Nanowires Science and Technology

Edited by

Nicoleta Lupu

Nanowires Science and Technology
Edited by Nicoleta Lupu

Republished by InTech
Janeza Trdine 9, 51000 Rijeka, Croatia

Technical Editor Teodora Smiljanic

Cover Designer Dino Smrekar

Additional hard copies can be obtained from orders@intechopen.com

Nanowires Science and Technology
Edited by Nicoleta Lupu
p. cm.
ISBN 978-953-7619-89-3

For EU product safety concerns:
IN TECH d.o.o., Prolaz Marije Krucifikse Kozulić 3, 51000 Rijeka, Croatia,
info@intechopen.com or visit our website at intechopen.com.

Preface

Nanowires can be defined as structures with thicknesses or diameters of tens of nanometers or less and unconstrained lengths. Many different types of nanowires exist, including metallic (e.g., Ni, Pt, Au or different alloys based on metals), semiconducting (e.g., Si, InP, GaN, etc.), insulating (e.g., SiO2, TiO2), and molecular nanowires (e.g. organic DNA or inorganic). Nanowires have many interesting properties that are not seen in bulk or 3-D materials.

There are two basic approaches of synthesizing nanowires: top-down and bottom-up approach. In a top-down approach a large piece of material is cut down to small pieces through different means such as lithography and electrophoresis. In a bottom-up approach the nanowire is synthesized by the combination of constituent ad-atoms. Most of the synthesis techniques nowadays are based on bottom-up approaches.

There are many applications where nanowires may become important: in electronic, opto-electronic, nanoelectromechanical devices, as additives in advanced composites, for metallic interconnects in nanoscale quantum devices, as field-emitters, as sensors or as leads for biomolecular (nano)sensors.

Nanowires still belong to the experimental world of laboratories. However, they may complement or replace carbon nanotubes in some applications. Some early experiments have shown how they can be used to build the next generation of computing devices.

The conductivity of a nanowire is expected to be much less than that of the corresponding bulk material. Nanowires also show other peculiar electrical properties due to their size. Unlike carbon nanotubes, whose motion of electrons can fall under the regime of ballistic transport (meaning the electrons can travel freely from one electrode to the other), nanowire conductivity is strongly influenced by edge effects. As a nanowire shrinks in size, the surface atoms become more numerous compared to the atoms within the nanowire, and edge effects become more important.

It is possible that semiconductor and magnetic nanowire crossings will be important to the future of digital computing. Though there are other uses for nanowires beyond these, the only ones that actually take advantage of physics in the nanometer regime are electronic.

Nanowires are being studied for use as photon ballistic waveguides as interconnects in quantum dot/quantum effect well photon logic arrays. Photons travel inside the tube, electrons travel on the outside shell. When two nanowires acting as photon waveguides cross each other the juncture acts as a quantum dot.

Because of their high Young's moduli, their use in mechanically enhancing composites is being investigated. Because nanowires appear in bundles, they may be used as tribological additives to improve friction characteristics and reliability of electronic transducers and actuators.

This book describes some nanowires fabrication and their potential applications, both as standing alone or complementing carbon nanotubes and polymers. Understanding the design and working principles of nanowires described here, requires a multidisciplinary background of physics, chemistry, materials science, electrical and optoelectronics engineering, bioengineering, etc.

This book is organized in eighteen chapters. In the first chapters, some considerations concerning the preparation of metallic and semiconductor nanowires are presented. Then, combinations of nanowires and carbon nanotubes are described and their properties connected with possible applications. After that, some polymer nanowires single or complementing metallic nanowires are reported. A new family of nanowires, the photoferroelectric ones, is presented in connection with their possible applications in non-volatile memory devices. Finally, some applications of nanowires in Magnetic Resonance Imaging, photoluminescence, light sensing and field-effect transistors are described.

The book offers new insights, solutions and ideas for the design of efficient nanowires and applications. While not pretending to be comprehensive, its wide coverage might be appropriate not only for researchers but also for experienced technical professionals.

Editor

Nicoleta LUPU

National Institute of Research and Development for Technical Physics, Iaşi
Romania

Contents

1

Nickel Silicide Nanowire Growth and Applications

Joondong Kim
Korea Institute of Machinery and Materials (KIMM)
Korea

1. Introduction

Due to the high potential and successful fabrication of one-dimensional nanomaterials such as carbon nanotubes and nanowires, intensive researches have been performed for practical applications (Kim. et al., 2009a).
Carbon nanotube is an ideal candidate for the high sensitive gas detection due to the peculiar hollow structure and a large surface area (Kim. et al., 2009b; Yun. et al., 2009). An electric conductive nickel silicide nanowire proved the high potential to be a functional microscopy tip, which may read the nanoscale structural and electrical information as well (Kim. et al., 2008a). Needle-shaped nanostructures would be utilized for field emitters, which may reduce the turn-on voltage by the enhanced electric field at the tips (Kim. et al., 2008b). Recently, semiconducting nanowires were applied as active light absorbers for Schottky solar cells (Kim. et al., 2009). Additionally, excellent electric conductive nanowires would provide a route to substitute the conventional copper interconnect and overcome the upcoming bottle neck of the current transport limit in a deep submicron integration (Kim. et al., 2005a; Kim & Anderson, 2006a).
Under Moore's law the semiconductor components have been scaled-down in every two years. A significant problem of conventional copper wire may cause an electromigration when current density exceeds 10^6 A/cm^2. Major industry leaders have predicted that the interconnect will be a significant issue for the device scale-down. ITRS (International Technology Roadmap for Semiconductors) declared that the increasing RC delay is one of the crucial problems for the device performance. A significant attention has given to the carbon nanotubes and nanowires as one-dimensional nanoscale interconnects in nanoelectronics.
There has been a remarkable interest and attention of carbon nanotubes and nanowires as one-dimensional nanoscale interconnects in nanoelectronics. Carbon nanotubes and metallic nanowires are considered as potential candidates to solve the general concerns in terms of electrical resistance and device speed. Metallic silicide nanowires have an advantage of compatibility to the Si technology over carbon nanotubes and perform uniformly. In this chapter, the growth of silicide nanowires is reviewed and the practical applications are presented. Electrical excellent silicide nanowires were applied for nanoscale interconnects and field emitters.

2. Silicide

The silicide is a compound of Si with an electropositive component (Kim & Anderson 2005b). Silicides are commonly used in silicon-based microelectronics to reduce resistivity of gate and local interconnect metallization. The popular silicide candidates, $CoSi_2$ and $TiSi_2$, have some limitations. $TiSi_2$ showed line width dependent sheet resistance and has difficulty in transformation of the C49 phase to the low resistive C54. $CoSi_2$ consumes more Si than $TiSi_2$ (Colgan et al., 1996). Nickel silicide is a promising material to substitute for those silicide materials providing several advantages; low resistivity, lower Si consumption and lower formation temperature (Kim et al., 2003). Recently, Ni silicide has emerged as an ideal electrical contact materials to the source, drain and gate in complementary metal oxide silicon devices and also shows an excellent scaling down behavior (Lavoie et al., 2003; Kittl et al, 2003; Morimoto et al., 1995).

2.1 Nanowire growth

Several nanowire growth mechanisms were reported, such as vapor-liquid-solid (VLS), solid-liquid-solid (SLS), and solid-solid (SS) types. The VLS type was first presented by Wagner and Treuting (Wagner & Ellis, 1964; Edwards et al., 1962) and is also the most popular method for growing nanowire today. The liquid catalyst acts as the energetically favored spot for absorbing gas-phase reactants (Morales et al., 1998). The high temperature for nanowire growth has been reduced to 320–600 °C by use of gas type Si sources such as $SiCl_4$ or SiH_4 with Au (Westwater et al., 1997; Wu et al, 2004; Zeng et al., 2003). Otherwise, a high temperature close to or above 5000 °C is needed to liquefy the catalyst and Si (Yu et al., 1998; Wang et al., 1999; Feng et al., 2000; Zhang et al. 1998; Geng et al. 2008).

Recently, SLS synthesis was presented. Metal catalyst coated Si prevents direct forming of vaporized Si atoms but results in liquid droplets of Si and metal, even at a high temperature of 900–950 °C (Chen et al., 2003; Yan et al., 2000). It was also claimed that SS synthesis can grow nanowires at 1050 °C by simple annealing in a CH_4:H_2 mixture gas. In this mechanism, the metal particles are observed on the tip of nanowires, different from the SLS mechanism (Lee et al., 2004).

Joondong Kim and professor Anderson (University at Buffalo, State University of New York) reported a unique mechanism of the nanowire growth in 2005 (Kim & Anderson 2005b). A unique nanowire growth mechanism is that of the metal-induced growth (MIG) method. The highly linear nanowires were grown by solid-state reaction of Ni and Si at 575 °C by sputtering method. The low-temperature process is desirable for applying nanowires as nanoscale interconnections with little or no damage on the fabricated structures.

Metal-diffusion growth (MDG) was also presented by Joondong Kim et al. in 2007 (Kim et al., 2007 a). It proved the uniform composed Ni silicide nanowires grown by Plasma-enhanced chemical vapor deposition method. The processing temperature (350 ^{o}C)was much reduced. It proved the similar electrical performance of each nanowire. Pre-patterned trench fabrication is also reported (Wang et al., 2007). SiO_2 patterning templates the nanowire shape and a focused ion beam milling was utilized. The summary of nanowire growth types and processing temperatures are presented in Table I.

Growth/synthesis types	Processing temperature [^{o}C]	Metal	References
VLS	300-600	Au	Westwater et al., 1997; Wu et al, 2004; Zeng et al., 2003
	1150-1200	Fe	Yu et al., 1998; Wang et al., 1999; Feng et al., 2000; Zhang et al. 1998
	5000	Ni	Geng et al. 2008
SLS	950	Ni	(JMR 15-16 Chen et al., 2003; Yan et al., 2000)
SS	1050	Ni	Lee et al., 2004
MIG	575	Ni	Kim & Anderson 2005b
MDG	350	Ni	Kim et al., 2007 a
Patterning	550	Ni	Wang et al., 2007

TABLE 1. Nanowire fabrication types and conditions

3. Metal-induced growth

3.1 Briefs

Metal-induced growth (MIG) is a spontaneous reaction of metal and solid-type sputtered Si. Professor Anderson (University at Buffalo, State University of New York) has initiated the concept to achieve quality crystalline Si for cost-effective thin film solar cells. Joondong Kim was joined his group in 2002 and had performed a research of thin film Si solar cells and observed interesting morphological changes. He and Professor Anderson revealed the unique mechanism of nanowire growth. They reported that MIG method is versatile to grow films or nanowires as well by modulating temperatures, catalysts, catalyst thickness and so on.

3.2 Crystalline film by MIG

One of the great advantages of MIG is to grow an epitaxial Si film, which adopts the concepts of solid phase crystallization and solid phase epitaxy. The deposited metal catalyst film, as a seed layer, interacts with sputtered Si to form a silicide layer, as a seed layer, to grow Si film above it. Ni and Co are common metal catalysts in the MIG method. The small lattice mismatch of $CoSi_2$ or $NiSi_2$ provides an excellent precursor layer to grow an epitaxial Si film as well as to render a spontaneously formed good ohmic contact layer. This MIG method is a fast Si crystallization method at a low processing temperature and would provide versatile approaches in Si application. More details may be found from the author's previous reports (Kim & Anderson 2007b; Kim et al., 2007c, 2008b).

3.3 Nanowire growth by physical method; MIG

MIG method provide a unique mechanism of nanowire growth. Unlike the MIG film growth mechanism, it utilizes the property of fast metal diffusion to Si. MIG nanowire growth mechanism is depicted in figure. 1.

The Ni deposited layer has been grooved and agglomerated during substrate temperature ramp up and Si sputtering. The first stage of nanowire growth is grooving and agglomeration. The process temperature of 575 ^{o}C is far below the Ni melting point of 1455 ^{o}C but it is at the recrystallization temperature known as 0.3 of the melting point in metallic materials. The grain growth occurs in the Ni layer on SiO_2 by thermal heating.

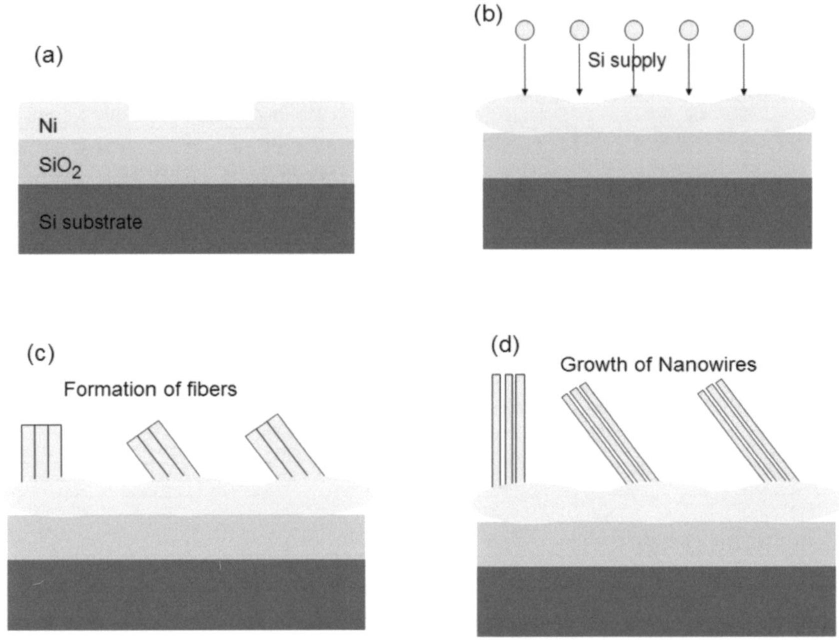

Fig. 1. Schematic diagram of MIG nanowire growth. (a) Groove and agglomeration of Ni layer by heating, (b) Clustering during Si supply, (c) Formation of nanofibers, and (d) Growth of nanowires.

The larger grains grow larger at the expense of small grain shrinkage or sacrifice resulting in grooves at the grain boundaries. It has been reported that agglomeration is driven by the reduction in surface and interfacial energy and starts with grain boundary grooving in the silicide layer (Colgan et al., 1996). Deep grooves finally lead to an agglomeration phenomenon in the Ni layer. After the grooving and agglomerating stage, two different Ni layer regions may be formed; a convex-thick Ni layer part and a concave-thin Ni layer part as shown in figure 1(a).

The clustering stage followed by grooving formation affects the different preferential growing directions. Each cluster may have a different orientation which also determines nanowire growth direction. The longest nanowire growing directions are vertical or parallel to the surface. Figure 1(b) shows the clustering stage. The Si sputtering onto the Ni layer reproduces the top morphology to form nickel silicide clusters due to heteroepitaxial growth. The kinetic energy comes from the thermal heating and transferring energy from the plasma gas (H_2/Ar). Because the concave region is a thinner Ni layer, this region may quickly become Si saturated resulting in no grown nanowires. The convex region with thicker Ni grows nanowires during the Si sputtering process. At times, clusters form nanofibers as in Figure 1(c). Figure 1(d) shows the segregation phenomenon resulting in several nanowires from one nanofiber. The segregation may be explained by the different thermal expansions of Ni and Si while forming Ni silicide. The thermal linear expansion coefficient for Ni is 13.3×10^{-6} C^{-1} while for Si is 3×10^{-6} C^{-1}. Schematic diagrams are provided in Figure 2 to show the simplified morphological changes at each stage. MIG nanowire growth formation is based on initiating NiSi formation at the onset and retaining the NiSi mechanism to lengthen the nanowire during the process. The substrate has been maintained at the processing temperature of 575 °C before and during the sputtering. Equation (1) shows the first reaction of Ni to the sputtered Si nanoparticles. Si sputtering on the grooved Ni layer leads to Ni diffusion into the accumulated Si to initiate the first NiSi formation.

$$2Ni + 2Si \text{ (sputtered from a target)} \rightarrow 2(NiSi) \qquad (1)$$

$$2\,(NiSi) \rightarrow Ni2Si + Si \text{ (accumulated Si)} \qquad (2)$$

$$Ni_2Si + Si \text{ (sputtered from a target)} \rightarrow (NiSi) \rightarrow NiSi \text{ nanowires} \qquad (3)$$

(Reprinted with permission from Kim et al., 2005b. Copyright 2005, Elsevier.)

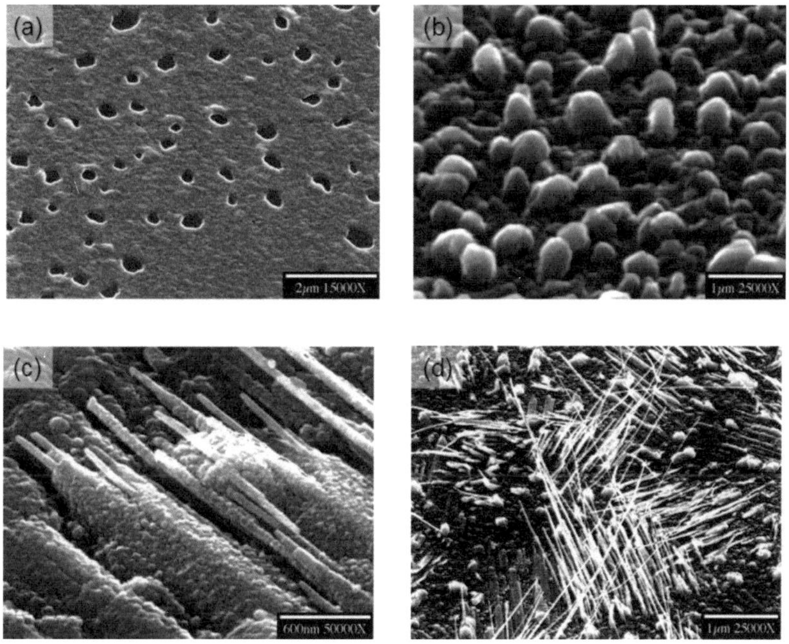

Fig. 2. Scanning electron microscopy (SEM) observation of the nanowire growth. (a) Ni layer was grooved by heating. The concave and convex regions were formed. (b) Ni silicide clusters were formed by reaction of Ni and Si. (c) Ni silicide fibers were formed. (d) Nanowires were grown. Some nanowires were segregated from a fiber

NiSi formation as shown in the Equations (2) and (3) may be a main source to continue the nanowire growth in the MIG method. Ni is a dominant moving species to initiate and continue the MIG nanowire growth mechanism. Ni diffuses from the Ni deposited layer to impinging Si from the target to form the NiSi layer shown in Equation (1). Ni also moves inside the nanowires to continue the NiSi mechanism of Equations (2) and (3). Ni diffusion may continue in the NiSi nanowire growth after reacting with sputtered Si. Some amount of NiSi contributes to lengthen the NiSi nanowires and some supplies the feedback mechanism to continue the process. Temperature is one of the most important factors to continue the nanowire growth in the NiSi nanowires growth mechanism and also prohibit the Si-rich formation like $NiSi_2$ which is the most stable composition in Ni silicides. The formation of NiSi is known as being diffusion controlled, opposite to a nucleation-controlled reaction (Ti and Co silicide). Ni is the dominant diffusing species in Ni silicide formation. If the temperature is low enough, Si is not significantly mobile (Lavoie et al., 2003). The optimum growth temperature of NiSi nanowires by the MIG method is 575 °C, which is close to the point of fastest Ni diffusion in Si. Short and thick nanofibers were grown at lower temperature. At higher temperature, the diameter of nanowires increased and the length was shortened. $NiSi_2$ may have formed at high temperature to limit the NiSi reaction. The most attractive point of MIG NiSi nanowires is straight line growth which facilitates nanobridge connections.

3.4 Catalyst effects

Ni-induced nanowire are in Figure 3(a). Ni is a major diffusing species in Ni monosilicide to form NiSi nanowires. No nanowires were grown by a Co catalyst as shown in Figure 3(b), or thick and short nanowhiskers were formed by Pd as in Figure 3(c). In the case of Co, Co is not a major moving species but Si moves in the CoSi phase (Bartur & Nicolet, 1983). Thus, Co did not grow nanowires. In the Pd case, the major moving species is a mixture of Pd and Si (Finstad et al., 1978). Thick and short nanowhiskers were grown with a low linearity.

Figure 4 shows a HRTEM image of a nanowire grown in [210]. The left side inset presents a prototype of a single nanowire, which consisted of 3 parts—root, stem, and tip. The Ni content was found from root to stem to tip by Ni to Si ratio as 1:0.8, 1:1.06, and 1:1.32, respectively. This result suggests that Ni diffused inside the nanowire. It could be considered that the end of nanowire growth at a tip is due to the deficiency of Ni. The right inset image is taken from the nanowire showing the NiSi structure. The cross-sectional transmission electron microscopy (TEM) analysis was performed to investigate the composition of the Ni silicide layer, which is a seed layer below the nanowire grown surface.

The cross-sectional view of the bottom of the Ni silicide layer underneath the nanowire grown surface is shown in Figure 5(a). The Ni silicide layer formed a crystalline structure shown in Figure 5(b). A Ni atomic map is shown in Figure 5(c).

It is clearly shown that the SiO_2 layer acted as a diffusion barrier of Ni to the Si substrate. Ni has diffused upward to react with sputtered Si to form a silicide layer. Electron diffraction analysis in the selected area of the cross section at the very bottom of the Ni silicide layer revealed a Ni-rich phase of Ni_3Si, as shown in Figure 5(d). The Ni diffusion in the silicide layer is similar to that of the nanowire body. Chemical analysis revealed that the Ni to Si ratio has a gradient from bottom to surface of the layer. The reaction of Ni and Si causes the compressive stress by volume changes at the silicide interface (Gergaud et al., 2003; Gambino & Colgan, 1998) as well as by the different mobility between the Ni and Si (Gambino & Colgan, 1998). The Si bonds are softened by intermixing of Si and Ni atoms and existing Ni in the interface of Ni and Si (Ottaviani, 1981). The induced stress can be relieved by rearranging of atoms due to volume shrinkage. It is considered that the nanowire structure is determined from these interactions to be epitaxially grown above the silicide layer, as shown in Figure 4.

3.3 Morphological changes by Ni thickness

Figure 6 (a) shows the different morphological changes by Ni film thickness, which has been achieved in Ni deposition procedure by tilting a specimen thinning from 150 nm. The nanowire growth region was clearly observed in the Figure 6(c) and presented a critical thickness of Ni film to be 60–80 nm.

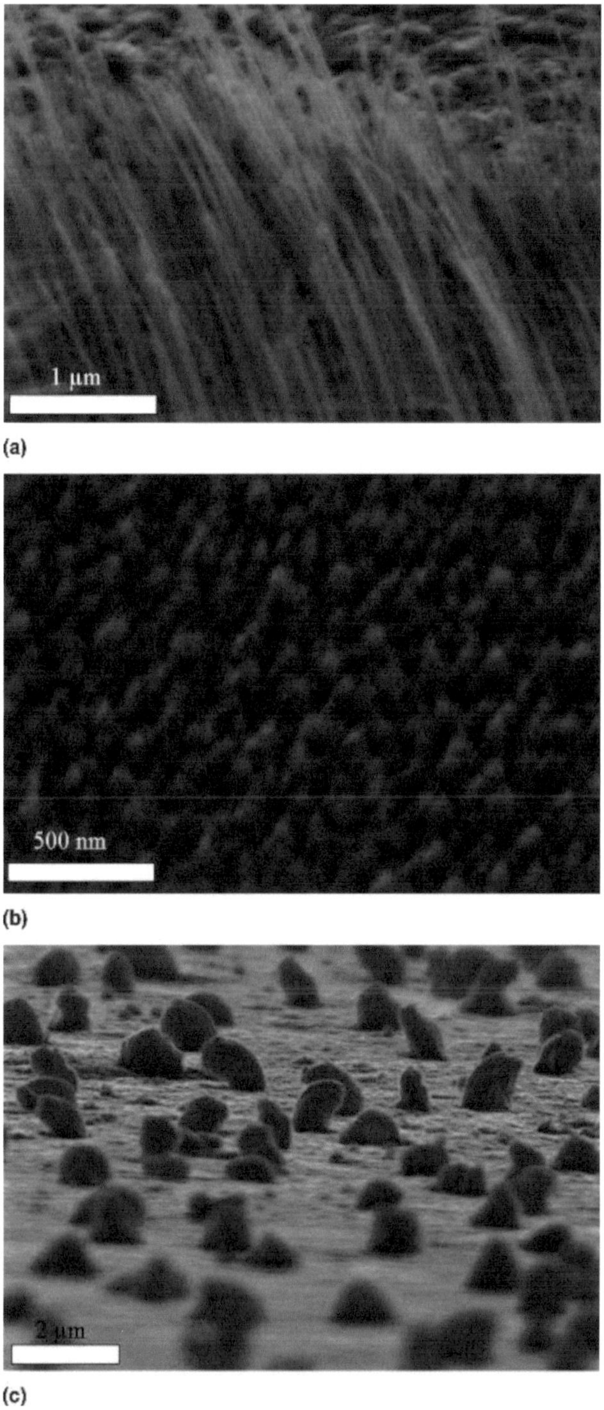

Fig. 3. MIG nanowire growth on a metal deposited layer: (a) Ni, (b)Co, and (c) Pd. Reprinted with permission from Kim et al., 2006b. Copyright 2006, Material Research Society.

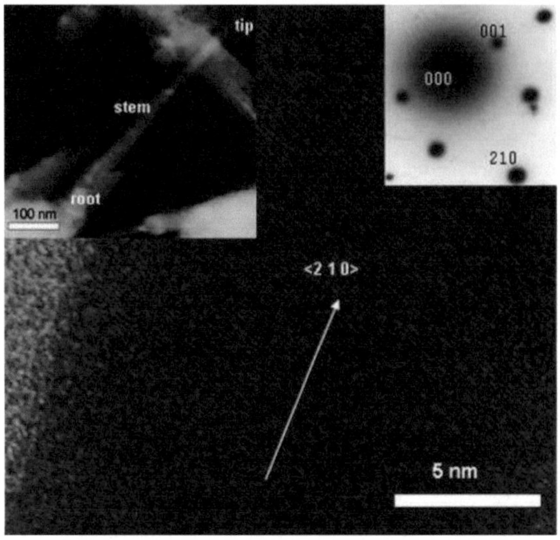

Fig. 4. Transmission electron microscopy of a single nanowire grown in the [210] direction. The left inset is a lower magnification showing root, stem, and tip of the nanowire, and the right inset is a diffraction pattern showing a NiSi structure. Reprinted with permission from Kim et al., 2006b. Copyright 2006, Material Research Society.

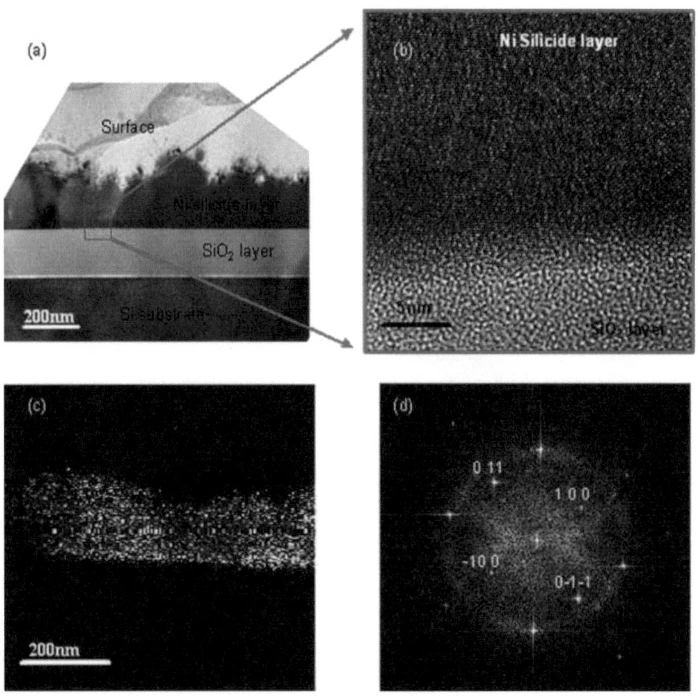

Fig. 5. Nanowire grown layer analyses: (a) low magnification of nanowire grown base layer, (b) high magnification of the square area of (a), (c) Ni mapping, and (d) electron diffraction analysis on the bottom layer. Reprinted with permission from Kim et al., 2006b. Copyright 2006, Material Research Society.

A thicker or a thinner film than the critical thickness formed larger silicide drops instead of growing nanowires as shown in Figure 6(b) and (d), respectively. This implies that the reaction of Ni and Si controls the silicide formation. Due to the equilibrium phase is determined by the ratio of Ni atoms to Si atoms (Gambino & Colgan, 1998), the thinner film tends to be a Si rich silicide and the thicker film prefers to be a Ni rich phase (Julies et al., 1999; Zheng et al., 1983).

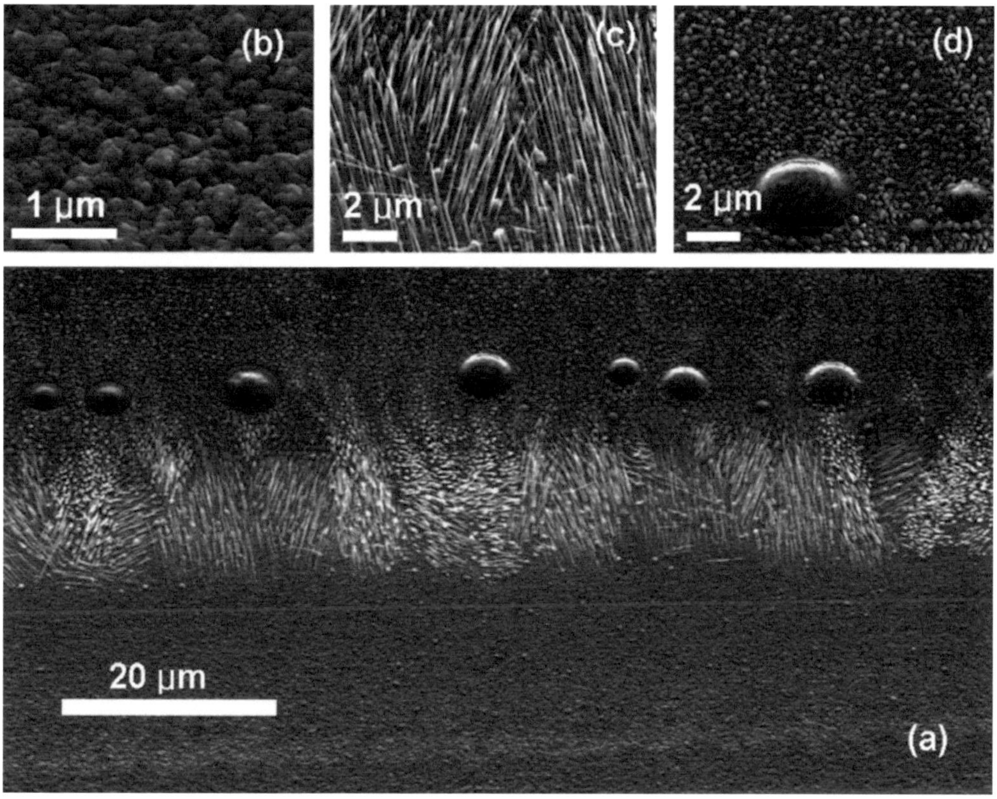

Fig. 6. (a) A SEM image of morphological changes with varying Ni thickness from 150 nm. Ni silicide nanowires were grown in a limited scope of the initial Ni thickness of 60–80 nm. Large images of (b) thicker, (c) moderate, (d) thinner Ni thickness region. Reprinted with permission from Kim et al., 2008b. Copyright 2008, Elsevier.

Thermodynamically, the large grains get larger by the tendency to reduce the surface free energy. The large grain formation is also observed in the thinner Ni region. The cross-sectional observations of Ni silicide nanowires on a SiO_2-coated Si substrate and a tungsten plate are presented in Figure 7. The optimum growth temperatures were found to be 575 °C for a SiO_2-coated Si substrate and 550 °C for a tungsten plate. The higher thermal conductivity of tungsten is attributed to reduce the processing temperature.

XRD was performed to investigate the formation of Ni silicide phases revealing a NiSi peak as well as Ni rich phase of Ni_3Si as shown in Figure 8. The reaction between Ni and Si atoms firstly forms a Ni rich phase and then further Si supply causes Ni diffusion to Si forming a Ni silicide nanowire. Due to the unique mechanism of Ni to Si, it has been found that the Ni silicide nanowires would be grown by the metal-induced growth method both in physical vapor deposition (Kim et al., 2005a, 2005b, 2006a, 2006b) and plasma-enhanced chemical vapor deposition (PECVD) (Kim et al., 2007a). By using the Ni diffusion to grow silicide nanowires, the Ni silicide formation is an important factor to form nanowires of NiSi in PVD and Ni_3Si_2 in PECVD process.

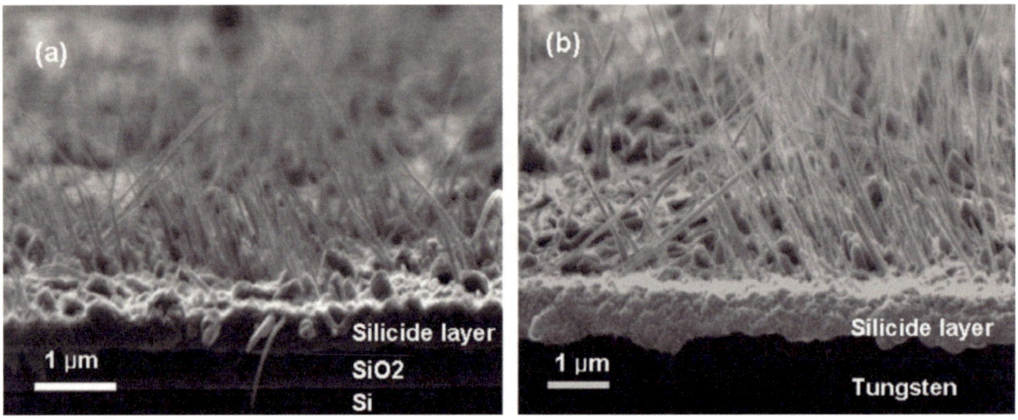

Fig. 7. Cross-sectional SEM images of Ni silicide nanowires grown on a SiO_2-coated Si substrate (a) and a tungsten plate (b). Reprinted with permission from Kim et al., 2008b. Copyright 2008, Elsevier.

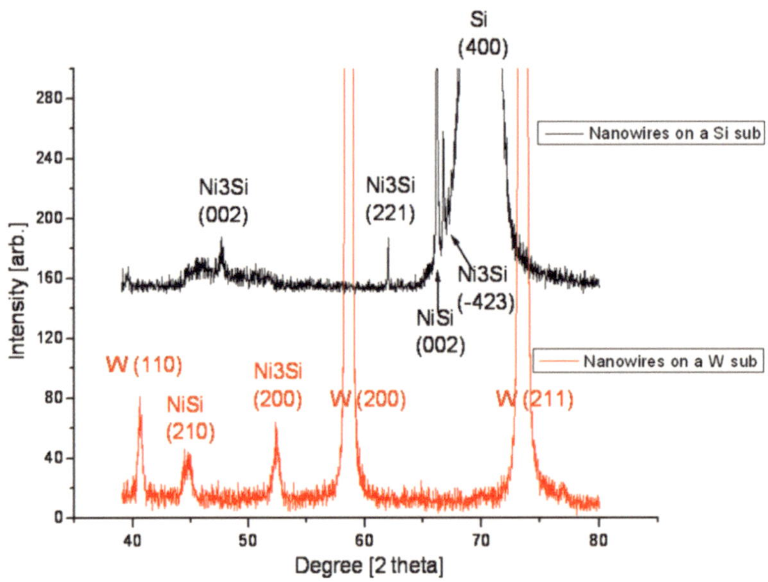

Fig. 8. XRD spectrums of the Ni silicide nanowires grown on a Si substrate and a tungsten plate. Reprinted with permission from Kim et al., 2008b. Copyright 2008, Elsevier.

The spontaneous reaction between Ni and Si firstly forms a silicide layer and then the further reaction causes the Ni diffusion to grow a unique linear nanostructure to be a nanowire above a silicide layer, which is a remarkable feature to form a spontaneous contact formation without an architectural electrode fabrication (Kim et al., 2005a, 2006a). Moreover, the suitable growth of Ni silicide nanowire on various materials may increase the practical uses. Various factors, such as a processing temperature, a metal thickness, and a Si supply amount determine the formation or transition of Ni silicide phases. In this research, Ni thickness is mainly investigated on the growth of nanowires in a specific condition by the balanced reaction of Ni and Si.

4. Ni silicide nanowire growth by chemical method

Beside the MIG methods, chemical vapor deposition (CVD) and DC arc-discharge approaches were presented (Kim et al., 2005a). The low growth temperature from 320 to 420 °C is accomplished by the decomposition of silane gas on Ni surfaces to make different composition (NiSi, Ni_2Si, and Ni_3Si_2) of nanowires (Decker et al., 2004). Similar growth was also reported but Ni gradient was appeared through a nanowires. (Kim et al., 2007).
The amorphous nanowires have been grown by the solid-liquid-solid mechanism at a high temperature of 950 °C to utilize the $NiSi_2$ eutectic liquid droplets.11 In our previous work, NiSi nanowires have been reported without using a gas-type silicon source by the MIG method.12,13 For the DC arc-discharge growth method, ultra high temperature above 5000 ^{o}C was applied to vaporize of Ni and Si (Geng et al., 2008) It has still remained as an assignment to grow single composed Ni silicide nanowires at a low temperature without a metal gradient through a nanowire, which may limit the nanowire-embedding system performance.

4.1 Metal diffusion growth

Joondong Kim has reported Ni silicide nanowire by Plasma-enhanced chemical vapor deposition (PECVD) growth method in 2007 (Kim et al., 2007a). Metal-diffusion growth (MDG) provides advantages in mass production and crystallinity of nanowires. It proved the uniform composed Ni silicide nanowires grown by Plasma-enhanced chemical vapor deposition method. The processing temperature (350 oC) was much reduced. This Ni silicide nanowire has a single composition of Ni and Si through a body, which will benefit the uniform performance of nanowire-embedding devices. Moreover, it has been revealed that the most PECVD grown nanowires have the same composition of Ni and Si.

4.2 Growth condition

The single crystalline Ni silicide nanowire were achieved at a low temperature of 350 °C. Ni as a catalyst was thermally evaporated onto a 500 nm SiO_2 coated Si substrate at a high vacuum level of 5×10^{-7} Torr before loading for PECVD. Silane (SiH_4) as a Si source was supplied to react the Ni. The gas flow rate was fixed at 50 standard cubic centimeter per minute (SCCM) for 10 min. PECVD system was operated at 13.6 MHz.

4.3 Growth mechanism

Typical Ni silicide nanowire grown morphology by MDG is shown in Figure 9. High dense, long, and thin nanowires were grown at 350 °C with a SiH_4 pressure of between 10 and 100 Torr. The length is most above 5 μm with 20 - 30 nm in diameter. The Ni–Si binary system has various phases of Ni_2Si, NiSi, or Ni_3Si_2 before forming the $NiSi_2$ phase, which is the most stable phase in the system and to be a seed to grow a crystalline thin Si film due to its little lattice mismatch to Si (Kim et al., 2007b, 2007c). The formation of NiSi and Ni_3Si_2 is known as the typical transitions phased by the reaction of Ni and Si. Ni is a fast diffuser to Si and there exists a thermodynamically unstable scope causes serious morphological changes resulting in clustering and impinging whiskers (Kim et al., 2005b). The amorphous Ni silicide is unstable and Ni continuously diffuses to Si rich region remaining crystalline Ni silicide nanoscale structure (Levenger & Thompson, 1990). Due to the use of gas type Si source, the Ni silicide nanowire growth temperature has been lowered to 350 °C comparing to the solid Si source case of 575 °C (Kim et al., 2005b).

Fig. 9. Ni silicide nanowire grown by metal-diffusion growth method in PECVD system.

4.4 MDG nanowire analysis

High resolution transmission electron microscopy (TEM) image shows that the nanowire has a single crystalline structure grown perpendicular to the (001) plane of Ni3Si2 orthorhombic structure, as shown in Figure 10(a). The measured spaces are 0.694 and 0.540

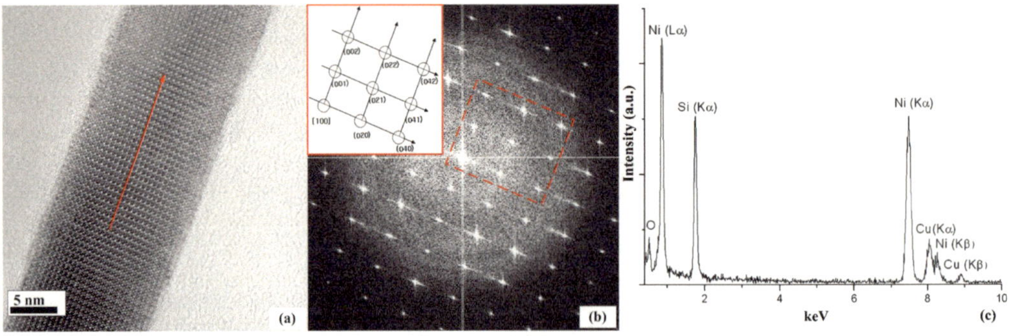

Fig. 10. (a) High resolution TEM micrograph of Ni silicide nanowire with diameter of 28.3 nm. Arrow indicates that the nanowire was grown in the (001) direction. (b) Diffraction pattern indicates that the single crystalline nanowire has an orthorhombic Ni_3Si_2 structure. The inset is an indexing of the diffraction pattern marked as a dotted line. (c) EDS analysis of the nanowire showing a Ni silicide formation. Reprinted with permission from Kim et al., 2007a. Copyright 2007, American Institute of Physics.

nm, consistent with the Ni_3Si_2 interplanar spaces of *d*001=0.692 nm and *d*020 =0.540 nm. The lattice spacings and angular relationships showed that a single crystalline Ni silicide nanowire has an orthorhombic Ni_3Si_2 structure (*a*=1.222 nm, *b*=1.080 nm, and *c*=0.6924 nm) indexing (001), (021) and (020) spots, as shown in Figure 10(b). Energy dispersive spectroscopy investigation also confirmed the Ni silicide formation of a single nanowire, as shown in Figure 10(c).

5. Applications

Due to the tiny scale and excellent electrical conductivity, Ni silicide nanowires have a high potential in the nanoscale electronics. In this chapter, two promising applications of Ni silicide nanowire are presented. First part discusses the nanoscale interconnect and last part presents Ni silicide field emitters.

5.1 Nanoscale interconnect

Under Moore's law, the semiconductor components have been shrunk every two years. The interconnect is one of the major issues of component scale-down to increase electrical resistance resulting in device performance loss or malfunction.

A significant problem of conventional copper wire may cause an electromigration when current density exceeds 10^6 A/cm^2. Intel and other companies predicted that the interconnect resistance will start to become a significant limit for the device scale-down. ITRS (International Technology Roadmap for Semiconductors) declared that the increasing RC delay is one of the crucial issues for the device performance.

There has been a significant interest and attention in carbon nanotubes and nanowires as one-dimensional building blocks for nanoscale interconnects. Carbon nanotubes and metallic nanowires are considered as potential candidates to solve the general concerns in terms of electrical resistance and device speed.

Carbon nanotube has been attracted as a possible replacement for copper wires in semiconductor devices. It may be able to pass high current of 10^9 A/cm^2 without failure and has a good mechanical stability. In spite of these benefits, carbon nanotubes are bearing a problem of uniformity. Some nanotubes are semiconductors, while others have metallic characteristics which may not ensure the uniform performance. Additionally, carbon is heterogeneous material to Si and it may require an additional process to Si technology.

In contrast, silicide nanowire is compatible to the Si technology. Ni silicide has been intensively researched for use as a contact material of gate and source/drain in complementary metal oxide-semiconductor (CMOS) devices. It is superior to other candidates, such as $TiSi_2$ and $CoSi_2$. $TiSi_2$ has difficulty in transformation of the C49 phase to the low resistive C54 phase. $CoSi_2$ is limited by high Si consumption and junction leakage. These merits prompt interest in Ni silicide nanowires to make a one-dimensional nanoscale building block.

5.1.1 Current density of Ni silicide nanowires

There has been a significant interest of other high transport nanomaterials. NiSi nanowire interconnect has proven a potential to deliver high level current. It also confirmed that the each NiSi nanowire delivered current uniformly. The ohmic contact behavior was obtained from the Ni silicide nanowire connected Pt electrodes, as shown in Figure 11(a).

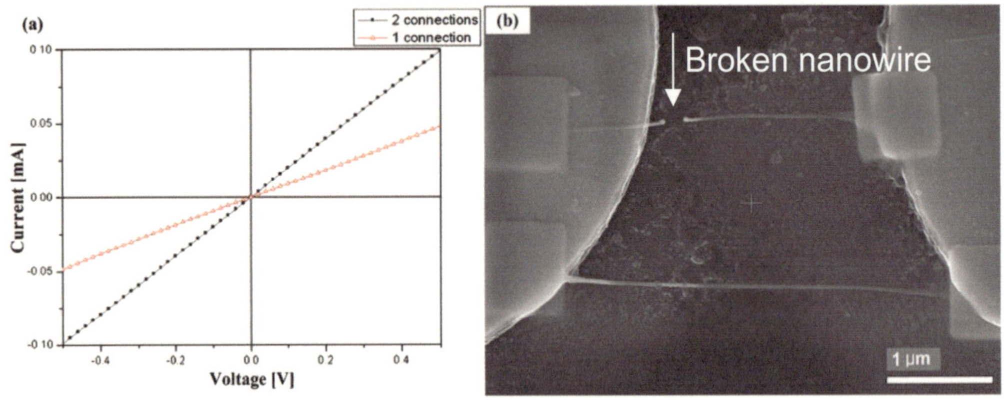

Fig. 11. (a) Electrical measurement was obtained from the nanowire connected Pt electrodes. Double connected nanowires gave 5.07 kΩ and single connected nanowire showed 10.44 kΩ. (b) SEM micrograph of a single nanowire connection. Reprinted with permission from Kim et al., 2007a. Copyright 2007, American Institute of Physics.

The resistance obtained from the two nanowire connections gave 5.07 kΩ. After then, a single nanowire connection was achieved by applying a high impulse voltage to break one nanowire, as shown in Figure 11(b). The *I-V* measurement was performed again to give 10.44 kΩ from the one nanowire connection. The resistance of the disconnected nanowire was obtained to be 9.84 kΩ (Kim et al., 2007a). It is worthy to remark that two nanowire connections have similar resistances indicating a uniform characteristic of each nanowire connection with little contact noise (Dong et al., 2005).

It also proved that the uniform current delivery performances of NiSi nanowires. The current density of the Ni silicide nanowire obtained to be 3.36×10^7 A/cm^2. Wu et al. reported a high failure current density of 3×10^8 A/cm^2 through a Ni silicide nanowire, which was synthesized by Ni coating on a Si nanowire and heat treatment (Wu et al., 2004).

Joondong Kim presented the spontaneous grown and connected parallel nanowire to be called as a nanobridge (Kim et al., 2007a). The nanobridge carries 2.58×10^8 A/cm^2 without breaking the connection implying even higher current is acceptable to transport. Other reports of Ni silicide nanowire current delivery are summarized in table 2.

Current density	Composition of nanowire	Growth or synthesis method	References
3×10^8 A/cm^2	NiSi	Ni coating on a Si nanowire	Wu et al., 2004
8×10^7 A/cm^2	NiSi, Ni_2Si, $Ni_{31}S_{12}$	Lithography	Zhang et al.,2006
2.58×10^8 A/cm^2 or higher	NiSi	MIG	Kim et al., 2006a
3.4×10^7 A/cm^2	NiSi	CVD	Kim et al., 2007
1×10^8 A/cm^2	NiSi	Focus ion beam milling	Wang et al., 2007
3.36×10^7 A/cm^2	Ni_3Si_2	MDG	Kim et al., 2007a

Table 2. Summaries of Ni silicide nanowire current density.

5.1.2 Nanowire interconnect: Nanobridge

Beyond nanowire growth, use of nanowires in nanoelectronics may be a breakthrough to overcome the difficulty in scaling down of integrated circuits. The first and inevitable step of nanoelectronics integration is to form contacts.

Many different approaches have been used to make contacts or control nanoscale entities, such as guided self-assembly using magnetic trapping (Tanase et al., 2002), SiO_2 patterning and cracking (Alaca et al., 2004), field emission-induced growth on a scanning tunneling microscopy tip, (Thong et al., 2002). ac or dc electric field (Fan et al., 2004; Smith et al., 2000), electron beam lithography (Nastaushev et al., 2002) and "grow-in-place" (Shan et al., 2004). To employ nanoscale structures in devices, it is desirable to make interconnections that are independent from the use of complex processes or multiple lithography steps.

The bottom-up approach provides cost-effective nanowires compared to the top-down approach and offers much promise in future integrated circuits. The self- and directed assembly method based on the bottom-up process will inaugurate the volume manufacturing of nanotechnology. The control and manipulation of self-assembled nanostructures may contribute to the advance of memory chips (Marrian et al., 2003).

The spontaneous nanowire interconnect (nanobridge) may provide an essential connection route as a one-dimensional building block in nanoelectronic devices. The nanowires have a desired property of linear propagation parallel to a substrate at a substrate edge. The deposited Ni has been grooved and agglomerated at substrate heating of 575 °C by the reduction in surface energy, which may determine the silicide islands formation. The silicide island formed at a trench edge may have a property to propagate parallel nanowires across the trench (Kim et al., 2006a).

This could explain the specific nanowire propagation at the edge of the trench. Si sputtering on the grooved Ni layer results in Ni diffusion into Si to form NiSi nanowires. The parallel propagated nanowire formed a spontaneous connection between Ni deposited electrodes shown in Figure 12(a). The MIG nanowire contact nanobridge (NB) is solid and strong with a small contact area that is the same as the diameter of the nanowire itself, about 58 nm.

The electrical transport in Figure 12 (b) gave a low resistance of about 147.9 Ω. The NB resistivity was obtained to be 10.8 μΩ·cm, comparable to the value for single crystalline20 (Meyer et al., 1997). NiSi of 10 μΩ·cm and NiSi thin film of 11 μΩ·cm. The value is also close to the reported resistivity of 9.5 μΩ·cm for the selective Ni-coating NiSi nanowire (Wu et al., 2004). Other results on NiSi nanowires were reported as about 370 μΩ·cm from Ti/Au deposition (Lee et al., 2004). or nonlinear high resistance characteristics from Pt deposition (Dong et al., 2005; Decker et al., 2004) implying that there is an effect of foreign metal use or poor adhesion between electrodes and nanowire.

The common method by others to characterize a nanoscale structure is to detach it to an insulation layer, such as SiO_2, then make contacts above it. The patterns for electrical leads are usually fabricated with the assistance of e-beam lithography (Wu et al., 2004; Lee et al., 2004; Lew et al., 2004). The electron beam assisted pattern fabrication needs complicated process steps as well as about 10 nm tolerances. Besides the delicate requirement, it has been currently reported about the Ti thickness effects in use with Ti-Au combination, which is a popular binary alloy to make electrodes in e-beam patterning. Ti thickness significantly affected resistance by almost 6 orders of magnitude variation by modulating Ti-Au within a fixed total metal thickness (Hwang et al., 2004). For those reasons, it is preferred that contacts for nanowire electrical testing be of a similar metal to the nanowire.

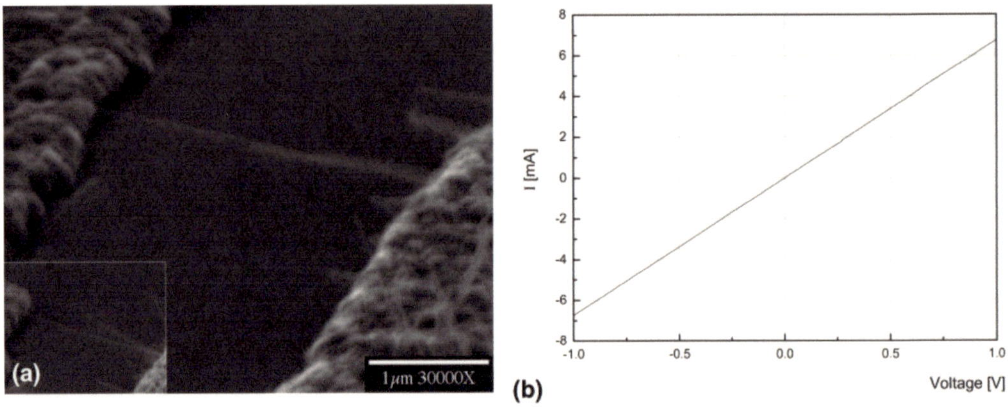

Fig. 12. (a) Nanobridge in a trench formed by a metal lift-off process (inset is a SEM image at a different angle). The NB is 58 nm in diameter and 2.9 μΩ in length. (b) I-V characteristic measured from two pads connected by a NB. Reprinted with permission from Kim et al., 2006a. Copyright 2006, American Chemical Society.

One advantage of the metal-induced grown NB is in providing compatible metallic pads for electrode use. The MIG pads are homogeneous to the NiSi NB to relieve an extrinsic effect caused by foreign metal use. The electric transport was also measured on a pad and substrate to see the effects on the nanowire measurement. Resistance from point to point on a pad was about 7 Ω providing a reliably low resistance compared to the resistance of the NB. The measured resistance from pad to pad on the substrate without a NB was 40.49 MΩ, the sign of a good insulation layer. It is considered that the pad gave little effect on the NB I-V result without interference from a high resistance substrate, graphically shown in Figure 13.

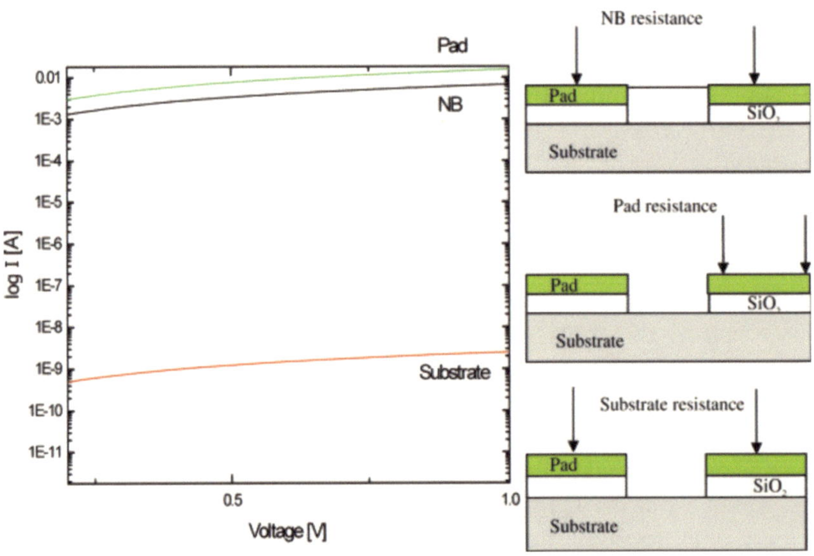

Fig. 13. Pad resistance of 7 Ω is low enough compared with an NB resistance of 147.9 Ω. Schematics illustrate points where resistance values were measured. Reprinted with permission from Kim et al., 2006a. Copyright 2006, American Chemical Society.

5.2 Ni silicide nanowire field emission properties

One dimensional nanostructures of nanotubes and nanowires have been emerged as effective cold cathodes, which may reduced the turn on voltage by an enhanced field at sharp tip structures. Field emission in vacuum has a significant benefit over the conventional electron-hole carrier transport in semiconductor without energy loss in medium. To realize a promising of nanostructure for a field emitter, the excellence in electrical conductivity is highly required.

Using a nanoscale structure as a field emitter, it requires a low turn-on voltage and a stable emission current. Various approaches have been performed in fabrication of emitting entities by Si (Kanemaru et al., 1996), diamond and molybdenum tips (Yoon et al., 1999). To enhance the electric conductivity, doping process (Kanemaru et al., 1996) or silicide coating on Si structures (Yoon et al., 1999) have been reported. In the fabrication aspects, a good electric nanoscale emitter is highly desirable. Silicides of Ni, Co, Ti have been utilized to be contact materials in Si technologies due to the compatibility to Si (Kim et al., 2007b, 2007c, 2008a) and low resistance driving intensive interests on forming a silicide nanowires (Wang et al., 2007; Wu et al., 2004; Lee et al., 2004; Decker et al., 2004; Kim et al., 2005a, 2005b, 2006a, 2007a) in applications of a nanoscale interconnection (Kim et al., 2005a, 2006a, 2007a) and a microscopy tip (Yoon et al., 1999). The field emission measurements of Ni silicide nanowires grown on a SiO_2-coated Si and a tungsten substrates were performed in a vacuum of 10^{-7} Torr. An anode electrode of ITO was spaced 300 μm to emitters. In the connection, it requires an extrinsic contact conductor for the Ni silicide nanowires grown on a SiO_2-coated Si substrate, otherwise tungsten plate was directly used a conductor. The electric excellence of Ni silicide nanowires has been reported in previous researches.

Although Ni silicide nanowire has been known as a good nanoscale conductor, limited researches have been performed on the nanowire growth observation and the field emission properties. As a one-dimensional conductor, Ni silicide nanowire has a high advantage to be a field emitting entity. Two different samples of were investigated for the Ni silicide nanowire field emission. The cross-sectional observations of Ni silicide nanowires on a SiO_2-coated Si substrate and a tungsten plate are presented in Figure 7.

Figure 14 presents that the emission properties of Ni silicide nanowires grown on a SiO_2-coated Si and a tungsten substrates was measured in a vacuum level of 10^{-7} Torr. As denoting a turn-on state as the emission current density to be 10 $\mu A/cm^2$, the inset electric field was achieved to be 3.23 V/μm at 970 V and 3.47 V/μm at 1040 V from Ni silicide nanowires grown on a tungsten plate and a SiO_2-coated Si substrate, respectively. Additionally, the current density from the tungsten plate use provided higher value of 172.5 $\mu A/cm^2$ than 76.5 $\mu A/cm^2$ of a SiO_2-coated Si case at 5 V/μm.

The lower field emission current density of a SiO_2-coated Si substrate may be explained by the structure of the metal-oxide-silicon (MOS), in which the potential drops across an oxide layer and a Si substrate to form the conduction state. Otherwise, no significant potential drop is considerable across a tungsten plate, which is a refractory metal and a good conductor. The Fowler-Nordheim (F-N) plots are presented in Figure 14(b). Current density (J) is related to the electric field (E) and describes as $J = aE^2\exp(-b/E)$. The constants of a and b are expressed by 1.56×10^{-10} β^2/φ and 6.83×10^{9} $\varphi^{1.5}/\beta$, respectively, where φ is the local work function and β is the field enhancement factor (Joo et al., 2006).

Due to the linear function of the F-N plot (Bonard et al., 2002). The b values were obtained to be 3180 and 3002 from the Ni silicide nanowires grown on a Si substrate and a tungsten plate. According to the similar b characteristics, the field emission properties of Ni silicide nanowires are considered to be uniform. Moreover, the use of a conducting substrate such as tungsten may enhance a current density and lower a turn-on voltage. The field enhancement factor of Ni silicide nanowires have provided much improved performance comparing to that of $NiSi_2$ nanorods (Ok et al., 2006) of 630, which is mainly due to the high aspect ratio nanowires.

In conclusion, The field emission of Ni silicide nanowires gave a high electron emission property at a reduced electric field. The onset electric fields, to give the current density of 10 $\mu A/cm^2$, were achieved to be 3.23 V/μm at 970 V and 3.47 V/um at 1040 V from Ni silicide nanowires grown on a tungsten plate and a SiO_2-coated Si substrate, respectively. The field enhancement factors were similar from different substrate materials to be 3180 from a tungsten plate and 3002 from a SiO_2-coated Si substrate. The current density from a tungsten plate use provided higher value of 172.5 $\mu A/cm^2$ than 76.5 $\mu A/cm^2$of a SiO_2-coated Si case at 5 V/μm indicating the improvement of emission current by using a conductive substrate. The Ni silicide nanowire having a high emission performance with a uniformity shows a potential to be a field emitter in various applications.

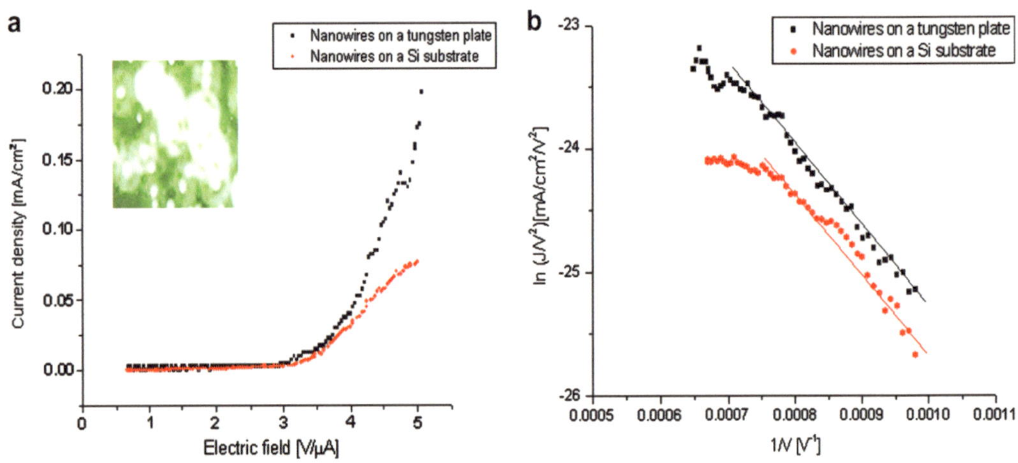

Fig. 14. Electron emission current characteristics: (a) plots of the current density to electric field (inset is emission spots monitored on a phosphor screen); and (b) plots of Fowler-Nordheim indicates linear slopes. Reprinted with permission from Kim et al., 2008b. Copyright 2008, Elsevier.

6. Summaries

In this chapter, we reviewed the growth mechanism of Ni silicide nanowires mainly focused on the metal-induced growth (MIG) and metal-diffusion growth (MDG). Although various techniques were reported to achieve the Ni silicide nanowires, metal-assisted growth (MIG or MDG) provides good electrical properties. Moreover, MDG method may guarantee the uniform composition through a single nanowire and group as well. This chapter discussed two applications of Ni silicide nanowires for nanoscale interconnects and field emitters.

Nanowires and carbon nanotubes have attracted high interests for high current delivery nanoscale interconnect. However carbon nanotube has shown the ability of high current transport, it may require an additional fabrication step to be adopted in Si devices. Moreover, the lack of uniformity limits the carbon nanotube application in nanoelectronics.

In this aspect, the Ni silicide nanowire may be one of the most promising one-dimensional nanoscale building blocks in terms of compatibility to the conventional Si technology and the excellent electrical conductivity. According to the device scale-down, it is inevitable to substitute the conventional copper interconnects, which are bearing a limit of an electromigration when the current density exceeds 10^6 A/cm^2. Several reports have proved the potential of high current delivery of Ni silicide nanowires.

Beyond the nanowire growth, use of nanowires in nanoelectronics may be a breakthrough to overcome the bottle-neck in scaling down. The bottom-up approaches would provide a route for the cost-effective and volume manufacturing scheme of nanotechnology. Spontaneously contacted nanowire connection (nanobridge) is one of the solution for the "grow-in-place" scheme, which may accomplish the nanoscale interconnect without complex processes. The Ni silicide nanobridge showed a high current density of 2.58×10^7 A/cm^2 or higher due mainly to the homogeneous contacts between pads to Ni silicide nanowire.

Additionally, one dimensional nanostructures of nanotubes and nanowires have been emerged as effective cold cathodes, which may reduced the turn on voltage by an enhanced field at sharp tip structures. Field emission in vacuum has a significant benefit over the conventional electron-hole carrier transport in semiconductor without energy loss in medium. To realize a promising of nanostructure for a field emitter, the excellence in electrical conductivity is highly required. Needle-shaped nanostructures would be utilized for field emitters to reduce the turn-on voltage by the enhanced electric field at the tip.

Due to the tiny scale and good electric properties, Ni silicide nanowires are being highly considered as to be entities. In this chapter, the properties of Ni silicide nanowire field emission was reviewed. The onset electric fields, to give the current density of 10 μA/cm^2, were achieved to be 3.23 V/μm at 970 V from Ni silicide nanowires grown on a conductive tungsten plate. High field enhancement factor of 3180 was achieved with reducing a turn-on voltage. The Ni silicide nanowire having a high emission performance with a uniformity shows a potential to be a field emitter in various applications.

Beside these above mentioned applications, nanostructures would provide a promise for various nanoelectronics applications such as, the sensitive gas sensors, functional microscopy tips, solar cells, and so on.

The author would like to finish this chapter with a sentence of "There's plenty of room at the bottom" by Richard P. Feynman.

7. Acknowledgement

The author acknowledge the financial support from National Research Foundation (NRF, 2009-0082018) by the Ministry of Education, Science and Technology (MEST) in Korea. Some parts of this chapter were referred from the author's previous reports (Kim et al., 2005b, 2006a, 2006b, 2007a, 2008b).

8. References

Alaca, B. E.; Sehitoglu, H. & Saif, T. (2004). Guided self-assembly of metallic nanowires and channels. *Applied Physics Letters, 84*, 4669-4671.

Bartur, M. & Nicolet, M.-A. (1983) Marker experiments for diffusion in the silicide during oxidation of PdSi, Pd2Si, CoSi2, and NiSi2 films on Si. *Journal of Applied Physics*, 54, 5404-5415.

Bonard, J. -M.; Dean, K. A. Coll, B. F. & Klinke, C. (2002). Field Emission of Individual Carbon Nanotubes in the Scanning Electron Microscope. *Physical Review Letters*, 89, 197602.

Chen, X.; Xing, Y. Xu, J. Xiang, J. & Yu, D. (2003) Rational growth of highly oriented amorphous silicon nanowire films. *Chemical Physics Letters*, 374, 626-630.

Colgan, E. G.; Gambino, J. P. & Cunningham, B. (1996). Nickel silicide thermal stability on polycrystalline and single crystalline silicon. *Materials Chemistry and Physics*, 46, 209-214.

Cui, Y.; Lauhon, L. J. Gudiksen, M. S. Wang, J. & Lieber, C. M. (2001). Diameter-controlled synthesis of single-crystal silicon nanowires. *Applied Physics Letters*, 78, 2214-2216.

Decker, C. A.; Solanki, R. Freeouf, J. L. Carruthers, J. R. & Evans, D. R. (2004). Directed growth of nickel silicide nanowires *Applied Physics Letters, 84, 1389-1391.*

Dong, L.; Bush, J. Chirayos, V. Solanki, R. Jiao, J. Ono, Y. Conley, J. F. & Ulrich, B. D. (2005). Dielectrophoretically Controlled Fabrication of Single-Crystal Nickel Silicide Nanowire Interconnects. *Nano Letters*, 5, 2112–2115.

Edwards, P. L. & Happel, R. J. (1962). Beryllium oxide whiskers and platelets. *Journal of Applied Physics*, 33, 943-948.

Fan, D. L.; Zhu, F. Q. Cammarata, R. C. & Chien, C. L. (2004). Manipulation of nanowires in suspension by ac electric fields. *Applied Physics Letters, 85, 4175- 4177.*

Feng, S. Q.; Yu, D. P. Zhang, H. Z. & Bai, Z. G. & Ding, Y. The growth mechanism of silicon nanowires and their quantum confinement effect. *Journal of Crystal Growth*, 209, 513-517.

Finstad, T. G.; Mayer, J. W. & Nicolet, M.-A. (1978). The formation of NiSi from Ni2Si studied with a platinum marker. *Thin Solid Films*, 51, 391-394.

Gambino, J. P. & Colgan, E. G. (1998). Silicides and ohmic contacts. *Materials Chemistry and Physics*, 52, 99-146.

Geng, Z. R.; Lu, Q. H. Yan, P. X. Yan, D. & Yue, G. H. (2008). Efficient preparation of NiSi nanowires by DC arc-discharge. *Physica* E, 41, 185-188.

Gergaud, P.; Megdiche, M. Thomas, O. & Chenevier, B. (2003): Influence of Si substrate orientation on stress development in Pd silicide films grown by solid-state reaction. *Applied Physics Letters*, 83, 1334-1336.

Hayzelden, C. & Batstone, J. L. (1993). Silicide formation and silicide-mediated crystallization of nickel-implanted amorphous silicon thin films. *Journal of Applied Physics*, 73, 8279-8289.

Hwang, J. S.; Ahn, D. Hong, S. H. Kim, H. K. Hwang, S. W. Jeon, B.-H. & Choi, J.-H. (2004). Effect of Ti thickness on contact resistance between GaN nanowires and Ti/Au electrodes. *Applied Physics Letters, 85*, 1636-1638.

Joo, J.; Lee, S. J. Park, D. H. Kim, Y. S. Lee, Y. Lee, C. J. & Lee, S.-R. (2006). Field emission characteristics of electrochemically synthesized nickel nanowires with oxygen plasma post-treatment. *Nanotechnology*, 17, 3506-3511.

Julies, B. A.; Knosesen, D. Pretorius, R. & Adams, D. (1999). A study of the NiSi to NiSi2 transition in the Ni-Si binary system. *Thin Solid Films*, 347, 201-207.

Kanemaru, S.; Hirano, T. Tanoue, H. & Itoh, J. (1996). Control of emission characteristics of silicon field emitter arrays by an ion implantation technique. *Journal of Vacuum Science and Technology* B, 14, 1885-1888.

Kim, C.-J.; Kang, K. Woo, Y. S. Ryu. K.-G. Moon, H. Kim, J.-M. Zang, D.-S. & Jo, M.-H. (2007). Spontaneous Chemical Vapor Growth of NiSi Nanowires and Their Metallic Properties. *Advanced Materials*, 19, 3637-3642.

Kim, G. B.; Yoo, D. -J. Baik, H. K. Myoung, J. -M. Lee, S. M. Oh, S. H. & Park, C. G. (2003). Improved thermal stability of Ni silicide on Si (100) through reactive deposition of Ni. *Journal of Vacuum Science and Technology* B, 21, 319-322.

Kim, J.; Anderson, W. A. Song, Y.-J. & Kim, G. B. (2005a). Self-assembled nanobridge formation and spontaneous growth of metal-induced nanowires. *Applied Physics Letters*, 86, 253101.

Kim, J. & Anderson W. A. (2005b). Spontaneous nickel monosilicide nanowire formation by metal induced growth. *Thin Solid Films*, 483, 60-65.

Kim, J. & Anderson W. A. (2006a). Direct Electrical Measurement of the Self-Assembled Nickel Silicide Nanowire. *Nano Letters*, 6, 1356-1359.

Kim, J.; Bae, J.-U. Anderson, W. A. Kim, H.-M. & Kim, K.-B. (2006b). Solid-state growth of nickel silicide nanowire by the metal-induced growth method. *Journal of Materials Researches*, 21, 2936-2940.

Kim, J.; Shin, D. Lee, E.-S. Han, C.-S. & Park, Y. C. (2007a). Electrical characteristics of single and doubly connected Ni silicide nanowire grown by plasma-enhanced chemical vapor deposition. *Applied Physics Letters*, 90, 253103.

Kim, J.& Anderson W. A. (2007). Metal silicide-medicated microcrystalline silicon thin-film growth for photovoltaics. *Solar Energy Materials and Solar Cells*, 91, 534-538.

Kim, J.; Piwowar, A. M. Nowak, R. Gardella Jr. J. A. & Anderson, W. A. (2007c). Cobalt-induced polycrystalline silicon film growth. *Applied Surface Science*, 253, 3053-3056.

Kim, J.; Shin Y.-H. Yun, J.-H. & Han, C.-S. (2008a). A nickel silicide nanowire microscopy tip obtains nanoscale information. *Nanotechnology* 19, 485713.

Kim, J.; Lee, E.-S. Han, C.-S. Kang, Y. Kim, D. & Anderson W. A. (2008b). Observation of Ni silicide formations and field emission properties of Ni silicide nanowires. *Microelectronic Engineering* 85, 1709-1712.

Kim, J.; Han, C.-S. Park, Y. C. & Anderson W. A. (2008c). Three-dimensional crystalline Si film growth by Ni silicide mediation. Applied Physics Letters, 92, 043501.

Kim, J.; Yun, J.-H. Han, C.-S. Cho, Y. Park, J. & Park, Y. C. (2009a). Multiple silicon nanowires-embedded Schottky solar cell. *Applied Physics Letters*, 95, 143112.

Kim, J.; Yun, J.-H. Song J.-W. & Han, C.-S. (2009b). The spontaneous metal-sitting structure on carbon nanotube arrays positioned by inkjet printing for wafer-scale production of high sensitive gas sensor units. *Sensors and Actuators* B, 135, 587-591.

Kittl, J. A.; Lauwers, A, Chamirian, O. Van Dal, M. Akheyar, A. De Potter, M. Lindsay, R. & Mzex, K. (2003) Ni- and Co-based silicides for advanced CMOS applications. *Microelectronic Engineering*, 70, 158.

Lavoie, C.; d'Heurle, F. M, Detavernier, C, & Cabral Jr. C, (2003). Towards implementation of a nickel silicide process for CMOS Technologies. Microelectronic Engineering, 70, 144-157.

Lee, K. S.; Mo, Y. H. Nahm, K. S. Shim, H. W. Suh, E. K. Kim, J. R. & Kim, J. J. (2004). Anomalous growth and characterization of carbon-coated nickel silicide nanowires. *Chemical Physics Letters, 384*, 215-218.

Lee, K. H.; Yang, H. S. Baik, K. H. Bang, J. Vanfleet, R. R. & Sigmund, W. (2004). Direct growth of amorphous silica nanowires by solid state transformation of SiO2 films. *Chemical Physics Letters*, 383, 380-384.

Levenger, A. & Thompson, C. V. (1990). Journal of Applied Physics, 67, 1325-1333.

Lew, K. K.; Pan, L. Bogart, T. E. Dilts, S. M. Dickey, E. C. Redwing, J. M. Wang, Y. Cabassi, M. & Mayer, T. S. (2004). Structural and electrical properties of trimethylboron-doped silicon nanowires. *Applied Physics Letters, 85, 3101-3103.*

Marrian C. R. K. & Tennant, D. M. (2003). Nanofabrication. *Journal of Vacuum Science and Technology* A, 21, S207-S215.

Meyer, B.; Gottlieb, U. Laborde, O. Yang, H. Lasjaunias, J. C. Sulpice, A. & R. Madar, R. (1997). Intrinsic properties of NiSi. Journal of Alloys and Compounds *262*, 235-237.

Morales, A. M. & Lieber, C. M. (1998). A laser ablation method for the synthesis of crystalline semiconductor nanowires. *Science* 279, 208-211.

Morimoto, T.; Ohguro, T. Momose, H. S. Iinuma, T. Kunishima, I. Suguro, K. Katakabe, I. Nakajima, H. Tsuchiaki, M. Ono, M. Katsumata, & Y. Iwai, H. (1995). Self-aligned nickel-mono-silicide technology for high-speed deep submicrometer logic CMOS ULSI. *IEEE Transaction on Electron Devices*, 42, 915-922.

Nastaushev, Y. V.; Cavrilova, T. Kachanova, M. Nenasheva, L. Kolosanov, V. Naumova, O. V. Popov, V. P. & Assev, A. L. (2002). 20-nm Resolution of electron lithography for the nano-devices on ultrathin SOI film. *Materials Science Engineering C, 19, 189-192.*

Ok, Y.-W.; Seong, T.-Y, Choi, C.-J. & Tu, K. N. (2006). Field emission from Ni-disilicide nanorods formed by using implantation of Ni in Si coupled with laser annealing. *Applied Physics Letters*, 88, 043106.

Ottaviani, G. (1981). Interface metallurgy and electronic properties of silicides. *Journal of Vacuum Science and Technology, 18*, 924-928.

Shan, Y.; Kalkan, A. K. Peng, C.-Y. & Fonash, S. J. (2004). From Si Source Gas Directly to Positioned, Electrically Contacted Si Nanowires: The Self-Assembling "Grow-in-Place" Approach. *Nano Letters, 4,* 2085-2089.

Smith, P. A.; Nordquist, C. D. Jackson, T. N. Mayer, T. S. Martin, B. R. Mbindyo, J. & Mallouk, T. E. (2000). Electric-field assisted assembly and alignment of metallic nanowires. *Applied Physics Letters, 77,* 1399-1401.

Tanase, M.; Silevitch, D. M. Hultgren, A. Bauer, L. A. Searson, P. C. Meyer, C. J. & Reich, D. H. (2002). Magnetic trapping and self-assembly of multicomponent nanowires. *Journal of Applied Physics*, 91, 8549-8551.

Thong, J. T. L.; Oon, C. H. Yeadon, M. & Zhang, W. D. (2002). Field-emission induced growth of nanowires. *Applied Physics Letters, 81,* 4823-4825.

Wagner, R.S. & Ellis, W.C. (1964). Vapor-liquid-solid mechanism of single crystal growth. *Applied Physics Letters,* 4, 89-90.

Wang, N.; Tang, Y. H. Zhang, Y. F. Lee, C. S. Bello, I. & Lee, S. T. (1999). Si nanowires grown from silicon oxide. *Chemical Physics Letters.* 299, 237-242.

Wang, Q.; Luo, Q. Gu. C. Z. (2007). Nickel silicide nanowires formed in pre-patterned SiO2 trenches and their electrical transport properties. Nanotechnology, 18, 195304.

Westwater, J.; Gosain, D. P. Tomiya, S. Usui, S. and Ruda, H. (1997). Growth of silicon nanowires via gold/silane vapor-liquid-solid reaction. *Journal of Vacuum Science and Technology B. 15, 554-557.*

Wu, Y.; Cui, Y. Huynh, L. Barrelet, C. J. Bell, D. C. & Lieber C. M. (2004). Controlled growth and structures of molecular-scale silicon nanowires. *Nano Letters,* 4, 433-436.

Wu, Y.; Xiang, J. Yang, C. Lu, W. & Lieber, C. M. (2004). Single-crystal metallic nanowirs and metal/semiconductor nanowire heterostructures. *Nature,* 430, 61-65.

Yan, H. F.; Xing, Y. J. Hang, Q. L. Yu, D. P. Wang, Y.P. Xu, J. Xi, Z. H. & Feng S. Q. (2000). Growth of amorphous silicon nanowires via a solid-liquid-solid mechanism. *Chemical Physics Letters* 323, 224-228.

Yoon, Y. J.; Kim, G. B. & Baik, H. K. (1999). Effects of phase and thickness of cobalt silicide on field emission properties of silicon emitters. *Journal of Vacuum Science and Technology* B, 17, 627-631.

Yu, D. P.; P. Bai, Z. G. Ding, Y. Hang, Q. L. Zhang, H. Z. Wang, J. J. Zou, Y. H. Qian, W. Xiong G. C. Zhou, H. T. & Feng, S. Q. (1998). Nanoscale silicon wires synthesized using simple physical evaporation. *Applied Physics Letters,* 72, 3458-3460.

Yun, J.-H.; Kim, J. Park, Y. C. Song, J.-W. Shin, D. H. & Han, C.-S. (2009) Highly sensitive carbon nanotube-embedding gas sensors operating at atmospheric pressure. *Nanotechnology,* 20, 055503.

Zeng, X. B.; Xu, Y. Y. Zhang, S. B. Hu, Z. H. Diao, H. W. Wang, Y. Q. Kong, G. L. & Liao, X. B. (2003). Silicon nanowires grown on a pre-annealed Si substrate. *Journal of Crystal Growth,* 247, 13-16.

Zhang, Z.; Hellstrom, P.-E. Ostling, Mikael & Zhang, S.-L. (2006). Electrically robust ultralong nanowires of NiSi, Ni2Si, and Ni31Si12. *Applied Physics Letters,* 88, 043104.

Zhang, H. Z.; Yu, D. P. Bai, Z. G. Hang, Q. L. & Feng S. Q. (1998). Dependence of the silicon nanowire diameter on ambient pressure. *Applied Physics Letters,* 73, 3396-3398.

Zheng, L. R.; Hung, L. S. Mayer, J. W. (1983). Silicide formation in lateral diffusion couples. *Journal of Vacuum Science and Technology* A, 1, 758-761.

2

Syntheses of Silver Nanowires in Liquid Phase

Xinling Tang[1] and Masaharu Tsuji[1,2]
[1]*Department of Applied Science for Electronics and Materials, Graduate School of Engineering Sciences, Kyushu University*
[2]*Institute for Materials Chemistry and Engineering, Kyushu University*
Japan

1. Introduction

Nanowires which were defined as having at least two dimensions between 1 and 100 nm, have received a great interest due to their unique optical, electrical, magnetic, and thermal properties with dimensionality and size confinement [1-10]. The intrinsic properties of nanowires are mainly determined by its size and composition. In order to study the size dependent properties, it is the crucial task to synthesize size-controlled nanowires. Among all these nanowires, the synthesis of silver (Ag) nanowires has been and continues to be an area of active research due to the high electrical and thermal conductivities of bulk silver, which is an important material in many fields. Moreover, silver nanowires have also been used as sacrificial templates to generate other nanostructures such as gold nanotubes, which are difficult to be fabricated [9,10]. The research progress in synthesis strategy is mandated by advancements in all areas of industry and technology. In the past ten years, owing to the efforts from many research groups, splendid strategies were developed for the synthesis of Ag nanowires with various levels of control over the growth parameters.

These synthetics strategies have been conveniently categorized into vapor phase approaches and liquid phase growth approaches. Vapor phase approaches mainly utilize physical methods such as an electronic-beam. Limited by the space of the chapter, we lay a strong emphasis on the introduction of liquid phase synthesis of Ag nanowires. Liquid phase syntheses were most widely used because these approaches have the advantages of nature of homogeneous reaction, wide range of solvents, simple monitoring technology, and low cost. In this article, we review some current research activities that center on Ag nanowires. Representative techniques are discussed and supply a basic understanding of the methods and mechanisms for preparing Ag nanowires.

2. Typical synthetic strategies of Ag nanowires

2.1 Polyol method

Polyol synthesis was originally introduced by Fievet et al. [11] as an excellent method for the synthesis of submicrometer-sized metallic nanoparticles. Xia and co-workers successfully developed this method to the preparation of single-crystal Ag nanoparticles with uniform size and shape using polyvinylpyrrolidone (PVP) as a protecting agent [12]. Now the polyol method has become widely used by many research groups for the synthesis of metal nanostructures. Xia et al. synthesized Ag nanowires with higher aspects ratios by injection of ethylene glycol (EG) solutions of $AgNO_3$ and PVP, added drop-by-drop, at a constant solution temperature of 160 °C [12]. In the polyol process, the introduction of an exotic reagent is considered to be the key factor that leads to the formation of wire-like structures. In their experiments, Ag nanowires are generated using a self-seeding process and EG acts as both solvent and reducing agent. For the formation of silver nanowires, low precursor concentrations and slow addition rates are necessary. By controlling the injection rate, multiple-twined particles (MTPs) formed at the initial stage of the reduction process could serve as seeds for the subsequent growth of silver nanowires, which is the so-called self-seeding process. At a lower precursor concentration, it is possible to reduce the chemical potential to a relatively low level so as to make MTPs thermodynamically stable because it is bound almost entirely by the lower energy {111} facets [13-15]. At the same time, due to the existence of twin defects at the borders of combined tetrahedron, silver atoms will preferentially deposit on those active sites. Once five-twinned particles have been formed, they will change easily from MTPs to silver nanowires. Figs. 1A-1D show the scanning electron microscopic (SEM) and transmission electron microscopic (TEM) images and XRD of the silver nanowires with a large ratio of length to width. It was found that the morphologies and aspect ratios of Ag nanowires strongly depend on the molar ratio between the repeating unit of PVP and $AgNO_3$. When the molar ratio between PVP and $AgNO_3$ was more than 15, the final product was essentially composed of silver nanoparticles. When the molar ratio decreased to 6, the resulted product contained many

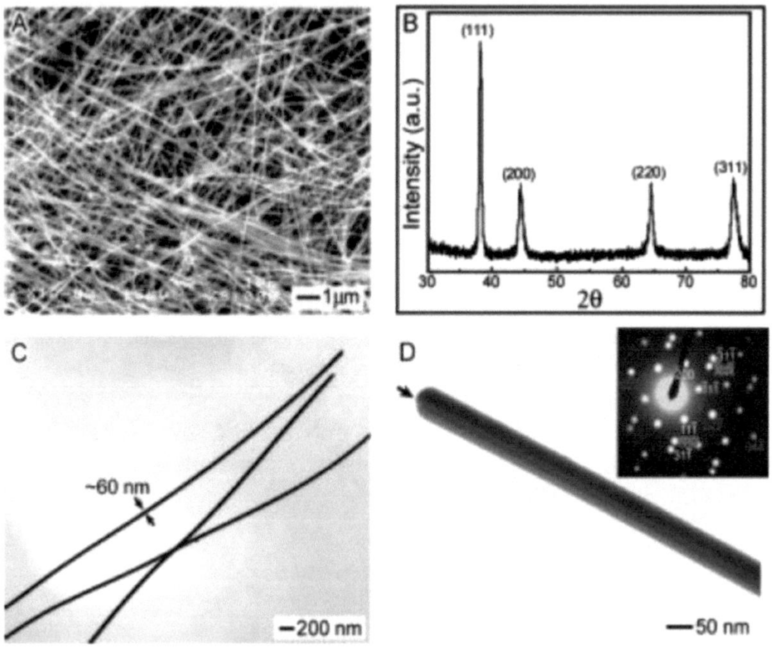

Fig. 1. A) SEM and C) TEM images of uniform silver nanowires which were synthesized via the self-seeding polyol process. B) XRD pattern of these nanowires, indicating the fcc structure of silver. D) The TEM image of an individual silver nanowire. The inset gives the microdiffraction pattern recorded by focusing the electron beam on this wire. Reprinted with permission from reference [12f]. Copyright 2002 ACS.

nanorods. Nanowires were synthesized with the molar ratio of 1.5. If the molar ratio was further decreased to 0.6, nanowires become significantly thicker and some of them were non-uniform along the longitudinal axis. It was generally accepted that for a given coordination reagent such as PVP, it can kinetically control the growth rates of various faces of metal through selective adsorption and desorption on these surfaces. An advantage of this method is that large-scaled synthesis of Ag nanowires is possible. It was believed that the synthetic approach described by Xia et al. [12] can be extended to many other metals by selecting appropriate coordination reagents.

2.2 Salt-mediated polyol method

Based on the PVP-assisted polyol method, Xia and co-workers also developed a salt-mediated polyol process to prepare silver nanowires [12,16]. The addition of a trace amount of salt, such as NaCl, $Fe(NO)_3$, $CuCl_2$ and CuCl, has been shown to influence the morphology of the final metal products. Usually salt-mediated synthesis strategy is a simple and effective method which is useful for the mass synthesis of silver nanowires [16]. For example, they synthesized silver nanowires in high yields by reducing $AgNO_3$ with EG heated to 148 °C in the presence of PVP and a trace amount of NaCl. It was found that oxygen must be removed from the reaction solution in the presence of Cl^- anions in order to obtain silver nanowires. Fig. 2 shows the proposed mechanism of oxidative etching and growth of MTPs. Five-twinned decahedral particles were proposed to be seeds of Ag nanowires. The defects among single-crystal tetrahedron subunits of decahedral particles provide high-energy sites for atomic addition, leading to one dimensional Ag products via anisotropic growth along the <110> direction. The addition of chloride causes enhanced oxidation and preferential etching of twinned particles, leaving only single-crystal particles (or seeds) to grow. Under the similar conditions, Xia's experiment has been carried out under argon which supplied an anoxic condition and the twinned particles that formed in the early stage of the reaction could grow into uniform nanowires [17].

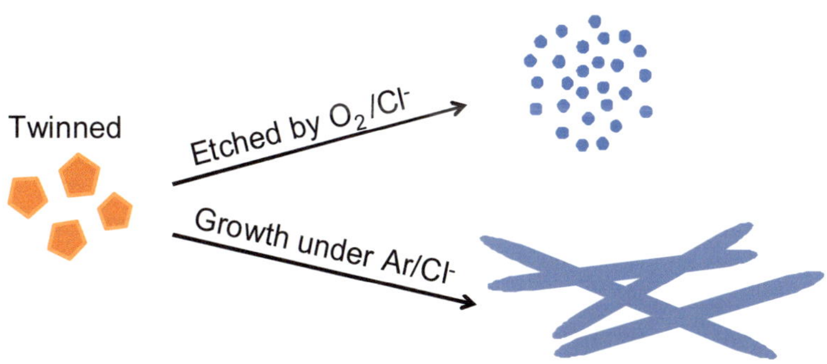

Fig. 2. Illustration of the proposed mechanism by which twinned decahedron was etched with the existence of O_2/Cl^- and silver nanowire was obtained under argon protection.

In another typical synthesis of silver nanowires by Xia et al. [16a], reduction reaction was carried out through a CuCl or $CuCl_2$ mediated polyol process but without introducing any inert gases. Both the cation and anion are necessary for the successful production of silver nanowires. It was found that oxidative etching effect was not observed even at a comparably high Cl^- concentration. This result was attributed to a decrease in the amount of dissolved O_2 present during the reduction reaction. Fig. 3 shows the role of Cu-containing salts in the polyol synthesis of Ag nanowires. O_2 can adsorb and dissociate on silver surfaces and block the {111} reactive sites from adding Ag atoms efficiently [18,19]. Cu(I) [Cu(II) can be reduced into Cu(I) by EG] effectively scavenge any adsorbed atomic oxygen from the surface of silver seeds and benefit the growth of silver nanowires. These results suggest that silver nanowires can be achieved more rapidly when oxygen is most effectively removed. A trace amount of Cl^- also acts an important role in the polyol synthesis of silver nanowires. Firstly, it provides electrostatic stabilization for the initially formed silver seeds. Secondly, Cl^- can reduce the concentration of free Ag^+ ions in the solution through the formation of AgCl nanocrystallites and supply a slow release process of Ag^+ which facilitates the high-yield formation of the thermodynamically more stable MTPs required for wire growth.

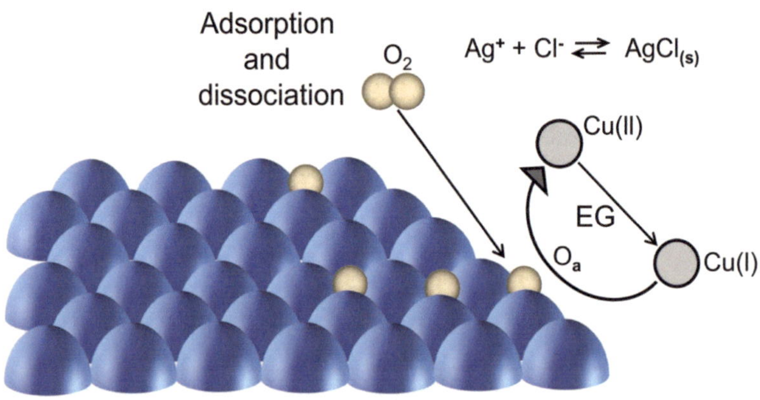

Fig. 3. A schematic illustration depicting the role of Cu-containing salts in the polyol synthesis of Ag nanowires. Molecular oxygen present during initial seed formation can adsorb and dissociate on the Ag seeds. Cu(I) rapidly scavenges this adsorbed atomic oxygen (O_a).

All previous reports of the synthesis of silver nanowires from an $AgNO_3/Cl^-/PVP/EG$ solution described that it is necessary to remove the oxygen dissolved in the reaction solution in order to achieve the growth of silver nanowires. Recently, our group [20-23] and Gou et al. [24] found that the mechanism described above is not applicable to the synthesis of Ag nanowires under rapid microwave(MW)-polyol method in the presence of Cl^- anions because one dimensional (1D) structures were produced preferentially from unstable decahedrons even in the presence of O_2. It was predicted that not only Cl^- from NaCl, but also O_2 dissolved in EG probably takes part in the shape-selective evolution of Ag 1D products.

We extended the MW-polyol method to an air-assisted method which is carried out in an oil-bath tank. In general, mixtures of nanowires and other spherical and cubic particles are prepared in the polyol synthesis. Therefore, centrifugal separation between nanowires and other products was necessary to obtain nanowires. We have recently prepared Ag nanostructures under bubbling air or N_2 gas [25]. Then, we succeeded in the synthesis of monodispersed silver nanowires in high yield (>90% without isolation) under bubbling air through a reagent solution under oil-bath heating from ca. 20 °C to the boiling point of EG (198 °C) for 20 min. Fig. 4 depicts typical TEM images of Ag nanowires obtained under bubbling air and N_2. Long nanowires of uniform diameters of ca. 40 nm and lengths of 5–30 μm could be obtained under bubbling air without centrifugal separation of other particles

(Figs. 4A and 4B). On the other hand, mixtures of cubes, bipyramids, and wires were prepared under bubbling N_2 (Figs. 4C and 4D). The yield of 1D particles is only 20% because of high yields of cubes and bipyramids under bubbling N_2. Similar Ag nanostructures were obtained when we prepared Ag nanostructures without gas bubbling. The yield of nanowires (75%) was an intermediate value between that obtained undr bubbling air (90%) and that under bubbling N_2 (20%). This result can be attributed to that a small amount of residual O_2 gas is dissolved in EG without gas bubbling. It takes part in the production of each Ag nanostructure.

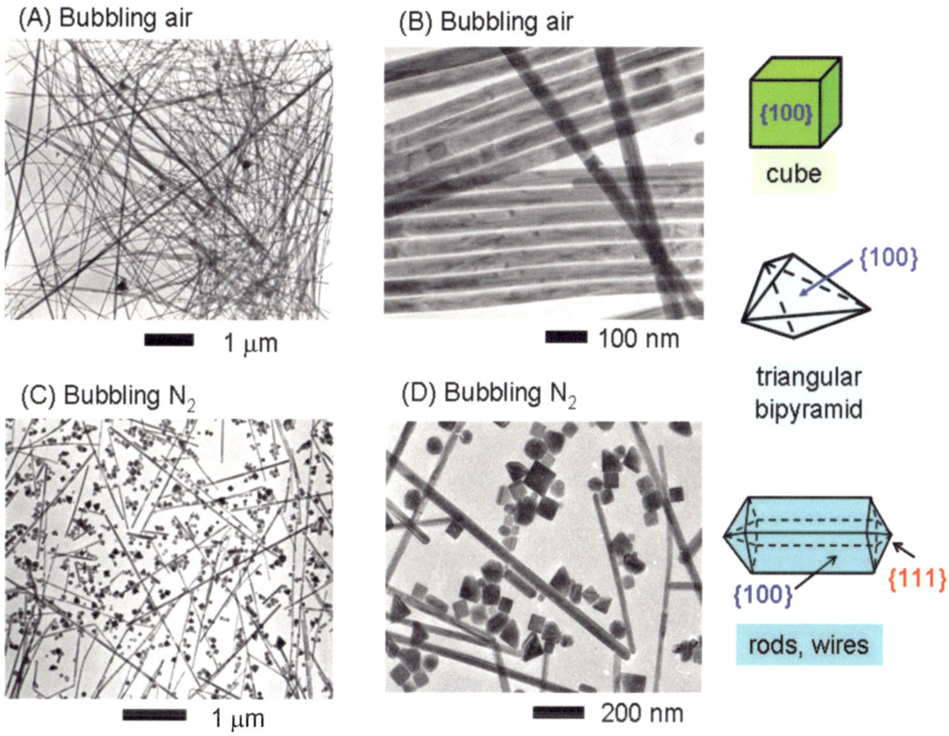

Fig. 4. Typical TEM images of Ag nanostructures prepared from $AgNO_3$ (46.5 mM)/NaCl (0.3 mM)/PVP (264 mM)/EG mixtures at 198 °C by bubbling of air (A and B) and N_2 (C and D). Also shown are crystal structures of 1D products, cubes, and triangular bipyramids.

The growth process of silver nanostructures was also studied by observing ultraviolet-visible (UV-Vis) spectra and TEM images at various temperatures. It is well documented that Ag nanostructures display different SPR bands with different frequencies depending on shapes and sizes, as reviewed by Mock et al. [26] and Wiley et al. [27]. The main SPR peaks for Ag nanorods and nanowires appear at ca. 350 and ca. 380 nm as shown in Fig. 5A at 198 °C. These peaks were attributed to the transversal modes of the 1D products with pentagonal cross sections, corresponding respectively to the out-of-plane quadrupole resonance and out-of plane dipole resonance modes. The SPR bands of cubic (ca. 80 nm edge length) and triangular bipyramidal (75 nm edge length) Ag crystals were observed respectively at 320–800 nm with a peak at ca. 470 and 320–900 nm with a peak at ca. 520 nm [12,26,27]. The peaks described above mutually overlap when mixtures of Ag nanostructures of various shapes and sizes were prepared.

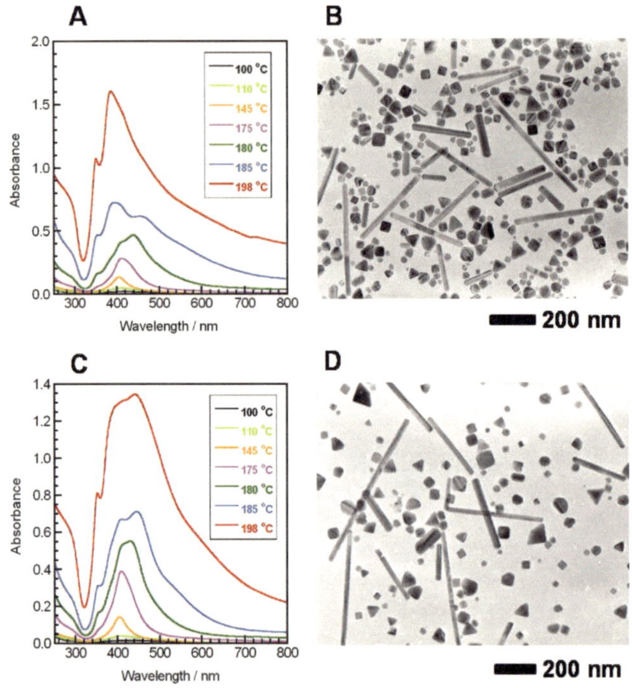

Fig. 5. UV-Vis spectra of Ag nanostructures prepared from $AgNO_3$/NaCl/PVP/EG mixtures under bubbling of (A)air and (C) N_2. Also shown are typical TEM images of Ag nanostructures obtained by bubbling of (B) air and (D) N_2 at 180 °C.

To examine the shape evolution of Ag nanostructures through bubbling air and N_2, the UV-Vis absorption spectra of the reagent solutions were measured at various solution temperatures (Figs. 5A and 5C). UV spectra obtained under bubbling air and N_2 are similar at low temperature. A symmetrical SPR band with a peak at ca. 410 nm appears at ca. 100 °C and becomes strong and broad up to ca. 175 °C. This suggests that the formation of spherical nanoseeds and the sizes become large with the increase in temperature. With increasing temperature from 175 to 180 °C, the SPR peaks become broad and three peaks appear at 355, 410, and 440 nm. The two former peaks are ascribable to SPR bands of 1D Ag nanoparticles; the last one can be assigned to SPR bands of quasi-cubic and quasi-bipyramidal structures. To confirm morphologies of Ag nanostructures at 180 °C, TEM images of products were measured for bubbling of N_2 and air (Figs. 5B and 5D). An outstanding finding is that there is no marked difference in yields of respective products obtained under bubbling air and N_2 at 180 °C. On the other hand, it was found that the growth rate of 1D product is much higher than those of cubes, bipyramids, and spheres in the 180–198 °C range for both bubbled gases; growth is especially rapid under bubbling air. Furthermore, the last peak of the cubes and bipyramids disappears at 198 °C, which indicates the yields of cubes, bipyramids, wires, and spheres are much lower under bubbling air in the 180–198 °C range.

Based on the above findings, we proposed the mechanism of the Ag nanostructure formation under bubbling N_2 and air (Fig. 6). The reaction process was separated into two stages. The first stage is from room temperature to ca. 180 °C and it need about 16 min to finish this process under oil-bath heating at 500 W; seed nucleation and formation take place at this low temperature range. The solubility constant of AgCl in EG is very small at a room temperature (ca. 20 °C). Therefore, there is little free Cl^- anion existed in the solution and almost all Cl^- from NaCl is present as small AgCl nanoparticles. These indiscerptible AgCl particles can serve as seeds of Ag anisotropic nanoparticles. With increasing the temperature from room temperature to 180°C, AgCl is dissolved gradually in EG and much Ag^+ and Cl^- are released to the reaction solution, which supply reducing materials to neutral Ag. In a separate experiment we have confirmed that Ag nanoparticles are actually produced directly from AgCl by heating an AgCl solution from room temperature to 180 °C. This result indicates that AgCl is dissolved and reduced to Ag. It finally provides small Ag nanoparticles in EG in this temperature region.

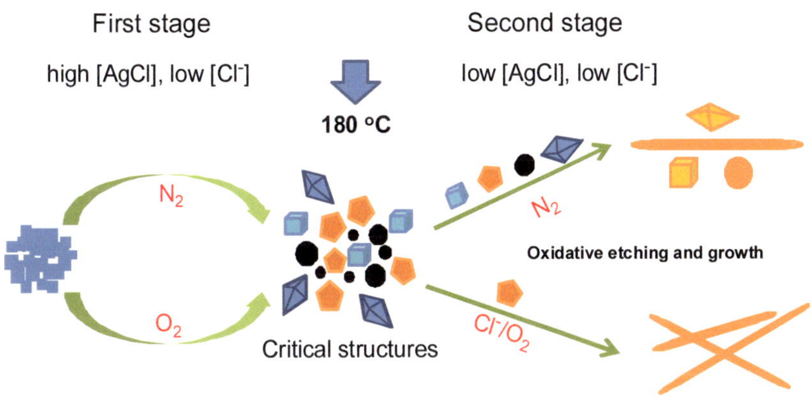

Fig. 6. Proposed mechanism of Ag nanostructures synthesized from $AgNO_3$/NaCl/PVP/EG mixtures using bubbling of N_2 and air.

In the first stage in Fig. 6, the Cl^- concentration is low so that oxidative etching rates are slow. Under such conditions, the etching rate is insufficiently high to etch small Ag particles. Therefore, effects of gas are unimportant, and mixtures of cubes, bipyramids, and rods are produced in both bubbling of N_2 and air at temperatures less than ca. 180 °C. The second stage is the high-temperatures from 180 to 198 °C. In this range, the etching rate increases with the increasing in the concentration of Cl^-. Thereby, oxidative etching rates become so high that small Ag particles can be etched in the presence of O_2. In this range, the growth rates of Ag seeds also increase considerably: rapid crystal growth and oxidative etching occur competitively. The relative rate of crystal growth and oxidative etching strongly depends on shapes of Ag particles. In this condition, the growth rate from decahedra to 1D products is fast, whereas the etching rates of other products are fast. Therefore, Ag nanowires are produced in a high yield.

In general, shapes of the final products depend not only on the stability of crystal structures against etching by Cl^-/O_2 but also the density of Ag^0 atoms around the growing nuclei. A high related ion concentration around the nuclei can create an environment with high twinning probability for the formation of MTPs from which pentagonal 1D rods and wires are grown. Thus, five-twinned 1D products are favorable at higher concentration of Ag^+ ions, whereas single crystal cubes are preferentially formed at lower concentration of Ag^+ ions.

In order to further confirm the effect of NaCl in the synthesis of silver nanowires, a series of experiments were carried out without gas bubbling [28]. First Ag nanostructures were

prepared at various NaCl concentrations from 0 to 0.45 mM. We found that the concentration of NaCl strongly affects the final shapes of products. A low concentration of NaCl (0.12 mM) is in favor of the formation of cubes and bipyramids, whereas a higher concentration of 0.3 mM is the best condition for the formation of NaCl nanowires.
We classified mechanisms for the formation of Ag nanostructures into three cases as shown in Fig. 7, depending on the molar ratios of [NaCl]/[$AgNO_3$]. Case (a) is [NaCl]/[$AgNO_3$] = 0. When the NaCl is absent, Ag^+ ions are reduced to Ag^0, nucleation occurs, and then the Ag seeds are formed and their growth leads to spherical particles formed randomly. Case (b) is [NaCl]/[$AgNO_3$] = 1. When equal molar concentrations of $AgNO_3$ and NaCl are mixed at room temperature, AgCl is formed due to the small solubility constant of AgCl. In this case, AgCl is dissolved as Ag^+ and Cl^- with increasing temperature. After then Ag^+ ions are reduced to Ag^0 and nucleation, seed formation, and crystal growth take place. Since the concentration of Ag^+ ions is low, twinning probability is low. Therefore, single crystal cubes and single-twin bipyramids are produced preferentially. Case (c) is an intermediate between the above two cases: [NaCl]/[$AgNO_3$] = 0.0065. In this case, although AgCl is formed, a large excess amount of $AgNO_3$ is present in the solution. The standard electrode potential of $AgCl + e^- \rightarrow Ag + Cl^-$ (+0.2233 V vs. SHE) is lower than that of $Ag^+ + e^- \rightarrow Ag$ (+0.799 V vs. SHE). Under such a condition, most of the Ag^0 atoms do not arise from AgCl but from free Ag^+. In this case, major products are five-twinned 1D particles because the concentration of Ag^+ ions is sufficiently enough to provide a high probability for twinning.

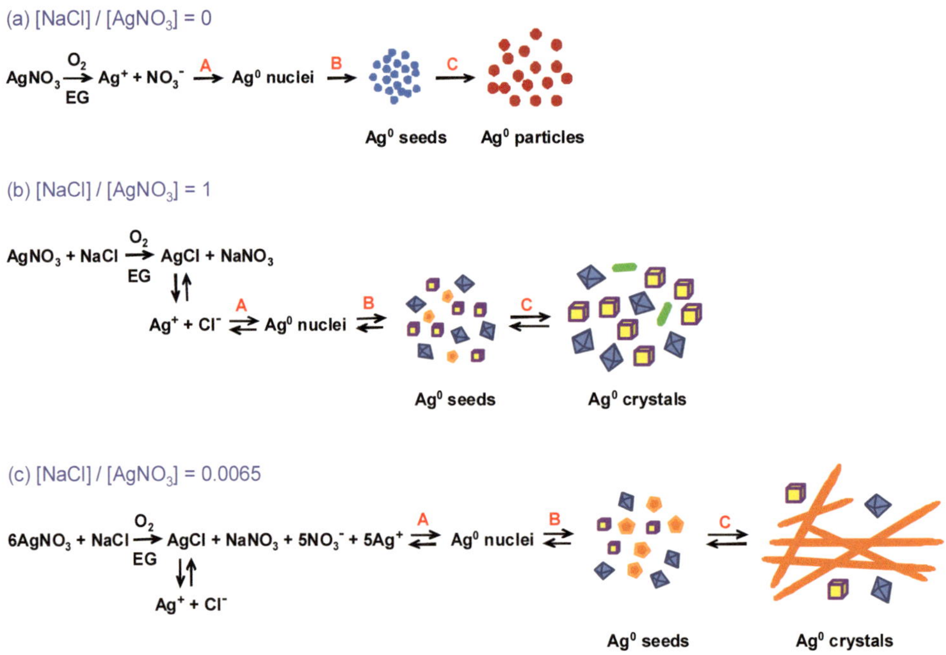

Fig. 7. Possible nucleation and growth mechanisms of Ag nanostructures at different molar ratios of [NaCl]/[$AgNO_3$].

In the case of the liquid phase method, previous studies on anisotropic growth of Ag nanomaterials have demonstrated that Pt seeds as exotic nuclei play a key role with PVP served as a capping reagent. Xia et al. have described a soft template assisted solution-phase approach for the production of 1D nanostructures by reducing $AgNO_3$ in the presence of PVP and $PtCl_2$ in EG at a high temperature using oil bath [29]. They have suggested that the formation of the 1D Ag nanostructures proceeds via two steps. The first step involve the formation of Pt nanoparticles with an average diameter of 5 nm by reducing $PtCl_2$ with EG .

$$2HOCH_2\text{-}CH_2OH \rightarrow 2CH_3CHO + 2H_2O \quad (1)$$

$$2CH_3CHO + PtCl_2 \rightarrow CH_3CO\text{-}COCH_3 + Pt + 2HCl \quad (2)$$

At second step, $AgNO_3$ and PVP are added into the reaction system drop-by-drop, leading to the nucleation and growth of Ag nanowires. They thought that the pre-synthesized Pt nanoparticles serve as seeds for the heterogeneous nucleation and growth of silver owing to the close match in crystal structure and lattice constants. Then Ag nanoparticles with diameters of 20-30 nm were created following the nucleation process. Many Ag atoms also nucleated through a homogeneous process and formed critical particles with diameter of 1-5 nm. At high temperature, critical particles progressively disappeared to benefit the growth of large particles via the famous Ostwald ripening [30]. With the assistance of PVP, some of the large particles grow into nanowires with diameters in the range of 30-40 nm. Usually the growth process would continue until all these critical silver particles with diameters less than 5 nm were completely consumed, and only nanowires and larger nanoparticles can survive stably.

Our group have conducted a further detailed study on the real role of Pt seeds as a nucleation regent for the preparation of 1D Ag nanostructures by using the MW-polyol method [23]. To examine effects of Pt seeds and Cl^- anions, a various kind of experiments were carried out. When Pt catalysts were prepared from two different agents, $H_2PtCl_6{\cdot}6H_2O$ and $Pt(acac)_2$, the 1D Ag nanostructures could be easily prepared only by using $H_2PtCl_6{\cdot}6H_2O$, although Pt seeds could be produced when using either of them. This result might indicate that Pt catalysts probably do not act as seeds for evolution of the 1D Ag nanostructures. In contrast, it is possible for Cl^- ions resulting from $H_2PtCl_6{\cdot}6H_2O$ to play an important role in assisting the formation of the 1D Ag nanostructures. So we used NaCl or KCl as a source of Cl^- ions to elucidate the role of Cl^- ions. We found that nanowires could be prepared when using either of them. In addition, besides the 1D Ag products, the formation of some well-crystallized cubic and bi-pyramid Ag nanocrystals with sharp edges has been found to be related to existence of Cl^- ions. We confirmed that the so-called Pt seeds-mediated process is in fact a salt-mediated polyol process.

2.3 Seed-mediated growth approach

Murphy et al. firstly reported the large-scaled preparation of silver nanowires with uniform diameter by seed-mediated growth approach in a rod-like micellar media [31]. In principle, two steps were necessary in order to achieve the formation of nanowires. Firstly, Ag nanoseeds with an average diameter of 4 nm were prepared by chemical reduction of $AgNO_3$ by $NaBH_4$ in the presence of trisodium citrate. Then $AgNO_3$ was reduced by ascorbic acid in the presence of Ag seed obtained in the first step, the micellar template cetyltrimethylammoniun bromide (CTAB), and NaOH in order to synthesize nanorods and nanowires with various aspect seed. This seed-mediated growth approach is prone to form silver nanorods and nanowires of controllable aspect ratio.

It is well documented that Ag nanowires were one of the earliest twinned Ag nanostructures reported and now have become an active area of research in Ag nanoparticles. But it was always a great challenge to achieve control over the length and width of Ag nanorods and nanowires. At the same time, silver nanorods and nanowires with pentagonal cross-section have been proposed to come from the evolvement of decahedra. Recently, Kitaev et al. synthesized monodisperse size-controlled faceted pentagonal silver nanorods by thermal regrowth of decahedral silver nanoparticle in aqueous solution at 95 °C, using citrate as a reducing agent [32]. Here we will introduce this method although it is a synthesis method of Ag nanorods. Firstly monodisperse decahedral Ag nanoparticles were synthesized by photochemical method and the size of decahedron can be controlled in a larger scale range by regrowth. Then, silver decahedra were added into a temperature-controlled solution and a thermal regrowth was carried out at 95 °C with the presence of PVP with citrate as a reducing agent. Fig. 8 shows the typical SEM images of the resulted monodisperse pentagonal Ag nanorods. These faceted pentagonal rod particles readily self-assemble into densely packed arrays upon drying of their dispersions owing to the excellent monodispersity and good stabilization in aqueous solution. It was found that PVP can improve the yield and stability of Ag nanoparticle, but not absolutely necessary for the shape control. Ag nanorods with various aspect ratio of length to diameter can be achieved by semicontinuous silver addition and the width of nanorods was determined by the size of decahedron. Due to the ease of preparation and excellent shape and size control, this method is well worth further studying and can be extended to the synthesis of Ag nanowires in the future. These nanorods will be attractive for diverse practical applications due to the reliable synthetic approach, high monodispersity, excellent size control.

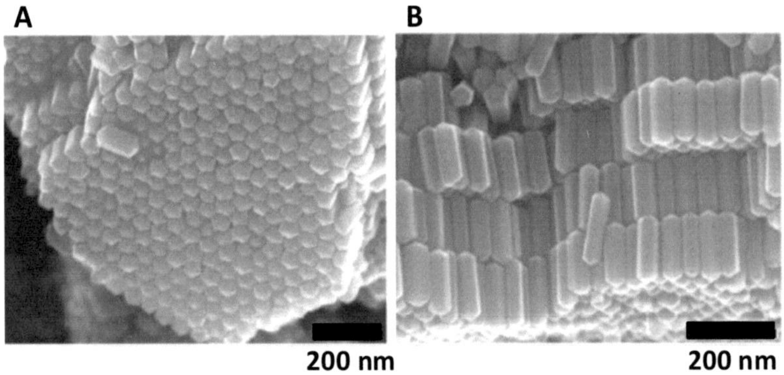

Fig. 8. SEM images of self-assembled packing of monodisperse pentagonal Ag nanorods. Reprinted with permission from renference [32]. Copyright 2009 ACS.

Before the work of Kitaev et al. [32], silver nanorods and nanowires have also been synthesized in our group using multiple twin decahedron (MTD) of gold as the seed by the MW-polyol method [33]. Because gold and silver have the same FCC crystal structures, and the lattice constants of Au (0.4079 nm) and Ag (0.4086 nm) are very similar, there are many kinds of Au@Ag core-shell structures can be prepared from Au/Ag reagents. The final shape of Ag outer shell was decided by the shape of Au inner core. The Au@Ag core-shell nanostructures were synthesized in two steps. The first step was the synthesis of an Au inner core, and the second step was the preparation of an Ag outer shell including nanorods and nanowires. Fortunately, the thin boundary layers of Au@Ag core-shell structures can easily be observed from TEM images, because Au cores and Ag shells appear as clear black and blight contrasts, respectively. Based on the TEM observation of Au@Ag particles prepared at different [Ag]/[Au] molar ratios (Fig. 9) and EDS data, the growth mechanism of Au@Ag nanocrystals overgrown on an MTD Au core was illustrated in Fig. 10. The MTD nanocrystals consist of ten {111} facets and crystal growth occurs in two opposite <110> directions with the same growth rates. Because PVP molecules selectively adsorb on the surface of five rectangular side {100} facets and, crystal growth occurs on the exposed active {111} facets. Thus Ag shells are continuously overgrown on MTD particles along <110> directions because active {111} facets with a constant total area remain during the crystal growth. During our studies on core-shell nanocrystals, we have found that Ag rods and wires are overgrown on the MTD Au crystals located in the center of the 1-D particles. This result provides the first definite experimental evidence that the previously proposed growth mechanism of 1-D Ag nanorods and nanowires is valid. It was found that optical properties of Au@Ag nanocrstals are different from that of pure Au core and Ag shell nanocrystals, they can be applied to a new optical device.

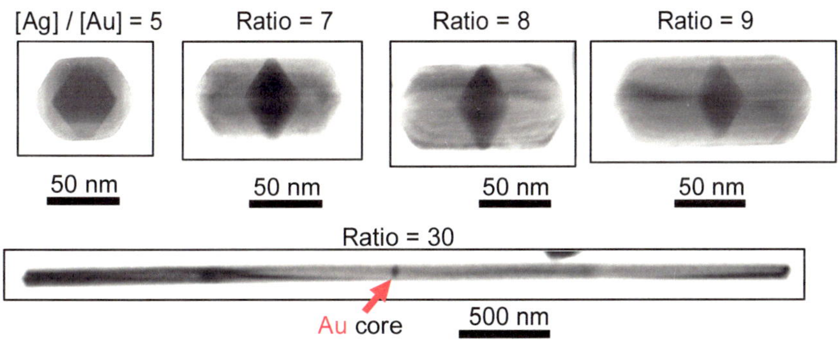

Fig. 9. TEM photographs of Au@Ag nanorods and nanowires prepared by addition of various amounts of $AgNO_3$ to Au cores.

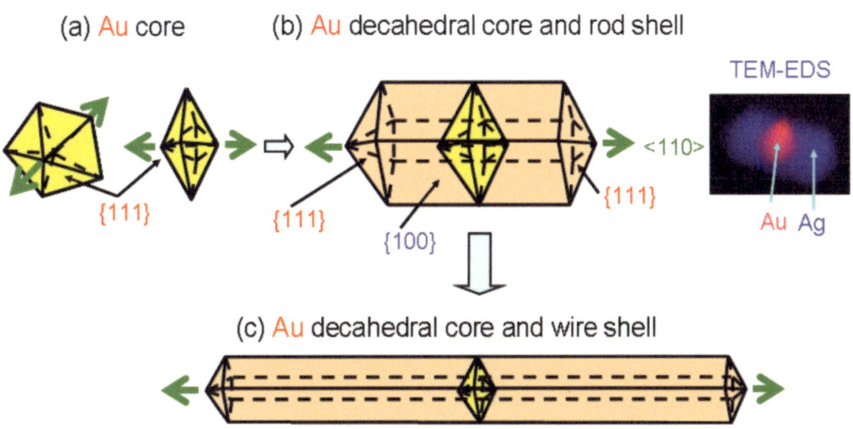

Fig. 10. Growth mechanism of Au@Ag nanorods and nanowires from the decahedral Au core.

2.4 Seedless and surfactantless wet chemical synthesis

Previous strategies to control the growth of Ag nanowires have involved either softer directing agents, including surfactants or polymers. It is hard work to remove the directing agent from nanowires surface because multiple washings are required. For some sensing applications, the surface-bound residue from the synthesis increases the difficulty of doing chemistry at the surface. Murphy and co-workers reported a method to make crystalline silver nanowires in water, in the absence of a surfactant or polymer to direct nanoparticle growth, and without externally added seed crystallites [34]. The reaction is one in which silver nitrate is reduced to silver metal at 100 °C by sodium citrate in the presence of NaOH. Hydroxide ion concentration was an important factor in determining the morphology of the final product. Ag nanowires prepared by this method is up to 12 microns long. In this synthesis, the citrate plays multiple roles. It not only strongly complexes with the silver ion but it is also responsible for the reduction to silver metal and acts as a capping agent to the silver metal. Although only small amount of hydroxide exists in the reaction solution and hydroxide weakly complexes with the silver ion, but it is enough to interfere with the capping ability of citrate and lead to the formation of Ag nanowires. It is a simple method to prepare nanowires of silver in high yield, with relatively few spherical nanoparticle byproducts. Most importantly, it does not require a surfactant or polymer template, or seeds. This method makes it possible to prepare metallic nanostructures with as "clean" surfaces in water.

2.5 Seedless and surfactant assisted synthesis

Surfactant assisted synthesis is one of the soft-template processes. Murphy et al. have successfully synthesized silver nanorods or nanowires with a rod-like micelle template in CTAB assisted process [31]. During their experiment procedure, seeds are also required in order to achieve the growth of nanorods or nanowires as described in 2.3, so the reaction system becomes complicated. For the surfactant approaches, when the concentration of surfactant is above the critical concentration, micelles with a certain shape will form and act as a soft template. Nanorods and nanowires with controllable aspect ratio are fabricated with the absence of seeds. If the surfactant concentration is too low to form any micelles, it only acts as a capping agent or dispersing agent. Murphy and coworkers also reported the seedless and surfactantless wet chemical synthesis of silver nanowires, in which tri-sodium citrate serves as a capping agent and reducing agent [34] as shown in 2.4. The limitation of surfactantless approach is that it can't well control the dimensions and morphologies of the product and nanowires obtained with this method are poorly dispersed.

Tian and co-workers developed a seedless surfactant process to synthesize high-quality Ag nanorods and nanowires in large quantities [35]. Silver nitrate is reduced by tri-sodium citrate ($Na_3C_6H_5O_7$) in the presence of sodium dodecylsulfonate (SDSN). In this process, tri-sodium citrate plays an important role, and SDSN only acts an assisting role as a capping agent in the formation of Ag nanowires. The concentration of SDSN actually used in the experiment process of Ag nanowires is only 1 mM, which is far lower than the critical concentration for the formation of spherical micelles (9.7 mM, 40 °C). Especially, the critical concentration of the micelle formation for SDSN increases with temperature rising. So there are no SDSN micelles formed to serve as a soft template during the whole process of forming nanostructures. The diameters and aspect ratios of Ag nanowires or nanorods can be controlled by changing the concentration of tri-sodium citrate. However, it does not mean that SDSN is unnecessary for the formation of nanorods or nanowires. Ag nanowires cannot generate under too-low or too-high concentration of SDSN. The key to form Ag nanowires is to well control the tri-sodium citrate concentration for a fixed SDSN concentration.

Fig. 11 shows the proposed schematic illustration for the growth of Ag nanorods and nanowires. At a lower concentration of tri-sodium citrate below 0.2 mM, Ag monomer concentration favors the formation of nanospheres according to Ostwald ripening and small particles dissolve into the solution and deposit on large particles. At a higher concentration of tri-sodium citrate beyond 0.2 mM, a large quantity of monomers is produced in the solution and the growth mode is diffusion-controlled. The monomer concentration in the stagnant solution keeps the solubility of the (110) facet by a rapid growth onto or dissolution from the facet in the assistance of SDSN. It is well documented that the activation energy of the (110) facet is lower than that of (100) and (111) owing to the relative high surface free energies for a FCC structured metal particle. Therefore, monomer particles grow up along <110> directions owing to the strong bonding abilities and chemical reactivities of (110) facet. Thus nanorods are obtained. At higher concentration tri-sodium citrate (>0.8 mM), nanowires with smaller diameter and higher aspect can be formed because of the higher Ag monomer concentration and high velocity of crystal growth.

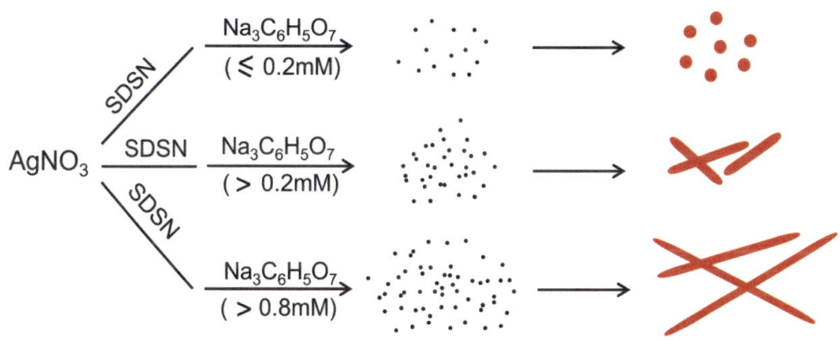

Fig. 11. Schematic illustration of the experimental mechanisms to generate spherical, rod- and wire-like nanoparticles.

3. Typical synthetic strategies of Ag nanowires of Coaxial nanocables

In recent years, coaxial nanocable as a new nanostructure has also received great attention because their functions could be further enhanced by fabricating the core and sheath from different materials, which will find use as interconnects in fabricating nanoelectronic devices. Ag-polymer nanocables became a major focus of cable-shaped nanostructures, due to its extremely high electrical conductivity in the bulk and its unusual optical properties depending on the size. It has also been predicted that Ag nanowires with polygonal or circular cross sectional symmetry would exhibit different resonances and a dielectric coating would lead to a red-shift of the SPR peak [36,37]. According to the traditional crystal-growth theory, it is impossible to form Ag nanowires with circular cross-section. Cable-like structures make it possible for the formation of nanowires with expected cross-sections. Then many approaches were developed to achieve Ag nanocables.

Xia's group first reported the preparation of Ag nanocables by directly coating bicrystalline silver nanowires with conformal sheaths of silica through a sol-gel process [38]. Yu et al. [39] introduced a simple large-scale hydrothermal synthesis of flexible nanocables with silver as cores and cross-linked poly(vinyl alcohol) (PVA) as shells. In a typical procedure, a clear solution containing 0.06 mM $AgNO_3$ and 5 mL PVA (3 wt %) solution was added into Teflon-lined stainless steel autoclave, which was sealed and maintained at 160 °C for 36 or 72 h and then cooled to room temperature naturally. The resulting product is composed of flexible fibers with a diameter of 0.7-4 μm and length up to 100 μm (Fig. 12a). All these silver nanowires show a composite structure of a smooth core and a surrounding sheath (Figs. 12b-12e). At the same time, it is well distinguished between surrounding polymer and inner core. The shell thickness of the obtained nanocables is decided by the reaction time and the temperature can also affect the thickness of the shell. It is well worth noting that PVA cannot become cross-linked in the absence $AgNO_3$ after a hydrothermal treatment and keep other parameters constant, which indicate that $AgNO_3$ is essential for the formation of cables. In fact, $AgNO_3$ acts as a catalyst for oxidation-reduction reactions and PVA can be oxidized by $AgNO_3$ at 160 °C into cross-linked PVA. It is promising to extend this method to the synthesis of other metal-polymer coaxial nanocables.

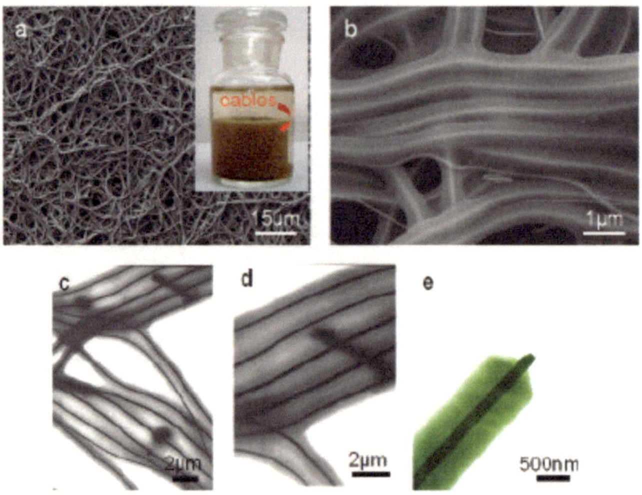

Fig. 12. SEM and TEM images of the nanocables. (a) General view of the cables. Inset is the sample obtained after reaction for 72 h. (b) Magnified SEM image of the cables, clearly showing the inner core-shell structure. (c-e) TEM images of the cables. Reprinted with permission from reference [39]. Copyright 2005 ACS.

Li et al. achieved the synthesis of cylindrical Ag nanowires by using an amorphous coating to modify the growth behaviour of Ag nanowires [40]. In their synthesis, Ag^+ ions are reduced by glucose under hydrothermal conditions with the presence of PVP in the temperature range of 140-180 °C. These resulted Ag nanowires at 180 °C show a narrow diameter distribution of 16 to 24 nm and the lengths usually reach tens of micrometers. Such slim nanowires show excellent flexibility, which indicates good machinability for future nanodevice fabrication. Further insight to the structure was obtained by high-resolution TEM, which showed Ag nanowires are coated by a multilayer sheath and cable-like structures are created. The sheath should be composed of carbonaceous layers according to previous study that carbonization of glucose could take place under similar conditions. For the formation of Ag@C nanocables, they proposed that Ag nanowires first grew up and then carbonaceous growth units stick on the edges of polygonal crystals owing to the higher surface tension and more dangling bonds. At a low reaction temperature of 140 °C, only well-crystallized pentagonal nanowires were obtained. This reason can be attributed to that carbonization process is significantly delayed. In summary, high-purity cylindrical Ag nanocables can be selectively prepared through kinetic control of the hydrothermal conditions.

4. Conclusion

Fabrication of silver nanowires has a special significance in many applications, since the optical properties of silver show superior properties compared to that of any other metals with respect to the losses in the localized surface plasmon resonance, or in the propagation of the one-dimensional surface plasmons. The fact that silver nanowires are excellent conductors of both electrons and photons opens up intriguing possibilities for electro-optical devices. Due to high electrical conductivity, silver nanowires are also promising materials as ultra large-scale circuits and quantum components, as well as for electron field emission sources in FEDs (field emission displays).

This article gave a simple introduction to the synthesis methods and growth mechanisms of Ag naowires, which were developed in the recent ten years by many research groups. Silver nanowires as the most typical metal nanowires has become an important branch of metal nanowires, and these methods used for the synthesis of Ag nanowires can supply reference for the synthesis of other metal nanowires. Every approach involved in this article has the merit and disadvantage, but all of them can give us some elicitation which will benefit to explore new strategies in the future.

5. Acknowledgment

This work was supported by Joint Project of Chemical Synthesis Core Research Institutions, and Grant-in-Aid for Scientific Research on Priority Areas "unequilibrium electromagnetic heating" and Grant-in-Aid for Scientific Research (B) from the Ministry of Education, Culture, Sports, Science, and Technology of Japan (Nos. 19033003 and 19310064).

6. References

[1] Wu, Y.; Xiang, J.; Yang, C.; Lu, W. & Lieber, C. M. (2004). Single-crystal metallic nanowires and metal/semiconductor nanowire heterostructures. *Nature,* Vol. 430, No. 7000, 704-704, ISSN: 0028-0836.

[2] Pascual, J. I.; Mendez, J.; Gomez-Herrero, J.; Baro, A. M.; Garcia, N.; Landman, U.; Luedtke, W. D.; Bogachek, E. N. & Cheng, H. P. (1995). Properties of metallic nanowires - from conductance quantization to localization, *Science,* Vol. 267, No. 5205, 1793-1795, ISSN: 0036-8075.

[3] Murphy, C. J.; Sau, T. K.; Gole, A. & Orendorff, C. J. (2005). Surfactant-directed synthesis and optical properties of one-dimensional plasmonic metallic nanostructures. *MRS Bulletin,* Vol. 30, No. 5, 349-355, ISSN: 0883-7694.

[4] Alivisatos, A. P.; Barbara, P. F.; Castleman, A. W.; Chang, J.; Dixon, D.A.; Klein, M. L.; McLendon, G. L.; Miller, J. S,; Ratner, M. A. & Rossky, P. J. (1998). Stupp SI &Thompson ME, From molecules to materials: Current trends and future directions. *Advanced Materials,* Vol. 10, No. 16, 1297-1336, 0935-9648.

[5] Thiaville, A. & Miltat, J. (1999). Magnetism - Small is beautiful. *Science,* Vol. 284, No. 5422, 1939-1940, ISSN: 0036-8075.

[6] Wang, Z. L. (2000). Characterizing the structure and properties of individual wire-like nanoentities. *Advanced Materials,* Vol. 12, No. 17, 1295-1298, ISSN: 0935-9648.

[7] Whitney, T. M.; Jiang, J. S.; Searson, P. C. & Chien, C. L. (1993). Fabrication and magnetic-properties of arrays of metallic nanowires, *Science,* Vol. 261, No. 5126, 1316-1319, ISSN: 0036-8075.

[8] Daniel, M. C. & Astruc, D. (2004). Gold nanoparticles: Assembly, supramolecular chemistry, quantum-size-related properties, and applications toward biology, catalysis, and nanotechnology. *Chemical Reviews,* Vol. 104, No. 1, 293-346, ISSN: 0009-2665.

[9] Braun, E.; Eichen, Y.; Sivan, U. & Ben-Yoseph, G. (1998). DNA-templated assembly and electrode attachment of a conducting silver wire. *Nature,* Vol. 39, No. 6669, 775-778, ISSN: 0028-0836.

[10] Sun, Y. G.; Mayers, B. T. & Xia, Y. N. (2002). Template-engaged replacement reaction: A one-step approach to the large-scale synthesis of metal nanostructures with hollow interiors. *Nano Letters,* Vol. 2, No. 5, 481-485, ISSN: 0028-0836.

[11] Fievet, F.; Lagier, J. P.; Blin, B.; Beaudoin, B. & Figlarz, M. (1989). Homogeneous and heterogeneous nucleations in the polyol process for the preparation of micron and sub-micron size metal particles, *Solid State Ionics,* Vol. 32, No. 3, 198-205, ISSN: 0167-2738.

[12] (a) Xia, Y.; Xiong, Y.; Lim, B. & Skrabalak, S. E. (2009). Shape-controlled synthesis of metal nanocrystals: Simple chemistry meets complex physics? *Angewandte Chemie – International Edition,* Vol. 48, No. 1, 60-103, ISSN: 1521-3773. (b) Wiley, B.; Sun, Y. G. & Xia, Y. N. (2007). Synthesis of silver nanostructures with controlled shapes and properties. *Accounts of Chemical Research,* Vol. 40, No. 10, 1067-1076, ISSN: 0001-4842. (c) Wiley, B.; Sun, Y. G.; Chen, J. Y.; Cang, H.; Li, Z.Y.; Li, X.D. & Xia, Y.N. (2005). Shape-controlled synthesis of silver and gold nanostructures. *MRS Bulletin,* Vol. 30, No. 5, 356-361, ISSN: 0883-7694. (d) Chen, J. Y.; Wiley, B. J. & Xia, Y. N. (2007). One-dimensional nanostructures of metals: Large-scale synthesis and some potential applications. *Langmuir,* Vol. 23, No. 8, 4120-4129, ISSN: 0743-7463. (e) Wiley, B.; Sun, Y. G.; Mayers, B. & Xia, Y. N. (2005). Shape-controlled synthesis of metal nanostructures: The case of silver. *Chemistry-A European Journal,* Vol. 11, No. 2, 454-463, ISSN: 0947-6539. (f) Sun, Y. G. & Xia, Y. N. (2002). Large-scale synthesis of uniform silver nanowires through a soft, self-seeding, polyol process. *Advanced Materials,* Vol. 14, No. 11, 833-837, ISSN: 0935-9648.

[13] Mullin, J. (1997). *Crystallization,* Oxford University Press, ISBN: 0750611294, New York.

[14] Marks, L. D. (1994). Experimental studies of small-particles structures. *Reports on Progress in Physics,* Vol. 57, No. 63, 603-649, ISSN: 0034-4885.

[15] Cleveland, C. L. & Landman, U. (1991). The energetic and structure of nickel clusters-size dependence. *Journal of Chemical Physics,* Vol. 94, No. 11, 7376-7396, ISSN: 0021-9606.

[16] (a) Korte, K. E.; Skrabalak, S. E. & Xia, Y. (2008). Rapid synthesis of silver nanowires through a $CuCl^-$ or $CuCl_2^-$ mediated polyol process. *Journal of Materials Chemistry,* Vol. 18, No. 4, 437-441, ISSN: 0959-9428. (b)Wiley, B.; Sun, Y. G. & Xia, Y. N. (2005).

Polyol synthesis of silver nanostructures: Control of product morphology with Fe(II) or Fe(III) species. *Langmuir,* Vol. 21, No. 8, 8077-8080, ISSN: 0743-7463.

[17] Wiley, B.; Herricks, T.; Sun, Y. G. & Xia, Y. N. 2004. Polyol synthesis of silver nanoparticles: Use of chloride and oxygen to promote the formation of single-crystal, truncated cubes and tetrahedrons. *Nano Letters,* Vol. 4, No. 9, 1733-1739, ISSN: 1530-6984.

[18] Campbell, C. T. (1985). Atomic and molecular-oxygen adsorption on Ag(111). *Surface Science,* Vol. 157, No. 1, 43-60, ISSN: 0039-6028.

[19] deMongeot, F. B.; Cupolillo, A.; Valbusa, U. & Rocca, M. (1997). O_2 dissociation on Ag(001): The role of kink sites. *Chemical Physics Letters,* Vol. 270, No. 3-4, 345-350, ISSN: 0009-2614.

[20] Tsuji, M.; Hashimoto, M.; Nishizawa, Y.; Kubokawa, M. & Tsuji, T. (2005). Microwave-assisted synthesis of metallic nanostructures in solution. *Chemistry-A European Journal,* Vol. 11, No. 2, 440-452, ISSN: 0947-6539.

[21] Tsuji, M.; Nishizawa, Y.; Matsumoto, K.; Miyamae, N.; Tsuji, T. & Zhang, X. (2007). Rapid synthesis of silver nanostructures by using microwave-polyol method with the assistance of Pt seeds and polyvinylpyrrolidone. *Colloids and Surfaces A: Physicochemical and Engineering Aspects,* Vol. 293, No. 1-3, 185-194, ISSN: 0927-7757.

[22] Tsuji, M.; Matsumoto, K.; Miyamae, N.; Tsuji, T. & Zhang, X. (2007). Rapid preparation of silver nanorods and nanowires by a microwave-polyol method in the presence of Pt catalyst and polyvinylpyrrolidone. *Crystal Growth & Design,* Vol. 7, No. 2, 311-320, ISSN: 1528-7483.

[23] Tsuji, M.; Matsumoto, K.; Jiang, P.; Matsuo, R.; Tang, X. L. & Karnarudin, K. S. N. (2008). Roles of Pt seeds and chloride anions in the preparation of silver nanorods and nanowires by microwave-polyol method. *Colloids and Surfaces A: Physicochemical and Engineering Aspects,* Vol. 316, No. 1-3, 266-277, ISSN: 0927-7757.

[24] Gou, L. F.; Chipara, M. & Zaleski, J. M. (2007). Convenient, rapid synthesis of Ag nanowires. *Chemistry of Materials,* Vol. 19, No. 7, 1755-1760, ISSN: 0897-4756.

[25] Tang, X. L.; Tsuji, M.; Jiang, P.; Nishio, M.; Jang, S. M. & Yoon, S. H. (2009). Rapid and high-yield synthesis of silver nanowires using air-assisted polyol method with chloride ions. *Colloids and Surfaces A: Physicochemical and Engineering Aspects,* Vol.338, No.1-3, 33-39, ISSN: 0927-7757.

[26] Mock, J. J.; Barbic, M.; Smith, D. R.; Schultz, D. A. & Schultz, S. (2002). Shape effects in plasmon resonance of individual colloidal silver nanoparticles. *Journal of Chemical Physics,* Vol. 116, No. 15, 6755-6759, ISSN: 0021-9606.

[27] Wiley, B. J.; Im, S. H.; Li, Z. Y.; McLellan, J.; Siekkinen, A. & Xia, Y. N. (2006). Maneuvering the surface plasmon resonance of silver nanostructures through shape-controlled synthesis. *Journal of Physical Chemistry B,* Vol. 110, No. 32, 15666-15675, ISSN: 1520-6106.

[28] Tang, X. L.; Tsuji, M.; Nishio, M. & Jiang, P. (2009). Roles of chloride anions in the shape evolution of anisotropic silver nanostructures in poly(vinylpyrrolidone) (PVP)-assisted polyol process. *Bulletin of the Chemical Society of Japan,* Vol. 82, No. 10, 1304–1312.

[29] Sun, Y. G.; Yin Y. D.; Mayers, B. T.; Herricks, T. & Xia, Y. N. (2002). Uniform silver nanowires synthesis by reducing $AgNO_3$ with ethylene glycol in the presence of

seeds and poly(vinyl pyrrolidone). *Chemistry of Materials*, Vol. 14, No. 11, 4736-4745, ISSN: 0897-4756.

[30] Matijevic, E. (1993). Preparation and properties of uniform size colloids. *Chemistry of Materials*, Vol. 5, No. 4, 412-426, ISSN: 0897-4756.

[31] Jana, N. R.; Gearheart, L. & Murphy, C. J. (2001). Wet chemical synthesis of silver nanorods and nanowires of controllable aspect ratio. *Chemical Communications*, No. 7, 617-618, ISSN: 1359-7345.

[32] Pietrobon, B.; McEachran, M. & Kitaev, V. (2009). Synthesis of size-controlled faceted pentagonal silver nanorods with tunable plasmonic properties and self-assembly of these nanorods. ACS Nano, Vol. 3, No. 1, 21-26, ISSN: 1936-0851.

[33] Tsuji, M.; Miyamae, N.; Lim, S.; Kimura, K.; Zhang, X.; Hikino, S. & Nishio, M. (2006). Crystal structures and growth mechanisms of Au@Ag core-shell nanoparticles prepared by the microwave-polyol method. *Crystal Growth & Design*, Vol. 6, No. 8, 1801-1807, ISSN: 1528-7483.

[34] Caswell, K. K.; Bender, C. M. & Murphy, C. J. (2003). Seedless, surfactantless wet chemical synthesis of silver nanowires. *Nano Letters*, Vol. 3, No. 5, 667-669, ISSN: 1530-6984.

[35] Hu, J. Q.; Chen, Q.; Xie, Z. X.; Han, G. B.; Wang, R. H.; Ren, B.; Zhang, Y.; Yang, Z. L. & Tian, Z. Q. (2004). A simple and effective route for the synthesis of crystalline silver nanorods and nanowires. *Advanced Functional Materials*, Vol. 14, No. 2, 183-189, ISSN: 1616-301X.

[36] (a) Kottmann, J. P.; Martin, O. J. F.; Smith, D. R. & Schultz, S. (2001). Plasmon resonances of silver nanowires with a nonregular cross section. *Physical Review B*, 235402, Vol. 64, No. 23, ISSN: 1098-0121. (b) Kottmann, J. P.; Martin, O. J. F.; Smith, D. R. & Schultz, S. (2001). Dramatic localized electromagnetic enhancement in plasmon resonant nanowires. *Chemical Physics Letters*, Vol. 341, No.1-2, 1-6, ISSN: 0009-2614.

[37] (a) Gray, S. K. & Kupka, T. (2003). Propagation of light in metallic nanowire arrays: Finite-difference time-domain studies of silver cylinders. *Physical Review B*, Vol. 68, No. 4, 045415, ISSN: 1098-0121 (b) Oliva, J. M. & Gray, S. K. (2003). Theoretical study of dielectrically coated metallic nanowires. *Chemical Physics Letters*, Vol. 379, No.3-4, 325-331, ISSN: 0009-2614.

[38] Yin, Y. D.; Lu, Y.; Sun, Y. G. & Xia, Y. N. (2002). Silver nanowires can be directly coated with amorphous silica to generate well-controlled coaxial nanocables of silver/silica. *Nano Letters*, Vol. 2, No. 4, 427-430, ISSN: 1530-6984.

[39] Luo, L. B.; Yu, S. H.; Qian H. S. & Zhou, T. (2005). Large-scale fabrication of flexible silver/cross-linked poly(vinyl alcohol) coaxial nanocables by a facile solution approach. *Journal of the American Chemical Society*, Vol. 127, No. 9, 2822-2823, ISSN: 0002-7863.

[40] Sun, X. M. & Li, Y. D. (2005). Cylindrical silver nanowires: Preparation, structure, and optical properties. Advanced Materials, Vol. 17, No. 21, 2626-2630, ISSN: 0935-9648.

3

Growth of Nanowire and Nanobelt Based Oxides by Thermal Oxidation with Gallium

Qing Yang[1], Takahito Yasuda[1], Hitonori Kukino[1], Miyoko Tanaka[2] and Hirokazu Tatsuoka[1]
[1]*Shizuoka University*
[2]*National Institute for Materials Science*
Japan

1. Introduction

Quasi-one-dimensional nanostructure based oxides, such as nanowires and nanobelts, have attracted much attention in recent years, which exhibit unusual optical, electronic or mechanical properties as compared to those of bulk materials, due to the significant surface related defects originating from the high surface to volume ratios. In addition, some important advantages such as good high temperature stability, oxidation resistance and stable electric properties, demonstrate their potential basic building blocks for new classes of environmentally conscious electronics. Much effort has been made to develop nanoscale optical and electronic devices for the generation of renewable energy using the nanowire and nanobelt based oxides. As a simple and scalable growth method, the oxidation behavior of metals has been typically conducted on the nanostructure growth of metal oxides over several decades (Chen et al., 2008; Gu et al., 2002; Jiang et al., 2002; Ren et al., 2007; Takagi, 1957; Yu et al., 2005; Yu et al., 2006). On the other hand, Pan et al. reported the SiO_x nanowire growth using gallium (Pan et al., 2002; 2003; 2005). In addition, a variety of oxide nanostructures have been synthesized by the thermal oxidation of silicide alloys with gallium (Ogino et al., 2007). In this paper, we report a variety of nanowire and nanobelt based oxides (ZnO, α-Fe_2O_3, β-Ga_2O_3) grown by the thermal oxidation with gallium in air.

2. Growth & characterization

The oxide nanostructures were synthesized by the thermal oxidation of the source substrates with gallium in air. Gallium was melted around 35~40 °C, then manually applied to the substrates using a stick. The substrates with gallium were loaded into a quartz tube which is open to the air. The oxide nanostructure growth was performed by exposure of the substrates with gallium to the air, and the oxide nanostructures were grown at elevated temperatures for several hours.
The as-grown oxide nanostructures were characterized by scanning electron microscopy (SEM), transmission electron microscopy (TEM) with selected area electron diffraction (SAED) and high-resolution transmission electron microscopy (HRTEM). The compositional analysis was made using energy dispersive X-ray spectroscopy (EDS).

3. Oxide nanowires & nanobelts

3.1 ZnO

ZnO is well known as one of the most promising wide band-gap semiconductor materials, and the ZnO quasi-one-dimensional nanostructures have attracted tremendous attention for its potential applications in nanoscale electronics and optoelectronics (Lu et al., 2006). ZnS has been adopted as a buffer or source material for the growth of various ZnO crystals. ZnO nanorods (Wang et al., 2004) and nanowires (Yuan et al., 2003) were grown by the thermal evaporation of ZnS powders; while using ZnS and $Fe(NO_3)_3$, ZnO whiskers were grown (Hu et al., 2001); ZnO nanobelts were formed by the oxidation of a ZnS nanobelt (Li et al., 2004). Miyake et al. reported a simple growth technique of epitaxial ZnO layers using ZnS substrates (Miyake et al., 2000). The growth of ZnO nanostructures is expected when the thermal oxidation with gallium is applied to the ZnO growth using ZnS substrates. In this section, we report the ZnO nanowires and nanobelts grown at temperatures of 700~900 °C for 5 h in air.

Fig. 1. SEM images of ZnO nanostructures with various morphologies.

Figure 1 shows the SEM images of ZnO nanowires with various morphologies. The diameters of the nanowires vary from 20 to 200 nanometers, and the maximum length extends to tens of micrometers. As shown in Figure 2, the EDS mapping of several nanowires revealed that the observed atomic ratio of Zn/O is near unity, while the impurity atoms of Ga and S are quite few.

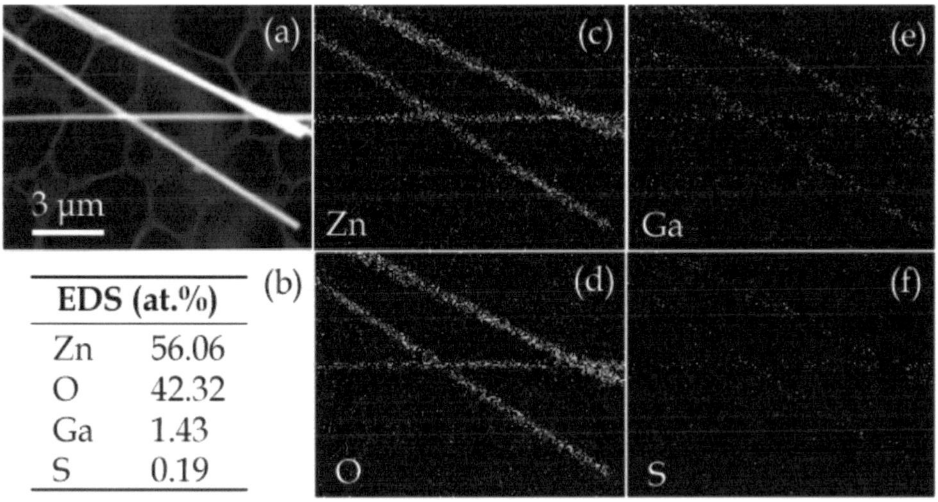

EDS (at.%)	
Zn	56.06
O	42.32
Ga	1.43
S	0.19

Fig. 2. (a) SEM image of several nanowires and corresponding EDS mapping images of (c) Zn, (d) O, (e) Ga and (f) S. The observed EDS result is shown in (b).

Figures 3 (a) and (b) show the TEM image of a nanowire and corresponding SAED pattern, respectively. The SAED pattern agrees with that of hexagonal ZnO with the incident electron beam parallel to the $[2\bar{1}\bar{1}0]$ direction, and the growth axis of the nanowire is parallel to the [0001] direction.

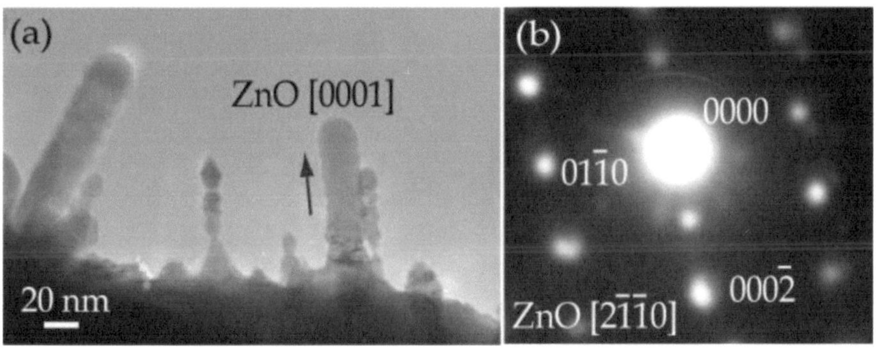

Fig. 3. (a) TEM image and (b) corresponding SAED pattern of one of the nanowires. (Yang et al., 2009)

Figure 4 (a) shows the enlarged TEM image of a nanowire. It was found that the surface of the nanostructure is covered by a thin layer. Figures 4 (b), (c) and (d) show the HRTEM images with SAED patterns in the insets, as marked by B, C, D in Fig. 4 (a), respectively. The plane spacing perpendicular to the growth direction is about 0.26 nm, which is consistent with that of the (0002) planes of the ZnO crystal. Stacking faults lying on the (0001) planes are observed perpendicular to the growth direction, and the streaking SAED pattern in the inset of Fig. 4 (c) is caused by the existence of the stacking faults. As shown in Fig. 4 (d), the average distance of the lattice planes is about 0.43 nm and 0.26 nm. The angle between them is about 75°. The spacing lattice conforms to that of the cubic $ZnGa_2O_4$ crystal (α = 0.8335 nm; JCPDS Card No. 38-1240). The SAED pattern in the inset of Fig. 4 (d) shows that the

$ZnGa_2O_4$ grew along the [010] direction. It was found that the growth direction of the nanowires is ZnO[0001], and which is parallel to $ZnGa_2O_4$[010]. In addition, ZnO$[2\bar{1}\bar{1}0]$ is nearly parallel to $ZnGa_2O_4$[301]. This relationship is roughly equivalent to the relationship as ZnO$(\bar{1}100)$ // $ZnGa_2O_4$(101) and ZnO$(11\bar{2}0)$ // $ZnGa_2O_4$$(\bar{1}01)$. In addition, the mismatch of the plane spacings between $d_{ZnO(\bar{1}100)}$ (0.2815 nm) and $d_{ZnGa2O4(202)}$ (0.2947 nm) is about 4.6%, and that of the plane spacings between $d_{ZnO(11\bar{2}0)}$ (0.1625 nm) and $d_{ZnGa2O4(\bar{4}04)}$ (0.1474 nm) about 9.3%. This crystallographic relationship shows a relatively lower lattice mismatch configuration for the ZnO and $ZnGa_2O_4$ couple. It is considered that the observed relationship is provided by the result of the ZnO$(\bar{1}100)$ // $ZnGa_2O_4$(101) and ZnO$(11\bar{2}0)$ // $ZnGa_2O_4$$(\bar{1}01)$ crystallographic orientation relationships with a lower lattice mismatch.

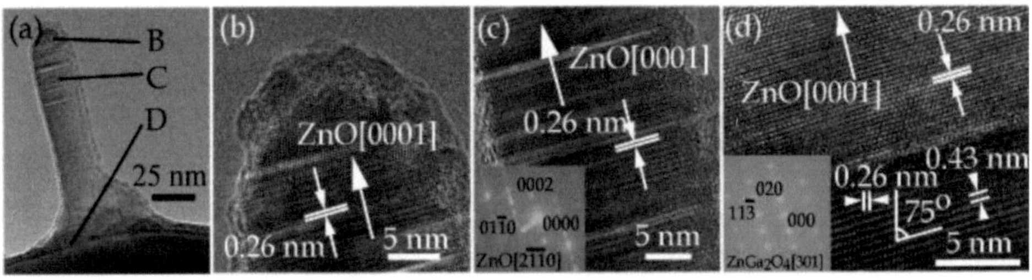

Fig. 4. (a) TEM image of one nanowire, (b,c,d) HRTEM images of the nanowire and corresponding SAED patterns in the insets, as marked by B,C,D in (a). (Yang et al., 2009)

Figures 5 (a) and (c, d, e, f) show the TEM image of the nanowire shown in Fig. 4 (a) and the corresponding Zn, O, Ga and S EDS mapping, respectively. The pictures reveal that the Ga atoms are highly concentrated at the substrate of the nanowire, while the Zn atoms are distributed in both the layer and nanowire. The EDS result reveals that the nanowire mainly contains zinc and oxygen, and the ratio of zinc to oxygen is about 54:45, indicating stoichiometric ZnO. It should be noted that small amounts of Ga or S remain in the ZnO layer. The observed stoichiometry of the substrate region does not strictly agree with that of $ZnGa_2O_4$, which might be caused by the co-existence of other Zn or Ga compounds in deeper regions of the substrate.

Figure 6 shows typical SEM images of the ZnO nano- and micro-belts with a variety of morphological features. Branched (Fig. 6b), stepped (Fig.6c and d) and irregular shaped (Fig. 6d and e) belt-like structures were observed. The morphology of the belt shown in Fig. 6 (e) is quite similar to that reported by (Jian et al., 2006; Zhao et al., 2004), but no droplet exists on the tip of the belt presented here. Meanwhile, several thick belts with cross sectional sides are shown in Fig. 6 (c). In comparison, thin belts were also observed, as shown in Fig. 6 (f, g) and Fig. 8 (d).

Figures 7 (a) and (b, c, d) show the HRTEM images of a stepped nanobelt and the corresponding fast Fourier transform (FFT) patterns, as marked by b, c, d in (a), respectively. As shown in Fig. 7 (a), two domains of the stepped nanobelt are marked by A and B, respectively, and domain A grew from domain B. The plane spacing (shown in the inset of Fig. 7a) parallel to the growth direction of domain A is about 0.15 nm, which corresponds to the $(02\bar{2}0)$ plane of hexagonal ZnO. The FFT patterns reveal that domain A grew along the [0001] direction with wide and narrow surfaces of $\pm(2\bar{1}\bar{1}0)$ and $\pm(0\bar{1}10)$, respectively; while the growth direction of domain B is along the $[0\bar{1}11]$ direction, which is parallel to the

[0001] growth direction of domain A, and both of them have the same flat surfaces of $\pm(2\bar{1}\bar{1}0)$. The domain B growth is quite similar to the nanoplatelet (Ye et al., 2005), which

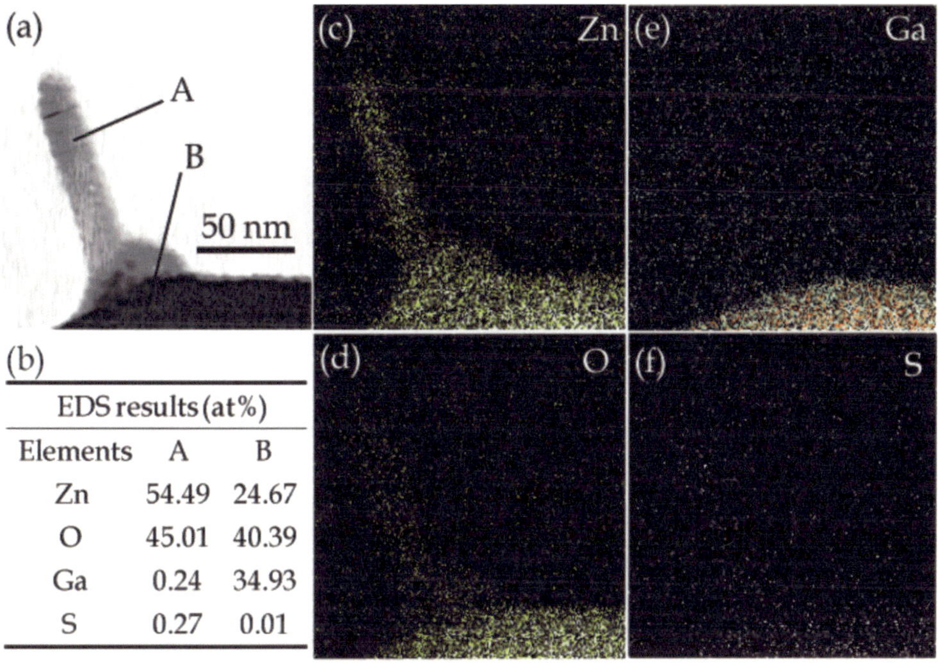

EDS results (at%)		
Elements	A	B
Zn	54.49	24.67
O	45.01	40.39
Ga	0.24	34.93
S	0.27	0.01

Fig. 5. (a) TEM image of the nanowire shown in Fig. 4 (a), and corresponding EDS mapping images of (c) Zn, (d) O, (e) Ga and (f) S. The observed EDS results are shown in (b). (Yang et al., 2009)

Fig. 6. (a,f) Low magnification SEM images of belt-like structures; typical high magnification SEM images of (b) branched, (c,d) stepped, (d,e) irregular shaped and (f,g) thin belts.

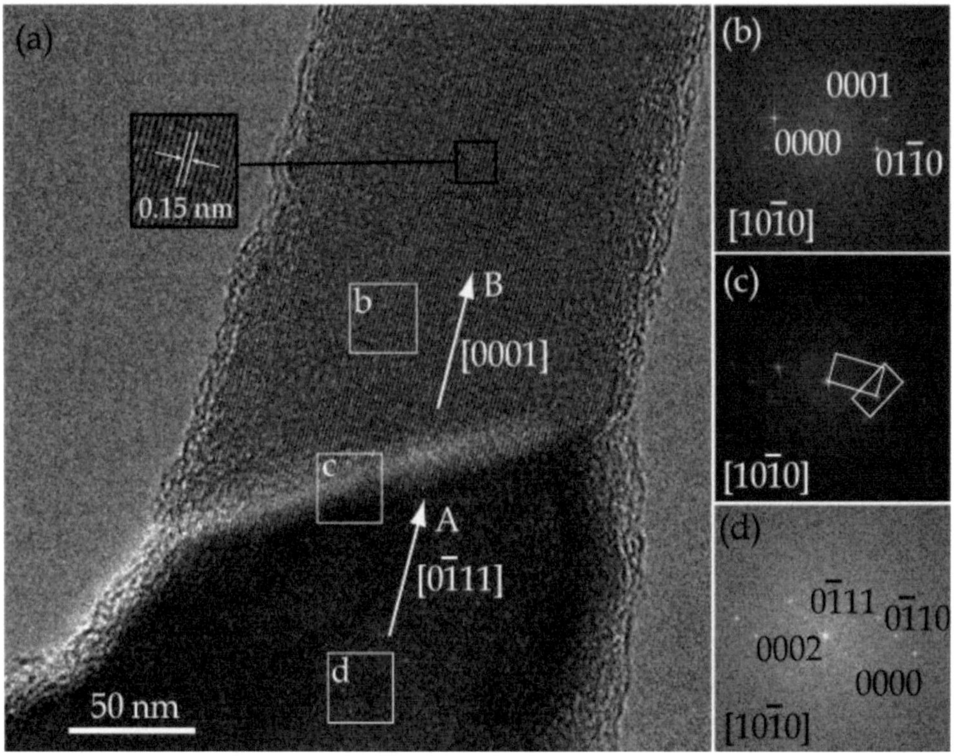

Fig. 7. (a) HRTEM images of a stepped nanobelt and (b,c,d) corresponding FFT patterns.

also grew along the $[01\bar{1}1]$ direction with flat surfaces of $\pm(\bar{2}\bar{1}10)$. The relationship of the two domains is roughly equivalent to the relationship as $(01\bar{1}0)_A // (10\bar{1}0)_B$ and $(1\bar{1}00)_A // (01\bar{1}1)_B$. The mismatch of the plane spacings between $d_{(01\bar{1}0)}$ (0.2814 nm) and $d_{(10\bar{1}0)}$ (0.2814 nm) is 0, and that of the plane spacings between $d_{(1\bar{1}00)}$ (0.2814 nm) and $d_{(01\bar{1}1)}$ (0.2476 nm) is about 12%. This crystallographic relationship shows a relatively lower lattice mismatch configuration for the two domains of the stepped nanobelt. It is considered that the observed relationship is provided by the result of the $(01\bar{1}0)_A // (10\bar{1}0)_B$ and $(1\bar{1}00)_A // (01\bar{1}1)_B$ crystallographic orientation relationships with a lower lattice mismatch.

Figures 8 (a, b) and (c) show the HRTEM images of a belt and corresponding FFT pattern, respectively. The belt shows two sets of lattice fringes with a spacing of about 0.25 nm and 0.17 nm, normal to each other, corresponding to the (0002) and $(11\bar{2}0)$ planes of hexagonal ZnO. The FFT pattern revealed that the nanobelt grew along the $[11\bar{2}0]$ direction with wide and narrow surfaces of $\pm(1\bar{1}00)$ and $\pm(0001)$, respectively.

Figures 8 (d) and (e) show the TEM image of several nanobelts and corresponding SAED pattern, respectively. The SAED pattern is recorded from the nanobelt as marked by E in Fig. 8 (d), and it can be indexed based on the hexagonal ZnO cell with an electron beam along $[\bar{1}2\bar{1}3]$. It demonstrated that the flat surfaces of the nanobelt is $\pm(\bar{1}2\bar{1}3)$, while the growth directions are perpendicular to $(11\bar{2}\bar{1})$.

ZnO crystals generally grow along the [0001] direction (Fig. 3a, Fig. 4a and Fig. 7a) because of the kinetically fast growth of the (0001) planes for the anisotropic crystal structure (Baxter et al., 2003). Meanwhile, the growth rate of the ZnO crystals along the [0001] direction is

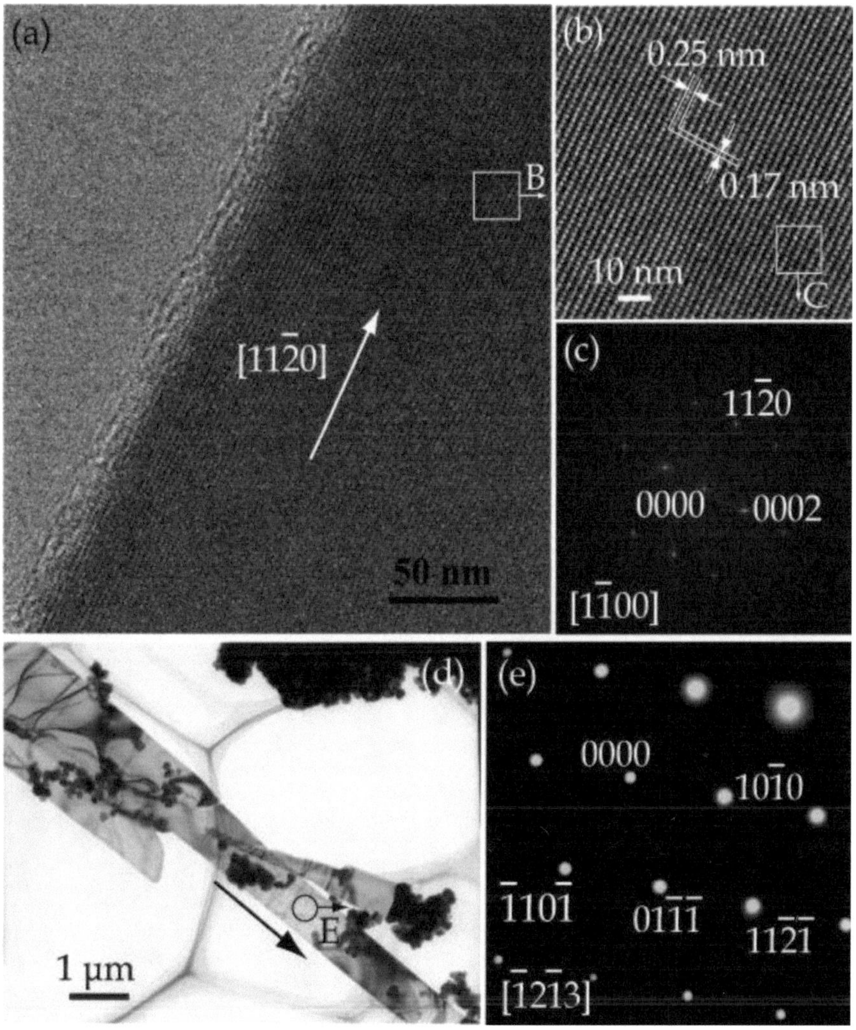

Fig. 8. (a,b) HRTEM images of a nanobelt and (c) corresponding FFT pattern. (d) TEM image of a nanobelt and (e) corresponding SAED pattern.

faster than along the $[0\bar{1}11]$ direction (Li et al., 1999). Therefore, the transfer of the growth direction from $[0\bar{1}11]$ to [0001] for the stepped nanobelt (Fig. 7a) is possibly determined by the kinetically driven growth and minimization of the surface energy (Ye et al., 2005). ZnO belts growing along the [0001] direction with wide and narrow surfaces of $\pm(2\bar{1}\bar{1}0)$ and $\pm(0\bar{1}10)$ (Fig. 7a) were reported by (Pan et al., 2001; Wang et al., 2004; Ye et al., 2005), and those growing along the $[11\bar{2}0]$ direction with wide and narrow surfaces of $\pm(1\bar{1}00)$ and ±(0001) (Fig. 8a) were reported by (Deng et al., 2005; Li et al., 2004). A few reports described the synthesis of ZnO belts with both morphologies (Ding et al., 2004). However, in the present work, we reported a new type of ZnO nanobelt with flat surfaces of $\pm(\bar{1}2\bar{1}3)$ and growth directions perpendicular to $(11\bar{2}\bar{1})$ (Fig. 8d). Meanwhile, all of the types of ZnO belts were synthesized by the simple growth method, which provides the possibility for the selective growth of ZnO nanostructures by controlling the growth conditions.

The characterizations of the above ZnO nanostructures show that the surface region of the substrate is composed of ZnO or $ZnGa_2O_4$, which indicates that the ZnO nanostructures grew on ZnO or $ZnGa_2O_4$ domains by the oxidization of the surface of ZnS substrate. It has been reported that the selective growth of a particular phase is mainly triggered by a specific interface composition, and is governed by the diffusion flux to the interface (Majni et al., 1981). In this sense, when ZnS is used as the starting material of the growth evolution, O diffusion into ZnS leads to the formation of ZnO near the surface. The enthalpy of formation of ZnO (41.9 kcal/g-atom) (Kubaschewski & Alcock, 1979) is greater than that of ZnS (24.5 kcal/g-atom) (Kubaschewski & Alcock, 1979). The formation of ZnO is thermodynamically favored as compared to that of ZnS. On the other hand, the enthalpy of formation of Ga_2O_3 (51.8 kcal/g-atom) (Kubaschewski & Alcock, 1979) is greater than that of ZnO. If the Ga droplets remained on the ZnS surface, Ga could be oxidized to form Ga_2O_3. It is assumed that Ga atoms are consumed to form $ZnGa_2O_4$, and possibly incorporated into ZnO as impurities.

3.2 α-Fe_2O_3

As one of the oxide semiconductors with a small band-gap of 2.1 eV, α-Fe_2O_3 is the most stable iron oxide under ambient conditions. It is nontoxic, magnetic, corrosion resistant, and

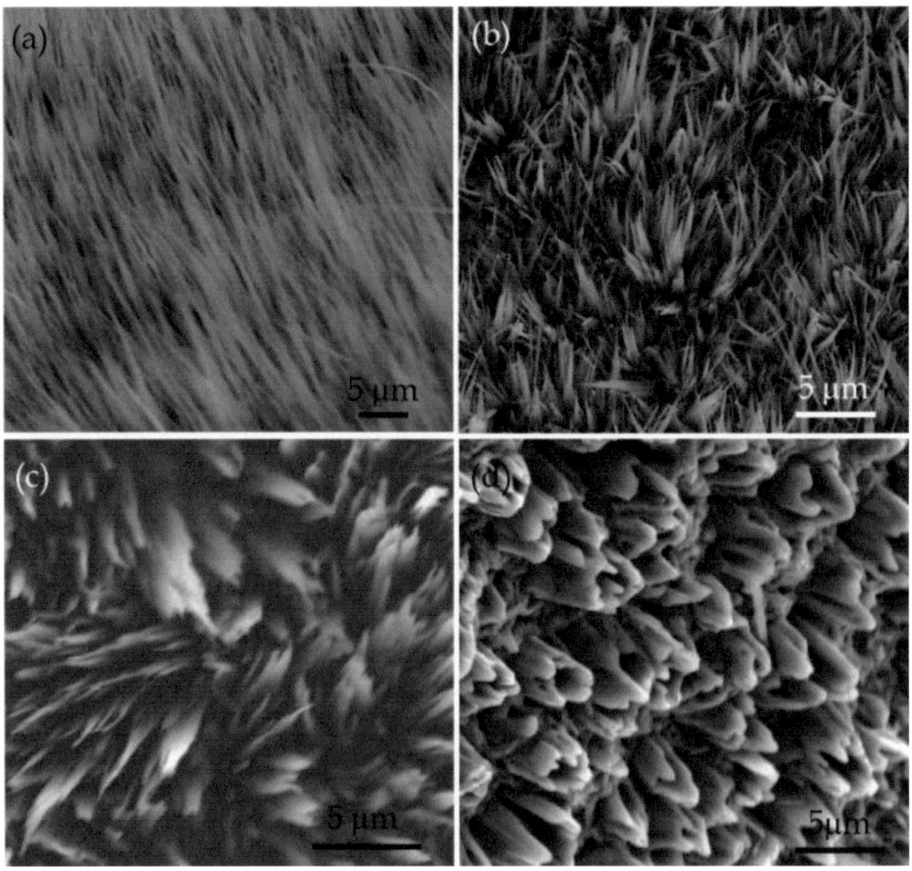

Fig. 9. SEM images of α-Fe_2O_3 nanostructures grown on iron substrates (a,c) without or (b,d) with gallium at (a,b) 700 and (c,d) 800 °C.

has promising applications in photocatalyst (Ohmori et al., 2000), sensor (Liao et al., 2008), magnetic storage media (Kim et al., 2006) and field emission (Yu et al., 2006). To date, various α-Fe_2O_3 nanostructures, including whiskers, nanowires and nanobelts, have been synthesized by the oxidation of iron during a wide temperature range using a hotplate (Yu et al., 2006), reactive oxygen plasma (Cvelbar et al., 2008), oxygen (Takagi, 1957; Wen et al., 2005), ozone (Srivastava et al., 2007), or a gas mixture (Fu et al., 2003; Wang et al., 2005; Xu et al., 2009). In this section, we report the simple synthesis of α-Fe_2O_3 nanowires and nanobelts, via the direct thermal oxidation of iron substrates with or without gallium at temperatures of 600~800 °C for 3 h in air.

Figure 9 shows the SEM images of α-Fe_2O_3 nanostructures grown at temperatures of (a, b) 700 and (c, d) 800 °C by the thermal oxidation of iron substrates (a, c) without or with (b, d) gallium in air. α-Fe_2O_3 nanobelts were grown on iron substrates at 800 °C rather than nanowires grown at 700 °C. On the other hand, α-Fe_2O_3 nanobelts and nanowires were grown on iron substrates with gallium at 700 °C. In addition, it should be pointed out that the morphologies of the two kinds of nanobelts are quite different, and the width of the nanobelts grown without gallium at 800 °C (Fig. 9c) is generally larger than that grown with gallium at 700 °C (Fig. 9b). The detailed structures of the nanowires (Figs. 9a&b) and nanobelts (Figs. 9b&c) were characterized by transmission electron microscopy, which are shown in Figures 10-12.

The TEM image and corresponding SAED pattern of one of the nanowires, grown by the thermal oxidation of iron substrates at 700 °C (Fig. 9a), are shown in Figures 10 (a) and (b), respectively. As shown in Fig. 10 (b), the SAED pattern is composed of two sets of diffraction patterns with zone axis parallel to the [001] and $[\bar{1}11]$ directions, respectively, as marked by the solid lines.

It is indicated that the nanowire is composed of at least two overlapped grains to form a bicrystal structure, and the two overlapped grains share the same growth direction along the [110]. Figure 10 (c) shows an enlarged TEM image of the area marked by circle C in Fig. 10 (a), and the crystal boundary is clearly seen. Figure 10 (d) shows the TEM image of the rectangular enclosed area of the right-side crystal in Fig. 10 (c), the fringe spacing along the growth direction is about 0.252 nm, which corresponds to the interplanar spacing of the (110) planes of α-Fe2O3 (JCPDS Card No. 71-5088), and the corresponding FFT pattern with the $[\bar{1}11]$ zone axis is shown in Figure 10 (e), which confirms the growth direction along [110]. In addition, as shown in Fig. 10 (b), several additional weak diffraction spots as marked by the arrows are observed, which belong to the diffraction patterns with zone axis parallel to the $[\bar{1}11]$ direction, as marked by the dotted lines. It is considered that the diffracted beam travelling through the grain viewed along the [001] direction is re-diffracted when it passes into the other grain viewed along the $[\bar{1}11]$ direction. The observed weak diffraction pattern should be the double diffraction pattern, which occurs at the interface between the thin grains. In addition, few similar double diffraction spots are also observed in the SAED patterns of the nanobelts characterized below.

Figure 11 shows the TEM images of the nanostructures grown by the thermal oxidation of iron substrates with gallium at 700 °C (Fig. 9b). The TEM image containing nanobelts and nanowires is shown in Figure 11 (a). Figure 11 (b) shows the TEM image of a nanobelt with the corresponding SAED pattern in the inset. Figure 11 (c) shows the TEM image of a nanowire, and an enlarged TEM image of the tip part is shown in the left-lower inset with the corresponding SAED pattern shown in the right-upper inset. The nanowire, with a

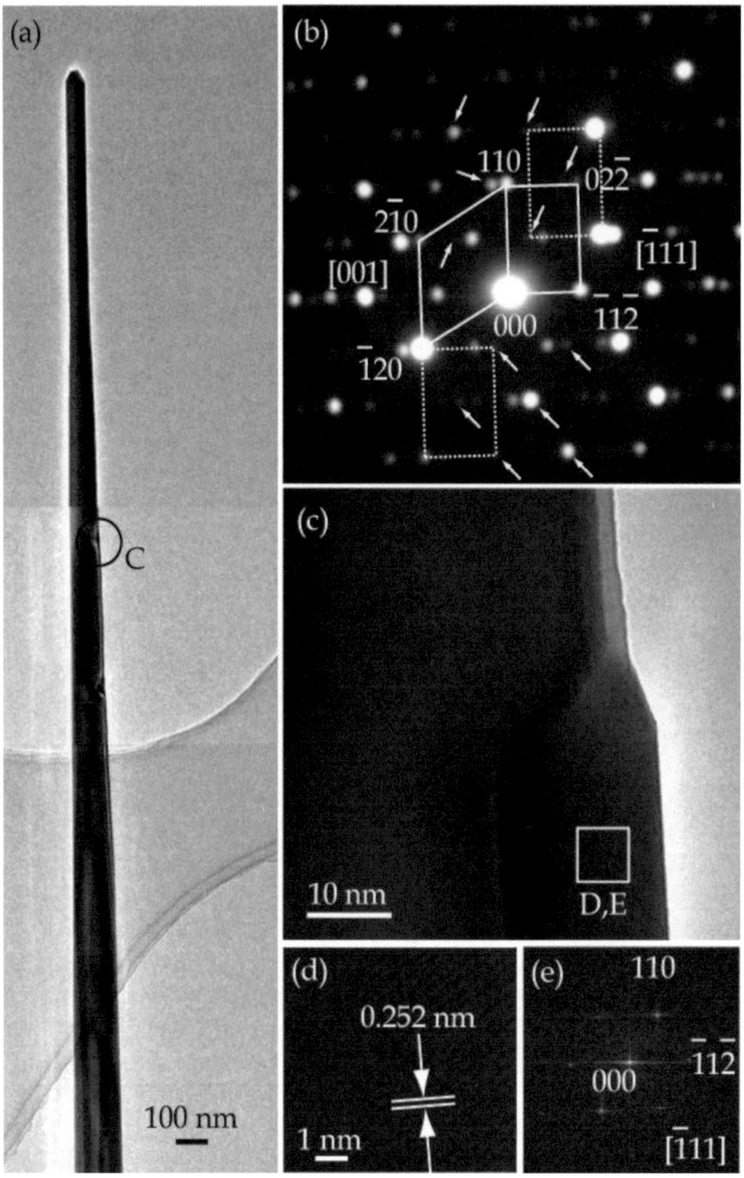

Fig. 10. (a) TEM image and (b) corresponding SAED pattern of one of the nanowires shown in Fig. 9 (a). (c) enlarged TEM image of the area marked by circle C in (a). (d) TEM image and (e) corresponding FFT pattern of the rectangular enclosed area in (c).

diameter of about 100 nm, is as small as that observed in Fig. 10 (a), but half as much as the width of the nanobelt. The SAED patterns reveal that either the nanobelt or the nanowire has a bicrystal structure with the growth direction along [110].

Figures 12 (a) and (e) show the TEM images of the nanobelts grown by the thermal oxidation of iron substrates at 800 °C (Fig. 9c). A nanobelt with a sharp tip is shown in Fig. 12 (a). Figures 12 (b-d) show the corresponding SAED patterns from the rectangular B, C and D enclosed areas in Fig. 12 (a), and Figures 12 (f-g) show the corresponding SAED

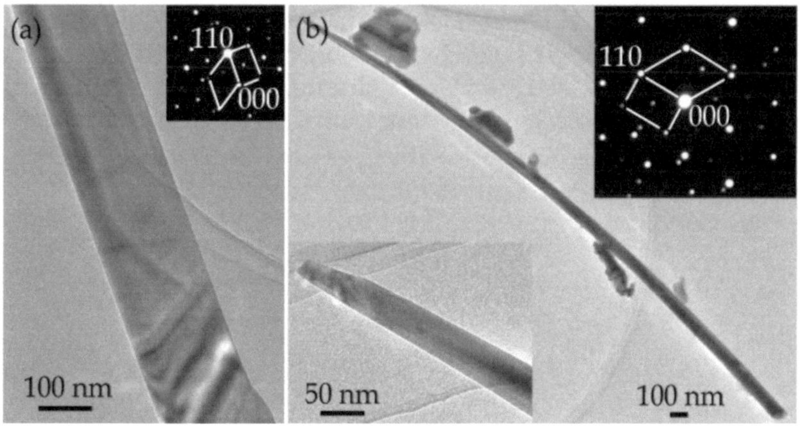

Fig. 11. (a) TEM image of the nanostructures shown in Fig. 9 (b). (b) TEM image of a nanobelt and corresponding SAED pattern in the inset. (c) TEM image of a nanowire with enlarged TEM image of the tip and corresponding SAED pattern in the insets, respectively.

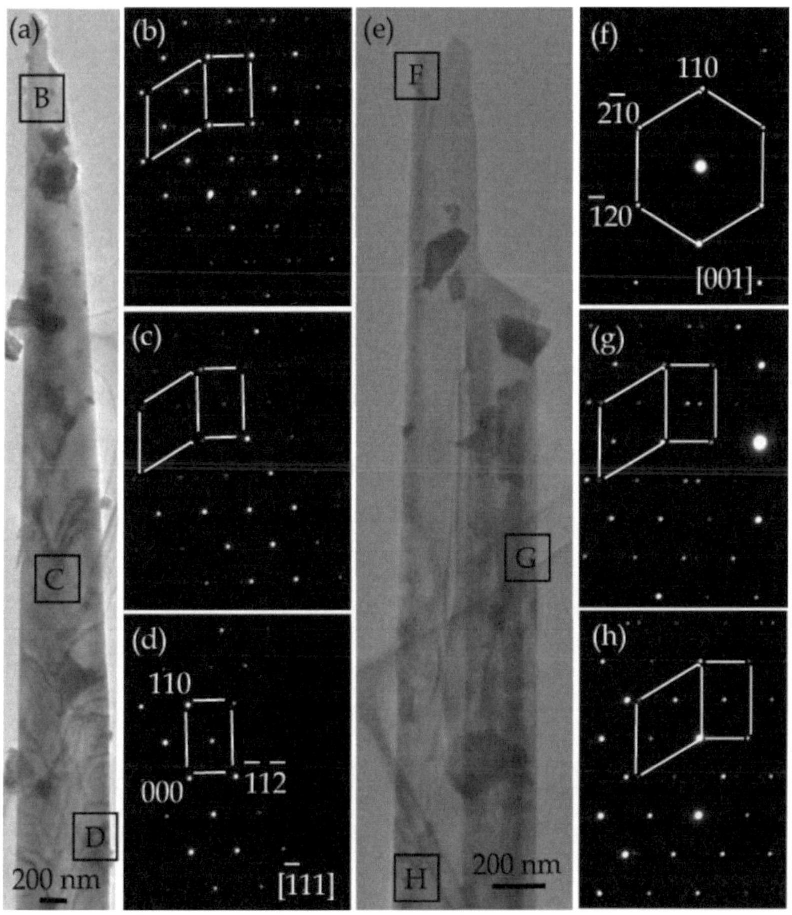

Fig. 12. (a,e) TEM images of the nanobelts shown in Fig. 9 (c). (b-d, f-h) corresponding SAED patterns recorded from rectangular D-E and F-H enclosed areas in (a) and (e), respectively.

patterns from the rectangular F, G and H enclosed areas in Fig. 12 (e). Although only one set of diffraction pattern is shown in Figs. 12 (d, f), actually some extra weak diffraction spots exist in the SAED patterns after careful examination. Thus, the nanobelts also have a bicrystal structure and grow along the [110] direction.

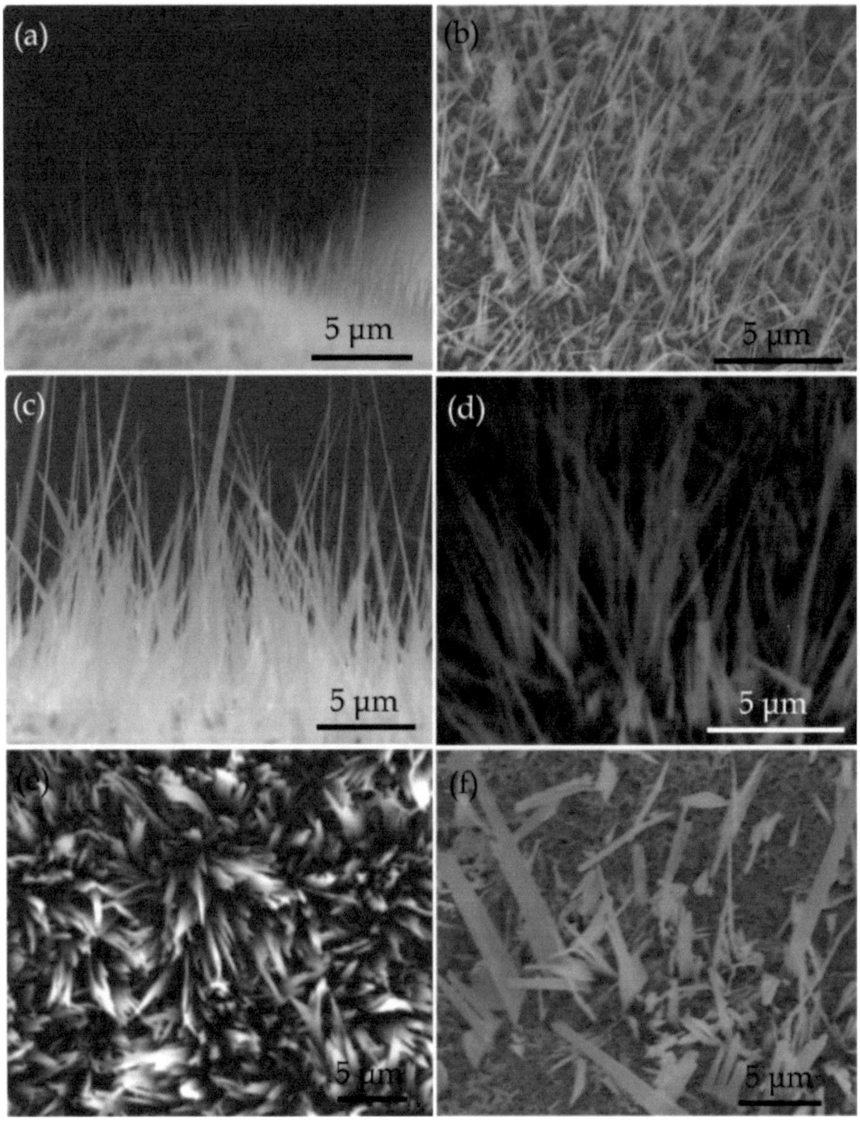

Fig. 13. SEM images of α-Fe_2O_3 nanostructures grown on iron substrates (a,c,e) without or (b,d,f) with gallium at (a,b) 600, (c,d) 650 and (e,f) 750 °C.

The surface energy is one of the major factors that affect the growth direction of the nanowire or nanobelt (Ye et al., 2005). The nanowires or nanobelts preferentially grow along the direction that minimizes their surface energy. Although the surface energy of the [110] orientation is not the highest (de Leeuw & Cooper, 2007), the α-Fe_2O_3 nanoscale crystals preferred to grow along the [110] direction (Cvelbar et al., 2008; Kim et al., 2006; Takagi,

1957). In the crystal structure of α-Fe_2O_3, the O atoms are close-packed in the (110) planes, the O-rich and Fe-deficient character on the (110) planes is considered to be the driving force for the preferential growth along the [110] direction (Srivastava et al., 2007; Wen et al., 2005). In addition, the bicrystal structure observed here did not affect the growth direction along [110].

In order to investigate the growth temperature dependence of the α-Fe_2O_3 nanostructure growth, the α-Fe_2O_3 nanostructure growth was tried by the thermal oxidation of iron substrates with or without gallium at temperatures of 600, 650 and 750 °C for 3 h in the air, as shown in Figure 13. By the thermal oxidation of iron substrates in the air, nanowires were grown at temperatures of 600~700 °C (Figs. 13a&c, 9a), while nanobelts were grown at temperatures of 750 (Fig. 13e) and 800 °C (Fig. 9c). The nanowires or nanobelts cover the surface of the irons and the density increases with the temperatures. However, nanobelts were grown by the thermal oxidation of iron substrates with gallium at temperatures of 600~750 °C, meanwhile, few nanowires were also observed (Figs. 13b,d&f, 9c). In addition, micro-crystals were grown at 800 °C, and nanostructures were seldom obtained (Fig. 9d). Most of the nanowires and nanobelts become narrow towards the tip, and the length extends to tens of micrometers. In addition, the nanowires are generally thinner than the nanobelts.

Due to the low Fe vapor pressure (Honig & Kramer, 1969), and no evidence of deposition was found anywhere except on the iron substrate, the evaporation can be negligible at the growth temperatures in the air. In consideration of the rough surface of the iron substrate as observed in the as-grown samples, it is assumed that nanoscale Fe droplets were formed on the surface of the iron substrate at elevated temperatures, and Fe and O were absorbed into the droplets from the iron substrate and the air, respectively. The α-Fe_2O_3 nanostructures would be formed on the iron surface when the droplets got supersaturated, and the growth would be terminated when the liquid is condensed into the solid state during cooling down (Kim et al., 2006; Yu et al., 2006). The α-Fe_2O_3 nanostructure growth should be controlled by the Fe surface diffusion and supersaturation (Lu & Ulrich, 2005; Takagi, 1957; Ye et al., 2005; Zhang et al., 2008).

In this study, the transformation from nanowire growth to nanobelt growth or even micro-crystal growth was observed in Figs. 9 and 13. It is generally suggested that high supersaturation favors the two-dimensional growth of the nanostructures (Lu & Ulrich, 2005; Ye et al., 2005; Zhang et al., 2008). At low temperatures (below 600 °C), few droplets were formed and the diffusion rate was very low, thus sparse nanowires were grown (Fig. 13a). More droplets were formed and the diffusion rate increased with the growth temperatures, thus the nanowires with high density were grown at 700 °C (Fig. 9a). When the temperature was higher than 750 °C (Figs. 13e&9c), the supersaturation was relatively high, two-dimensional nucleation would be initiated, and the increased diffusion rate accelerated the nanobelt growth (Wen et al., 2005). On the other hand, the presence of gallium would significantly decrease the formation temperature of the droplets by forming eutectic (Okamoto, 2000), and a high supersaturation could be achieved at low temperatures, which would facilitate the nanobelt growth, as compared with the nanowires grown by the thermal oxidization of iron. The nanobelts with a high density were grown at 700 °C (Fig. 9b), while micro-crystal growth rather than nanobelt growth occurred at 800 °C (Fig. 9d) due to the very high supersaturation and diffusion rate. In addition, the Fe diffusion from the bottom to tip is a diffusion limited process (Han et al., 2006), which is considered to result in the narrowing of the nanowires (Figs. 10a&11c) and nanobelts towards the tip (Fig. 12a).

3.3 β-Ga_2O_3

β-Ga_2O_3 is chemically and thermally stable with a wide band-gap of 4.9 eV, thus has potential applications in optoelectronic devices (Fu et al., 2003) and high temperature stable gas sensors (Ogita et al., 1999). So far, a variety of β-Ga_2O_3 nanostructures has been widely synthesized through chemical vapor deposition (Zhang et al., 2008), thermal evaporation (Zhang et al., 1999) or catalytic assisted growth (Choi et al., 2009) using gallium as one of the starting materials. As comparison, β-Ga_2O_3 nanostructure growth was carried out by the thermal oxidation of gallium at temperatures of 700~900 °C for 24 h in air, and the β-Ga_2O_3 nanostructures were observed on the surface of molten gallium. As shown in Figure 14, a series of SEM images reveal the growth evolution with the increase in the growth temperatures. β-Ga_2O_3 nanowires with a length less than 1 μm were grown at 700 °C (Fig. 14a), while β-Ga_2O_3 nanowires with a high density were grown at 800 °C (Fig. 14b), and the density and length increased with the temperature (Fig. 14c). It is indicated that a high growth temperature favored the nucleation and nanowire growth of the β-Ga_2O_3 crystals.

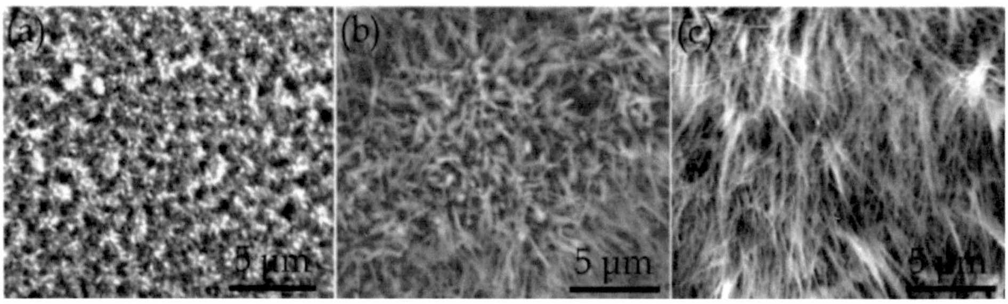

Fig. 14. SEM images of β-Ga_2O_3 nanowires grown at (a) 700, (b) 800 and (c) 900 °C.

4. Conclusion

In this study, we showed a variety of nanowire and nanobelt based oxides, including ZnO, α-Fe_2O_3 and β-Ga_2O_3, grown by the thermal oxidation with gallium in air. These results revealed the availability of such a simple growth method for the oxide nanostructure growth. In addition, the effect of gallium on the nanostructure growth (e.g., ZnO nanowires & nanobelts) and morphology transformation (e.g., α-Fe_2O_3 nanowires & nanobelts) was demonstrated, which provided the possibility of the selective growth of oxide nanostructures. Various other oxide nanostructures are expected when considering the anisotropic surface energy and nonhomogeneous supersaturation conditions, as well as the growth condition dependence of the selective growth.

5. References

Baxter, J. B.; Bessems, R. E. M. W. & Aydil, E. S. (2003). Growth and characterization of ZnO nanowires. *Materials Research Society Symposium Proceedings*, Vol.776, pp.Q7.9.1-Q7.9.6

Chen, J. T.; Zhang, F.; Wang, J.; Zhang, G. A.; Miao, B. B.; Fan, X. Y.; Yan, D. & Yan, P. X. (2008). CuO nanowires synthesized by thermal oxidation route. *Journal of Alloys and Compounds*, Vol.454, No.1-2, pp.268-273

Choi, K. H.; Cho, K. K.; Kim, K. W.; Cho, G. B.; Ahn, H. J. & Nam, T. H. (2009). Catalytic growth and structural characterization of semiconducting β-Ga_2O_3 nanowires. *Journal of Nanoscience and Nanotechnology*, Vol.9, No.6, pp.3728-3733

Cvelbar, U.; Chen, Z.; Sunkara, M. K. & Mozetic, M. (2008). Spontaneous growth of superstructure α-Fe_2O_3 nanowire and nanobelt arrays in reactive oxygen plasma. *Small*, Vol.4, No.10, pp.1610-1614

de Leeuw, N. H. & Cooper, T. G. (2007). Surface simulation studies of the hydration of white rust $Fe(OH)_2$, goethite α-FeO(OH) and hematite α-Fe_2O_3. *Geochimica et Cosmochimica Acta*, Vol.71, No.7, pp.1655-1673

Deng, G.; Ding, A.; Cheng, W.; Zheng, X. & Qiu, P. (2005). Two-dimensional zinc oxide nanostructure. *Solid State Communications*, Vol.134, No.4, pp.283-286

Ding, Y.; Gao, P. X. & Wang, Z. L. (2004). Catalyst-nanostructure interfacial lattice mismatch in determining the shape of VLS grown nanowires and nanobelts: a case of Sn/ZnO. *Journal of The American Chemical Society*, Vol.126, No.7, pp.2066-2072

Fu, L.; Liu, Y.; Hu, P.; Xiao, K.; Yu, G. & Zhu, D. (2003). Ga_2O_3 nanoribbons: synthesis, characterization, and electronic properties. *Chemistry of Materials*, Vol.15, No.22, pp.4287-4291

Fu, Y. Y.; Wang, R. M.; Xu, J.; Chen, J.; Yan, Y.; Narlikar, A. V. and Zhang, H. (2003). Synthesis of large arrays of aligned α-Fe_2O_3 nanowires. *Chemical Physics Letters*, Vol.379, No.3-4, pp.373-379

Gu, G.; Zheng, B.; Han, W. Q.; Roth, S. & Liu, J. (2002). Tungsten oxide nanowires on tungsten substrates. *Nano Letters*, Vol.2, No.8, pp.849-851

Han, Q.; Xu, Y. Y.; Fu, Y. Y.; Zhang, H.; Wang, R. M.; Wang, T. M. & Chen, Z. Y. (2006). Defects and growing mechanisms of α-Fe_2O_3 nanowires. *Chemical Physics Letters*, Vol.431, No.1-3, pp.100-103

Honig, R. E. & Kramer, D. A. (1969). Vapor pressure data for the solid and liquid elements. *RCA Review*, Vol.30, pp.285-305

Hu, J. Q.; Ma, X. L.; Xie, Z. Y.; Wong, N. B.; Lee, C. S. & Lee, S. T. (2001). Characterization of zinc oxide crystal whiskers grown by thermal evaporation. *Chemical Physics Letters*, Vol.344, No.1-2, pp.97-100

Jian, J. K.; Wang, C.; Zhang, Z. H.; Chen, X. L.; Xu, L. H. & Wang, T. M. (2006). Necktie-like ZnO nanobelts grown by a self-catalytic VLS process. *Materials Letters*, Vol.60, No.29-30, pp.3809-3812

Jiang, X.; Herricks, T. & Xia, Y. (2002). CuO nanowires can be synthesized by heating copper substrates in air. *Nano Letters*, Vol.2, No.12, pp.1333-1338

Kim, C. H.; Chun, H. J.; Kim, D. S.; Kim, S. Y.; Park, J.; Moon, J. Y.; Lee, G.; Yoon, J.; Jo, Y.; Jung, M. H.; Jung, S. I. & Lee, C. J. (2006). Magnetic anisotropy of vertically aligned α-Fe_2O_3 nanowire array. *Applied Physics Letters*, Vol.89, No.22, pp.223103

Kubaschewski, O. & Alcock, C. B. (1979). *International Series on Materials Science and Technology: Metallurgical Thermo-Chemistry*, Pergamon Press, Oxford

Li, W. J.; Shi, E. W.; Zhong, W. Z. & Yin, Z. W. (1999). Growth mechanism and growth habit of oxide crystals. *Journal of Crystal Growth*, Vol.203, No.1-2, pp.186-196

Li, Y.; You, L.; Duan, R.; Shi, P. & Qin, G. (2004). Oxidation of a ZnS nanobelt into a ZnO nanotwin belt or double single-crystalline ZnO nanobelts. *Solid State Communications*, Vol.129, No.4, pp.233-238

Liao, L.; Zheng, Z.; Yan, B.; Zhang, J. X.; Gong, H.; Li, J. C.; Liu, C.; Shen, Z. X. & Yu, T. (2008). Morphology controllable synthesis of α-Fe_2O_3 1D nanostructures: growth mechanism and nanodevice based on single nanowire. *The Journal of Physical Chemistry C*, Vol.112, No.29, pp.10784-10788

Lu, J. G.; Chang, P. & Fan, Z. (2006). Quasi-one-dimensional metal oxide materials - synthesis, properties and applications. *Materials Science and Engineering R*, Vol.52, No.1-3, pp.49-91

Lu, J. J. & Ulrich, J. (2005). The influence of supersaturation on crystal morphology - experimental and theoretical study. *Crystal Research and Technology*, Vol.40, No.9, pp.839-846

Majni, G.; Nobili, C.; Ottaviani, G.; Costaato, M. & Galli, E. (1981). Gold-aluminum thin-film interactions and compound formation. *Journal of Applied Physics*, Vol.52, No.6, pp.4047-4054

Miyake, A.; Kominami, H.; Tatsuoka, H.; Kuwabara, H.; Nakanishi, Y. & Hatanaka, Y. (2000). Growth of epitaxial ZnO thin film by oxidation of epitaxial ZnS film on Si(111) substrate. *Japanese Journal of Applied Physics*, Vol.39, No.11B, pp.L1186-L1187

Ogino, K.; Honda, S.; Yasuda, T.; Tatsuoka, H.; Inaba, T.; Kominami, H.; Nakanishi, Y. & Murakami, K. (2007). Simple synthesis of a variety of nano-structures using silicide alloys with Ga droplets. *ECS Transactions*, Vol.11, No.8, pp.77-81

Ogita, M.; Saika, N.; Nakanishi, Y. & Hatanaka, Y. (1999). Ga_2O_3 thin films for high-temperature gas sensors. *Applied Surface Science*, Vol.142, No.1-4, pp.188-191

Ohmori, T.; Takahashi, H.; Mametsuka, H. & Suzuki, E. (2000). Photocatalytic oxygen evolution on α-Fe_2O_3 films using Fe^{3+} ion as a sacrificial oxidizing agent. *Physical Chemistry Chemical Physics*, Vol.2, No.14, pp.3519-3522

Okamoto, H. (2000). *in Desk Handbook: Phase Diagrams for Binary Alloys*, ASM International, Materials Park, OH

Pan, Z. W.; Dai, Z. R. & Wang, Z. L. (2001). Nanobelts of semiconducting oxides. *Science*, Vol.291. No.5510, pp.1947-1949

Pan, Z. W.; Dai, Z. R.; Ma, C. & Wang, Z. L. (2002). Molten gallium as a catalyst for the large-scale growth of highly aligned silica nanowires. *Journal of the American Chemical Society*, Vol.124, No.8, pp.1817-1822

Pan, Z. W.; Dai, S.; Beach, D. B. & Lowndes, D. H. (2003). Temperature dependence of morphologies of aligned silicon oxide nanowire assemblies catalyzed by molten gallium. *Nano Letters*, Vol.3, No.9, pp.1279-1284

Pan, Z. W.; Dai, S. & Lowndes, D. H. (2005). Gallium-catalyzed silicon oxide nanowire growth. *Tsinghua Science and Technology*, Vol.10, No.6, pp.718-728

Ren, S.; Bai, Y. F.; Chen, J.; Deng, S. Z.; Xu, N. S.; Wu, Q. B. & Yang, S. (2007). Catalyst-free synthesis of ZnO nanowire arrays on zinc substrate by low temperature thermal oxidation. *Materials Letters*, Vol.61, No.3, pp.666-670

Srivastava, H.; Tiwari, P.; Srivastava, A. K. & Nandedkar, R. V. (2007). Growth and characterization of α-Fe_2O_3 nanowires. *Journal of Applied Physics*, Vol.102, No.5, pp.054303

Takagi, R. (1957). Growth of oxide whiskers on metals at high temperature. *Journal of The Physical Society of Japan*, Vol.12, No.11, pp.1212-1218

Wang, G. Z.; Ma, N. G.; Deng, C. J.; Yu, P.; To, C. Y.; Hung, N. C.; Aravind, M. & Dickon H. L. Ng (2004). Large-scale synthesis of aligned hexagonal ZnO nanorods using chemical vapor deposition. *Materials Letters*, Vol.58, No.16, pp.2195-2198

Wang, R.; Chen, Y.; Fu, Y.; Zhang, H. & Kisielowski, C. (2005). Bicrystalline hematite nanowires. *The Journal of Physical Chemistry B*, Vol.109, No.25, pp.12245-12249

Wang, X.; Ding, Y.; Summers, C. J. & Wang, Z. L. (2004). Large-scale synthesis of six-nanometer-wide ZnO nanobelts. *The Journal of Physical Chemistry B*, Vol.108, No.26, pp.8773-8777

Wen, X.; Wang, S.; Ding, Y.; Wang, Z. L. & Yang, S. (2005). Controlled growth of large-area, uniform, vertically aligned arrays of α-Fe_2O_3 nanobelts and nanowires. *The Journal of Physical Chemistry B*, Vol.109, No.1, pp.215-220

Xu, Y. Y.; Zhao, D.; Zhang, X. J.; Jin, W. T.; Kashkarov, P. & Zhang, H. (2009). Synthesis and characterization of single-crystalline α-Fe_2O_3 nanoleaves. *Physica E*, Vol.41, No.5, pp.806-811

Yang, Q.; Tanaka, M.; Yasuda, T. & Tatsuoka, H. (2009). Growth of ZnO nanowires using ZnS substrates with Ga droplets. *e-Journal of Surface Science and Nanotechnology*, Vol. 7, pp.25-28

Ye, C.; Fang, X.; Hao, Y.; Teng, X. & Zhang, L. (2005). Zinc oxide nanostructures: morphology derivation and evolution. *The Journal of Physical Chemistry B*, Vol.109, No.42, pp.19758-19765

Yu, T.; Zhu, Y.; Xu, X.; Shen, Z.; Chen, P.; Lim, C. T.; Thong, J. T. L. & Sow, C. H. (2005). Controlled growth and field-emission properties of cobalt oxide Nanowalls. *Advanced Materials*, Vol.17, No.13, pp.1595-1599

Yu, T.; Zhu, Y.; Xu, X.; Yeong, K. S.; Shen, Z.; Chen, P.; Lim, C. T.; Thong, J. T. L. & Sow, C. H. (2006). Substrate-friendly synthesis of metal oxide nanostructures using a hotplate. *Small*, Vol.2, No.1, pp.80-84

Yuan, H. J.; Xie, S. S.; Liu, D. F.; Yan, X. Q.; Zhou, Z. P.; Ci, L. J.; Wang, J. X.; Gao, Y.; Song, L.; Liu, L. F.; Zhou, W. Y. & Wang, G. (2003). Characterization of zinc oxide crystal nanowires grown by thermal evaporation of ZnS powders. *Chemical Physics Letters*, Vol.371, No.3-4, pp.337-341

Zhang, H. Z.; Kong, Y. C.; Wang, Y. Z.; Du, X.; Bai, Z. G.; Wang, J. J.; Yu, D. P.; Ding, Y.; Hang, Q. L. & Feng, S. Q. (1999). Ga_2O_3 nanowires prepared by physical evaporation. *Solid State Communications*, Vol.109, No.11, pp.677–682

Zhang, X.; Liu, Z. & Hark, S. (2008). Synthesis and cathodoluminescence of□β-Ga_2O_3 nanowires with holes. *Journal of Nanoscience and Nanotechnology*, Vol.8, No.3, pp.1284-1287

Zhang, Y.; Li, R.; Zhou, X.; Cai, M. & Sun, X. (2008). Selective growth of α-Al_2O_3 nanowires and nanobelts. *Journal of Nanomaterials*, Vol.2008, pp.250370

Zhao, D.; Andreazza, C.; Andreazza, P.; Ma, J.; Liu, Y. & Shen, D. (2004). Temperature-dependent growth mode and photoluminescence properties of ZnO nanostructures. *Chemical Physics Letters*, Vol.399, No.4-6, pp.522-526

4

Nano-cones Formed on a Surface of Semiconductors by Laser Radiation: Technology, Model and Properties

Artur Medvid'
Riga Technical University,
Latvia

1. Introduction

Nowadays, nanostructures are some of the most investigated objects in semiconductor physics, especially Quantum confinement effect (QCE) in quantum dots (QDs) (Alivisatos, 1996), quantum wires (QWs) (Xia et al., 2003) and quantum wells (Fowler et al., 1966). In the case of nano-size structures the energy band diagram of semiconductor changes strongly. This leads to a crucial change of semiconductor properties such as: electrical, due to the change of free charge carrier concentration and electrons' and holes' mobility; optical, such as: absorption coefficient, reflectivity index, radiative recombination efficiency (Emel'yanov et al., 2005); mechanical and heating properties.

The growth of investigations of the homogeneous structures, such as well-defined one-dimensional axial heterostructures, multiheterostructures and core-shell nano-wire heterostructures generates a great interest for their potential applications. Conglomeration of the QDs in a line leads to formation of QWs. If diameter of QWs changes monotonously then cones like structure is formed. The cone nanostructure properties have not only quantitative, but also strongly qualitative characteristics and new interesting properties, for example, formation of graded band gap structure in elementary semiconductors (Medvid' et al., 2008).

Fabrication of nanostructures without lithographic process, based on the self-assembling processes is very promising for future nano-electronics. In this case the self-assembling processes utilize the microscopic structures on the surface or the strain induced by lattice mismatch.

Today we have only some very well elaborated methods for formation of nanostructures (NSs) in semiconductors. They are: molecular beam epitaxy (Talochkin et al., 2005), ion implantation (Zhu et al., 1997) chemical vapor deposition method (Hartmann et al., 2005), and laser ablation (Morales et al., 1998, Yoshida et al., 1998) with followed by thermal annealing in furnace. A lot of time and high vacuum or special environment, for example, inert gas Ar is needed for nano-crystals growth using these methods. As a result, nano-crystals grow with broad distribution in size. Therefore, elaboration of new methods for growth of NSs in semiconductors is a very important task for nano-electronics and optoelectronics. A significant amount of effort has been dedicated to the production of nano-structured Si-based systems (Werva et al., 1996). Several studies have used ion implantation of group IV elements in SiO_2 and heat treatment to obtain NSs that exhibit photoluminescence (PL). Si is the basic material in microelectronics because it is cheap and its technology is very well developed.

Photoluminescence in IR, red, and blue-violet region of spectra of Si and Ge nano-particles implanted into SiO_2 layer have been found by several scientific groups. Strong PL in the blue-violet region was achieved, with the values of excitation and emission energy being independent of the annealing temperature and consequently of the cluster mean size (Werva et al., 1996; Rebohle et al., 2000). Different explanations of this PL mechanism have been proposed. One part of scientists explain the PL by QCE (Rebohle et al., 2000)in Si and Ge nano-particles, but the second part (Fernandez et al., 2002) - by local Si-O and Ge-O vibration at Si-SiO_2 and Ge SiO_2 - interface. Phonon confinement effect model was developed by Campbell and Fauchet (Campbell et al., 1986) for Si nano-crystals. It was shown that using simultaneously micro-Raman back scattering spectrum and PL spectrum it is possible to determine the origin of the PL. It means if "blue shift" in PL spectrum and "read shift" of LO line in Raman back scattering spectrum are simultaneously present QCE takes please in NSs. A new flexible possibility is predicted to change the semiconductor basics parameters into QDs of Si_xGe_{1-x}/Si solid solution both by change of x (concentration of Ge) and QDs diameter. Increase both content of Ge atoms - x, and diameter of QDs leads to the same effective shift of PL spectrum toward low energy of spectrum, so called "red shift". It has been shown that increase of x from 0.096 to 0.52 leads to shift of maximum position in IR part of PL spectrum toward low energy (0.3 eV) (Sun et al., 2005). The same, "red shift" of PL spectrum on 0.7 eV has been observed for nano-particles with diameter 5-50 nm and $x = 0.237 - 0.75$ in visible part of spectrum (Mooney et al., 1995). Authors explain this result by the incorporation of the Ge atoms into Si nano-particles and associate it with surface state. Another effect has been observed for pure i-Ge crystal. Decrease of QDs diameter till 4 nm leads to "blue shift" of PL spectrum with maximum position up to 1.65 eV (Medvid' et al., 2007) in comparison with PL spectrum of bulk crystal. Therefore in this paper we will show that the main role in control of PL spectrum and its intensity is QCE with small influence of Ge content.

Compound semiconductor crystals such as GaAs, 6H-SiC, CdTe and ternary solid solutions CdZnTe are very useful materials in performance of the semiconductor laser, **x** and γ ray detectors (Gnatyuk et al., 2006) and recently it was found that they can be used for construction of solar cells (Vigil-Galan et al., 2005). New possibility for these devises construction was opened using QCE resulting in, for example, increase of laser radiation intensity and increase of efficiency for the third generation solar cells (Green et al., 2004), construction of light sources with controllable wave length i.e. Therefore, research of possibility of NSs formation on a surface of the semiconductors by laser radiation and study of their optical properties are the main tasks of the study. With this aim it is necessary to know the mechanism of the powerful laser radiation (LR) interaction with a semiconductor. The basic model used today for description of LR effects in semiconductor is the thermal model (Beigelsen et al., 1985), at list for nanosecond laser pulls duration. It implies that energy of light is transformed into thermal energy. But it is only the first step in the understanding of this process. Different models have been proposed for explanation of self-assemble of NSs on a surface of a semiconductor by LR. One of them is photo-thermo-deformation (PTD) model (Emel'janov et al., 1997). According to the PTD model conversion of the light to the heat and deformation of the crystalline lattice of a semiconductor takes place due to inhomogeneous absorption of the light. And formation of the periodical

structure on the surface arises due to redistribution of interstitials and vacancies. Disadvantage of the RTD model are following: impossibility to explain of the NSs formation in semiconductors, for example, Ge (Medvid' et al., 2005) and 6H-SiC (Medvid' et al., 2004) at high intensity of LR when phase transition from solid state to liquid phase takes place, accumulation and saturation effects (Medvid' et al., 2002; Medvid'et al., 1999). It was shown that at high absorption of the powerful LR in a semiconductor high gradient of temperature occurs which causes impurities atoms and intrinsic defects, interstitial and vacancies, drift toward temperature gradient, so called Thermogradient effect (TGE) (Medvid'et al., 1999). According to TGE theory atoms which have bigger effective diameter than atoms of basic semiconductor material drift toward the maximum of temperature, but atoms with smaller effective diameter toward the minimum of temperature. As a result in semiconductor arises compressive mechanical stress on the irradiated surface and tensile mechanical stress in the balk of a semiconductor. An evidence of the TGE presence at these conditions is formation of p-n junction on a surface of p-Si (Mada et al., 1986), p-InSb (Fujisawa, 1980; Medvid' et al., 2001), p-CdTe (Medvid' et al., 2001) and p-InAs (Kurbatov et al., 1983).
A new laser method elaborated for cone like nanostructure (diameter of the nano-cone is increased gradually from top of cone till a substrate) formation on a surface of semiconductors is reported. Model of the nanostructures formation and their optical properties are proposed.

2. Experiments and discussion

2.1 Experiments on elementary semiconductors Ge, Si their solid solution Si_xGe_{1-x} and compound 6H-SiC single crystals.

Experiments were performed in ambient atmosphere at pressure of 1 atm, T = 20^{0}C, and 60% humidity. Radiation from a pulsed Nd:YAG laser for Ge single crystals and Si_xGe_{1-x}/Si solid solution basic frequency with following parameters: pulse duration τ=15 ns, wavelength λ = 1.06 μm, pulse rate 12.5 Hz, power P = 1.0 MW and for Si single crystals with SiO_2 cover layer second harmonic with τ =10 ns and λ = 532 nm were used. SiO_2 cover layer, in experiments with Si, GaAs and CdZnTe, for privet evaporation of material is used. Usually laser beam was directed normally to the irradiated surface of sample. Samples of Ge(1 1 1) or Ge(001) i-type single crystal with sizes 1.0x0.5x0.5 cm^3 and resistivity ρ = 45 Ω·cm were used in experiments. The samples were polished mechanically and etched in CP-4A solution to ensure the minimum surface recombination velocity S_{min} = 100 cm/s on all the surfaces. The spot of laser beam of 3 mm diameter was scanned over the sample surface by a two coordinate manipulator in 1 or 2 mm steps. N_2 laser with following parameters: λ = 337 nm, τ =10 ns and P = 0.16 MW in experiments with 6H-SiC (0001) single crystals doped with N and B atoms was used. The surface morphology was studied by atomic force microscope (AFM) and electron scanning microscope and electron scanning microscope (ESM). Optical properties of the irradiated and no irradiated samples were studied by photoluminescence (PL) and back scattering Raman methods. For PL the 488nm line of a He-Cd laser and for micro-Raman backs catering a Ar+ laser with λ=514.5nm were used.
The AFM study of Ge surface morphology after irradiation by basic frequency of the Nd:YAG laser is shown in Fig.1
The most interesting results were found at increasing the LR intensity up to 28.0 MW/cm^2, when nano-cones arise on the irradiated surface of Ge, which are self-organized into a 2D lattice. The 2D picture of the irradiated surface of a Ge sample as seen under ESM is shown

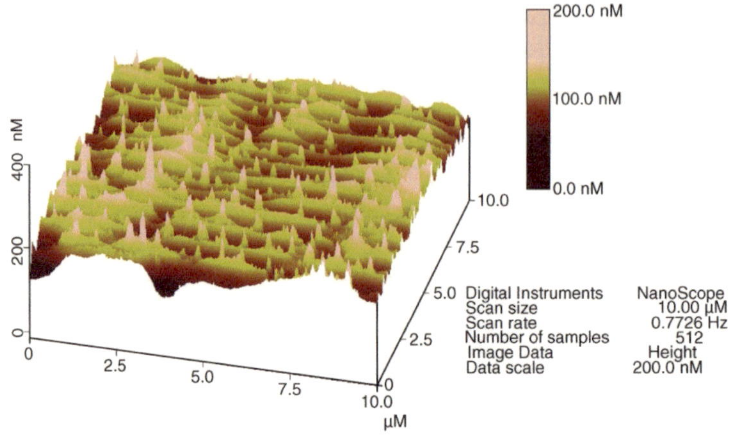

Fig. 1. Three-dimensional AFM image of self-organized nano-structures formed under Nd: YAG laser radiation at intensity of 28 MW/cm^2.

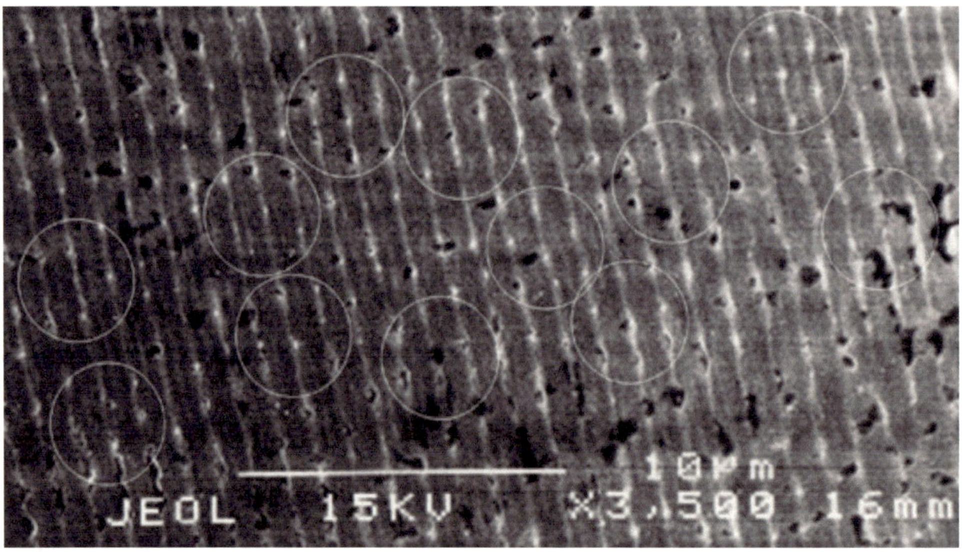

Fig. 2. SEM image of the Ge single crystal surface irradiated by Nd:YAG laser at intensity 28 MW/cm^2. C6i patterns are marked by white circles.

in Fig. 2. The 2D lattice is characterized by translation symmetry along and perpendicular to periodic lines with a pattern of C point group symmetry and repetition period of 1 µm. C_{6i} patterns are marked by white circles. Patterns orientation and their symmetry depending on orientation of Ge surface were not observed. An explanation of the phenomenon was sought in calculations of the time-dependent distribution of temperature in the bulk of the Ge sample using the heat diffusion equation with values for Ge parameters from (Okhotin et al., 1972; Vorobyev et al., 1996) (Fig. 3). As seen from the results, a close to adiabatic overheating of the crystalline lattice (Lin et al., 1982; Von der Linde et al., 1982) under conditions of the experiment; at LR intensity exceeding 28.0 MW/ cm^2. A thermal gradient of $3 \cdot 10^8$ K/m is reached during the first 10 ns. According to synergy ideas, a turbulent fluency and self-organized stationary structures of hexagonal symmetry, the so-called Bernar's cells, may arise at the presence of non-equilibrium liquid phase at a high gradient of temperature.

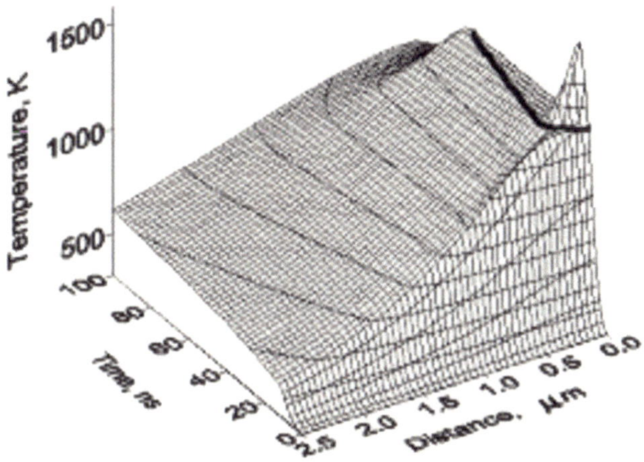

Fig. 3. Calculations of time-dependent distribution of temperature in the bulk of Ge single crystal at Nd:YAG laser radiation intensity of 28 MW/cm^2. Black line is isotherm of Ge melting temperature.

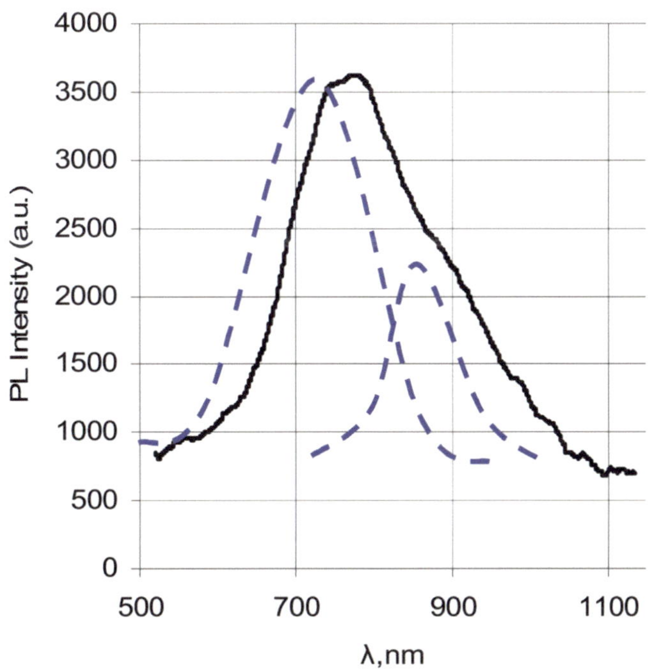

Fig. 4. PL spectrum of the surface of Ge after irradiation by Nd:YAG laser at intensity up to 28 MW/cm^2.

The characteristic size of the pattern is determined by thickness of the liquid layer being approximately 1 μm in our experiments. Unusual PL spectrum from the irradiated surfaces of Ge was found in the visible range of spectrum with maximum at 750 nm (1,65eV), as shown in Fig. 4. PL spectrum is usual situated at 2 μm and intensity of PL is very too low due to indirect band structure of Ge. This "blue shift" of PL spectrum we explain by presents of QCF on the top of nano-cones where ball radius is equal or less than Bohr's radii of electron and hole. Our calculation of the ball diameter on the top of nano-cone using formula (1) from paper (Efors et al., 1982) and band gap shift from PL bands with maximums at 1.65 eV and 1.3 eV (Fowler et al., 1966) at parameters of Ge: m_e=0.12 m_0 and m_h=0.379 m_0 for electron and hole effective masses, respectively, gives diameters of balls 4 nm and 6 nm.

$$E_g = E^0{}_g + \frac{(\pi\hbar)^2}{2d^2}\left(\frac{1}{m_e^*} + \frac{1}{m_h^*}\right) \tag{1}$$

An evidence of our suggestion is Raman back scattering spectra of the no irradiated (black curve) and of the irradiated (red curve) surfaces of Ge crystal by the laser, as shown in Fig.5.

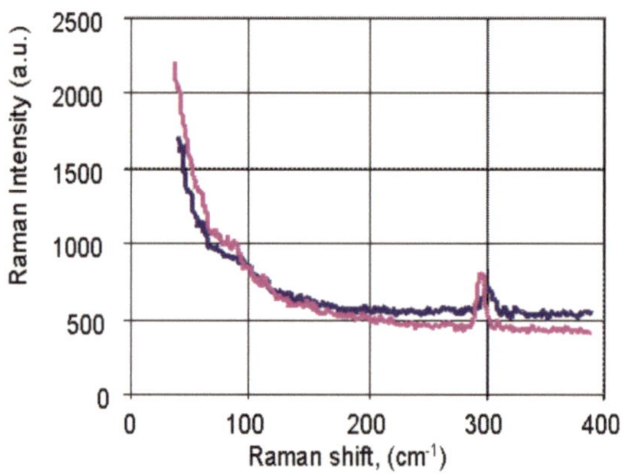

Fig. 5. Micro Raman spectrum of the no irradiated (black curve) and of the irradiated surfaces of Ge single crystal by Nd:YAG laser (red curve) at intensity up to 28 MW/cm^2.

Line at 300 cm^{-1}of the no irradiated surface of Ge attributed to bulk Ge (Ge-Ge vibration, LO line). Red shift of the LO line in Raman back scattering spectra on 6 cm^{-1} after irradiation of Ge surface takes place. Calculated line width and peak frequency Raman spectrum as a function of average crystal size (d_{ave}) for spherical Ge particles from paper (Kartopu et al., 2004) are shown in Fig.6. "Read shift" of the LO line on 6 cm^{-1} in the Raman spectrum correspond to 4 nm diameter of Ge nano-ball on the top of nano-cone. This value is in good agreement with our precious calculation from formula (1) using the PL spectrum.

An AFM 3D image of Si surface after irradiation by second harmonic of Nd :YAG laser at I=2.0 MW/cm^2 of SiO_2/Si structure, and the same AFM 3D image of Si surface after subsequent chemical etching by HF are shown in figure 7a and figure 7b, respectively. Photoluminescence spectra of the irradiated (curves 1 and 2) and non-irradiated (curve 3) surface of SiO_2/Si at intensity of laser radiation up to 2.0 MW/cm^2 are shown in figure 8. The surface morphology of SiO_2 layer is smooth "stone-block" like, but really under SiO_2

layer are very sharp Si nano-cones (Fig. 7b), which arise on the SiO_2/Si interface after irradiation by the laser. SiO_2 layer was fully removed by HF acid from SiO_2/Si structure. Photoluminescence of the SiO_2/Si structure in visible range of spectrum with maximum at 2.05 eV (600 nm) obtained after irradiation by the laser at intensity I =2.0 MW/cm^2, is shown in figure 8. PL of this structure after removing of SiO_2 layer by chemical etching in HF acid is similar and is obtained in the same range of spectrum and having the same positions of maximums. It means that PL is not connected with local Si-O vibration at Si-SiO_2 interface (Fernandez et al., 2002). Therefore, we explain our results by Quantum confinement effect in nano-cones. Decrease of the PL intensity can be explained by increase of reflection index of the structure after removing of SiO_2 layer. We can see that the visible PL spectrum of

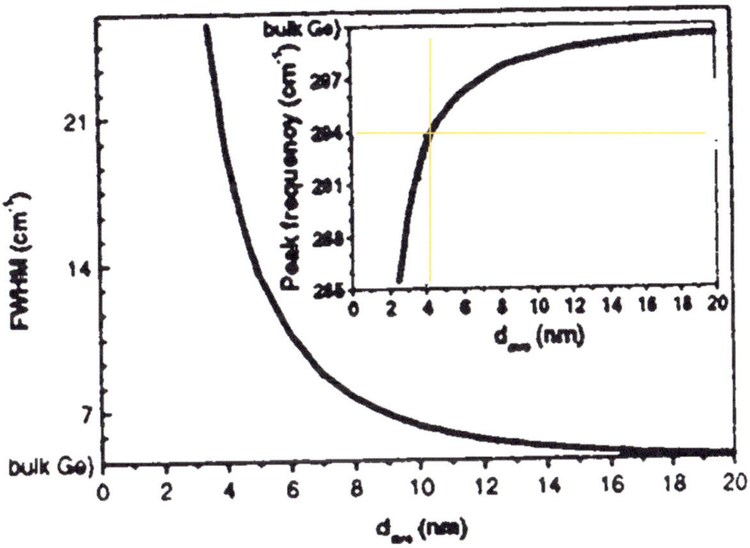

Fig. 6. Calculated line width (FWHM) and peak frequency Raman spectrum as a function of average crystal size (dave) for spherical Ge particles [4].

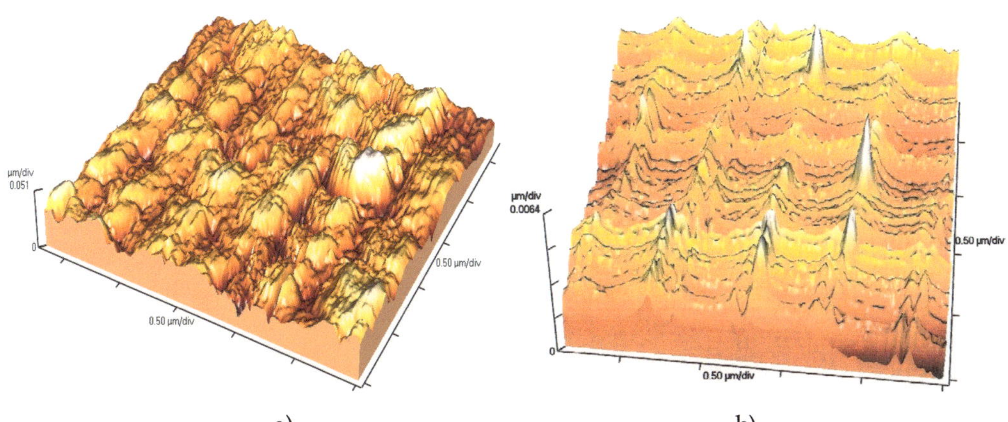

Fig. 7. AFM 3D images of: (a) SiO_2 surface after irradiation of SiO_2/Si structure by Nd:YAG laser at I=2.0 MW/cm^2 and (b) Si surface after subsequent removing of SiO_2 by HF acid.

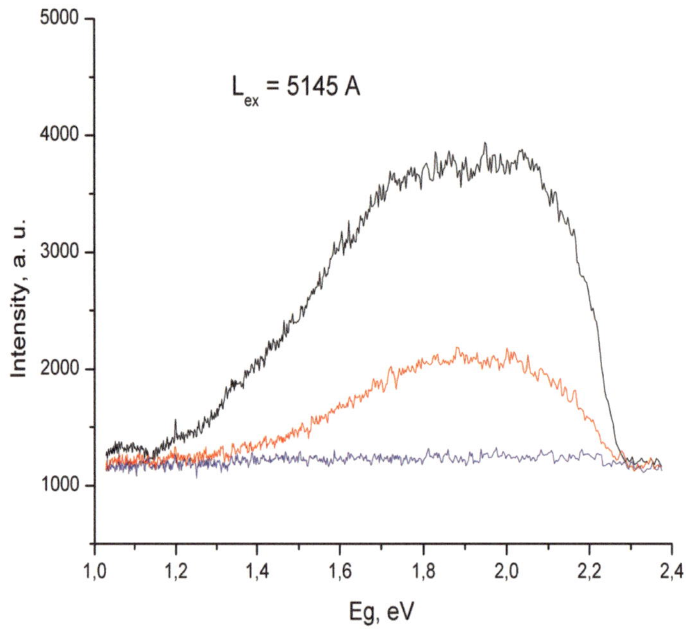

Fig. 8. Photoluminescence spectra of the SiO_2/Si structure irradiated by the laser at intensity 2.0 MW/cm^2 (red and black curves), after removing of SiO_2 layer by chemical etching in HF acid (black curve). Blue curve correspond to PL of the no irradiated surface.

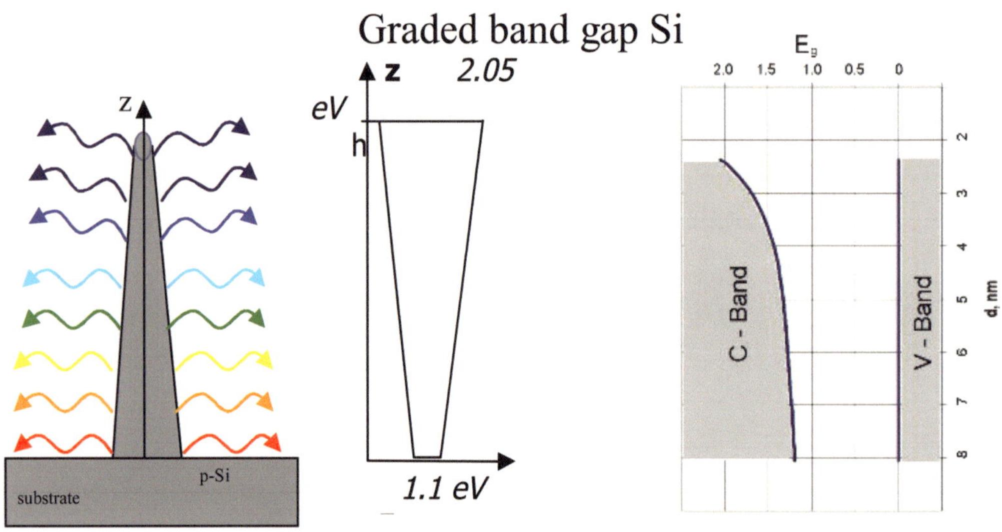

Fig. 9. A schematic image of a nano-hill with a gradually decreasing diameter from p-Si substrate till top, formed by laser radiation - a. and band gap Si structure b. c. A calculated band gap structure of Si as a function of nano-wires diameter using formula (2) from ref. [37].

SiO_2/Si structure is wide and asymmetric with gradual decrease of intensity in IR range of the spectra. It is typical for graded band gap structure. These results present a dramatic rise of PL with energy much higher than the indirect band gap of Si. Schematic image of a nano-cone with graduated decrease of diameter from p-Si substrate till top is shown in Fig.9.
Increase of energy of a radiation quantum from substrate till top of the Si single crystal at photoluminescence of nano-cone takes place due to Quantum confinement effect in nano-wire, according to formula (Li et al., 2004)

$$\Delta E_g = \frac{2\hbar^2 \zeta^2}{m^* d^2}, \tag{2}$$

where $1/(m^*) = 1/(m_e^*) + 1/(m_h^*)$, ($m_e^*$ and m_h^* are electron and hole effective-masses, respectively) and ***d*** is the diameter. For QWs, ζ =2.4048. In our case the diameter of nano-cone/nano-wires is a function of height ***d(z)***, therefore, it is graded band gap semiconductor. Our calculation of Si band gap as a function of nano-wires ***d*** from PL spectrum using formula (2) from paper (Li et al., 2004) is shown in Fig.9. We can see that the dependence is nonlinear and decreasing function of diameter, especially very rapidly at the small size of diameters. In our case the maximum of band gap is 2.05 eV which corresponds to the minimal diameter 2.3 nm on the top of nano-cones /nano-wire.
We have found a new method for formation of the graded band gap in elementary semiconductor. Graded change of band gap arises due to Quantum confinement effect. Usually graded band gap semiconductor structure is formed by conventional method - molecular beam epitaxy, changing molecular components concentration layer by layer.
Crystal $Si_{1-x}Ge_x$ alloys were grown on Si(100) wafers by Molecular Beam Epitaxy (MBE). $Si_{1-x}Ge_x$ films were grown by MBE on top of a 150 nm thick Si buffer layer on Si. Alloys containing 30% Ge were used in the experiments. The surface of a Si Ge /Si structure was irradiated by basic frequency of the Nd:YAG laser. The three-dimensional surface morphology of $Si_{1-x}Ge_x$/Si hetero-epitaxial structure recorded by AFM measurements after irradiation by the Nd:YAG laser at intensities of 7.0 MW/cm^2 (a) and 20.0 MW/cm^2 (b) is shown in Fig. 10.
In Fig. 10(a) are seen the nano-cones of the average height of 11 nm formed by laser radiation at the intensity of 7.0 MW/cm^2. Similar nano-cones of the average height of 27 nm seen in Fig. 10(b) have been obtained by irradiation intensity of 20 MW/cm^2. Due to higher irradiation intensity they are more compact in diameter and higher. After irradiation of the Si Ge/Si hetero-epitaxial structure by the laser at intensity of 7.0 MW/cm^2 the surface structure begins to look as spots on unwetting material, for example, it looks like water spots on a glass, Fig.10(c). It means that laser radiation induces segregation of Ge phases at the irradiated surface of the material. This conclusion is in agreement with data from paper [38] where it was shown that Ge phase starts formation at 50% concentration of Ge atoms in SiGe solid solution. According to the TGE (Medvid' et al., 2002), it is supposed that laser radiation initiates the drift of Ge atoms toward the irradiated surface of the hetero-epitaxial structure (Medvid' et al., 2009). PL spectra of the $Si_{1-x}Ge_x$/Si hetero-epitaxial structures with the maxima at 1.60 -1.72 eV obtained after laser irradiation at intensities of 2.0 MW/cm^2, 7.0 MW/cm^2 and 20.0 MW/cm^2 are shown in Fig. 11. The spectra are unique and unusual for the material, because, depending on Ge concentration, the band gap of SiGe is situated

between 0.67 eV and 1.12 eV (Sun et al., 2005). As seen from Fig. 11, the Si_{1-x} Ge_x structure emits light in the visible range of spectrum and the intensity of PL increases with the intensity of irradiation. The maximum of the PL band at 1.70 eV is explained by the QCE (Efors et al., 1982). Position of the observed PL peak compared with the bulk material shows a significant "blue shift". The maxima of PL spectra of the Si_{1-x} Ge_x /Si hetero-epitaxial structure slightly shift to higher energy when the laser intensity increases from 2.0 MW/cm^2 to 20.0 MW/cm^2, which is consistent with the QCE too. Our suggestions, concerning to Ge phase formation, are supported by the back scattering Raman spectra are shown in Fig.12.

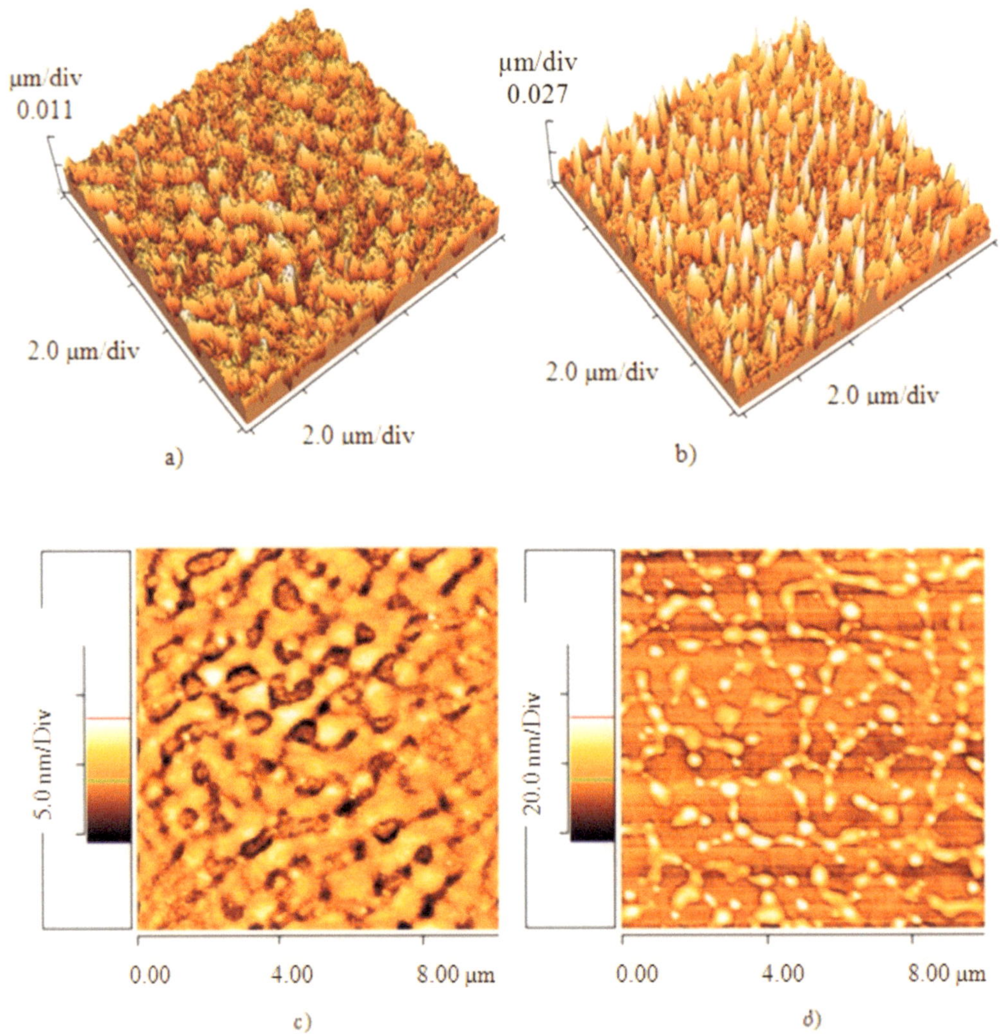

Fig. 10. Three-dimensional AFM images of $Si_{1-x}Ge_x$/Si surfaces irradiated by the Nd:YAG laser at intensity a)7 MW/ cm^2 and b) 20 MW/cm^2 and two-dimensional surface morphology of the same spots of structure at intensities: (c) 7.0 MW/cm^2 and (d) 20.0 MW/cm^2.

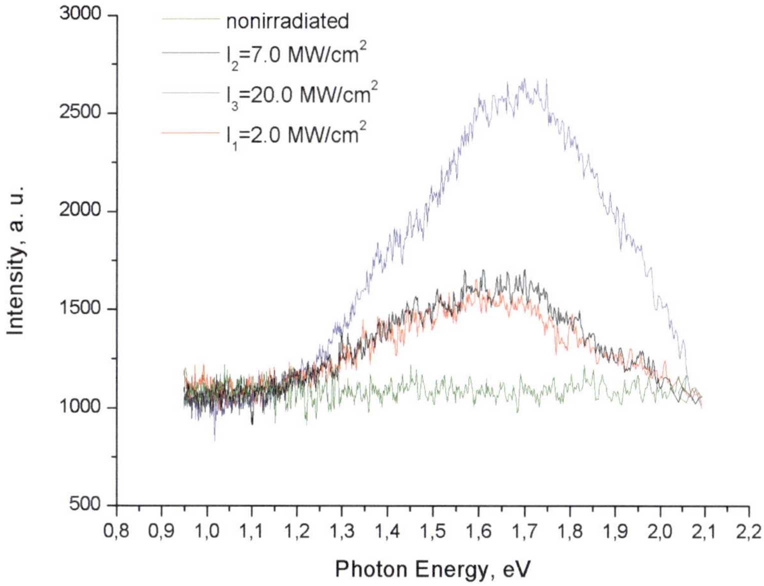

Fig. 11. PL spectra of $Si_{0.7}Ge_{0.3}/Si$ heteroepitaxial structures before and after irradiation by Nd:YAG laser

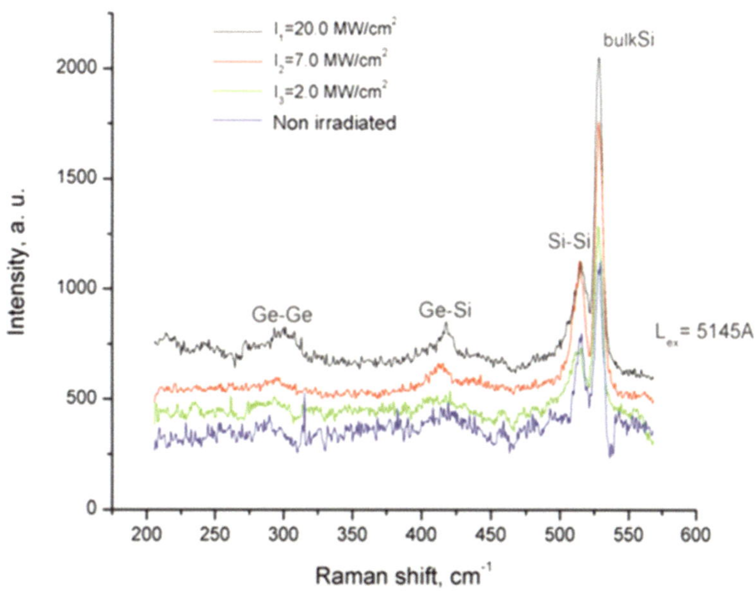

Fig. 12. Back scattering Raman spectra of $Si_{1-x}Ge_x/Si$ heteroepitaxial structure before (blue curve) and after irradiation by the laser. Appearance of the 300cm^{-1} Ge-Ge vibration band in Raman spectra is explained by the new phase formation in $Si_{1-x}Ge_x/Si$ heteroepitaxial structure.

After laser irradiation at the intensity of 20.0 MW/cm^2 a Raman band at 300 cm^{-1} appears in the spectrum. This band is attributed to the Ge-Ge vibration and is explained by formation of a new Ge phase (Kamenev et al., 2005) in the $Si_{1-x}Ge_x$/Si hetero-epitaxial structure. There is proposed to use modified formula (3) from paper (Efors et al., 1982) for determination of concentration x in the nano-cones of $Si_{1-x}Ge_x$/Si hetero-epitaxial structure formed by laser radiation using PL spectra.
Determination of x for diameter of nano-dot d = 4.2 nm on the top of nano-cone from AFM measurements and band gap from maximums of PL spectra at E_{g1} =1.74 eV, E_{g2}=1.69 eV and E_{g3}=1.60 eV, and E^0_g= 0.95eV for 30% of Ge in Si (Hogarth, 1965) were found Ge concentration in the nano-cones are x_1 =34%, x_2 =55% and x_3 =66%, respectively, where E^0_g is band gap of bulk material, $m_{e,h}^{*Ge,*Si}$ are the electron and hole effective mass for Ge and Si, respectively. In this preliminary analysis we used the same value of d.

$$E_g = E^0{}_g + \frac{(\pi\hbar)^2}{2d^2}\left[x_{Ge}\left(\frac{1}{m_e^{*Ge}}+\frac{1}{m_h^{*Ge}}\right)+(1-x)_{Si}\left(\frac{1}{m_e^{*Si}}+\frac{1}{m_h^{*Si}}\right)\right] \quad (3)$$

The following model is proposed for explanation of dynamics of nanostructures formation.

Model

Irradiation of SiGe/Si heterostructure by Nd:YAG laser initiates Ge atoms drift to the irradiated surface due to gradient of temperature - Thermogradient effect. Concentration of Ge atoms is increased at the irradiated surface. Ge atoms are localized at the surface of Si like a thin film. A mismatch of Si and Ge crystal lattices leads to compressed stress of Ge layer. The stress relaxation takes place by plastic deformation of the top Ge layer and creation of nanostructures on the irradiated surface according to the modified Stransky-Krastanov' mode.
In our experiments on SiC the nano-cones is formed on the surface of 6H-SiC after irradiation by the pulsed N_2 laser with following parameters: wavelength 337 nm, pulse duration 10 ns, energy of pulse $1.6\cdot10^{-3}$ J. These nano-cones are situated along the circular line with diameter about 400 nm as is shown in Fig.13, which is approximately 20 times smaller than the diameter of the spot of the focused laser beam. Usually the focused laser beam forms a crater of a cone shape on the surface of a semiconductor or a metal due to rapid spattering of the melted matter (Kelly et al., 1985). We have used crystals of 6H-SiC, produced by the Lely method doped with N and B or N with following concentrations: n_N = $2\cdot10^{18}$cm^{-3} and n_B= $5\cdot10^{18}$cm^{-3}. The experiments were carried out at room temperature and atmospheric pressure. The PL and the Friction Force Microscope (FFM) were used to detect the laser-induced changes in the chemical composition of the irradiated surface, while the AFM was used for studies of the surface morphology. A C surface of samples was irradiated by the laser. The threshold of average intensity of LR for formation of nanostructures is estimated as $<I_{th}> \approx 5$GW/cm^2. The FFM studies of the 6H-SiC(N) samples have shown that the chemical composition of nano-cones, which had arisen after the laser irradiation, differs from that of the no irradiated surface.
Luminescence spectra of the laser irradiated and no irradiated 6H-SiC(N) samples are shown in Fig.14, curves 3 and 4, correspondingly. Arising of the 2.8 eV band after laser irradiation is observed. As it is known this luminescence is assigned to the N_C centres (a N atom substituting for C in the lattice) (Gorban' et al., 2001). The observed phenomenon speaks in favor of increase of the nitrogen concentration in laser irradiated surface including the nano-cones.

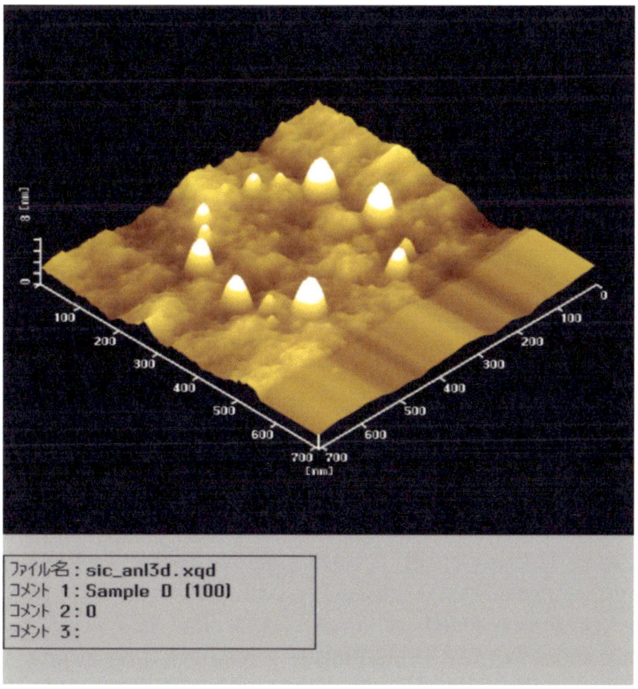

Fig. 13. Morphology of the 6H-SiC(N,B) the C surface after irradiation by N_2 laser at average intensity 5 GW/cm^2.

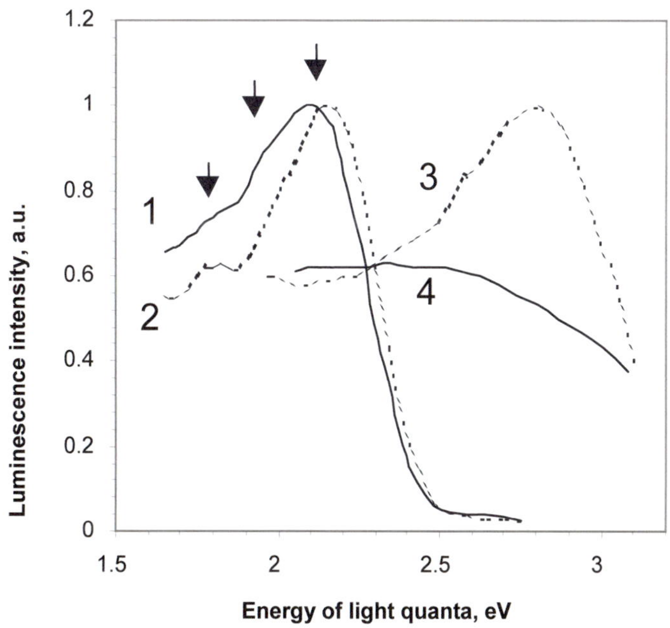

Fig. 14. Spectral distribution of photoluminescence of 6H-SiC samples: curves 1, 2 with impurities N, B and curves 3, 4 with impurity N; solid curves - no irradiated surface, dashed curves - after irradiation by N_2 laser at average intensity up to 5 GW/cm^2 .

The simultaneous formation of nano-cones with the changed chemical composition together with the emergence of the 2.8 eV luminescence band, ascribed to N_C centres allows us to suppose that these luminescence centers could be mainly located in the nano-cones. Consequently, it allows us to propose that the luminescence from the nano-cones is much more intensive than that of the total surface of the sample. Therefore, light emission from the nano-cones could be very bright. Normalized spectra of photoluminescence from no irradiated and laser irradiated 6H-SiC(N, B) sample are shown at Fig.14, curves 1 and 2, correspondingly. As it is seen from this picture the exposure of the surface of the sample to LR causes the decrease of both the 1.9 eV band, which is known to be originated by the recombination of charge carriers in donor-acceptor (D-A) pairs and 1.78 eV band (of the unknown origin) in comparison with the 2.1 eV band corresponding to electron transition from the conduction band to the_boron acceptor level (E_B= 0.65 eV) [42]. "Blue shift of PL spectrum and its maximum on 120 meV after irradiation of the sample by N_2 laser at intensity up to 5MW/cm^2 take place, as can see in Fig.14. This effect is explained by QCE on the top of nano-cones.
The threshold character of the effect together with the necessary condition for the energy of the laser light quantum hν>E_g (where the E_g is a band gap) for appearance of nano-cones, as well as the increase of N band on PL spectra and the decrease of D-A band of PL under numerous irradiation pulses (80-100 pulses at a point) allow us to suppose the presence of the TGE in these conditions (Medvid' et al., 2002). The TGE leads to redistribution of impurity atoms, vacancies (V) and interstitials as a result of their motion in the temperature gradient field of the crystalline lattice. According to the TGE in the case of SiC(N) the atoms of nitrogen are shifted towards the irradiated surface of the sample, because their covalent radii are larger than that of carbon atom. Due to their motion the N atoms fill the carbon vacancies V_C, resulting in the rise of the 2.8 eV luminescence band. In the SiC(N, B) crystals atoms of boron are shifted towards the opposite direction, because the radius of boron atom is smaller, than that of carbon. As a result, D-A pairs are disarranged and the 1.9 eV luminescence band decreases. We explain the appearance of nano-cones on the surface of 6H-SiC under exposure of a focused N_2 laser radiation by the so-called «lid effect».

"Lid effect" model

The subsurface area of the 6H-SiC crystal could melt in the region of the I_{max} of LR (Fig.15) because the temperature of the surface is lower than that in the bulk material. Such distribution of the temperature is caused by the low sublimation energy for SiC. The pressure of light is distributed along the surface according to the Gauss function, in Fig.15 dashed curve, and the average pressure is <P>=1.5 atm. The pressure of liquid phase is balanced by the pressure of light only in the area of $I_{max.}$ Beyond this area liquid matter is extruded on the surface in the form of nano-cones presumably along dislocations.
X0Y crossection of the semiconductor on the focused spot of the laser beam is shown together with the distribution of the laser beam intensity I along X coordinate-dashed curve. The dark areas on the surface of the semiconductor are nano-hills and the big dark ring is liquid matter. The black arrows show the pressure of light and the light arrows - the pressure of liquid matter. ⟨P⟩ = 1.5 atm is average pressure of light.
An evidence of the presence of the "lid effect" in these conditions can be gained from temperature distribution in the depth of the SiC crystal at irradiation by N_2 laser with intensity I= $4{\cdot}10^8$ W/cm^2, which is shown in Fig.16. As shown in Fig.16, maximum of temperature is situated in the bulk of the SiC crystal. This black line is isotherm of the melting temperature.

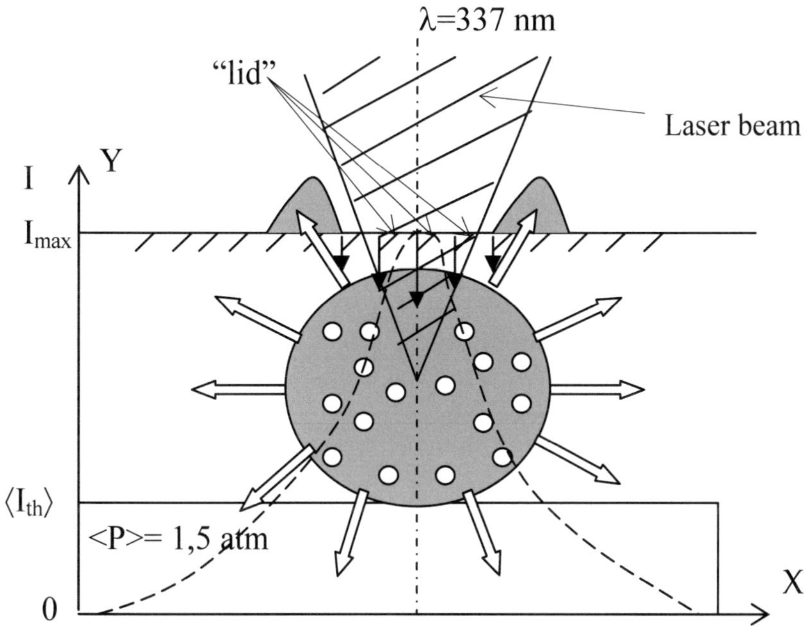

Fig. 15. Scheme of physical model for nano-hills formation on a surface of 6H-SiC caused by N_2 laser radiation λ=337 nm.

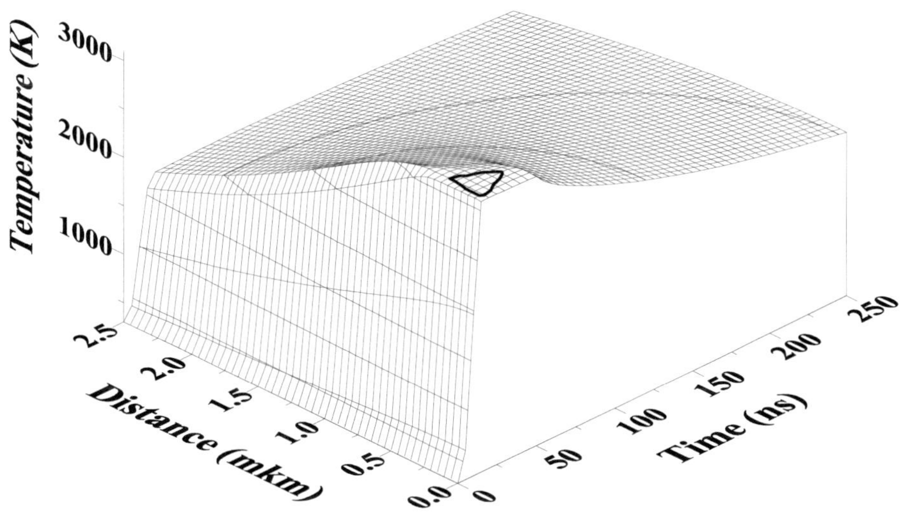

Fig. 16. The calculated distribution of temperature in the depth of SiC crystal after radiation by N laser at intensity I=0.4 GW/cm[2].

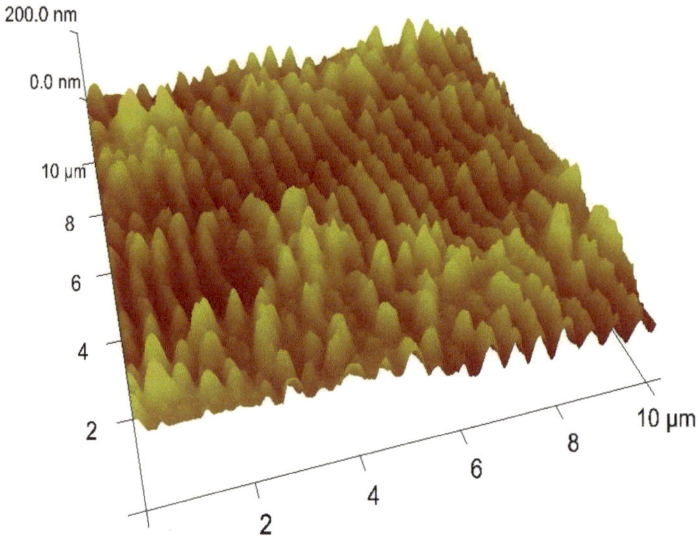

Fig. 17. AFM 3D image of GaAs surface after irradiation by YAG:Nd laser at I=5.5 MW/cm^2.

2.2 Experiments on binary GaAs and ternary $Cd_{1-x}Zn_xTe$ compound semiconductors

The same methods for study nano-cones on a surface of binary compound GaAs and ternary compound $Cd_{1-x}Zn_xTe$ semiconductors were studded. Nano- The increase of exciton energy on 2.5 meV proves the presence of the exciton quantum confinement effect at the top of nano-hills. The energy of band gap of the $Cd_{1-x}Zn_xTe$ crystal increases along the axis of the nano-hill perpendicular to the sample surface. Thus, a graded band gap structure with optical window is formed in the nano-hill ere formed on the surface of GaAs and $Cd_{1-x}Zn_xTe$ with x = 0.1 by the second harmonics of Nd:YAG laser radiation at intensity within 4.0 - 12.0MW/cm^2.

Morphology of the irradiated surface of the GaAs single crystal has shown formation of very sharp self-organized nano-cones, as shown in Fig. 17, which arise on the irradiated surface after irradiation by the laser with intensity I = 5.5 MW/cm^2. In the PL spectrum of irradiated GaAs single crystals (Fig.18) with maximum of spectrum at 750 nm we have also observed band gap blue shift on 100 nm. Intensity of PL at the maximum of the irradiated surface is 4 times more that intensity of PL no irradiated surface. But its intensity is much lower comparing to PL of Ge quasi QDs. The evidence of QCE presence in nanostructures on the irradiated surface of GaAs single crystal is back scattering Raman spectra, as shown in Fig. 19. We can see that spectral LO phonon line at 292 cm^{-1} is characterized by "red shift" after irradiation of the surface by the laser. Calculations of quasi quantum dots diameter on the top of nano-cones using band gap shift from PL spectra of GaAs and formula (1) gives d = 11 nm.

Irradiation of $Cd_{1-x}Zn_xTe$ (x=0.1) crystals by the Nd:YAG laser radiation at intensities below the threshold intensity of 4 MW/cm^2 does not change the surface morphology(Fig.20,a). The nanostructures begin forming at intensities $I \geq 4$ MW/cm^2 on the irradiated surface, as are shown in Fig.20.a and b, correspondently (Medvid' et al., 2008).

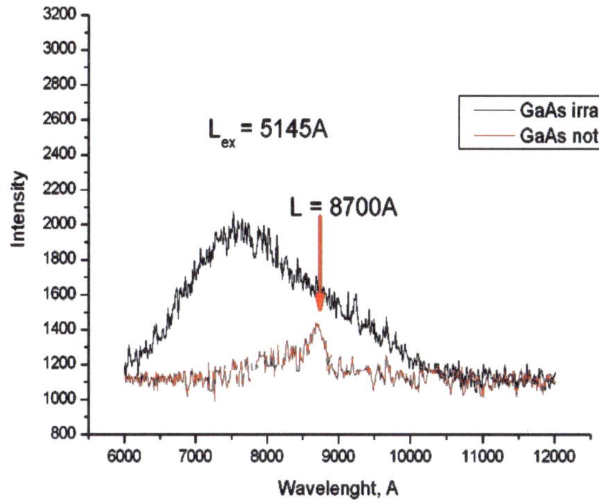

Fig. 18. PL spectra of the irradiated and nonirradiated surface of Si at intensity of LR up to 7.5 MW/cm[2].

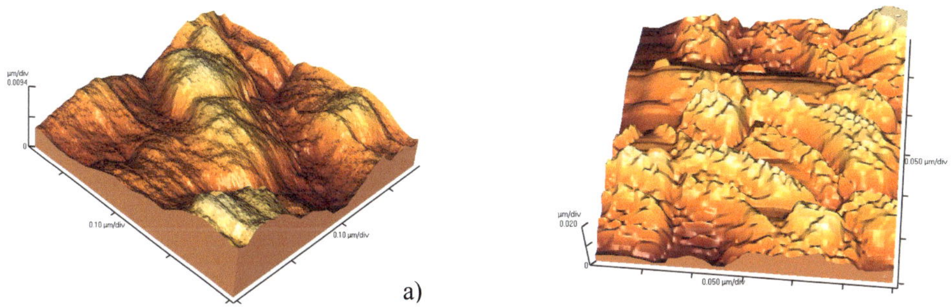

Fig. 20. Atomic force images of the $Cd_{1-x}Zn_xTe$ (x=0.1) surface (a) before irradiation and (b) after irradiation by the laser at intensity of 12 MW/cm[2].

The "blue shift" of exciton bands in PL spectra of the irradiated $Cd_{1-x}Zn_xTe$ is explained by Exciton quantum confinement effect in nano-cones (Brus, 1984). A new PL band at 1.88 eV is found. Appearance of the PL band is explained by formation of $CdTe/Cd_{1-x}Zn_xTe$ heterostructure in the bulk of the semiconductor with **x** =0.7 due to TGE (Medvid' et al., 2002). The increase of exciton energy on 2.5 meV proves the presence of the Exciton quantum confinement effect at the top of nano-cones. The energy of band gap of the $Cd_{1-x}Zn_xTe$ crystal increases along the axis of the nano-cone perpendicular to the sample surface. Thus, a graded band gap structure with optical window is formed in the nano-cone. The energy of band gap of the $Cd_{1-x}Zn_xTe$ crystal increases along the ZO axis of the nano-cone perpendicular to the irradiated surface of the sample, as s shown in Fig 22. Thus, a graded band gap structure with optical window is formed in the nano-cone.

The Thermogradient effect plays the main role in the redistribution of Cd and Zn atoms at the irradiated surface of $Cd_{1-x}Zn_xTe$ at low intensities of LR from 0.2 MW/cm[2] till 4.0 MW/cm[2]. Two layers are formed near the irradiated surface of semiconductor: the top layer consists of mostly CdTe crystal but the underlying layer - ZnTe crystal, as is shown in Fig.22.a. A mismatch of lattices of CdTe and ZnTe crystals is equal up to 5.8%. This

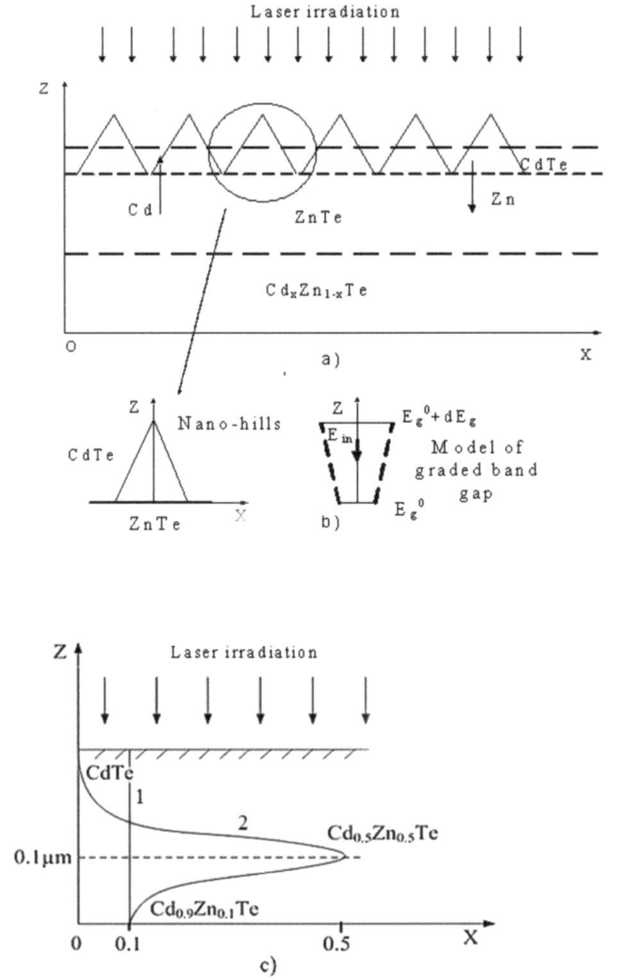

Fig. 22. a) Scheme of physical model for nano-cones formation on a surface of $Cd_{1-x}Zn_xTe$ crystal caused by Nd:YAG laser radiation λ=532 nm, the laser beam direction, Cd and Zn atoms drift in gradient T field and crossection of the irradiated sample; b) The energy of band gap of the $Cd_{1-x}Zn_xTe$ crystal increases along the symmetry axis Z of the nano-cone perpendicular to the irradiated surface of the sample. Thus, a graded band gap structure with optical window is formed in the nano-cone due to QCE; c) Distribution of Zn and Cd atoms in Zo direction in the sample before (curve 1) and after irradiation by the laser (curve 2).

plastically deformation of the top layer leads to creation of nanostructures of the irradiated surface according to the modified Stransky-Krastanov' mode.This result is in a good agreement with PL measurement data, as shown in Fig.21.b. A built-in quasi electric field, generated by graded band gap, is directed in the bulk of the semiconductor as a result decrease of surface recombination velocity. The photoconductivity spectra of CdZnTe samples were recorded at room temperature before and after irradiation by Nd:YAG laser, as shown in Fig.23. The photoconductivity spectra show, that at laser intensity up to 4MW/cm^2 the shift of maximum spectrum to the longer wavelength, "red shift", of spectra and increase of photocurrent at short wavelength take place, as shown in Fig.23, green curve. This effect is explained by decrease of surface recombination velocity. The irradiation of the sample by higher intensity of the laser causes the "red shift" of spectra and the total increase of photocurrent up to 2 times, as shown in Fig.23, red and black curves, are explained by formation of graded band gap structure on the top of nano-cones and increase of electron-hole pares life time due to increase concentration of D-A pares after irradiation by the laser.

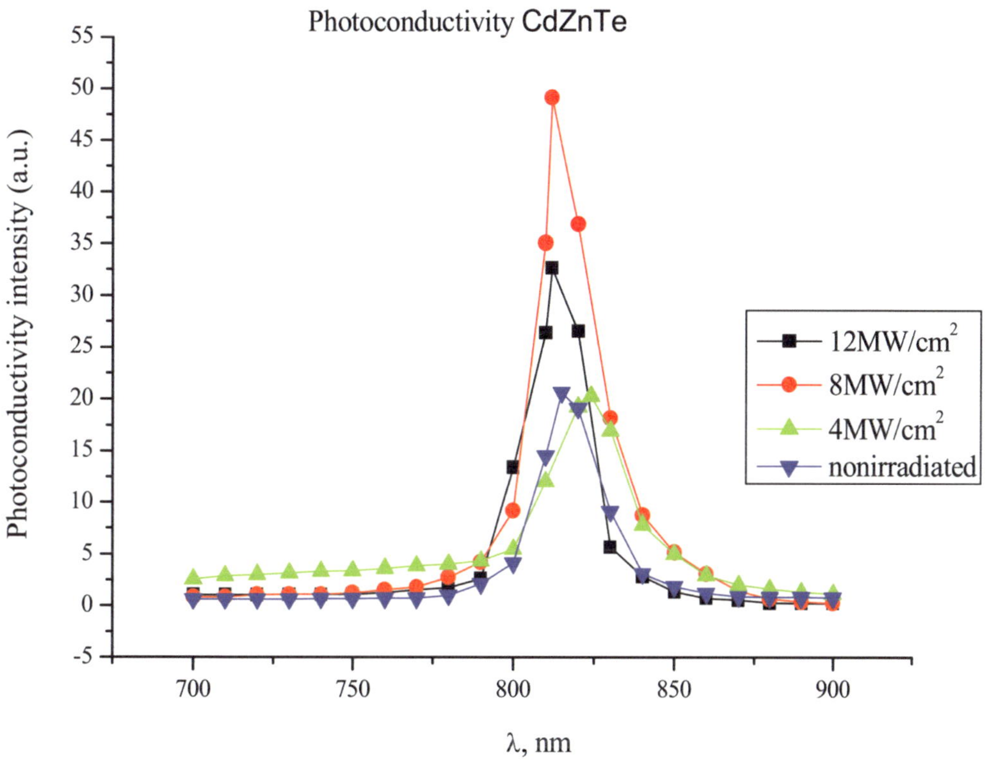

Fig. 23. Photoconductivity spectra of the irradiated surface of $Cd_{1-x}Zn_xTe$ before and after irradiation by the laser.

3. Conclusion

For the first time was shown the possibility of graded band gap structure formation in elementary semiconductors due to the presence of Quantum confinement effect in cone-like nanostructure.

Thermogradient effect plays a main role in initial stage of nano-cones and graded band gap structure formation by laser radiation in semiconductors.

The new laser method for nano-cones formation on a surface of semiconductors Si, Ge, GaAs, 6H-SiC crystals, SiGe and CdZnTe solid solutions is elaborated.

The model of cone-like nanostructure formation on the SiGe surface has been proposed. According to this model, irradiation the semiconductor by strongly absorbed laser radiation, leads to a huge temperature gradient (10^8K/m). It causes the drift of Ge atoms towards the irradiated surface. Concentration of Ge atoms is increased at the irradiated surface. Ge atoms are localized at the surface of Si like a thin film. A mismatch of Si and Ge crystal lattices leads to compressed stress of Ge layer. The stress relaxation by plastic deformation of the top Ge layer and creation of nanostructures on the irradiated surface according to the modified Stransky-Krastanov' mode takes place.

4. Acknowledgements

The author gratefully acknowledges financial support in part by Europe Project in the framework FR7-218000 Nr. L7477, by the Latvian Council of Science according to the Grant No. L7306, L7455 and by the Riga Technical University projects: FLPP-2009/32, R7355, R7234, U6968 & U7094.

5. References

Alivisatos, A. P. (1996). Semiconductor Clusters, Nanocrystals, and Quantum Dots. *Science,* Vol. 271, pp.933-937, ISSN: 0036-8075;

Beigelsen, D.K., Rozgonyi G.A,.Shank, C.V. (1985). *Energy Beam-Solid Interactions and Transient Thermal Processing* ,Vol.5, Material Research Society, Pittsburgh;

Brus, L.E. (1984). Electron–electron and electron-hole interactions in small semiconductor crystallites: The size dependence of the lowest excited electronic state, *J.Chem.Phys.,* Vol. 80, No.9, pp.4403- 44, ISSN:0021-9606 ;

Campbell, H. & Fauchet P.M. (1986). The effects of microcrystal size and shape on the one phonon Raman spectra of crystalline semiconductors, *Solid State Commun.,* Vol.58 pp.739-74, ISSN:0038-1098;

Efors, Al.L. & Efors, A.L. (1982). Band-band absorption of light in semiconductor ball. *Phys. and Techn.of Semicond.,* Vol.16, pp. 1209-121 ;

Emel'janov, V.I. & Panin, I.M. (1997). Defect-deformational self-organization and nanostructuring of solid surfaces, *Solid State Physics,* Vol.39, pp.2029-2035, ISSN:0081-1947;

Emel'yanov, A.M. Sobolev, N.A., Mel'nikova, T.M., Abrosimov, N.V. (2005). Point and extended defects engineering, *Solid state phenomena,* Vol.108-109, pp.761-764;

Fernandez, B.G., Lopez, M., Garcia, C., Perez-Rodriguez, A., Morante J.R., Bonafos, C., Carrada, M.& Claverie, A. (2002). *J.Appl.Phys.*Vol.91 pp. 789-7, ISSN:0021-8979;

Fowler, A. B., Fang, F. F., Howard, W. E. & Stiles P. J. (1966). Magneto-Oscillatory Conductance in Silicon Surfaces, *Phys. Rev. Lett.,* Vol.16, pp.901-903, ISSN:0031-9007;

Fujisawa, I. (1980). Type Conversion of InSb from p to n by Ion bombardment and laser Irradiation. *Jpn J. Appl. Phys.,* Vol. 19, pp. 2137-2140, ISSN:0021-4922;

Gnatyuk, V.A., Aoki, T., Hatanaka, Y. & Vlasenko O.I. (2006). Defect formation in CdTe during laser-induced doping and application to the manufacturing nuclear radiation detectors, *Phys. stat. sol. (c)* Vol.1, pp.121-12, ISSN: 1610-1634;

Gorban', I.S. & Krokhmal', A.P. (2001). The impurity optical absorption and conduction band structure in 6*H*-Si. *Semicond.,* Vol.35, pp.1242-1248, ISSN: 1063-7826;

Green, Martin A. (2004). *Third generation photovoltaics: advanced solar emerge conversion,* Springer-Verlag, Berlin;

Hartmann, J.M., Bertin, F., Rolland, G., Semeria, M.N. & Bremond, G. (2005). Effects of the temperature and of the amount of Ge on the morphology of Ge islands grown by reduced pressure–chemical vapor deposition, *Thin Solid Films,* Nr. 479, pp.113–120, ISSN: 0040-6090;

Kamenev, B.V., Baribeau, J.-M., Lockwood, D.J., & Tsybekov, A. (2005).Optical properties of Stranski-Krastanov grown three-dimensional Si/Si Ge nanostructures, *Physica E,* Vol. 26, pp. 174- 17, ISSN:1386-9477;

Kartopu, G., Bayliss S.C., Hummel, R.E.& Ekinci, Y. (2004). Report on the origin of the orange PL emission band. *J.Appl.Phys.,*Vol.957, pp. 2466-2472 , ISSN: 0021-8979.

Kelly, R., Cuomo, J.J., Leary, P.A., Rothenberg, J., Rraren, B.E.& Aliotta, C.F. (1985). Laser sputtering Part I: On the eksistence of rapid laser sputtering at 193 nm. *Nuclear Inst. and Meth. in Phys. Research,* Vol.B9, pp.329-340; ISSN: 0168-583x

Kurbatov, L., Stojanova I., Trohimchuk,P.P.&Trohin,A.S. (1983), Laser anieleing of $A^{III}B^{V}$ compound. *Rep.Acad.Sc.USSR,* Vol.268, pp.594- 597, (in Russian);

Kuwabara, H. & Yamada, S. (1975). Free-to-bound transition in β-SiC doped with boron. *Phys. Stat. Sol.(a,)* Vol.30, pp.739-74, ISSN:0031-8965;

Lin, J.M., Kurz, H.& Blomborgen, N. (1982). Picosecond time-resolved plasma and temperature-induced changes of reflectivity and transmission in silicon. *Appl. Phys. Lett.* , Vol.41, pp.643-64, ISSN:0003-6951;

Li, J. & Lin-Wang, (2004). Comparison between quantum confinement effects of quantum wires and quantum dots. *Chem.Mater.*, Vol.16, pp. 4012-4015, ISSN:0897-4756;

Mada, Y.& Ione, N. (1986). p-n junction formation using laser induced donors in silicon, *Appl. Phys. Lett.*, Vol.48, pp.1205-1207, ISSN:0003-6951;

Medvid', A. & Fedorenko, L. (1999). Generation of Donor Centers in p-InSb by Laser Radiation, *Materials Science Forum*, Vols. 297-298, pp. 311-314, ISSN:0255-5476;

Medvid', A., Litovchenko, V.G., Korbutjak, D., Krilyk, S.G., Fedorenko, L.L.& Hatanaka ,Y. (2001). Influence of laser radiation on photoluminescence of CdTe, Influence of laser radiation on photoluminescence of CdTe *Radiation Measurements,* Vol. 33, pp. 725-72, ISSN:1350-4487;

Medvid', A., (2002). Redistribution of the Point Defects in Crystalline Lattice of Semiconductor in Nonhomogeneous Temperature Field. *Defects and Diffusion Forum*, Vols. 210- 212 , pp. 89-101, ISSN:1012-0386;

Medvid', A. & Lytvyn, P. (2004). Dynamics of Laser Ablation in SiC, *Materials Science Forum,* Vols. 457-460, pp. 411-414, ISSN:0255-5476;

Medvid', Artur, Dmytruk, Igor, Onufrijevs, Pavels & Pundyk, Iryna. (2007). Quantum Confinement Effect in Nanohills Formed on a Surface of Ge by Laser Radiation. *Phys . Stat. Sol.(c)*, Vol.4, pp.3066-1069, ISSN: 1610-1634;

Medvid' A., Onufrijevs, P., Dmitruk, I.& Pundyk, I. (2008). Properties of Nanostructure Formed on SiO_2/Si Interface by Laser Radiation, *Solid State Phenomena,* Vols. 131-133 pp. 559-562, ISSN: 1012-0394;

Medvid', A., Mychko, A., Strilchyuk, O., Litovchenko, N., Naseka, Yu., Onufrijevs, P. & Pludonis, A. (2008).Exciton quantum confinement effect in nanostructures formed by laser radiation on the the surface of CdZnTe ternary compound, *Phys. Stat.Sol.(c)*, Vol.6, pp.209-212, ISSN: 1610-1634;

Medvid' A., Onufrijevs, P., Lyutovich, K., Oehme, M., Kasper, E., Dmitruk, N., Kondratenko, O.,I.Dmitruk & Pundyk, I. (2009).Self-assembly *of* nano-hills in Si_xGe_{1-x}/Si by laser radiation, *J. Nanoscience and Nanotechn*, (in press), ISSN:1533-4880, ISSN:1533-4880;

Medvid', A., Mychko, A., Gnatyuk, V. A., Levytskui, S.& Naseka, Y. Mechanism of nanostructure formation on a surface of CdZnTe crystal by laser irradiation. *J. Automation, Mobile Robotics & Intelligent Systems.*, Vol.3 pp.127-9. ISSN 1897-8649

Mooney, P. M.; Jordan-Sweet J. L.; Ismail, K.; Chu, J. O.; Feenstra R. M. & LeGoues F. K. Relaxed $Si_{0.7}Ge_{0.3}$ buffer layers for high-mobility devices. (1995). *Appl. Phys. Lett.*, Vol.67, pp.2373-237, ISSN:0003-6951;

Morales, A.M. & Lieber C.M. (1998). *Science*, Vol. 279, pp. 208–2, ISSN:0036-8075;

Okhotin, A.S., Pushkarskii, A.S. & Gorbachev, V.V. (1972).*Thermophysical Properties of Semiconductors,* Atom, Moscow, (in Russian).

Rebohle, L., von Borany, Frob, J.H. & Skorupa, W. (2000). *Appl.Phys.B: Lasers Opt.* Vol.B70 p. 131-13, ISSN: 0946-2171;

Reno, J. & Jones, E. (1992). Determination of the dependence of the band-gap energy on composition for $Cd_{1-x}Zn_xTe$. *Phys. Rev. B*, Vol.45, pp.1440–1442, ISSN:1350-4487;

Sun, K.W., Sue, S.H. & Liu, C.W. (2005). Physica E. Vol.28, pp.525-52, ISSN: 1386-9477;

Talochkin, A. B., Teys, S. A. & Suprun, S. P. (2005). Resonance Raman scattering by optical phonons in unstrained germanium quantum dots, Phys. Rev. Vol.B 72, p.115416-11154, ISSN: 1098-0121;

Vigil-Galán, O.; Arias-Carbajal, A.; Mendoza-Pérez, R.; Santana-Rodríguez, G.; Sastre-Hernández, J; Alonso, J C; Moreno- García, E.; Contreras-Puente, G. & Morales-Acevedo, A. (2005). Improving the efficiency of CdS/CdTe solar cells by varying the thiourea/$CdCl_2$ ratio in the CdS chemical bath. *Semicond. Sci. Technol.* Vol.20, pp.819–822, ISSN 0268-1242;

Von der Linde, D. & Fabricius, N. (1982). Observation of an electronic plasma in picosecond laser annealing of silicon. Appl. Phys. Lett., Vol. 41, pp.991-993, ISSN: 0003-6951;

Vorobyev, L.E., (1996). Semiconductor Parameters, In: Handbook Series, M. Levinshtein, S. Rumyantsev, M. Shur (Eds.), Vol.1, p. 232 ,World Scientific, ISBN: 978-981-02-2934-4, London;

Werwa, E., Seraphin, A. A., Chiu, L. A., Zhou, C., & Kolenbrander K. D. (1996). Synthesis and processing of silicon nanocrystallites using a pulsed laser ablation supersonic expansion method. *Appl. Phys. Lett.* Vol.64, pp.1821-182, ISSN: 0003-6951;

Wu, X. L.; Gao, T.; Bao, X. M; Yan, F.; Jiang, S. S. & Feng, D. (1997). Annealing temperature dependence of Raman scattering in Ge^+-implanted SiO_2 films. *J. Appl. Phys.* Vol. 82 pp. 2704, ISSN:0003-6951;

Yoshida, T., Yamada, Y.& Orii, T. (1998). Electroluminescence of silicon nanocrystallites prepared by pulsed laser ablation in reduced pressure inert gas. *J. Appl. Phys.* Vol. 83(10), pp.5427–32, ISSN: 0021-8979;

Xia, Y. & Yang, Y. (2003). Chemistry and physics of nanowires. *Advanced Materials*, Vol. 15, pp.351-353, ISSN: 0935-9648.

5

Magnetic Properties of Nanowires guided by Carbon Nanotubes

Miguel A. Correa-Duarte and Veronica Salgueirino
Universidade de Vigo
Spain

1. Introduction

The physical properties of one-dimensional (1D) nanostructures of magnetic materials are presently the subject of intensive research, taking into account the considerable attention they have recently received and the few cases reported. [1-4] Much of the early work was concerned with exploratory issues, such as establishing an easy axis for typical preparation conditions and the essential involvement of shape anisotropy, as opposed to magnetocrystalline anisotropy. More recently, attention has shifted towards the understanding of magnetization processes and related applications. Particularly interesting problems are the magnetic hysteresis of the wires and the time dependence of the magnetic reversal. Thus, magnetic nanowires have provided a highly successful test ground for understanding the microscopic mechanisms that determine macroscopically important parameters in the different applications where they can be used. [5]

On the other hand, these building blocks, as in the case of spherical nanoparticles, are at the border between the solid and molecular state displaying the novel effects that can now be exploited. Therefore, it becomes imperative to take into account the fact that the properties of materials composed of magnetic nanostructures are a result of both the intrinsic properties of the small building blocks and the interactions in between. [6]

This chapter is not meant as a survey of the present state and future developments of magnetic nanowires and since only two examples are considered, is far from being complete. The purpose of this chapter is three fold: a) an introductory level overview about magnetic colloids, the basic physics in the magnetism at the nanoscale; in terms of superparamagnetism, the concept of magnetic anisotropy and the dynamics of these systems. We have emphasized the dominant role of the surface effects on the intrinsic properties at the nanoscale and the competition with the interactions in the case of assemblies, leading to a characteristic magnetic behavior termed as *spin-glass.* Additionally, a brief introduction referred to carbon nanotubes (CNTs) is included. b) Characteristic examples of magnetic nanowires whose morphology was achieved by taking advantage of CNTs and exploiting wet-chemistry methods, and c) a complete analysis of the magnetic behavior displayed in both examples.

1.1 Colloids

Different preparation methods lead to magnetic nanostructures with differences in crystalline structure, surface chemistry, shape, etc. Hence, the fabrication technique has a great influence on the magnetic properties of the materials obtained. Numerous physical and chemical methods have been employed to produce magnetic nanostructures, such as *molecular beam epitaxy, chemical vapor deposition, normal incident pulsed laser deposition, sputtering* or *electrodeposition*. The magnetic colloidal nanostructures or colloids are remarkably different if compared to nanostructures formed by these methods, as they are chemically synthesized using wet chemistry and are free-standing nanocrystals grown in solution. [7] The magnetic colloids are thus a subgroup of a broader class of the magnetic materials that can be synthesized at the nanoscale level but using wet chemical methods. In this fabrication of colloidal nanocrystals, the reaction chamber is a reactor containing a liquid mixture of compounds that control the nucleation and the growth. In general, each of the atomic species that will be part of the nanostructures is introduced into the reactor in the form of a precursor. A precursor is a molecule or a complex containing one or more atomic species required for growing the nanocrystals. Once the precursors are introduced into the reaction flask they decompose, forming new reactive species (the monomers) that will cause the nucleation and growth of the nanocrystals. The liquid in the reactor provides the energy required to decompose the precursors, either by thermal collisions or by a chemical reaction between the liquid medium and the precursors, or by a combination of these two mechanisms. [8] The key parameter in the controlled growth of colloidal nanocrystals is the presence of one or more molecular species in the reactor that we will term as *ligands* hereafter. A ligand is a molecule that is dynamically adsorbed to the surface of the growing structure under the reaction conditions. It must be mobile enough to provide access for the addition of monomer units, while stable enough to prevent the aggregation of nanocrystals in solution. Figure 1 provides a schematic illustration of a nanostructure with an example of a mobile ligand adsorbed on its surface. Due to the increased surface-to-volume ratio, surface chemistry is of great importance for the chemical and physical properties of the colloids likewise synthesized, generally being caused by these molecular ligands stabilizing them.

Fig. 1. Schematic illustration of a nanostructure with an example of a mobile ligand adsorbed on its surface.

The preparation of nanostructures of desired sizes is the first and very important step, being a prerequisite of their further investigation and use. Narrow size distribution or monodispersity is highly desirable because magnetic properties become strongly size dependent in the nanometer size range. Additionally, the magnetic behavior of nanostructures depends not only on their chemical composition and size, but also on their crystalline modification and the presence of structural defects like stacking faults or twinned planes. Consequently, Murray *et al.* proposed the use of the term "*nanocrystal*" for crystalline particles with low concentrations of defects, while the more general term "*nanoparticle*" also includes particles containing gross internal grain boundaries, fractures, or internal disorder. [9] Another important characteristic of the wet-chemistry methods refers to the fact that the colloids are dispersed in solution, and so that, they can be produced in large quantities in a reaction flask and can later be transferred to any desired substrate.

1.2 Superparamagnetism

It is well known that a magnetic body has a structure that is divided into uniformly magnetized regions (domains) separated by domain or Bloch walls. This distribution of regions or domains permits the magnetic body to minimize its magnetostatic energy. Since it is the total energy that requires to be minimized, it will be a final balance of energies (including the magnetostatic and the exchange terms, the different anisotropies that come into play and the domain walls contribution) what will determines the domain structure. In order to work at the nanoscale, there will be an important reduction in dimensions, so that, the size of the domains will also be reduced. At this point, the domain structure will consequently change, but due to the domain wall formation energy cost, the balance with the magnetostatic energy will limit the subdivision in domains to a certain optimum size. Consequently, there is a size limit below which the nanostructure can no longer gain an energy favorable configuration by breaking up into domains. Hence, it remains with only one domain. The critical size, or single domain size D_s, below which a particle will not form domains, is where these two energies (energy cost of domain walls formation and magnetostatic energy) become equal, and the typical values for D_s range from 10 to 100 nm, with elongated nanostructures tending to have larger Ds.
Much of the behavior of these single-domain magnetic nanostructures can be described by assuming that all the atomic moments are rigidly aligned as a single "*giant*" spin. This is in essence the theory of superparamagnetism. Below the Curie temperature of a ferromagnet or ferrimagnet, all the spins are coupled together and so cooperate to yield a large total moment μ (the giant spin). This moment is bound rigidly to the nanostructure by one or more of the variety of anisotropies (that will be later discussed). The energy of this bond is KV, where K is the effective magnetic anisotropy (that takes into account the variety of anisotropies that come into play) and V is the volume of the particle. With decreasing size, the KV magnitude decreases and the thermal energy kT can disrupt the bonding of the total magnetic moment to the nanostructures. The magnetic moment in this situation is free to move and respond to an applied field independently of the nanostructure itself. An applied field would tend to align it, but kT would fight the alignment just as it does in a paramagnet (justifying therefore the use of the term superparamagnet). [10]
The phenomenon of superparamagnetism is, in fact, timescale-dependent due to the stochastic nature of the thermal energy kT. The anisotropy energy KV represents an energy barrier to the total spin reorientation, hence the probability for jumping this barrier is proportional to the Boltzmann factor exp($-KV/kT$). This can be made quantitative by introducing an attempt timescale, τ_0 which describes the timescale over which μ attempts to jump the KV barrier. Then, the timescale for a successful jump is

$$\tau = \tau_0 e^{-KV/kT} \tag{1}$$

The attempt timescale is about 10^{-9} s. The typical experiment with a magnetometer takes 10 to 100 s; and if M_S reverses at times shorter than the experimental timescale, the system appears superparamagnetic. Thus, using $\tau \approx 100$ s and $\tau_0 = 10^{-9}$ s, the critical volume becomes:

$$V_{sp} = \frac{25kT}{K} \tag{2}$$

$$T_B = \frac{KV}{25k} \tag{3}$$

The equation (2) can be rearranged to yield equation (3) and gives the blocking temperature T_B. Below T_B the free movement of $\mu = M_s V$ is blocked by the anisotropy; above T_B, kT permits the magnetic moment (μ) to fluctuate freely, so that the system appears superparamagnetic.

1.3 Magnetic anisotropy

The magnetic anisotropy concept describes the fact that the energy of the ground state of a magnetic system depends on the direction of the magnetization. The effect occurs either by rotations of the magnetic moment μ with respect to the external shape of the specimen (shape anisotropy) or by rotations relative to the crystallographic axes (intrinsic or magneto-crystalline anisotropy). The direction(s) with minimum energy, i.e. into which the magnetic moment μ points in the absence of external fields are called easy directions. The direction(s) with maximum energy are called hard directions. The magnetic anisotropy energy (MAE) between two crystallographic directions is given by the work needed to rotate the magnetic moment μ from an easy direction into the other directions. The MAE, that corresponds to a small contribution to the total energy of a bulk crystal, becomes more and more important as decreasing the size of the magnetic material until reaching the nanoscale, depending on different issues.

In bulk materials, magnetocrystalline and magnetostatic energies are the main sources of anisotropy, but in nanosized structures, other types of anisotropy can be of the same order of magnitude. As the properties are stated by the relaxation time τ of the magnetic moment μ on the nanostructures, τ being itself governed by the energy barrier KV (directly dependent on the effective magnetic anisotropy K), it is important to know all the possible sources of anisotropies and their contribution to the total energy barrier.

There are fundamentally two sources of magnetic anisotropy; (i) spin-orbit (LS) interaction and (ii) the magnetic dipole-dipole interaction. The more important interaction is the spin-orbit coupling, which couples the spin to the charge (orbital) density distribution in the crystal. Thus, the magnetic moment μ *"gets the feel"* of the crystal via the orbital motion of the magnetic electrons. The orbital motion is coupled to the lattice by an electric field (that indeed reflects the symmetry of this lattice). This field K is given by the sum of the electrostatic potentials $\phi(r_i)$ over the nearest neighbors at sites r_i. [11]

$$K = \sum_i^N e \cdot \phi(|\vec{r}_i|) \tag{4}$$

The first source (spin-orbit) of magnetic anisotropy includes the so-called magneto-crystalline and magneto-elastic contributions while the second one (from magnetic dipole-dipole interactions) includes contributions termed shape or magneto-static anisotropy. [12]
Accordingly, when considering the nanoparticulate systems, there are main contributions referred to the magnetocrystalline anisotropy whose energy can show various symmetries (uniaxial and cubic forms cover the majority of the cases), to the surface anisotropy (related to surface effects that stem from the fact that the existence of the surface represents a discontinuity for magnetic interactions) and to magnetostatic energy. Additionally, stress anisotropy should be taken into account as well since there is a second effect due to the surface, related to strains, because of a magnetostriction effect. [13]
If we refer to the magnetocrystalline anisotropy, the direction of the spontaneous magnetization of crystalline samples is oriented along certain directions. Typically, for bulk samples of bcc-Fe, fcc-Ni, and hcp-Co, these are the cubic ⟨100⟩, ⟨111⟩ and hexagonal ⟨0001⟩ directions, respectively. These directions are called easy magnetization directions and the magnetization along any other direction requires an excess energy. [14] The magnetic anisotropy energy density is indeed the excess work that needs to be put into the system to achieve saturation magnetization along a non-easy axis of magnetization. This excess work depends on the orientation of the magnetizing field with respect to the sample, and its magnitude will be different in general for magnetization in the different directions of the crystalline lattice. In sharp contrast to the small crystalline anisotropy of bulk samples, ultrathin films and nanostructures often exhibit an effective magnetic anisotropy that is orders of magnitude larger than the respective bulk value. The deviation from the respective bulk values has been ascribed in most cases to a magnetocrystalline surface anisotropy of the Néel type. [15]
In general, one can always apply the rule that indicates that the lower the symmetry of the crystals or of the local electrostatic potential around the magnetic moment, the larger the MAE is. [16] The mentioned increase of magnetic anisotropy for ultrathin films and nanostructures is often ascribed to so-called interface and surface anisotropy contributions that are attributed to the different atomic environment of the interface and surface atoms, where the symmetry of the crystals is much lower than in internal positions. [12]
The important impact of symmetry on the resulting magnetic anisotropy can also be understood in the crystal field description, which has been applied successfully to describe the magnetic anisotropy of 3d ions. [17] The energetically degenerate 3d levels of a free atom split into two groups, e_g (d_{z^2}, $d_{x^2-y^2}$ orbitals) and t_{2g} (d_{xy}, d_{xz} and d_{yz} orbitals) in the presence of the electric field of the surrounding atoms. The energy separation and the relative positioning of the e_g and t_{2g} levels depend on the symmetry of the atomic arrangement [18] and a tetragonal distortion lifts the degeneracy of these levels. [19] Thus, a change of symmetry, e.g. induced by strain, may lead to a change of the relative occupancy of the different d-orbitals, which in turn leads via spin-orbit coupling to a change of the magnetic anisotropy.
In order to illustrate the role of the symmetry (and surface effects) we can consider the following case. At each lattice point within for example, a cubic lattice, the dipole fields of all neighbors cancel out. This is, however, only true for an infinite system. As soon as surfaces are present, magnetic poles develop and thus, the dipole-dipole interaction leads to anisotropy. Since the shape of the sample determines this anisotropy (even for modest shape ratios c/a, the shape anisotropy can be very large), one usually calls it shape anisotropy. As a

result, in systems with reduced dimensions as nanostructures, the shape anisotropy might be even the dominating contribution to the overall MAE. [16]

Another important contribution comes from a well-known experimental fact derived from the coupling between magnetism and lattice strain, the magnetostriction, which origins the magnetoelastic anisotropy. A sample lowers its total energy upon magnetization by a lattice strain that depends on the magnetization direction with respect to the crystalline lattice. The underlying principle of the so-called magnetoelastic coupling can be described as the strain dependence of the magnetic anisotropy energy density. [20] For thin films and nanostructures, since are generally under considerable strain, this contribution can also determine the magnetic anisotropy.

Cantilever bending experiments and nanoindentation techniques are sensitive and accurate tools for the measurement of mechanical stress in atomic layers, at surfaces and on nanoparticles. The idea of the measurement in the former case is to detect the stress-induced change in curvature of a thin substrate. The curvature change is directly proportional to the film stress, integrated over the film thickness. The second case, the nanoindentation, has now long been used to study the elastic, plastic, and fracture properties of surfaces of bulk samples, and in the last decade, it has become possible to perform controlled compression and bending tests on nanostructures smaller than a micron, such as NPs, [20-24] nanowires [25-26] and nano-pillars. [27-28] In both cases, the elastic properties of the magnetic structures are given by the Young's modulus and the Poisson ratio and the magnetoelastic coupling coefficients can then be derived.

Although the important aspect of magnetic domain formation is not discussed here since we have restricted the scope of the introduction to theoretical single-domain nanostructures behavior, we will actually take into account the interplay of magnetostatic energy, exchange energy and the effective magnetic anisotropy that can lead to an energetically favored multi-domain state. Hence, aspects of configurational magnetic anisotropy that appear in submicron-sized magnets due to small deviations from the uniform state [29] and the so-called exchange anisotropy are considered.

The phenomenon of exchange anisotropy is the result of an interfacial exchange interaction between ferromagnetic (FM) and antiferromagnetic (AFM) materials, and only recently have the required experimental and analytical tools for dealing with interfacial behavior at the atomic level become available. [30] The observation (below room temperature) of a hysteresis loop shifted along the field axis, after cooling a nanostructured system in an applied field, indicates an exchange interaction across the interface between a FM and an AFM materials composing the sample. This loop shift was demonstrated to be equivalent to the assumption of a unidirectional anisotropy energy in the expression for the free energy at $T=0K$ of single-domain spherical particles with uniaxial anisotropy, aligned with their easy axis in the direction of the field, H, which is applied anti-parallel to any particle's magnetization M_s:

$$F = HM_s \cos\theta - K_u \cos\theta + K_1 \sin^2\theta \tag{5}$$

where θ is the angle between the easy direction and the direction of magnetization, and K_u and K_1 are the unidirectional and uniaxial anisotropy energy constants, respectively. Solutions of this equation are readily expressed in terms of an effective field H',

$$H' = H - \frac{K_u}{M_s} \tag{6}$$

that offers the hysteresis loop displaced by K_u/M_s, on the *H*-axis. Thus, an explanation of the loop shift is equivalent to explaining the unidirectional anisotropy.
AFM materials appear magnetically ordered below their Néel temperatures T_N, and have a zero net moment since have parallel and anti-parallel spins in a preferred direction. However, at the interface with a FM, there are localized net moments that arise from several sources. Indeed, in AFM nanostructures with compensated interfacial spin planes, there can be unequal numbers of parallel and antiparallel spins at the surface of the nanostructure, due to various origins such as its size, shape or roughness, generating localized net AFM moments. Since the FM is ordered at its Curie temperature T_C, greater than the T_N of the AFM, a field applied to couple FM-AFM systems at $T > T_N$ will align the FM magnetization in the field direction, while the AFM spins remain paramagnetic. As the temperature is lowered through T_N, the ordering net localized AFM spins will couple to the aligned FM spins, sharing their general spin direction. For high AFM magnetocrystalline anisotropy, if the interfacial AFM spins are strongly coupled to the AFM lattice, they will not be substantially rotated out of their alignment direction by fields applied at temperatures below T_N. However, since the localized uncompensated AFM spins are coupled to FM spins at the interface, they exert a strong torque on these FM spins, tending to keep them aligned in the direction of the cooling field, i.e., a unidirectional anisotropy.
Skumryev *et al.* demonstrated a magnetic coupling of FM nanoparticles with an AFM matrix as a source of a large effective additional anisotropy that led to a marked improvement in the thermal stability of the moments of the FM nanoparticles. The mechanism provides a way to beat the "*superparamagnetic limit*" in isolated particles so that, with the right choice of FM and AFM components, exchange anisotropy coupling could ultimately allow magnetically stable nanostructures. These nanostructures, only a few nanometers in size, would be able to surpass the storage-density goal of 1 *Tbit* in^{-2}, as set by the magnetic storage industry. [31]
As stated, the surface anisotropy is expected to contribute decisively for systems dominated by their surface properties, e.g. nanostructures offering an increased surface/volume ratio as it is the case of the magnetic colloids discussed here. Indeed, when it comes to magnetic nanostructures, the dominant surface contributions stem from the decisive role of ligands coverage onto the surface of the nanoparticles in order to render them dispersable in different solvents (colloidal chemistry), and therefore, became relevant for the resulting magnetic anisotropy. Experimental and theoretical reports have identified this decisive role of the ligands or molecular compounds adsorbed on the surface of different nanostructures in their magnetic anisotropy. [32] Since small structural relaxations can be induced by these ligands due to an induced spin reorientation related to the magnetoelastic coupling, it can be concluded that this magnetoelastic coupling opens the way to influence the magnetic anisotropy by even subtle structural and chemical changes at the surface. [33-36] Thus, surface magnetic properties result basically from the breaking of symmetry of the lattice, [37,38] which leads to site-specific, generally uniaxial, surface anisotropy, and from broken exchange bonds, which inevitably lead to surface spin disorder and frustration (most prominently in oxydic ferro-, antiferro- and ferrimagnets). [39,40]
On the other hand, the coordination of the ligands to the nanostructures surface should not alter the intrinsic specific physical properties of the particles nor those induced by their nanometer size. This latter point is important in order to take advantage of both the intrinsic and collective properties for future applications. Dumestre *et al.* have reported an

organometallic route towards the synthesis of metallic nanoparticles that can be applied to magnetic ones, based on the decomposition of an olefinic complex under a controlled atmosphere of H_2 in mild conditions of pressure. [41] Respaud *et al.* were able to demonstrate that the surface of this type of nanoparticles is free of contaminating agents so that, the magnetic properties are identical to those observed for nanoparticles produced and studied in ultra high vacuum. [42]

The magnetic anisotropy of magnetic nanostructures can therefore be tuned by a proper selection of the combination between the magnetic material itself and the molecular ligands attached to their surface to stabilize them, and by adjusting the growth parameters in the synthetic process used (herein we refer specifically to colloidal chemistry methods). Thus, the link between growth parameters and magnetism is given by the correlation between the nanostructures morphology, structure and the magnetic properties themselves. This correlation is given by the magnetoelastic coupling, and by the length scale of structure and morphological variations as compared to the length scale that describes magnetic properties. Such a magnetic length scale is for example, the domain wall, the exchange stiffness or the magnetic anisotropy. [12]

Additionally, despite the fact that we have just introduced single-domain nanostructures, if we consider the magnetization reversal by domain motion, an impact of structural and magnetic anisotropy variations on the coercivity will be expected. These variations should occur on a length scale larger than the domain wall width otherwise the magnetic inhomogeneities will be average out by the magnetic exchange interactions, and only little impact will result on the coercivity. [43]

1.4 Collective magnetic behavior

One of the most attractive topics in the field of condensed matter physics is slow dynamics such as nonexponential relaxation, aging (a waiting time dependence of observables) and memory effects. In the field of spin glasses, slow dynamics has been studied widely both experimentally and theoretically to examine the validity of novel concepts, such as a hierarchical organization of states and temperature chaos. These extensive studies have revealed interesting effects like memory and rejuvenation [44, 45] that led to also study slow dynamics in dense magnetic nanoparticulate systems by using experimental protocols already developed.

In the case of magnetic nanostructures systems, there are two possible origins of slow dynamics. The first one is a broad distribution of relaxation times originating solely from that of the anisotropy energy barriers of each nanoparticle moment. This is the only source of slow dynamics for sparse (weakly interacting) magnetic nanoparticle systems, in which the nanoparticles are fixed in space. However, for dense magnetic nanoparticle systems, there is a second possible origin of slow dynamics, namely, cooperative spin-glass dynamics due to frustration caused by strong dipolar interactions among the particles and randomness in the particle positions and anisotropy axis orientations. [46, 47] The 3D random distributions and random orientations of anisotropy axes of such nanostructures in an insulating matrix with high enough packing density and sufficiently narrow size distribution will create a competition of different spin alignments. [48] Which of the two is relevant depends essentially on the concentration of the particles. Therefore, in order to understand appropriately slow dynamics in magnetic nanoparticle systems, it is desirable to clarify both.

The comparison of the phenomena observed in relation to slow dynamics reveals some properties peculiar only to spin glasses, e.g. the flatness of the field-cooled magnetization below the critical temperature and memory effects in the zero-field-cooled magnetization. These two effects reflect the instability of the spin-glass phase under a static magnetic field of any strength meaning that it is indeed far from equilibrium. [49]
The intriguing properties of magnetic nanostructures including superparamagnetic and spin glass behavior are often ascribed to the delicate interplay between intrinsic properties and magnetostatic interactions.[50] When comparing between superparamagnets and spin-glass-like behavior of nanostructures, one significant difference is seen in the M_{FC} without intermittent stops (when cooling the system). While in the case of superparamagnets this M_{FC} increases as decreasing temperature, in the second case, the nearly constant M_{FC} is actually considered to be a typical property of ordinary spin glasses. Additionally, a further important phenomenon that is peculiar to superspin-glasses is the memory effect in the genuine ZFC protocol. This memory experiment in a ZFC protocol permits to differentiate between spin-glass phase and superparamagnets interacting. In this experiment, M_{ZFC} is measured with and without stops and in both cases cooling (and reheating) rate has to be the same. The stopping temperatures must be well below the blocking temperature T_B of the system measured. The results of the experiment will give us no significant difference in M_{ZFC} at the stopping temperatures and below when considering magnetic nanostructures weakly interacting. In this case then, the origin of slow dynamics is without doubt the broad distribution of relaxation times originating solely from the distribution of magnetic anisotropy energies of the nanostructures considered. On the contrary, if there is a significant difference in M_{ZFC} at the stopping temperatures and below, we must now considered our system of magnetic nanostructures as offering a spin-glass-like behavior.
It is important to carry out this latter experiment considering the M_{ZFC} and not the M_{FC} since only a spin glass left unperturbed by external fields at constant T, rearranges its spin configuration through a very slow process to reduce the domain wall energy. [51] Indeed, all measurements should be done in quite low magnetic fields in order to avoid nonlinear effects.[48] Moreover, Jonsson *et al.* reported that the collective behavior due to dipole-dipole interactions in concentrated samples extended the magnetic relaxation towards longer times and at the same time suppressed the relaxation at short observation times, a behavior that conforms to characteristic spin glass dynamics. [52]
One more point to be concerned about is the fact that collective magnetic excitations account for the precession or oscillations of the magnetic moment in the nanostructure about its magnetic easy axis, triggered at low temperatures by thermal energies insufficient to induce spin flips between opposite directions of the anisotropy axis, for $k_BT< K_{eff}V$. However, the particle surface-energy landscape can accommodate additional local minima due to lower coordination of the surface atoms, surface strain and spin canting associated with the surface itself. Many studies indicate that the breakage of superexchange bonds results in the creation of a surface shell within which spin disorder leads to spin-glass-like phase at the surface with closely spaced equilibrium states. [39, 53]

1.5 Carbon nanotubes

A great example representing the ever-increasing nanoscale-based research refers to the sp^2-bonded carbon nanotubes (CNTs) discovered in 1991. [54] CNTs have been pointed out as a paradigm material when talking about Nanotechnology. This new form in the carbon family with remarkable structure-dependent electronic, mechanical, optical and magnetic properties [55] has been the focus of an intensive study directed to numerous applications on many different fields, [56, 57] including synthetic routes where CNTs are useful for biological applications. [58, 59] Therefore, CNTs are expected to be controllably assembled into designed architectures as integral components of composites and/or supramolecular structures.

CNTs are hollow cylinders consisting of single or multiple sheets of wrapped graphite. According with the number of layers they are classified as single-walled carbon nanotubes (SWNTs) or multi-walled carbon nanotubes (MWNTs). Usually the structure of CNTs can be characterized using a pair of integers (n and m), which give us the rolling-up direction of the carbon sheet and the nanotube diameter. Thus, depending on the different rolling-up modes, the CNTs can be named as armchair with $n=m$, zigzag with $n=0$ or $m=0$, or generally chiral when any other n and m. Many of the potential applications of SWNTs and MWNTs are highly dependent on their electronic properties and in this context, their ballistic transport behavior and long electron mean free path have shown their potential as molecular wires. [60]

On the other hand, CNTs exhibit excellent structural flexibility and fluidity and can be bent, collapsed, or deformed into various shapes such as buckles, rings, or fullerene onions, providing a variety of shape-controlled physical properties of the nanostructures. Also by varying the geometric structure of the CNTs one can control electronic properties such as electrical conductivity or electron emission properties, thus providing modified electronic characteristics of the functional devices based on the CNTs.

Organic compounds have been used as templates for the generation of inorganic, organic or biological structures and materials rendering themselves the object of an increasing interest over the last years. In this way, taking into account that CNTs have been synthesized as an array of unprecedented structural, mechanical, and electronic properties, together with their high aspect ratio and surface area, these features have pointed them out as ideal templates for the deposition of a number of different materials on the search of new composite structures with promising properties and applications, offering the structural support that most of the inorganic, organic or biological materials generally lack. [61,62]

This ability to shape materials on a microscopic level is always desired but usually deficient. The formation of these hybrid structures is challenging due to the great application potential they display in many different fields and also from the scientific viewpoint. The application of these materials in useful processes and devices is ensured as soon as their production will be accomplished in a precise, reproducible manner, and if possible at reasonable costs.

Most of the applications proposed for CNTs have been shown to be strongly dependent on the development of strategies for functionalizing, processing and/or assembly of the CNTs themselves, mainly because their surface is rather inert, rendering very difficult any type of the mentioned mechanisms or techniques.

This is the case of a material deposition process onto CNTs, since becomes very difficult to control the final homogeneity. It is therefore important to explore feasible techniques whereby a surface modification would guarantee this material deposition onto the surface of the CNTs as a prior functionalization. Several approaches for functionalization of CNTs that have been developed can be classified as defect-site chemistry, covalent side-wall functionalization and non-covalent functionalization (figure 2).

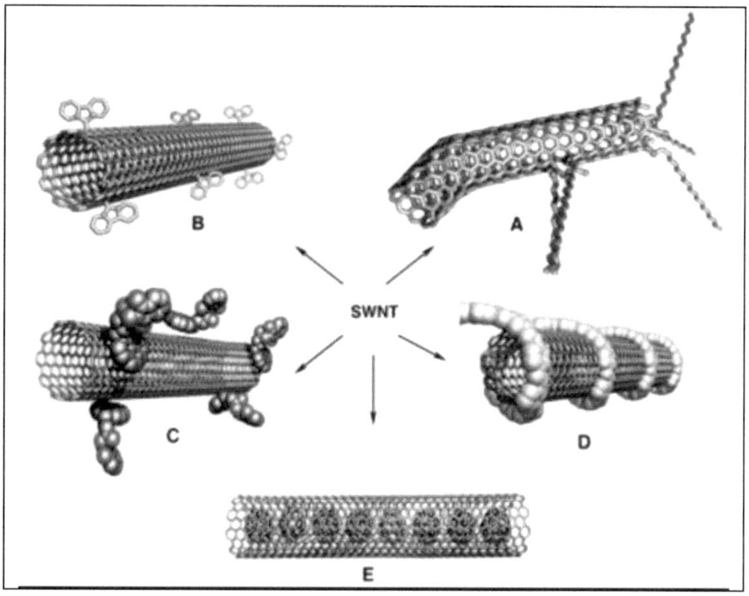

Fig. 2. Different types of SWNTs funtionalization: defect-site functionalization (A), covalent sidewall functionalization (B), non-covalent functionalization through, surfactants (C) or wrapping with polymers (D), and endohedral functionalization (E). Reprinted with permission from reference number 64. Copyright 2002 Wiley-VCH.

Defect-site chemistry Defect-side chemistry exploits the intrinsic defect sites both at the end and on the sidewalls of the CNTs as a result of the synthetic process. Additionally, the purification process of the CNTs involves the use of strong acids to remove the catalytic particles necessary for the synthesis, due to their oxidative action. This process ends up in holes with oxygenated functional groups like carboxylic acids or alcohol groups among others, which are promising starting points for the attachment of particles, molecular moieties or for further coordination chemistry at these sites.

Covalent-sidewall functionalization Covalent-sidewall functionalization is based on the chemical reactivity of the CNTs, related with the pyramidalization of the sp^2-hybridized carbon atoms and the π-orbital misalignment between adjacent carbon atoms. This pyramidalization and misalignment are scaled inversely with tube diameter, becoming more reactive tubes as decreasing their diameter. This agrees with the fact that fullerenes have a higher reactivity surface (which depends strongly on their curvature) compared to SWNTs which have no strongly curved regions that could serve for direct additions. This statement also explains why side-functionalization of SWNTs by covalent-bond formation needs highly reactive reagent.

Non-covalent functionalization Non-covalent functionalization comprises the dispersion of CNTs in aqueous solution, by means of surface active molecules as sodium dodecylsulfate (SDS) or by wrapping them with polymers. While the first one accommodates the CNTs in their hydrophobic interiors (sometimes by strong π-π -stacking interactions with the CNT sidewall if the hydrophobic part contains an aromatic group), the second one implies an association of the polymers with the sides of the CNTs based on the hydrophobic thermodynamic preference of CNT-polymer interactions compared to CNT-water interactions, thereby suppressing the hydrophobic surface of the CNTs.

Endohedral functionalization Endohedral functionalization comprises the use of the inner cavity of the CNTs for the storage of molecules or compounds since their interaction takes place with the inner surface of the sidewalls, very convenient for confined reactions inside the CNTs.

In summary, defect-side functionalization preserves the electronic structure of the CNTs, since the nanotubes can tolerate a number of defects before losing their unique electronic and mechanical properties. Covalent-sidewall functionalization generates a high degree of functionalization rendering this method very useful for composites formation. However, the destruction of the sp^2-hybridized structure may result in a loss of thermal conductivity, reducing the maximum buckling force or changing their electronic properties, displaying a semiconductor instead of a metallic behaviour.[63] Finally, the non-covalent functionalization has the main advantage since preserves intact the electronic properties and structure of the CNTs by maintaining the intrinsic nanotube sp^2-hybridization. [64]

2. Nanowires magnetic properties guided by carbon nanotubes

Several methods have been exploited for the synthesis of magnetic colloids on which the magnetic anisotropy can be tuned; relatively simple variations in surfactant composition used to selectively control the growth rates of different faces (in similar procedures as those concerning semiconductor materials [65,66]), [1] assembling previously synthesized magnetic nanoparticles as chains or necklaces, [2-4] [67-69] exploiting electrostatic interactions between the surface charge of magnetic nanoparticles and previously modified carbon nanotubes (CNTs), [70] or depositing the magnetic material on the surface of CNTs in a step-by-step procedure, giving place to homogeneous outer shells.[71] In order to investigate the possibility of obtaining nanowires with a very narrow size distribution and without chemical bonding at the surface we have chosen the third and fourth options schematically illustrated in figure 3. Thus, we have recently demonstrated that driving iron oxide nanoparticles or the direct reduction of nickel salt onto the surface of CNTs (in the second case in the presence of Pt nanoparticles) leads to very homogeneous magnetic nanowires.

2.1 Synthetic strategies

a. Fe_3O_4/γ-Fe_2O_3-coated CNTs

For this first strategy designed, an assembly of a compact layer of magnetic nanoparticles onto CNTs was taken under consideration. This option was carried out following a procedure that combines the polymer wrapping and LbL self-assembly techniques allowing the non-covalent attachment of iron oxide nanoparticles onto the CNTs and thus leaving intact their structure and electronic properties (see figure 2, D functionalization of CNTs by the polymer wrapping). [72] Poly (Sodium 4-styrene sulfonate) (PSS) was used as the initial wrapping polymer that permits to provide remarkably stable aqueous dispersions of multi-wall carbon nanotubes (MWNTs). Due to the high density of sulfonate groups on this negatively charged polyelectrolyte, the PSS acts as a primer onto the CNTs surface. So that, the subsequent and homogeneous absorption through electrostatic interactions of a cationic polyelectrolyte, the poly-(dimethyldiallylammonium chloride) (PDDA) is possible, supplying a homogeneous distribution of positive charges. These positive charges ensure the efficient adsorption (exploiting the same phenomenon, by means again of electrostatic interactions) of negatively charged magnetic nanoparticles onto the surface of CNTs.

Although the adsorption of nanoparticles (diameter D = 6-10 nm) onto CNTs is often problematic due to the extremely high curvature (D_{MWCNT} = 15-30 nm) that hinders the formation of dense coatings, the LbL self-assembly technique overcomes these difficulties, basically because of the electrostatic nature of the responsible interactions. [73, 74]

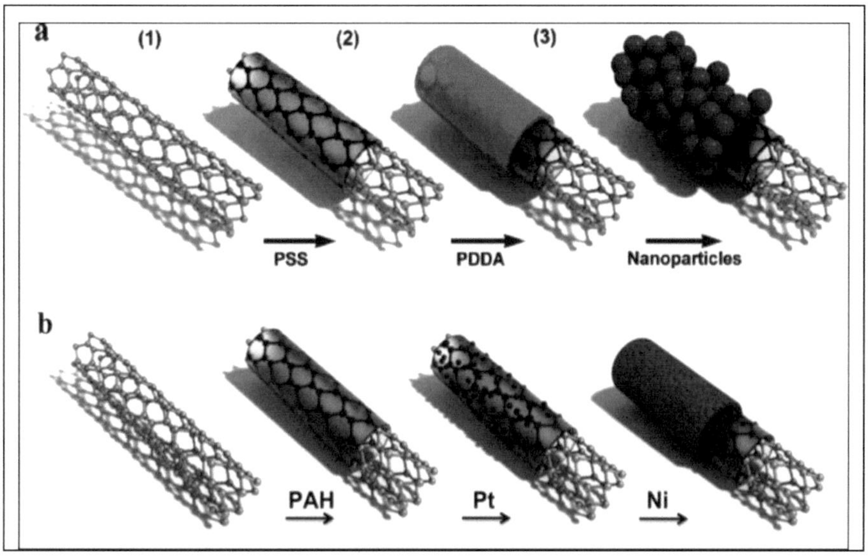

Fig. 3. Schematic illustrations of the synthetic processes used to produce magnetic nanowires using CNTs as guides and supports; a) by driven iron oxide nanoparticles and b) by reducing Ni atoms in a step-by-step process helped by the presence of catalytic Pt sites.

The magnetic nanoparticles prepared in aqueous solution (basic pH) are as mentioned, negatively charged and therefore electrostatically attracted to the positively charged PDDA outer layer adsorbed onto the CNTs. The pH for the most efficient adsorption of the $Fe_3O_4/\gamma\text{-}Fe_2O_3$ nanoparticles on polyelectrolyte was found to be 11.9-12.0 while at higher pH values no adsorption was observed. A uniform coating of magnetic particles onto the CNTs was achieved, as shown in the TEM images of Figure 4 (a and b) where long (> 5 μm) CNTs appeared completely covered with a dense layer of magnetic nanoparticles. These aqueous dispersions of magnetite/maghemite-coated CNTs were found to be very stable for several days or even weeks in the case of dilute solutions when (after centrifugation/washing) TMAOH (tetramethyl ammonium hydroxide) was added to the solution. Since the early work by Massart, [75] TMAOH has been a popular stabilizer for the preparation of aqueous ferrofluids, mainly comprising iron oxide nanoparticles. The stabilization of these magnetic CNTs in water takes place analogously through the formation of a double layer, with negative hydroxide ions fixed on the surface of the magnetic composites and positive tetramethylammonium as counterions in the basic environment.

Once coated, the magnetic response of these CNTs-based composites was easily and quickly visualized by holding the sample close to a small magnet. Indeed, when a drop of the dispersion was dried on a Si wafer without an external field, the magnetic CNTs were found randomly oriented on the silicon substrate (figure 5a). However, the magnetic CNTs can be oriented in the plane of a silicon wafer by using the mentioned external magnetic field. The magnetite/maghemite-coated CNTs were aligned as long chain structures by means of a

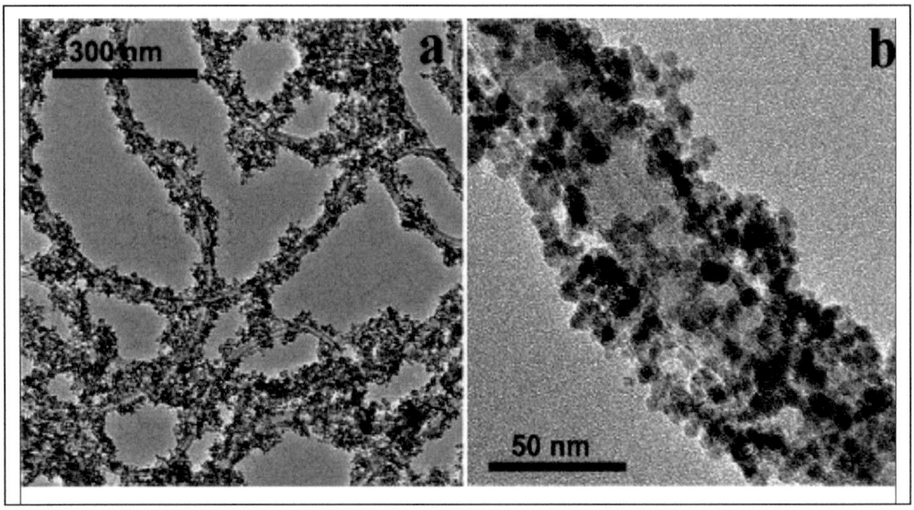

Fig. 4. TEM images at lower (a) and higher (b) magnifications of iron oxide-coated CNTs.

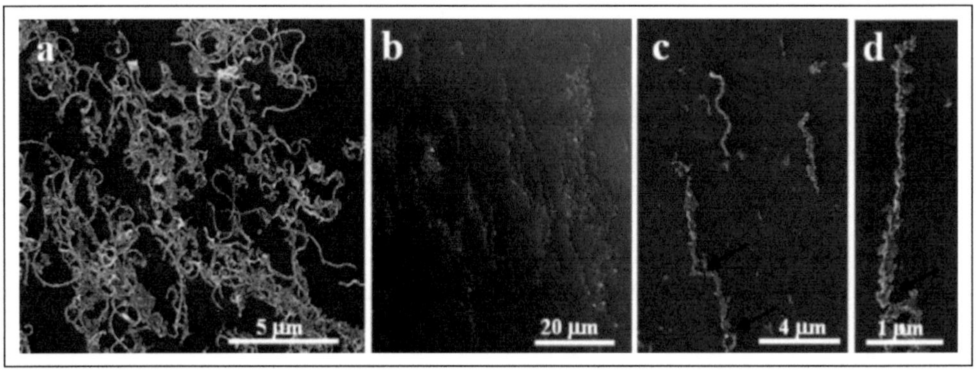

Fig. 5. SEM images of Fe_3O_4-coated CNTs in the absence (a) and in the presence (b,c,d) of an external magnetic field.

magnetophoretic deposition process on the silicon substrate at 300 K, since magnetic CNTs suspended in a liquid align parallel to the direction of the applied magnetic field (in this case under a magnetic field of 0.2T). This fact becomes quite easy to understand as considering that in zero field, the magnetic moments of the iron oxide nanoparticles randomly point in different directions resulting in a vanishing net magnetization (due to the thermal fluctuations able to influence the magnetic moments of the nanoparticle at this temperature). However, if a sufficiently large homogeneous magnetic field is applied, the magnetic moments of the nanoparticles align in parallel and the resulting dipolar interactions are sufficiently large to overcome thermal motion and to reorient the magnetic CNTs favouring the formation of chains of aligned carbon nanotubes. These chain-like structures are formed by connecting the magnetic CNTs in line, touching each other in a head to tail fashion - i.e. north to south pole (at positions indicated by arrows in figure 5c and d).

This behaviour has been reported for different magnetic nanoparticles systems. It takes place due to the anisotropic nature of the dipolar interaction. When comes into play, the north and south poles of the dipolar nanomagnets attract each other while particles coming close to each other side by side with the magnetization direction parallel will be repelled, thus favoring the formation of nanoparticle chains. [76] Therefore, at first glance one could find astonishing that some magnetic CNTs were found connected in parallel chains (figure 5b). This can be explained by either capillary and van der Waals forces or the fact that long magnetic chains whose end points are not next to each other will also attract each other forming double or triple chains which are parallel and close to each other but are offset to each other along the long axis.

b. Ni/NiO-coated CNTs

In the second strategy designed (figure 3b), the synthesis of well-defined, anisotropic magnetic nanotubes, using again the CNTs as templates, was carried out. Again, the sp^2 carbon structure was preserved obtaining in this case a ferromagnetic-like behavior at room temperature. Ni/NiO-coated CNTs/Pt nanocomposites were produced exploiting a stepwise reduction of $NiCl_2$ using hydrazine, following a procedure that starts with nanocomposites of CNTs previously functionalized with Pt nanoparticles. [77] With this step in mind, a previous CNTs polyelectrolyte functionalization needs to be carried out using another polyelectrolyte, the polyallylamine hydrochloride (PAH). The advantages in this case stem from the fact that ensure the presence of well-dispersed, individual nanotubes and additionally offers a homogeneous distribution of positive charges. This latter distribution of charges drives Pt nanoparticles once present in solution to be deposited onto CNTs.

These Pt nanoparticles were used as catalytic islands so that, the magnetic material can be grown directly on the CNT outer surface in a process that is mediated by this assembled layer of presynthesized, catalytic Pt nanoparticles. This intermediate step yields organic-inorganic hybrid composites that serve as 1D substrates for the preparation of magnetic CNT-supported Ni/NiO nanotubes.

Suspensions of Ni nanowires have also been proposed as magneto-optical switches because of their ability to scatter light that is perpendicularly incident to the wire axis. [78] These nanowires, when ferromagnetic, have large remnant magnetization owing to their large aspect ratios (increased contribution of the shape anisotropy) and hence can be used in low-field environments where superparamagnetic beads do not perform at all. [79] Additionally, since these composite structures involve a Ni/NiO antiferromagnetic/ferromagnetic interface, an exchange bias effect is expected which could find promising applications in magnetoresistive devices. [80]

The deposition of a very uniform and homogenous layer of Ni, without surfactants or other stabilizers in water solution represents a relevant issue since their use would induce a perturbation of the magnetic properties displayed. [42, 81]

Figure 6 shows representative TEM and HRTEM images of the samples produced. These images reflect the homogenous coating of individual CNTs and reveal the multidomain and crystalline nature of the Ni/NiO layers on the final composite. By Fourier transform analysis (of dark field images, not shown), the following intershell and interplanar distances were determined; MWNTs (3.36 Å), Pt (2.19 Å) and Ni (2.03 Å and 1.72 Å), corroborating the envisioned structure of CNTs@Pt/Ni/NiO nanocomposites. The reader has to take into account that CNTs@Pt/Ni nanocomposites were exposed to an oxygen-rich environment during the magnetic material deposition process (aqueous solution) promoting the formation of the chemically stable NiO outer layer around each Ni nanocrystal/nanoshell. [82] This surface passivation process provides the samples with an additional surface stabilization and simultaneously protects the inner metallic Ni from further oxidation. At this point it can also be underlined the importance of having a continuous shell of a magnetic material onto the CNTs, contrary to the nanocomposites produced on the first strategy, in which the individuality of the iron oxide nanoparticles was kept once deposited onto the CNTs. This has indeed consequences on the magnetic properties displayed, as detailed later on in this chapter.

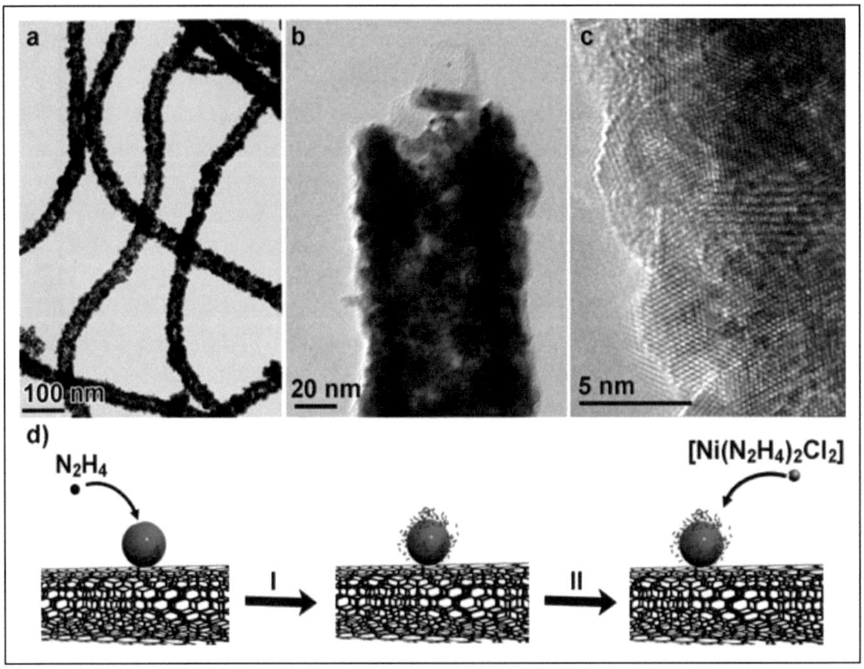

Fig. 6. Representative TEM images of the CNTs@Pt structures coated with a uniform outer layer of Ni/NiO. HRTEM image reflecting the polycrystalline nature of the Ni/NiO layer deposited onto the CNTs@Pt nanocomposites. d) Illustration of Ni reduction on the CNT/Pt side-walls. Step I involves the decomposition of hydrazine on the surface of Pt nanoparticles, which results in a charged surface, and step II the reduction of the hydrazine/Ni complex on the charged Pt surface.

The catalytic behavior of the small Pt nanoparticles play a crucial role in the formation of the metallic Ni on the walls of the CNTs by catalyzing the reduction of the Ni/hydrazine complex formed in aqueous solution, as described below. This reaction allows the CNT/Pt nanocomposites to be coated with a uniform and homogeneous Ni layer (ca. 10 nm thick), with no need for surfactants or other stabilizers in aqueous solution. This feature is relevant since the use of surfactants or other stabilizers usually hinders subsequent manipulation and implementation in different applications. The CNT/Pt@Ni composites are stable in solution (most likely because of the presence of a negatively charged NiO surface layer) and their surface is free of surfactants, thus allowing further functionalization if required. The proposed mechanism for the formation of metallic Ni on the surface of the Pt nanoparticles is depicted schematically in Figure 6d. Complexes between transition-metal ions and hydrazine are easily formed in water. [83] Such complexes can be decomposed in the presence of hydroxy groups, [84] in a process where Ni(II) complexes are reduced to Ni^0. However, the catalytic decomposition of hydrazine in the presence of Pt nanoparticles can take place on their surface by an electrophilic addition, that is, by forming electrophilic radicals that can then react with other hydrazine molecules from solution. The presence of metallic platinum nanoparticles therefore implies the reduction of nickel complexes without the need for additional hydroxy groups present in solution. In fact, we observed that reduction of Ni^{2+} did not occur in experiments using PAH-functionalized CNTs in the absence of Pt nanoparticles. Thus, the mechanism of Ni nanotube formation can be explained in the following terms: the excess hydrazine that is not complexed with Ni^{2+} ions can be catalytically decomposed on the surface of the platinum nanoparticles supported on the CNTs. This step generates a charged metallic surface (step I, Figure 6d) which promotes the reduction of the Ni(II) complex into Ni^0 (step II, Figure 6d). Subsequent decomposition of hydrazine on the surface of the reduced metal is facilitated, thus favoring further Ni reduction and growth of a homogeneous shell. The process terminates when all the Ni(II) has been reduced. The advantage of this surface-catalyzed reduction of Ni(II) is the formation of a continuous (though polycrystalline) Ni layer rather than an outer shell composed of nanoparticles, with an envisioned improvement of the resulting magnetic properties. Since the CNT/Pt@Ni nanocomposites formed are exposed to an oxygen-containing environment during the deposition of the magnetic material, a stable NiO outer layer is expected to form, passivating the Ni shell and preventing a full oxidation of the magnetic material.[82] However, the presence of nickel oxides on the outer surface of the nickel nanotube could not be confirmed by Fourier transform analysis. To determine whether nickel oxides (and/or hydroxides) were formed on the surface of the CNT/Pt@Ni nanocomposites, the samples were examined by X-ray photoelectron spectroscopy (XPS). Analysis of the Ni 2p peaks revealed a main peak for metallic Ni (852.8 eV) with contributions from different Ni oxidation states, presumably corresponding to the binding energies of NiO and $Ni(OH)_2$ (854.4 and 856.5 eV, respectively). These main lines are accompanied by satellite lines with binding energies that are 6 eV higher, which suggests the presence of these Ni oxides at the outer surface of the nickel wires. Apart from passivation, this surface oxidation process leads to ferromagnetic/antiferromagnetic (FM/AFM) interfaces (Ni/NiO), which give rise to an exchange bias effect that increases the potencial applications of these nanocomposites, as seen later in the chapter.

2.2 Magnetic properties

a. Fe_3O_4/γ-Fe_2O_3-coated CNTs

In order to check and understand the influence of using CNTs as supports to generate one-dimensional magnetic structures and of the CNTs themselves, the magnetic properties of magnetite/maghemite nanoparticles and the compositess were recorded in a Quantum Design Vibrating Sample Magnetometer (VSM). Figure 7 shows hysteresis curves of both structures, collected at 5 K and 300 K, that summarize the typical superparamagnetic behavior. Both iron oxide nanoparticles and magnetic CNTs show the same superparamagnetic behavior, i.e. the same coercive field (Hc = 280 Oe at 5 K) and no remanence or coercitivity at room temperature.

At first sight magnetic CNTs seem to have a larger magnetic moment per gram of sample. The saturation magnetization of the original Fe_3O_4/γ-Fe_2O_3 nanoparticles, whose relative concentration cannot be quantified due to the similarity of γ-Fe_2O_3 and Fe_3O_4 XRD spectra present a lower value than the bulk magnetization of γ-Fe_2O_3 (60-80 Am^2/kg). This was reported to be dependent on the particle size and on the different degree of vacancy ordering of the particles which is directly related to the sample preparation method.[85] Nevertheless, when assembled onto the surface of MWCNTs the saturation magnetization is increased by approximately 17 % compared to the initial particle powder and agrees rather well with the bulk magnetization of γ-Fe_2O_3. The origin of this magnetization increase is not clear and needs further investigation but the presence of nickel traces from the catalytic Ni surfaces used to grow the CNTs should also be taken into account. The error bar of the sample weight (less than 0.1 mg) could represent a maximum deviation of 1% on the magnetization, far away from the obtained ≈17%.

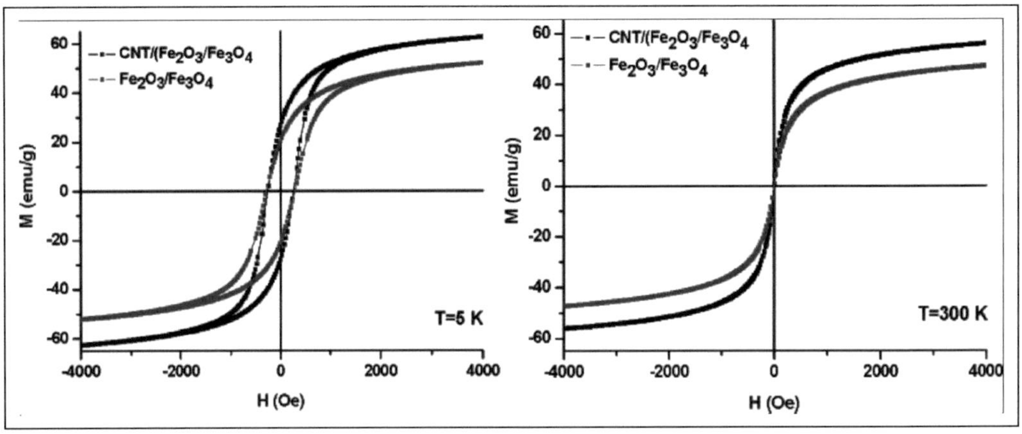

Fig. 7. Magnetic hysteresis cycles for γ - Fe_2O_3/Fe_3O_4-coated CNTs (black lines) and γ-Fe_2O_3/Fe_3O_4 nanoparticles (grey lines), showing M-H data at 5 (a) and 300 (b) K.

b. Ni/NiO-coated CNTs

The overall magnetic behaviour of these nanowires, comprising magnetic nanocrystals of Ni and NiO, is a result of both the properties of the individual constituents (dependent on intrinsic parameters like the size, shape and structure), the morphology they have acquired (anisotropic nanowires) and their interparticle interactions, with a negligible magnetic contribution from the CNTs. Some characteristics of a collective superspin-glass state that can be considered as counterparts of a conventional spin glass are reported, as a consequence of random orientations of the easy axes of the nanocrystals in every wire and the long-range character of the dipolar interactions. In the absence of an external magnetic field the magnetic moment of these nanocrystals constituting the composites would be in a blocked state, unless thermal activation would be able to overcome the anisotropy energy barrier and induce flipping of the moment between easy directions (superparamagnetism). [13] However, these assemblies of nanocrystals forming wires and containing an interface between a ferromagnet (FM) (Ni) and an antiferromagnet (AFM) (NiO) exhibit an additional unidirectional anisotropy due to magnetic coupling at the FM/AFM interface (exchange bias) which leads to magnetization stability. [30]

Exchange biasing has important technological applications, such as in giant-magnetoresistance-based spin valves that are used in hard drive read heads and other spintronics applications.[31] In nanoscale systems, the key factor in this magnetic interaction would be controlling the volume, the shape, crystallinity and structure of both FM and AFM materials. [80, 82, 86-90] Ni/NiO nanowires offer therefore an attractive approach to control

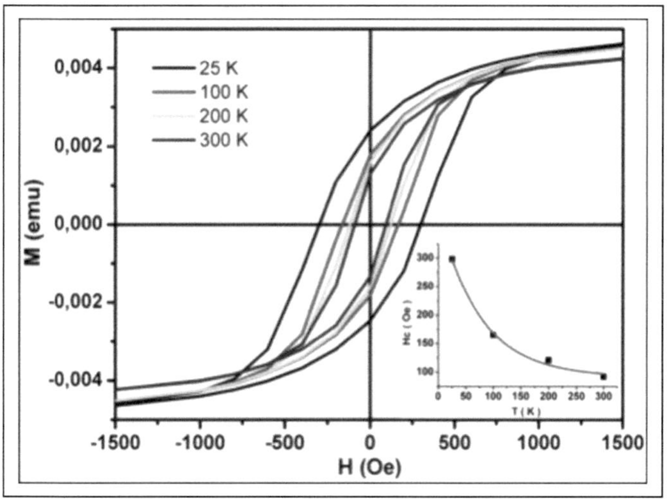

Fig. 8. Hysteresis loops at 25, 100, 200 and 300 K of Ni/NiO-coated CNTs/Pt nanocomposites. The inset shows the temperature dependence of coercivity (H_C).

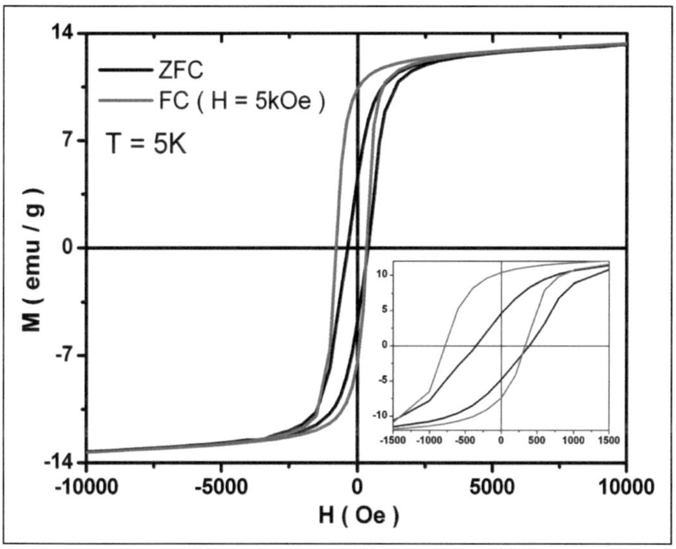

Fig. 9. a) Magnetic hysteresis loops at T1/45 K (zero-field cooled (ZFC) and field cooled (FC) (5 kOe)). b) Magnified view at smaller fields.

exchange coupling, an anisotropic morphology and spin-glass-like behaviour. Significant technological impacts based on this type of nanosized assemblies can thus be expected.

The main results obtained on the magnetic properties of CNTs@Pt/Ni/NiO nanocomposites were also measured in the solid state as powders, but in this case using a superconducting quantum device magnetometer (SQUID). The hysteresis loops, measured at 25, 100, 200 and 300 K are plotted in figure 8. The large values of the coercive field, even at room temperature, indicate that the magnetic moments are in a blocked state in the homogeneous Ni/NiO shell around every CNT. Each nanowire can be considered as a heterogeneous magnetic system consisting of FM Ni nanocrystals surrounded by AFM NiO shells,

homogeneously coating the CNTs/Pt nanotubes. The ferromagnetic component is confirmed by hysteresis (present at 300 K and lower temperatures), with coercivities (H_C) and remnant magnetizations. The antiferromagnetic NiO was identified by XPS (X-ray photoemission spectroscopy) and STEM (scanning transmission electron microscopy) analysis. The oxygen-rich environment of the nanocomposites (in aqueous solution) causes the formation of the stable NiO outer shells and favors therefore the appearance of interfacial FM/AFM exchange anisotropy. Indeed, when the system is cooled under an applied magnetic field of magnitude H_{FC} = 5kOe, the hysteresis loops of the nanocomposites are shifted with respect to the field axis (figure 9), confirming the presence of unidirectional anisotropy due to the exchange coupling at the interface of the FM/AFM materials.[30]
Figure 9 shows hysteresis curves collected at 5 K (zero-field cooled (ZFC) and FC (5kOe)) displaying this shift along the direction of the cooling field with coercivities $H_C=|H_{C1}-H_{C2}|/2$ = 370 Oe (ZFC) and H_C= 560 Oe (FC) and exchange bias field $H_E= -(H_{C1}+H_{C2})/2$ = 225 Oe (in the FC curve). The relatively large value of the exchange bias field H_E indicates an unidirectional exchange anisotropy due to the exchange interaction between the uncompensated surface spins of NiO and metallic Ni in the Ni/NiO-coated CNTs@Pt nanocomposites. This shift of the hysteresis loop can be established either by cooling the FM/AFM material in a magnetic saturation field below the Néel temperature of the AFM (5 kOe in this case) or by depositing both materials in an external magnetic field.[80]
Exchange bias at FM/AFM interfaces is also characterized by a coercivity enhancement, revealing the induced unidirectional anisotropy. The coercive field H_C = 90 Oe at 300 K of the Ni/NiO-coated CNTs/Pt nanocomposites is larger than typical values reported for bulk Ni. This can be explained taking into account an influence of the interface anisotropy, which through exchange coupling can modify the magnetism of the Ni/NiO nanocrystals. Roy and co-workers have anticipated that the NiO shell (in Ni/NiO core-shell nanoparticles) can act as a pinning layer, pinning the core spins near the interface of the Ni inner shell and the NiO outer shell via exchange interactions. The core spins are thus prevented from rotating freely and follow the applied field, thereby leading to the observed large coercivities.[82] In fact, the loop shifted along the magnetization axis indicates that after the field cooling process a fraction of the uncompensated moments is pinned due to a very high local anisotropy and cannot reverse at the available magnetic field strength. [91]
Summarizing these two first points concerning the magnetic properties of the second type of magnetic nanowires, in both cases there are important contributions from different anisotropies governing different mechanisms. This corresponds to the general contribution dealing with the magnetic anisotropy. This is indeed the quantity that determines the easy magnetization directions of the materials and it is also decisive for the magnetization reversal in external fields. [12]
In both cases the magnetocrystalline anisotropy concerning the crystalline structures of magnetite and maghemite in the first case and metallic Ni, nickel oxide and hydroxides contribute similarly, and there is actually no sense in a comparison as taking into account the nature of different materials.
There are however other three main contributions concerning the shape anisotropy, the surface anisotropy and the unidirectional/uniaxial anisotropy which may become dominant depending on the case, helping to understand therefore the different behavior.
Although having monodimensional nanostructures in both cases, we can consider the discontinuity in the magnetic materials if the outer magnetic shell around the CNTs is formed by individual nanoparticles. On the second case however, as forming the magnetic

shell by reducing the Ni^{2+} ions in a step-by-step process, we end up with continuous shells of Ni (that become Ni/NiO) onto the CNTs. This is not a trivial question when taking into account the shape anisotropy and surface anisotropy contributions, since in the first case we have spherical magnetic nanoparticles (assembled as a cylindrical shell) and cylindrical structures in the second case.

The third and additional contribution is present only in the second case of Ni/NiO coated CNTs/Pt nanocomposites. Due to the displayed morphology in which a ferromagnetic/antiferromagnetic interface is established. Thus, these composites were found to exhibit characteristic features of an exchange biased system. By this phenomenon, the magnetic behaviour of the ferromagnetic component in the nanocomposites is markedly affected by the outer shell of NiO, as was reflected by the open hysteresis loops. This interface generates an unidirectional anisotropy which increases the total anisotropy K so that the thermal energy at room temperature is overcome and the magnetic moments of the composites remain magnetically stable in the time window of the measurement.

2.2.3 Collective dynamics at low temperatures

The study of the magnetic properties of these nanoscaled systems of CNTs-based wires emphasize the dominant role of the surface/interface effects on the intrinsic properties of nanoscale systems and the competition with the interactions in the case of assemblies leading to characteristics magnetic behaviors. A further step in these fundamental studies will concern the dynamic aspect of the phenomena.

Magnetization versus temperature measurements at two different magnetic fields (H= 100 and 1000 Oe) for the Ni/NiO-coated CNTs/Pt are included in figure 10. ZFC and FC magnetization curves split below T = 300K and T = 100K at the applied magnetic fields and in both cases the ZFC magnetization curve exhibits a peak around T = 40 K. This temperature indicates a collective freezing of magnetic moments (blocking temperature T_B).

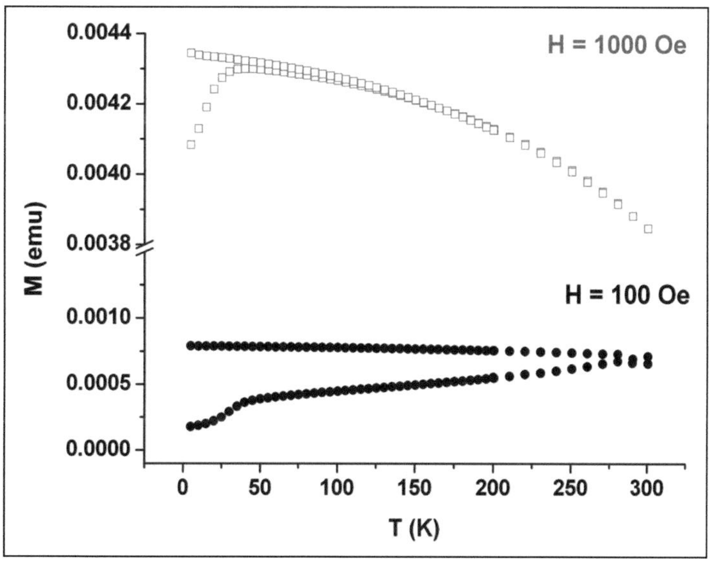

Fig. 10. Field cooled (FC) (5 kOe) and zero-field-cooled (ZFC) magnetization measurements of Ni/NiO-coated CNTs/Pt nanocomposites.

The irreversibility shown is strongly dependent on the magnitude of the applied magnetic field and can be presumably associated to slow relaxation processes for an assembly of interacting nanocrystals. This behaviour can be attributed to a random distribution of strong magnetic dipole-dipole interactions and surface anisotropy effects between the Ni cores and the NiO shells in the nanocrystals. The fact that the temperature at which the irreversibility takes place decreases with increasing magnetic fields [92] can be explained if the effects of the anisotropy field and the dipole-dipole interaction are overcome by the applied magnetic field. [93]

Indeed, one of the most challenging questions in nanoparticulated systems concerns the collective dynamics at low temperatures. In a dilute system, the magnetic dipole-dipole interaction between the particles is negligible compared to the anisotropy energy of an individual nanoparticle. The dynamics follow the predictions of the Néel-Brown model [10] and the system is considered as purely superparamagnetic. However, in a concentrated system the dipole-dipole interaction can be of the order of the particle anisotropy energy and strongly affects the low-temperature dynamics. Three-dimensional random distributions and random orientation of anisotropy axes of such nanoparticles in an insulating matrix with high enough packing density will create a competition of different spin alignments. [94] Despite sophisticated experimental work [95-98] and Monte Carlo simulations [99-101] supporting collective dynamics at low temperature, there are also contradictory results in favor of superparamagnetism behaviour and non-collective blocking. If a low-temperature collective superspin-glass state is formed, typical properties of an ordinary atomic spin glass should be observed in this phase. The collective behaviour of the Ni/NiO nanowires is not exclusive of conventional spin glasses and indeed has been reported in systems of concentrated magnetic particles [52] and other nanostructured magnetic materials, where dipolar interactions introduce a collective state and magnetic relaxation dependence. In order to label the nanowires collective behaviour as *spin-glass like,* we report memory effects by means of a ZFC magnetization experiment with a 2h stop during cooling at zero field. [51]

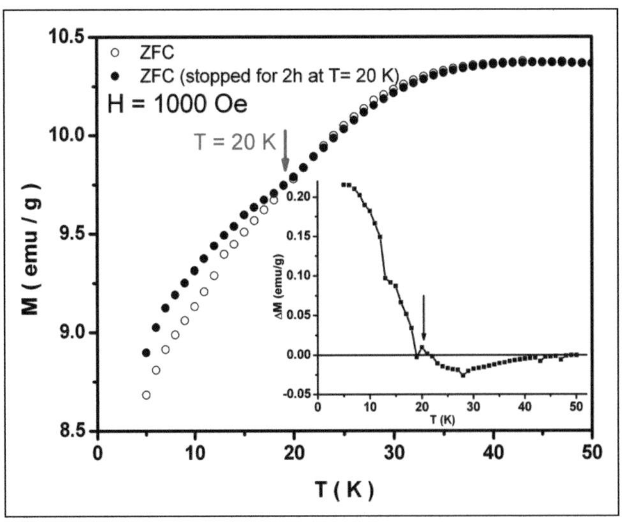

Fig. 11. ZFC magnetization curves measured with and without waiting time (t_w=2h) showing memory effect.

Figure 11 illustrates the memory effect of the ZFC magnetization after this stop-and-wait protocol performed below the blocking temperature. The sample was zero field cooled down to 5K twice, after a waiting time t_w=2h at 20K (T_w<T_B=40K) in the second case. The "stop-and-wait" magnetization data (solid symbols) lie significantly above the conventional ZFC magnetization at temperatures lower than 20 K (temperature at which the stop-and-wait experiment took place). This difference below T_w indicates that the magnetic moment configuration spontaneously rearranges towards equilibrium as increasing their correlation length. This implies that the correlation between the nanocrystals moments develops in a similar way as the correlation between the spins in spin glasses. Therefore, when interpreting the dynamic behaviour of these interacting Ni/NiO nanocrystals in the nanowires, the effects of the spin-glass-like correlations may have to be taken into account. The dynamic magnetic properties observed in the Ni/NiO-coated CNTs/Pt nanocomposites seem to favour the spin-glass hypothesis explaining the observed collectivity and glassiness.

3. Conclusion

We have reported the effective magnetic functionalization of CNTs by coating them with iron oxide nanoparticles using the polymer wrapping and LbL assembly techniques, or in a step-by-step process reducing Ni on their surface. Upon such functionalization, the alignment of CNT chains in relatively small external magnetic fields was demonstrated. The resulting magnetic CNTs-based structures are excellent candidates to be used as building blocks for the fabrication of novel composite materials with a preferential orientation of the magnetic CNTs.

4. Acknowledgments

The authors acknowledge the financial support of this work under Project MAT2008-06126 by the Spanish Ministerio de Ciencia e Innovación, under project INCITE08PXIB209007PR by the regional government (Xunta de Galicia, Spain) and by the L'Oréal-UNESCO *Woman in Science* Program. V. S. acknowledges the *Ramón y Cajal* Program fellowship (Ministerio de Ciencia e Innovación).

5. References

[1] Puntes, V. F.; Krishnan, K. M. & Alivisatos, A. P. (2001). Colloidal nanocrystals shape and size control: The case of Cobalt. *Science* 291, 2115-2117, ISSN: 0036-8075.

[2] Dumestre, F.; Chaudret, B.; Amiens, C.; Fromen, M.-C.; Casanove, M.-J.; Renaud, P. & Zurcher, P. (2002) Shape Control of Thermodynamically Stable Cobalt Nanorods throught Organometallic Chemistry, *Angew. Chem. Int. Ed.* 2002, 41, 4286-4289, ISSN: 0044-8249.

[3] Dumestre, F.; Chaudret, B.; Amiens, C.; Respaud, M.; Fejes, P.; Renaud, P. & Zurcher, P. (2003). Unprecedented Crystalline Super-Lattices of Monodisperse Cobalt Nanorods. *Angew. Chem. Int. Ed.* 42, 5213-5216, ISSN: 0044-8249.

[4] Goubault, C.; Leal-Calderon, F.; Viovy, J.-L. & Bibette, L (2005). Self-Assembled Magnetic Nanowires Made Irreversible by Polymer Bridging. *Langmuir*, 21, 3725-3729, ISSN: 0743-7463.

[5] Sellmyer, D. J.; Zheng, M. & Skomski, R., (2001). Magnetism of Fe, Co and Ni nanowires in self-assembled arrays, *J. Phys.: Condens. Matter* 13, R433-R460, ISSN: 0053-8984.

[6] Kodama, R. H. (1999). Magnetic nanoparticles. *J. Magn. Magn. Mater.*, 200, 359-372, ISSN: 0304-8853.

[7] Parak, W. J.; Manna, L.; Simel, F. C.; Gerion, D. & Alivisatos A. P. (2004) Quantum Dots, In: *Nanoparticles*, Schmid, G. (Ed.), 4-49, Wiley-VCH, ISBN: 3-527-30507-6, Weinheim.

[8] Murray, C. B.; Norris, D. J. & Bawendi, M. G. (1993). Synthesis and Characterization of Nearly Monodisperse CdE (E= Sulfur, Selenium, Tellurium) Semiconductor Nanocrystallites. *J. Am. Chem. Soc.* 115, 8706-8715, ISSN: 0002-7863.

[9] Murray, C. B.; Sun, S.; Doyle, H. & Betley, T. (2001).Monodisperse 3d transtion-metal (Co, Ni, Fe) nanoparticles and their assembly into nanoparticle superlattices. *MRS Bull*, 26, 985-991, ISSN: 0883-7694.

[10] Sorensen, C. M. (2001) Magnetism, In: *Nanoscale Materials in Chemistry*, Klabunde, K. J. (Ed.), 169-221, John Wiley and Sons, ISBN: 0-471-38395-3, New York.

[11] Farle, M. (2003). Magnetic thin films, In: *Nanoscale Materials*, Liz-Marzán, L. M. & Kamat, P. (Ed.) 395-422, Kluwer Academic Publishers, ISBN: 1-4020-7366-6, Dordrecht.

[12] Sander, D. (2004). The magnetic anisotropy and spin reorientation of nanostructures and nanoscale films. *J. Phys.: Condens. Matter*, 16, (R603-636), ISSN: 0053-8984.

[13] Dormann, J. L.; Fiorani, D. & Tronc, (1997) Magnetic Relaxation in Fine-Particle Systems, In: *Advances in Chemical Physics*, Prigogine, I. & Rice, S. A. (Eds.) John Wiley and Sons, 0-471-16285-X, New York.

[14] Gradmann, U., (1993). Magnetism in Ultrathin Transition metal Films, In: *Handbook of Magnetic Materials*, Buschow, K. H. J. (Ed.) vol. 7, chapter 1, pp 1-96, Elsevier Science, ISBN: 78-4-444-89853-1, Amsterdam.

[15] Eriksson, O., (2001). First principles theory of magnetism for materials with reduced dimensionality, In: *Lecture Notes in Physics: Band-Ferromagnetism*, Baberschke, K.; Donath, M. & Nolting, W. (Eds) 243-266, Springer, ISBN: 97-83-5404-2389-8, Berlin.

[16] Lindner, J. & Farle, M. (2007). Magnetic Anisotropy of Heterostructures, In: *Springer Tracks in Modern Physics*, 45-96, Springer-Verlag, ISBN: 978-3-540-73462-8-2, Berlin Heidelberg.

[17] Skomski, R. & Coey J. M. D. (1999) *Permanent Magnetism*, Institute of Physcis Publishing, ISBN: 0750304782, Bristol.

[18] Blundell, S. (2001). *Magnetism in Condensed Matter*, Oxford University Press, ISBN: 019 8505922, Oxford.

[19] Burkert, T.; Eriksson, O.; James, P.; Simak, S. I.; Johansson, B. & Nordström, L. (2004). Calculation of uniaxial magnetic anisotropy energy of tetragonal and trigonal Fe, Co, Ni, *Phys. Rev. B*, 69, 104426/1-4, ISSN: 0163-1829.

[20] Wu, R. (2001). First principles determination of magnetic anisotropy and magnetostriction in transition metal alloys, In: *Lecture Notes in Physics: Band-Ferromagnetism*, Baberschke, K.; Donath, M. & Nolting, W. (Eds) 46-71, Springer, ISBN: 97-83-5404-2389-8, Berlin.

[21] Mook, W. M.; Nowak, J. D.; Perrey, C. R.; Carter, C. B.; Mukherjee, R.; Girshick, S. L.; McMurry, P. H. & Gerberich, W. W. (2007). Compressive Stress Effects on Nanoparticle Modulus and Fracture. *Phys. Rev. B* 75, 214112/1-4, ISSN: 0163-1829.

[22] Carlton, C. E.; Lourie, O. & Ferreira, P. J. (2007). In-situ TEM Nano-Indentation of Individual Single-Crystal Nanoparticles, *Microsc. Microanal.* 13, suppl2, ISSN: 1431-9276.

[23] Zou, M. & Yang, D. (2006). Nanoindentation of silica nanoparticles attached to a silicon substrate, *Tribology Letters*, 22, 189-196, ISSN: 1023-8883.

[24] Rinaldi, A.; Licoccia, S.; Traversa, E.; Sieradzki, K.; Peralta, P.; Dávila-Ibáñez,A. B.; Correa-Duarte, M. A. & Salgueirino, V. (2009). Novel Plastic Behavior and Super-hard of Amorphous and Magnetic Composites Nanostructures, in preparation.

[25] Bin, W.; Heidelberg, A. & Boland, J. J. (2005). Mechanical properties of ultrahigh-strength gold nanowires, *Nature Materials*, 4, 526-529, ISBN: 1476-1122.

[26] Varghese, B.; Zhang, Y.; Dai, L.; Tan, V. B. C.; Lim, C. T. & Sow, C.-H. (2008). Structure-Mechanical Property of Individual Cobalt Oxide Nanowires, *Nano Letters*, 8, 3226-3232, ISSN: 1530-6984.

[27] Greer, J. R. & Nix, W. D. (2006). Nanoscale gold pillars strengthened through dislocation starvation. *Phys. Rev. B*, 73, 245410/1, ISSN: 0163-1829.

[28] Shan, W.; Mishra, R. K.; Asif, S. A. S.; Warren, O. & Minor, A. M. (2008). Mechanical annealing and source-limited deformation in submicrometer-diameter Ni crystals, *Nature Materials*, 7, 115-119, ISBN: 1476-1122.

[29] Cowburn, R. P.; Adeyeye, A. O.; & Welland, M. E. (1998). Configuration Anisotropy in Nanomagnets. *Phys. Rev. Lett.* 81, 5414/1-4. ISSN: 0031-9007.

[30] Berkowitz, A. E. & Takano, K. (1999) Exchange Anisotropy-a review, *J. Magn. Magn. Mater.* 200, 552-570, ISSN: 0304-8853.

[31] Skumryev, V.; Stoyanov, S.; Zhang, Y.; Hadjipanayis, G.; Givord, D. & Nogués, J. (2003). Beating the superparamagnetic limit with exchange bias, *Nature*, 423, 850-853, ISSN: 0028-0836.

[32] Ujfalussy, B.; Szunyogh, L.; Bruno, P. & Weinberger, P. (1996). First-principles calculation of the anomalous perpendicular anisotropy in a Co monolayer on Au (111), *Phys. Rev. Lett.*, 77, 1805-1808, ISSN: 0031-9007.

[33] Berkowitz, A. E.; Lahut, J. A. & VanBuren, C. E. (1980). Properties of Magnetic Fluid Particles, *IEEE Trans. Magn.*, 16, 184-190, ISSN:0018-9464.

[34] Ngo, A. T.; Bonville, P. & Pileni, M. P. (1999). Nanoparticles of $CoFe_2O_4$: syntheses and properties, *Eur. Phys. J. B.*, 9, 583-592, ISSN: 1434-6028.

[35] Vestal, C. R., Zhang, Z. J., (2003). Effects of Surface Coordination Chemistry on the Magnetic Properties of $MnFe_2O_4$ Spinel Ferrite Nanoparticles, *J. Am. Chem. Soc.* 125, 9828-9833, ISSN: 0002-7863.

[36] Shukla, N., Liu, C. Jones, P. M., Weller, D. (2003). FTIR study of surfactant bonding to FePt nanoparticles, *J. Magn. Magn. Mater.* 2003, 266, 178-184. ISSN: 0304-8853.

[37] L. Néel (1954). Anisotropie Magnétique superficielle et surstructures d'orientation, *J. Phys. Radium*, 15, 225-239.

[38] Buschow, K. H. J. (1993). *Handbook of Magnetic Materials*, Elsevier, ISBN: 0444514597, Amsterdam.

[39] Kodama, R. H.; Berkowitz, A. E.; McNiff Jr., E. J. & Foner, S. (1996). Surface Spin Disorder in $NiFe_2O_4$ Nanoparticles, *Phys. Rev. Lett.*, 77, 394-397, ISSN: 0031-9007.

[40] Iglesias, O. & Labarta, A. (2001). Finite-size and surface effects in maghemite nanoparticles: Monte Carlo simulations, *Phys. Rev. B*, 63, 184416/1-11, ISSN: 0163-1829.

[41] Dumestre, F.; Martinez, S.; Zitoun, D.; Fromen, M.-C.; Casanove, M.-J.; Lecante, P.; Respaud, M.; Serres, A.; Benfield, R. E.; Amiens, C.; Chaudret, B. (2004) Magnetic nanoparticles through organometallic synthesis: evolution of the magnetic properties from isolated nanoparticles to organized nanostructures, *Faraday Discuss.*, 125, 265-278, ISSN: 0301-7249 .

[42] Respaud, M.; Broto, J.-M.; Rakoto, H.; Fert, A.; Verelst, M.; Snoeck, E.; Lecante, P.; Mosset, A.; Thomas, L.; Barbara, B.; Osuna, J.; Ould Ely, T.; Amiens, C. & Chaudret, B. (1998). Surface effects on the magnetic properties of ultrafine cobalt particles, *Phys. Rev. B*, 57, 2925-2935, ISSN: 0163-1829.

[43] Wulfhekel, W.; Zavaliche, F.; Hertel, R.; Bodea, S.; Steierl, G.; Liu, G.; Oepen H. P. & Kirschner, J. (2003). Growth structure and magnetism of Fe nanostructures on W(001), *Phys. Rev. B*, 68, 144416/1-9, ISSN: 0163-1829.

[44] Nordblad, P. & Svedlindh, P. (1998). Spin glasses and Random Fields, In: *Spin glasses and Random Fields*, Young A. P. (Ed.) 1-27 World Scientific, ISBN: 9810231830, Singapore.

[45] Jonason, K.; Vincent, E.; Hammann, J.; Bouchaud, J. P. & Nordbald, P. (1998). Memory and Chaos Effects in Spin Glasses, *Phys. Rev. Lett.* 81, 3243-3246, ISSN: 0031-9007.

[46] Luo, W., Nagel, S. R., Rosenbaum, T. F., Rosensweig, R. E. (1991). Dipole Interactions with random anisotropy in a frozen ferrofluid, *Phys. Rev. Lett.* 67, 2721-2724, ISSN: 0031-9007.

[47] Sasaki, M.; Jönsson, P. E.; Takayama, H. & Mamiya, H. (2005) Aging and memory effects in superparamganets and superspin glasses, *Phys. Rev. B*, 71, 104405/1-9, ISSN: 0163-1829.

[48] Sahoo, S.; Petracic, O.; Kleemann, W.; Nordblad, P.; Cardoso, S. & Freitas, P. P. (2003). *Aging and memory in a superspin glass, Phys. Rev. B* 67, 214422-1/5, ISSN: 0163-1829.

[49] Takayama, H. & Hukushima, K. (2004). Field-Shift Aging Protocol on 3D Ising Spin-Glass Model: Dynamical Crossover between the Spin-Glass and Paramagnetic States, *J. Phys. Soc. Jpn.* 73, 2077-2080, ISSN: 0031-9015.

[50] Yang, H. T.; Hasegawa, D.; Takahashi, M. & Ogawa, T. (2009). Achieving a noninteracting magnetic nanoparticle system through direct control of interparticle spacing, *Appl. Phys. Lett.* 94, 013103/1-3, ISSN: 0003-6951.

[51] Rivadulla, F.; López-Quintela, M. A. & Rivas, J. (2004). Origin of the Glassy magntic Behavior of the Phase Segregated State of the Perovskites, *Phys. Rev. Lett.* 93, 167206-1/4, ISSN: 0031-9007.

[52] Jonsson, T.; Mattsson, J.; Djurberg, C.; Khan, F. A.; Nordblad, P. & Svedlindh, P. (1995). Aging in a Magnetic Particle System, *Phys. Rev. Lett.* 75, 4138-4141, ISSN: 0031-9007.

[53] Hendriksen, S.; Linderoth, S. & Lindgard, P.-A. (1993) Finite-size modifications of the magnetic properties of clusters, *Phys. Rev. B* 48, 7259-7273, ISSN: 0163-1829.

[54] S. Iijima, (1991). Helical microtubules of graphitic carbon, *Nature,* 354, 56-58, ISSN: 0028-0836.

[55] Wildoer, J. W. G.; Venema, L. C.; Rinzier, A. G.; Smalley, R. E. & Dekker, C. (1998). Electronic structure of atomically resolved carbon nanotubes, *Nature,* 391, 59-62, ISSN: 0028-0836.

[56] Baughman, R. H.; Zakhidov, A. A. & de Heer, W. A. (2002). Carbon Nanotubes-the Route toward Applications, *Science*, 297, 787-792, ISSN:0036-8075.

[57] Tomanek, D.; Jorio, A.; Dresselhaus, M. S. & Dresselhaus, G. (2001). Introduction to the Important and Exciting Aspects of Carbon-Nanotube Science and Technology. In: *Carbon Nanotubes Synthesis, Structure, Properties, and Applications,* Dresselhaus, M. S. & Dresselhaus, G. (Eds), Springer, 1-12, ISBN: 3540728643, Berlin Heidelberg.

[58] Correa-Duarte, M. A.; Wagner, N.; Rojas-Chapana, J.; Morsczeck, C.; Thie, M. & Giersig, M. (2004). Fabrication and Biocompatibility of Carbon Nanotube-based 3D Networks as Scaffolds for Cell Seeding and Growth, *Nano Lett. 4,* 2233-2236, ISSN: 1530-6984.

[59] Rojas-Chapana, J. A.; Correa-Duarte, M. A.; Ren, Z.; Kempa, K. & Giersig, M. (2004). Enhanced Introduction of Gold Nanoparticles into Vital Acidothiobacillus ferrooxidans by Carbon Nanotube-based Microwave Electroporation, *Nano Lett. 4,* 985-988, ISSN: 1530-6984.

[60] White, C. T. & Todorov, T. N. (1998). Armchair carbon nanotubes as long ballistic conductors, *Nature,* 393, 240-243, ISSN: 0028-0836.

[61] Wildgoose, G. G.; Banks, C. E. & Compton, R. G., (2006). Metal Nanoparticles and Related Materials Supported on Carbon Nanotubes: Methods and Applications. *Small*, 2, 182-193, ISSN: 1613-6810.

[62] Correa-Duarte, M. A. & Liz-Marzan, L. M. (2006). Carbon Nanotubes as templates for one-dimensional nanoparticle assemblies, *J. Mater. Chem.* 16, 22-25, ISSN: 1364-5501.

[63] Kamaras, K.; Itkis, M. E.; Hu, H.; Zhao, B. & Haddon, R. C. (2003). Covalent Bond Formation to a Carbon Nanotube Metal, *Science*, 301, 1501-1501.

[64] Hirsch, A. (2002). Functionalization of Single-Walled Carbon Nanotubes, *Angew. Chem.* 41, 1853-1859, ISSN: 0044-8249.

[65] Zhang, Z.; Tang, Z; Kotov, N. A. & Glotzer, S. C. (2007). Simulations and Analysis of Self-Assembly of CdTe Nanoparticles into Wires and Sheets, *Nano Lett.*, 7,1670-1675, ISSN: 1530-6984.

[66] Tang, Z.; Wang, Y.; Shanbhang, S.; Giersig, M. & Kotov, N. A. (2006). Spontaneous Transformation of CdTe Nanoparticles into Angled Te Nanocrystals: From Particles and Rods to Checkmarks, X-Marks, and Other unusual Shapes, *J. Am. Chem. Soc.*, 128, 6730-6736, ISSN: 0002-7863.

[67] Salgueirino-Maceira, V.; Correa-Duarte, M. A.; Hucht, A. & Farle, M. (2006). One-dimensional assemblies of silica-coated cobalt nanoparticles. Magnetic pearl necklaces. *J. Magn. Magn. Mater.* 303, 163-166, ISSN: 0304-8853.

[68] Scanna, S. & Philipse, A. P. (2006). Preparation and Properties of Monodisperse Latex Spheres with Controlled Magnetic Moment for Field-induced Colloidal Crystallization and (bipolar) Chain Formation, *Langmuir*, 22, 10209-10216, ISSN: 0743-7463.

[69] Klokkenburg, M.; Dullens, R. P. A.; Kegel, W. K.; Erné, B. H. & Philipse, A. (2006). Quantitative Real-Space Analysis of Self-Assembled Structures of Magnetic Dipolar Colloids, *Phys. Rev. Lett.*, 96, 037203/1-4, ISSN: 0031-9007.

[70] Correa-Duarte, M. A.; Grzelczak, M.; Salgueirino-Maceira, V.; Giersig, M.; Liz-Marzan, L. M.; Farle, M.; Sierazdki K. & Diaz, R. (2005). Alignment of Carbon Nanotubes under Low Magnetic Fields through Attachment of Magnetic Nanoparticles, *J. Phys. Chem. B*, 109, 19060-19063, ISSN: 1520-6106.

[71] Grzelczak, M.; Correa-Duarte, M. A.; Salgueirino-Maceira, V.; Rodrıguez-Gonzalez, B.; Rivas, J. & Liz-Marzan, L. M. (2007). Pt-Catalyzed Formation of Ni Nanoshells on Carbon Nanotubes, *Angew. Chem. Int. Ed.*, 46, 7026-7030, ISSN: 0044-8249.

[72] Correa-Duarte, M. A.; Sobal, N.; Liz-Marzan, L. M. & Giersig, M. (2004). Linear Assemblies of Silica-coated gold nanoparticles using carbon nanotubes, *Adv. Mater.*, 16, 2179-2184, ISSN: 0935-9648.

[73] Ostrander, J. W.; Mamedov, A. A. & Kotov, N. A. (2001). Two modes of Linear Layer-by-layer Growth of Nanoparticle-Polyelectrolyte Multilayers and Different Interactions in the Layer-by-layer Deposition, *J. Am. Chem. Soc.*, 123, 1101-1110, ISSN: 0002-7863.

[74] Spasova, M.; Salgueiriño-Maceira, V.; Schlachter, A.; Hilgendorff, M.; Giersig M.; Liz-Marzan, L. M. & Farle, M. (2005). Magnetic and Optical tunable microspheres with a magnetit/gold nanoparticle shell, *J. Mater. Chem.*, 15, 2095-2098, ISSN: 1364-5501.

[75] R. Massart (1981). Preparation of aqueous magnetic liquids in alkaline and acidic media, *IEEE Trans. Magn.* 1981, MAG-17, 1247-1248, ISSN:0018-9464.

[76] K. Butter, P. H. Bomans, P. M. Frederik, G. J. Vroege, A. P. Philipse, A. P. (2003). Direct observation of dipolar chains in ferrofluids in zero field using cryogenic electron microscopy, *J. Phys.: Condens. Matter* 2003, 15, S1451-S1470, ISSN: 0053-8984.

[77] Ahrenstorf, K.; Albrecht, O.; Heller, H.; Kornowski, A.; GSrlitz, D. & Weller, H. (2007). Colloidal Synthesis of Ni_xPt_{1-x} Nanoparticles with Tuneable Composition and Size, *Small*, 3, 271-274, ISSN: 1613-6810.

[78] Bentley A. K.; Ellis, A. B.; Lisensky, G. C. & Crone, W. C. (2005). Suspensions of nickel nanowires as magneto-optical switches, *Nanotechnology*, 16, 2193-2197, ISSN: 0957-4484.

[79] Reich, D. H.; Tanase, M.; Hultgren, A.; Bauer, L. A.; Chen, C. S. & Meyer, G. J. (2003). Biological applications of multifunctional magnetic nanowires. *J. Appl. Phys.*, 93, 7275-7280, ISSN: 0021-8979.

[80] Fraune, M.; Rudiger, U.; Guntherodt, G.; Cardoso, S. & Freitas, P. (2000). Size dependence of the Exchange bias field in NiO/Ni nanostructures, *Appl. Phys. Lett.*, 77, 3815-3817, ISSN: 0003-6951.

[81] van Leeuwen, D. A.; Van Ruitenbeek, J. M.; de Jongh, L. J.; Ceriotti, A.; Pacchioni, G.; Harberlen, O. D.& Rosch, N. (1994). Quenching of Magnetic Moments by Ligand-Metal Interactions in Nanosized Magnetic Metal Clusters, *Phys. Rev. Lett.*, 73, 1432-1435, ISSN: 0031-9007.

[82] Roy, A.; Srinivas, V.; Ram, S.; de Toro, J. A. & Mizutani, U. (2005). Structure and magnetic Properties of oxygen-stabilized tetragonal Ni nanoparticles prepared by borohydride reduction method, *Phys. Rev. B*, *71*, 184443/1-10, ISSN: 0163-1829.

[83] F. Bottomley (1970). The reactions of Hydrazine with Transition-metal Complexes. *Q. Rev. Chem. Soc.*, 24, 617-638, ISSN: 0009-2681.
[84] Wu, S.-H. & Chen, D.-H. (2004). Synthesis and Stabilization of Ni Nanoparticles in a Pure Aqueous CTAB Solution, *Chem. Lett.*, 33, 406-410, ISSN: 0366-7022.
[85] Morales, M. P.; Veintemillas-Verdaguer, S.; Montero, M. I. & Serna, C. J. (1999). Surface and Internal Spin Canting in γ-Fe_2O_3 nanoparticles, *Chem. Mater.*, 11, 3058-3064 , ISSN: 0897-4756.
[86] Makhlouf, S. A.; Parker, F. T.; Spada, F. E. & Berkowitz, A. E. (1997). Magnetic Anomalies in NiO nanoparticles, *J. Appl. Phys.*, 81, 5561-5564, ISSN: 0021-8979.
[87] Tracy, J. B.; Weiss, D. N.; Dinega, D. P. & Bawendi, M. G. (2005). Exchange Biasing and Magnetic Properties of Partially and Fully Oxidized Colloidal Cobalt Nanoparticles, *Phys. Rev. B*, 72, 064404/1-8, ISSN: 0163-1829.
[88] Kodama, R. H.; Makhlouf, S. A. & Berkowitz, A. E. (1997). Finite Size Effects in Antiferromagnetic NiO Nanoparticles, *Phys. Rev. Lett.*, 79, 1393-1396, ISSN: 0031-9007.
[89] Spasova, M.; Wiedwald, U.; Farle, M.; Radetic, T.; Dahmen, U.; Hilgendorff, M. & Giersig, M. (2004). Temperature dependence of Exchange Anisotropy in Monodisperse Cobalt Nanoparticles with a Cobalt Oxide Shell, *J. Magn. Magn. Mater.*, 272–276, 1508-1509, ISSN: 0304-8853.
[90] Salgueirino-Maceira, V.; Spasova, M. & Farle, M. (2005). Water-Stable, Magnetic Silica-Cobalt-Silica Multishell Submicron Spheres, *Adv. Funct. Mater.*, 15, 1036-1040, ISSN: 1616-301X.
[91] Tomou, A.; Gournis, D.; Panagiotopoulos, I.; Huang, Y.; Hadjipanayis, G. C. & Kooi, B. J. (2006). Weak Ferromagnetism and exchange biasing in cobalt oxide nanoparticle systems, *J. Appl. Phys.*, 99, 123915/1-5, ISSN: 0021-8979
[92] Antoniak, C.; Lindner, J.; Salgueiriño-Maceira, V. & Farle, M. (2006). Multifrenquency magnetic resonance and blocking behavior of FexPt1-x nanoparticles *Phys. Status Solidi (a)*, 203, 2968-2973, ISSN: 1862-6319.
[93] Zhang, P.; Zuo, F.; Urban III, F. K.; Khabari, A.; Griffiths, P. & Hosseini-Tehrani, A. (2001). Irreversible Magnetization in Nickel Nanoparticles, *J. Magn. Magn. Mater.*, 225, 337-345, ISSN: 0304-8853.
[94] Brown, Jr. W. F. (1963). Thermal Fluctuations of a Single-Domain Particle, *Phys Rev, 130,* 1677-1686, ISSN: 0031-9007.
[95] Dormann, J. L.; Fiorani, D.; Cherkaoui, R.; Tronc, E.; Lucari, F.; D'Orazio, F.; Spinu, L.; Nogués, M.; Kachkachi, H. & Jolivet, J. P., (1999). From pure superparamagnetism to glass collective state in γ-Fe_2O_3 nanoparticle assemblies, *J. Magn. Magn. Mater., 203,* 23-27, ISSN: 0304-8853.
[96] Djurberg, C.; Svedlindh, P.; Nordblad, P.; Hansen, M. F.; Bodker, F. & Morup, S. (1997). Dynamics of an Interacting Particle System: Evidence of Critical Slowing Down, *Phys. Rev. Lett., 79,* 5154-5157, ISSN: 0031-9007.
[97] Johsson, T.; Svedlindh, P. & Hansen, M. F. (1998). Static Scaling on an Interacting Magnetic Nanoparticle System, *Phys. Rev. Lett.*, 81, 3976-3979, ISSN: 0031-9007.
[98] Kleeman, W. Petracic, O. Binek, Ch. Kakazei, G. N. Pogorelov, Y. G. Sousa, J. B. Cardoso, S. & Freitas, P. P. (2001). Interacting ferromagnetic nanoparticles in

discontinouos Co80Fe30/Al2O3 multilayers: from superspin glass to reentrant superferromagnetism, *Phys. Rev. B*, *63*, 134423/1-5, ISSN: 0163-1829.

[99] García-Otero, J.; Porto, M.; Rivas, J. & Bunde, A. (2000). Influence of Dipolar Interaction on Magnetic Properties of Ultrafine Ferromagnetic Particles, *Phys. Rev. Lett.*, *84*, 167-170, ISSN: 0031-9007.

[100] Ulrich, M. García-Otero, J. Rivas, J. Bunde, A. (2003). Slow relaxation in ferromagnetic nanoparticles: Indication of spin-glass behavior, *Phys. Rev. B*, 67, 024416/1-4, ISSN: 0163-1829.

[101] Andersson, J. O.; Djurberg, C.; Jonsson, T., Svedlindh, P. & Nordblad, P. (1997). Monte Carlo studies of the dynamics of an interacting monodispersive magnetic-particle system, *Phys. Rev. B*, *56*, 13983-13988. ISSN: 0163-1829.

6

Synthesis of Germanium/Multi-walled Carbon Nanotube Core-Sheath Structures via Chemical Vapor Deposition

Dali Qian[1], Mark Crocker[1], A. Pandurangan[1], Cedric Morin[1] and Rodney Andrews[1,2]
[1]*Center for Applied Energy Research, University of Kentucky, Lexington, KY 40511,*
[2]*Department of Chemical and Materials Engineering, University of Kentucky, Lexington, KY 40506, USA*

1. Introduction

One-dimensional (1D) nanostructures such as nanotubes, nanowires, and nanobelts have been the focus of much recent attention, owing to the novel electronic and optical properties intrinsically associated with their low dimensionality and the quantum confinement effect. Such 1D nanostructures have potential applications in nanoelectronics, advanced composites, field emission devices, sensors, probes, optics and optoelectronics (Baughman et al., 2002; Agarwal & Lieber, 2006). Silicon nanowires have been preferentially studied since Si is of great technological importance in microelectronics (Morales & Lieber, 1998). Silicon nanowires exhibit significant differences in physical (Cui & Lieber 2001; Ma, et al., 2003; Sun et al., 2001) and chemical properties (Sun et al., 2003; Chen et al., 2005) from bulk Si, which have been exploited to fabricate nanoelectronic devices such as logic circuits (Huang et al., 2001), field effect transistors (Lieber, 2003), and sensors (Cui et al., 2001). Compared to Si, Ge nanostructures are of particular interest, since the exciton Bohr radius of bulk Ge (24.3 nm) (Maeda et al., 1991) is larger than that of Si (4.9 nm) (Cullis et al., 1997), resulting in more prominent quantum confinement effects. Ge also offers the advantage of lower processing temperatures with easier integration into conventional devices. Furthermore, Ge has much higher electron and hole mobility than Si (Sze, 1981), which is especially required when electronic devices are scaled down to the sub-100 nm regime.

Several growth methods have been developed for the synthesis of Ge nanowires, including laser ablation (Morales & Lieber, 1998; Zhang et al., 2000), thermal evaporation (Gu et al., 2001; Nguyen et al., 2005; Sun et al., 2006; Das et al., 2007; Sutter et al., 2008), supercritical-fluid synthesis (Ryan et al., 2003; Polyakov et al., 2006; Ziegler et al., 2004; Erts et al., 2006), liquid-state synthesis (Heath & LeGoues, 1993; Song et al., 2009), molecular beam epitaxy (Omi & Ogino, 1997), and chemical vapor deposition (CVD) (Kodambaka et al., 2007; Ryan et al., 2003). CVD has been the most widely employed of these synthesis methods, with the aim of synthesizing Ge nanowires in a controllable way via the selection of suitable Ge precursors and catalysts. To date a number of Ge-containing precursors have been used to grow Ge nanowires, including GeH_4 (Kamins et al., 2004; Wang & Dai, 2002; Jin et al., 2006; Kang et al., 2008; Wang & Dai, 2006), Ge_2H_6 (Kodambaka et al., 2007), $Ge(C_5H_5)_2$ (Mathur et al., 2004; Mathur & Barth, 2008), $C_{12}H_{12}Ge$ (diphenyl germane) (Ryan et al., 2003; Hanrath & Korgel, 2002; Polyakov et al., 2006; Ziegler et al., 2004; Erts et al., 2006), Ge-GeO_2 (Yin et al., 2007), and Ge-GeI_4 (Wu & Yang, 2000). In most CVD methods, as well as methods employing thermal evaporation of Ge powders (Gu et al., 2001; Nguyen et al., 2005; Sun et al., 2006; Das et al., 2007; Sutter et al., 2008), Au nanoparticles have been typically selected as the catalyst for Ge nanowire growth via the vapour-liquid-solid (VLS) mechanism (Wagner & Ellis, 1964) due to the low eutectic temperature of the Ge–Au alloy (360 °C). Other low melting point metals/alloys that have been employed as the catalyst include Al (Wang et al., 2006), Cu (Yao & Fan, 2007), Cu-Ni (Kang et al., 2008) and Fe(Ge) (Mathur, et al., 2004). Recently, the vapor-solid-solid (VSS) mechanism (Kamins et al. 2001; Bootsma & Gassen, 1971) has also been implicated, based on experimental observations that the catalyst state can be either liquid or solid depending on the thermal history, Ge_2H_6 pressure, and temperature. The supersaturation of Ge in the alloy catalyst caused by the growth process appears to be essential in stabilizing the liquid below the eutectic temperature; therefore, Ge nanowires can grow by both the VLS and VSS mechanisms, although at different rates (Kodambaka et al., 2007).

Although highly ordered metallic and semiconductor nanowires can be used in electronic devices, in many cases these nanowires are sensitive to oxygen and water vapor, resulting in degradation of the nanodevice performance (Zhang et al., 2007). Consequently, in some cases it is necessary to avoid any contact of these materials with the external environment. This is particularly important for Ge, which forms oxide coatings possessing unfavorable electronic properties. Untreated Ge nanowires are reported to oxidize upon exposure to ambient conditions to form a 1-2 nm thick GeO_{2-x} shell, which continues to grow to reach a self-limited thickness of ~4 nm over the course of ~24 h (Hanrath & Korgel, 2004). A promising solution to this problem is the coating of metallic or semiconductor nanowires with organic monolayer coatings such as those based on alkanethiols (Wang et al., 2005) or alkyl groups (Hanrath & Korgel, 2004), or with a protective shell of carbon nanotubes (CNTs) (Sutter & Sutter, 2006; Dai et al., 1996; Loiseau et al., 2000; Wu & Yang, 2000) or amorphous carbon (Huang et al., 2005).

Carbon-encapsulated Ge nanowires have previously been prepared using a chemical vapor deposition (CVD) technique (Sutter & Sutter, 2006). A supported Au catalyst was employed to catalyze the growth of Ge nanowires, which were then dispersed on an amorphous carbon film and annealed, resulting in the encapsulation of the nanowires by well defined, curved graphene sheets. The presence of gold nanoparticles (residual catalyst) on the surface of the nanowires was indicated as being instrumental in initiating graphene sheet formation. Other methods which have been employed for the preparation of Ge/C core-sheath nanostructures include the arc discharge synthesis of CNTs in the presence of Ge metal (Dai et al., 1996; Loiseau et al., 2000), and the deposition of graphitic coatings on pre-formed Ge nanowires by their treatment with organic vapors at 700-900 °C (Wu & Yang, 2000).

For practical circuit device fabrication, e.g., for photoresistors (Polyakov et al., 2006), photoluminescence devices (Ryan et al., 2003) and gas sensors (Rajaputra et al., 2008), ordered nanowire arrays are required. To this end, Holmes's research group has employed ordered mesoporous aluminosilicate thin films (Ryan et al., 2003; Ziegler et al., 2004) and AAO membranes (Polyakov et al., 2006; Erts et al., 2006) as templates to grow Ge nanowire arrays by the degradation of diphenylgermane ($C_{12}H_{12}Ge$) in supercritical CO_2. The experimental protocol involved placing the ordered porous films or AAO membranes inside a 25 mL high-pressure reaction cell with diphenylgermane placed inside an open top quartz glass boat adjacent to the membranes under an inert atmosphere. A typical synthesis at 600 °C under 37.5 MPa for 30 min resulted in conductive Ge nanowires which filled the template pores (Ziegler et al., 2004).

To meet the demands for practical device fabrication, we have studied the synthesis of Ge nanowires protected through encapsulation in multi-walled carbon nanotubes (MWCNTs) using a simple one-step synthesis method (Pandurangan et al., 2009). In this approach, a CVD method employing neat phenyltrimethylgermane (PTMG, boiling point of b.p. 183 °C) was used, the PTMG functioning as both the Ge and C source. Although transition metal catalysts have typically been employed for the synthesis of CNTs, it has recently been shown that semiconductor nanoparticles of SiC, Ge and Si can act as templates for the production of single-walled and double-walled carbon nanotubes during CVD synthesis with ethanol (Takagi et al., 2007). In this study, pure PTMG acts as a precursor for the formation of Ge nanoparticles which act as templates for the formation of MWCNT-encapsulated Ge nanowires. The effect of employing a floating Fe catalyst, using ferrocene as a precursor, has also been investigated. To tailor the derived Ge nanowire diameter and orientation, the use of anodized aluminum oxide (AAO) templates has also been evaluated.

2. Synthesis of Ge nanowires encapsulated within multiwalled carbon nanotubes

Previous work in our laboratory (Andrews et al., 1999) and others (Öncel & Yürüm, 2006) has established chemical vapor deposition (CVD) as an efficient means of preparing multiwalled carbon nanotubes (MWCNTs). In addition to the relatively simple equipment required, the use of CVD enables the process to be readily scaled, such that MWCNTs can be produced in quantities ranging from milligrams to kilograms. Consequently, to prepare MWCNT-encapsulated Ge nanowires, we elected to employ the same basic CVD technique. However, rather than using a floating catalyst such as ferrocene (required to catalyze MWCNT growth) dissolved in a carbon precursor such as xylene, we elected to employ neat PTMG as both the Ge and C source. In effect, PTMG acts as a precursor for the formation of Ge nanoparticles which, in turn, function as templates for the formation of MWCNT-encapsulated Ge nanowires (Pandurangan et al., 2009).

The two-zone furnace employed for the CVD preparation was similar to that described previously for the growth of pristine MWCNTs, and consisted of a quartz tube reactor with a flat quartz slide inserted at the reaction zone for additional deposition surface (Andrews et al., 1999). Neat PTMG was injected into the preheat zone of the reactor at a rate of 1 ml/h. After volatilization in the preheat zone (~250 °C), the PTMG was carried into the reaction zone of the furnace, maintained at 800 °C, by a 10% H_2/Ar sweep gas. The entire system was maintained at a slight positive pressure (~0.75 kPa) versus atmospheric pressure. Typical CVD runs were of 2 h duration, after which the material deposited on the quartz slide was collected for analysis.

Representative scanning electron micrographs of the material prepared at 800 °C are shown in Fig. 1. These show the sample to consist mainly of nanowires, possessing diameters of less than 500 nm (see Fig. 1b). The nanowires appear smooth, with little evidence of metal cluster formation. Furthermore, from Fig. 1 and other micrographs, it is apparent that the as-synthesized nanowires are largely free from amorphous carbon and possess large aspect ratios, the typical length being in the order of 10 μm. In contrast, at 700 °C decomposition of the precursor appears to be incomplete; consequently, only a few short wires are formed. Increase of the temperature above 800 °C results in the formation of amorphous carbon, with Ge nanoparticle formation being favored over the formation of nanowires.

Fig. 1. (a) Typical SEM image of Ge-filled MWCNTs synthesized at 800 °C, and (b) higher magnification image showing MWCNT tips.

As revealed by TEM, the product obtained at 800 °C (hereafter denoted as Ge@MWCNT/800) consists of crystalline Ge nanowires, encapsulated in a thin MWCNT sheath. Typical TEM images at low magnification (see Fig. 2a) reveal a pin-like morphology. TEM images at higher magnification (Fig. 2b-d) show that the Ge nanowires consist of well crystallized Ge cores which are completely encapsulated by the sheath-like MWCNTs possessing a thickness of 5-10 nm. Ge incorporation within the graphene sheets of the MWCNTs is not observed according to EDS analysis under STEM mode. Quantitative EDS analysis of the tube cores indicates a composition of 95% Ge and 5% C (atom %). Given the existence of 5-10 nm thick of graphene (C) layers covering the Ge core at the analytical probe position, it can be concluded that the core consists of pure Ge, as supported by HRTEM observations that show clear fringes corresponding to single crystal Ge (see Fig. 2c, 2d). Furthermore, EDS analysis confirmed that the near spherical heads of the nanowires consist of over 97 atom % Ge (the balance being C).

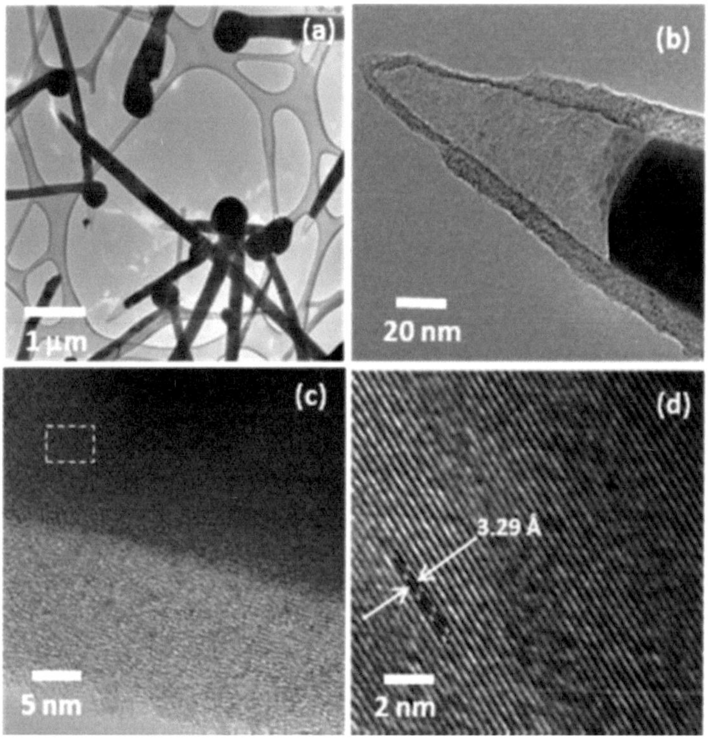

Fig. 2. (a) typical TEM images of Ge@MWCNT/800 at low magnification; (b) and (c) high magnification image showing Ge nanowire and MWCNT sheath; (d) higher magnification view of boxed area in (c) showing single crystal Ge.

Powder X-ray diffraction data confirm the crystalline nature of the nanowires (Fig. 3a). The positions of the observed diffraction peaks (2θ = 27.43, 45.45, 53.83, 66.13, 73.11, 83.79°) are in good agreement with literature values (Miikin, 1961) for the crystalline face centered cubic phase of Ge with lattice parameter a = 5.660 Å. Furthermore, the neighbor interlayer spacing indexed in HRTEM images (see Fig. 2d) is approximately 3.29Å, which is very close to the calculated distance between neighboring (111) planes of 3.27 Å. Thus, the as-produced Ge@MWCNT/800 consists primarily of FCC Ge.

As shown in Fig. 3b, the X-ray photoelectron spectrum of Ge@MWCNT/800 contains two Ge 3d photoelectron lines. The signal corresponding to a binding energy of 29.7 eV can be assigned to Ge metal, albeit that it is slightly shifted (~0.3 eV) compared to 3d binding energies typically observed for bulk Ge. The other signal at 32.5 eV is characteristic of GeO_2. Deconvolution of the signals and integration indicates a Ge(0):Ge(IV) atomic ratio of 78:22, suggesting that the Ge nanowires are largely, although not completely, protected from oxidation. Additionally, it is likely that amorphous, partially oxidized Ge is present, which contributes to the GeO_2 signal. While SEM analysis of Ge@MWCNT/800 suggests that nanowires constitute at least 80% of the sample, the presence of irregularly shaped Ge particles can be observed in some SEM images (as shown in Fig. 5c below).

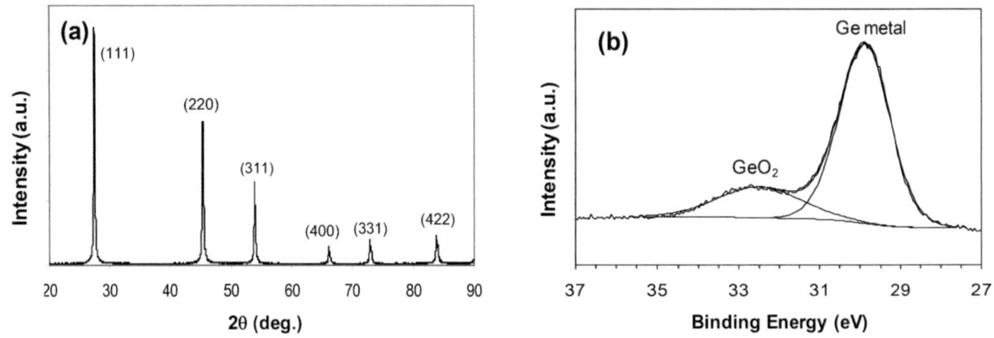

Fig. 3. Characterization of Ge@MWCNT/800: (a) powder XRD pattern and (b) XPS spectrum.

The results of thermogravimetric analysis performed on Ge@MWCNT/800 in air are shown in Fig. 4a. Noteworthy is the observation that a pronounced increase in sample weight commences at around 550 °C, consistent with the oxidation of Ge to GeO_2. This temperature corresponds to the initiation of MWCNT combustion (Bom et al., 2002), suggesting that Ge oxidation proceeds as the nanowires' outer carbon shells are oxidized and the Ge cores are exposed. Allowing for the initial carbon content of 3 wt.% in the sample (by elemental analysis), and assuming that all of the Ge present is oxidized to GeO_2, the observed gain in sample weight (37%) corresponds to an initial molar Ge:GeO_2 ratio of 86:14 (in reasonable agreement with the value determined by XPS). As shown in Fig. 4b, differential scanning calorimetry (DSC) performed under N_2 gave a melting point of 934 °C for the Ge nanowires.

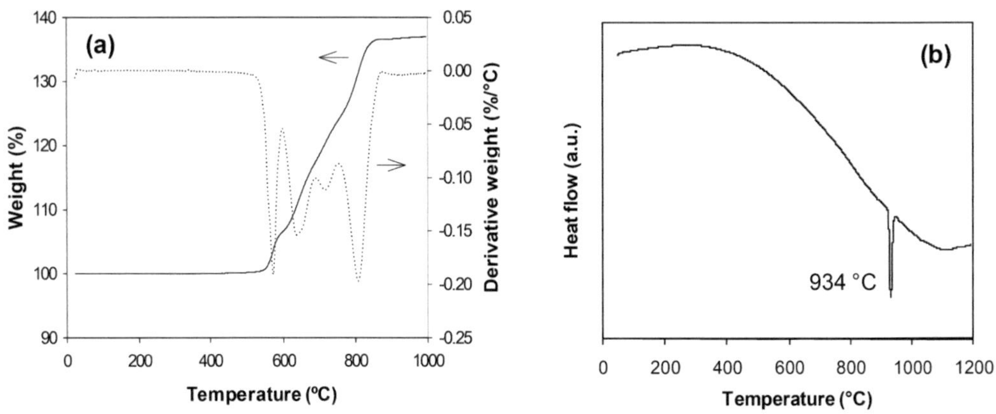

Fig. 4. Characterization of Ge@MWCNT/800: (a) thermogravimetric analysis in air and (b) differential scanning calorimetry in N_2 (10 °C/min heating rate used in both cases).

In the absence of an external carbon source, the carbon constituting the MWCNTs derives from the phenyltrimethylgermane precursor. In principle, an atomic Ge:C ratio of 1:9 is feasible in the product if decomposition of the precursor proceeds according to: $C_6H_5Ge(CH_3)_3 \rightarrow Ge + 9C + 7H_2$. In fact, elemental analysis of the product typically showed a Ge:C ratio of *ca.* 5:1, indicating that most of the carbon is lost in the form of volatile species (methane and ethene).

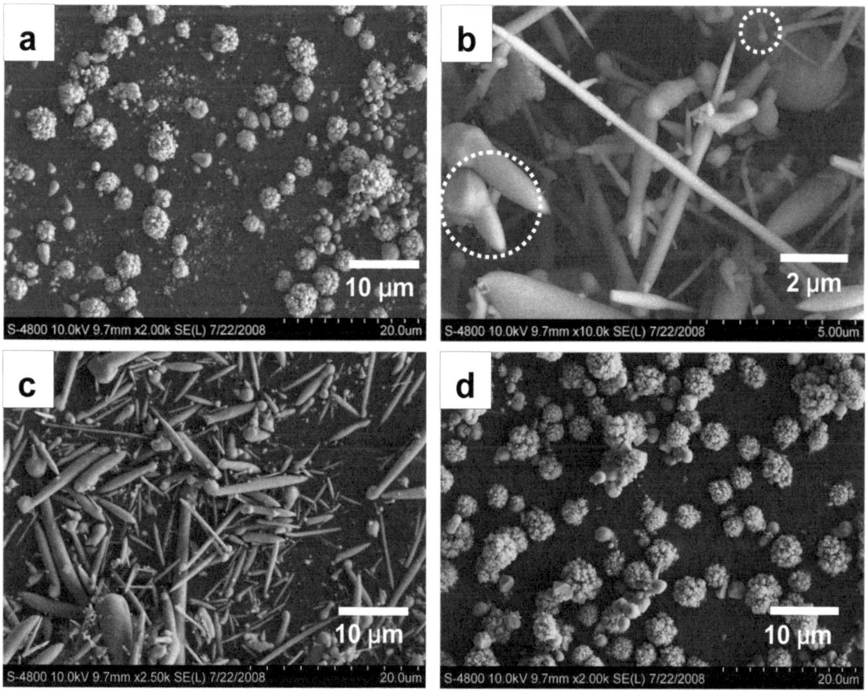

Fig. 5. (a)-(d) SEM images of CVD product obtained using neat PTMG: (a) 3″, (b) 4″, (c) 5″ and (d) 7″ from inlet; the circled regions in image b show Ge particles at an early stage of nanowire formation.

Insight into the factors controlling nanowire formation is provided by the results of CVD experiments in which the deposited solid was sampled from quartz slides placed at fixed distances from the inlet, i.e., along the length of the tube reactor. The reaction zone was again held at 800 °C. SEM images of these materials are shown in Figs. 5 and 6. Images 5a-5d correspond to a CVD run performed using neat PTMG as the precursor, while images 6a-6d correspond to the product obtained using PTMG diluted to 50 wt% with xylene. For both runs, light deposits of mainly amorphous Ge particles were obtained closest to the inlet. For the run using neat PTMG, nanowires could be found in a localized region of the reactor, between 4″ and 5″ from the inlet. Further away from the inlet, Ge particles were obtained (Fig. 5d). Evidently, while carbon is available beginning at 4″ from the inlet from cracking of the PTMG, the partial pressure of carbon in the atmosphere is insufficient to produce Ge-filled nanotubes along the entire length of the reactor. In contrast, when the PTMG/xylene mixture was used, not only was the total yield of Ge-filled nanotubes increased (estimated at >50% based on Ge), Ge-filled nanotubes were observed in the entire region 4″-7″ from the inlet. Furthermore, the dimensions of the filled nanotubes were rather uniform, with diameters in the 200-300 nm range and lengths of 6-10 µm. These results suggest that the form of the germanium deposit obtained is regulated by the partial pressure of carbon in the atmosphere.

For the runs performed with both neat PTMG and PTMG/xylene, structures could be observed which appear to correspond to the early stages of nanotube/nanowire growth. Examples of this are shown in the circled regions of Fig. 5b. CNT growth evidently occurs

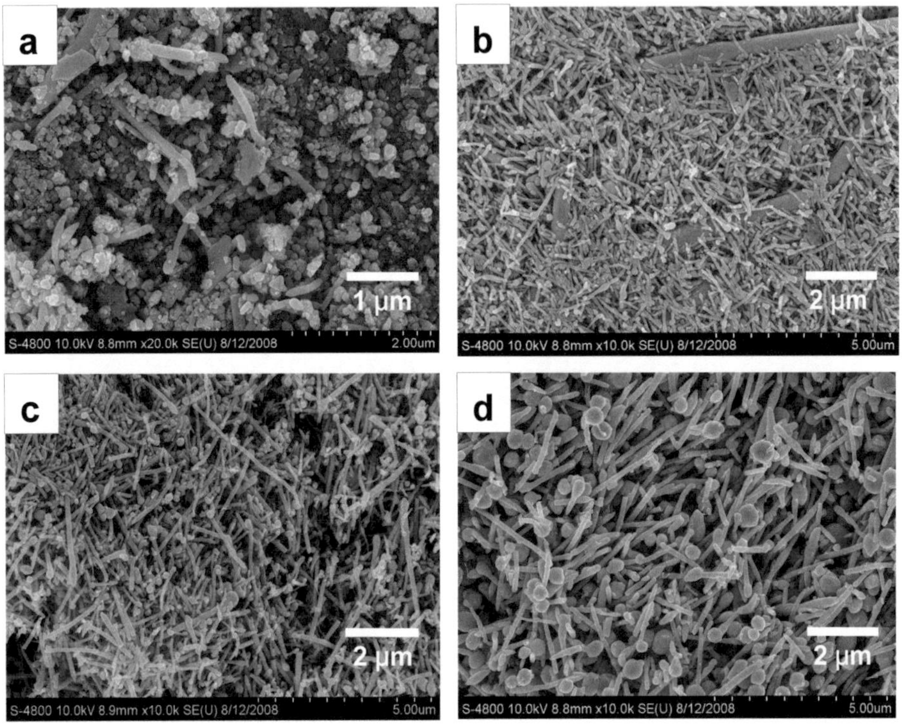

Fig. 6. SEM images of CVD product obtained using 50 wt% PTMG in xylene: (a) 3", (b) 4", (c) 5" and (d) 7" from inlet.

from the Ge particle (root growth mechanism) and appears to carry Ge as a nanowire away from the initial particle. Presumably, this synthesis mode is the result of wetting at the interface of the Ge and the growing carbon nanotube. This is illustrated in Fig. 2c, which shows the high degree of association between the carbon sheath and the Ge nanowire. It should be noted that the synthesis temperature used (800 °C) was more than 100 °C lower than the melting point of Ge. As shown in Fig. 4b, a m.p. of 934 °C was measured for the Ge nanowires. While this is only slightly lower than the melting point for bulk Ge (938 °C), it has been reported that smaller nanoscale metal structures can exhibit much greater melting point depression (Buffat & Borel, 1976; Couchman & Jesser, 1977). Hence, melting point depression can be expected during the initial stages of nanowire growth. This effect, together with possible capillary action (Dujardin et al., 1994), can explain the apparently low Ge viscosity and the tendency for CNT filling by Ge.

Finally, it is worth noting that initial attempts to prepare MWCNT-encapsulated Si, Sn or Bi nanowires using this approach have proved unsuccessful. In the case of silicon, this can be attributed to its high melting point (1410 °C), such that any Si nanoparticles formed will be solid at the synthesis temperature. In the case of Sn and Bi, the melting points are much lower (232 and 271 °C, respectively). In this case, the observation of metallic deposits in the pre-heat zone is consistent with the relatively low decomposition temperatures of the organometallic precursors used ($Sn(C_6H_5)(CH_3)_3$ and $Bi(C_6H_5)_3$), i.e., deposition occurs in a region where the temperature is too low for carbon nanotube formation to occur. Consequently the need for precursors which are less thermally labile is indicated.

3. Synthesis of Ge-Fe nanowires encapsulated by multiwalled carbon nanotubes

Recent work has shown that the properties of nanotubes formed by assembly of metal encapsulated silicon and germanium clusters can be tailored by a suitable choice of metal atoms (Pokropivnyi, 2001). Fe-doped ferromagnetic Si nanotubes and Mn-doped Si nanotubes with nearly degenerate ferromagnetic and antiferromagnetic states have been obtained with high local magnetic moments (Singh et al., 2004), making them attractive for nanomagnetic and spintronic devices. Although Mn doping in bulk Ge has found to give rise to weak ferromagnetic behavior (Park et al., 2002; Cho et al., 2002), in low dimensional systems such as nanowires the magnetic behavior could be very rich depending upon the nanostructure and doping of metals. From this it follows that the synthesis of MWCNT-encapsulated Ge-Fe nanowires is of interest.
To prepare Ge-Fe nanowires, the same one-step CVD process was used as successfully employed for the growth of encapsulated Ge nanowires, employing PTMG, pyridine and ferrocene as precursor materials. Ferrocene (b.p. 249 °C) has been shown to be an excellent precursor for producing metallic iron catalyst particles which can seed carbon nanotube growth. Pyridine was used as the solvent and additionally functions as the primary carbon source. In a typical run 1.15g of ferrocene was dissolved in 10 g pyridine, to which 1.0 g of

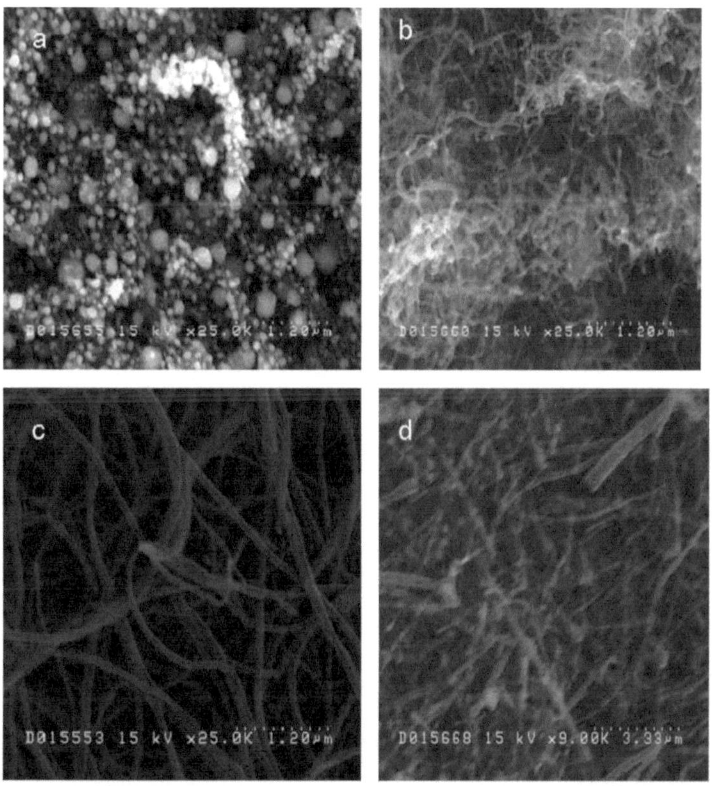

Fig. 7. Typical SEM images of Ge-Fe filled MWCNTs produced by CVD at (a) 700 °C, (b) 800 °C, (c) 900 °C, (d) 1000 °C showing nanotube formation. The micrometer-sized particles in (a) are Ge-Fe clusters as confirmed by EDS mapping.

PTMG was added. The homogeneous solution was injected into the preheat zone of the reactor (~250 °C) at an addition rate of 1 mL/h and the resulting vapor was swept into the reaction zone of the furnace by a 10% H_2/Ar carrier gas.

After a typical CVD run of 2 h duration, the black deposit obtained was recovered and analyzed by SEM. Figure 7a-7d shows SEM images of the product formed at temperatures varying between 700 and 1000 °C. From Figure 7b and other micrographs it is apparent that the nanotube bundles obtained at 800 °C are free from other carbonaceous materials and possess large aspect ratios, typical nanotube length being in the order of 10 μm. In contrast, at 700 °C only a few short MWCNTs are formed (Figure 7a). The majority of the carbon present is amorphous in form, while the metals are observed as particles which are agglomerated on the surface of the carbon. At 800 °C, carbon nanotube production improves significantly, while there is an absence of metal particles. As for Ge@MWCNTS, these findings suggest that the growth kinetics of the MWCNTs at these different temperatures significantly impact the form of the deposited metals. Further increase of the temperature results in decreased MWCNT purity and increased amounts of pyrolytic forms of carbon.

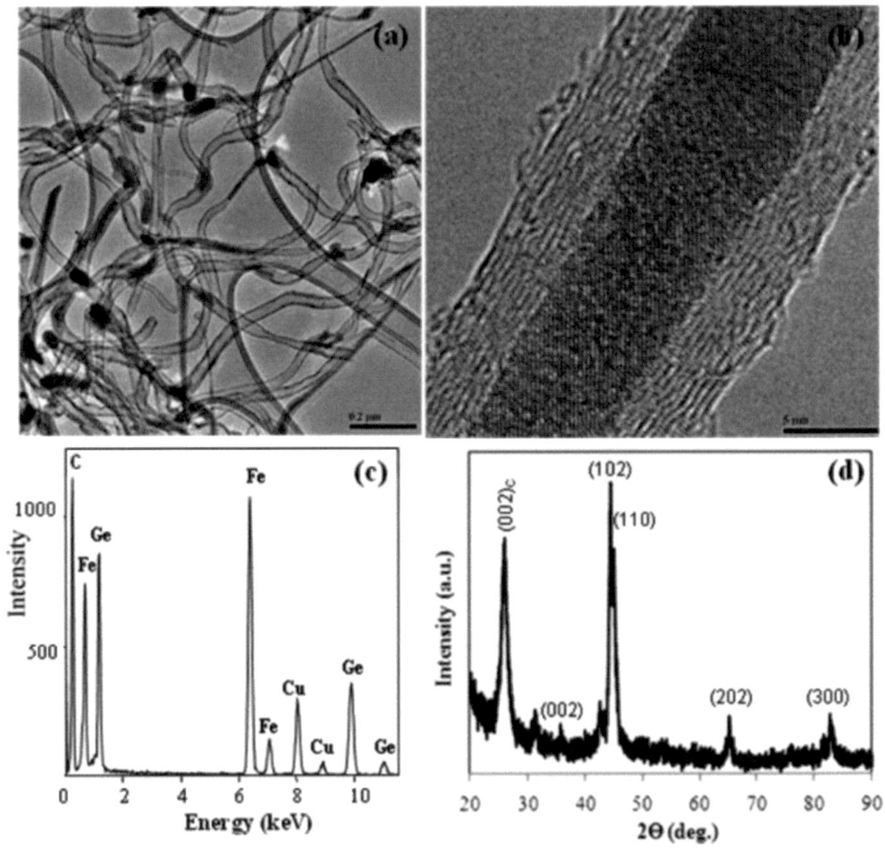

Fig. 8. (a) Low magnification TEM image of Ge-Fe nanowires encapsulated by MWCNTs (produced by CVD at 800 °C); (b) HRTEM image of isolated Ge-Fe filled MWCNT; (c) EDS spectrum of the Ge-Fe core in (b); (d) XRD of bulk Ge-Fe filled MWCNT sample indexed as $Fe_{1.67}Ge$; the additional diffraction peak at 26.0° corresponds to the carbon (002) reflection.

As revealed by TEM, the CVD-derived material prepared at 800 °C (Figure 8) consists of pure MWCNTs possessing cores composed of a Ge-Fe alloy. As shown in Figures 8a and 8b, the Ge-Fe nanowires consist of a uniform single crystal core with the exception of an enlarged head and tail portion, and are completely encapsulated by the MWCNTs. Figure 8b shows a high magnification view of a completely filled MWCNT. From this and other images it is apparent that these nanowires are indeed crystalline and without an amorphous coating of any kind, as is sometimes observed for metallic nanowires (Morales & Lieber, 1998). According to TEM, the outer diameter of the Fe-Ge filled MWCNTs lies in the 20-70 nm range, with the diameter of the Ge-Fe cores falling in the range 5-15 nm. Electron energy loss spectroscopy (EELS) analysis under STEM mode shows a sharp C-K edge at 284 eV, indicating that the carbon atoms are in the sp^2-hybridized state. The EDS spectrum obtained from the MWCNT-sheathed nanowires show the presence of Ge, Fe and C (Figure 8c), while XRD (Figure 8d) shows diffraction peaks which are close to those characteristic for the phase $Fe_{1.67}Ge$ (PDF# 00-017-0232).

Examination of Ge binary phase diagrams shows that the Ge-rich region of the Ge-Fe diagram (Mathur et al., 2004) is similar to that of the Si-Fe diagram (Morales & Lieber, 1998; Moffatt, 1976); above 838 °C the phases are $FeGe_{x(l)}$ + $Ge_{(s)}$, and below this temperature they are β-$FeGe_{2(s)}$ + $Ge_{(s)}$. The mechanism of Ge-Fe nanowire formation is therefore likely to follow that proposed above for the growth of Ge nanowires, i.e., involving Ge-Fe nanoparticle formation, followed by carbon nanotube growth and nanotube filling by the mobile Ge-Fe phase. In comparison with the Ge nanowire synthesis, the elemental C:metal ratio used in the Ge-Fe nanowire synthesis is much higher, i.e., C:(Ge+Fe) = 112, as opposed to the C:Ge ratio of 24 used in the optimized Ge nanowire synthesis (employing 50 wt% PTMG in xylene). Consequently, there appears to be insufficient Ge-Fe alloy to completely fill the nanotubes, as evidenced by the TEM image in Fig. 8a. Finally, we note that these results contrast with MWCNTs grown using a simple pyridine-ferrocene feedstock via the same floating catalyst CVD method (Qian et al., 2003). In the latter case, conical Fe_3C catalyst particles are observed in the product, located at each nanotube root (being consistent with a typical root growth model), without any filling of the tube cores by Fe. This supports the idea that the active catalyst in the Ge-Fe system is indeed a Ge-Fe alloy, rather than an iron carbide type catalyst.

4. Growth of MWCNT-encapsulated Ge nanowires with uniform diameters

In previous work, vertically aligned carbon nanotube arrays have been grown in the pores of AAO templates using a CVD method similar to the general method described for the synthesis of MWCNTs (Andrews et al., 1999), with the exception that pure xylene was used as the hydrocarbon source without any ferrocene catalyst precursor in the feed (Rajaputra et al., 2008). The resulting aligned MWCNT arrays have been used to fabricate gas sensors (Rajaputra et al., 2008); by monitoring the electrical resistance response, low concentrations of NH_3 and NO_2 can be detected by these arrays.

In this study, porous AAO templates were fabricated by a two-step anodization process of high purity aluminium tape in a 0.3 M oxalic acid medium as reported earlier (Rajaputra et al., 2008). By selecting suitable electrochemistry parameters, AAO membranes with uniform pore size of ~45 nm and thickness of 50 μm were prepared. Several pieces of these AAO templates were placed on a quartz plate that was inserted in a quartz tube (CVD reactor) at 800 °C, while undiluted PTMG was fed into reactor at a rate of 1 ml/h for 2 h. After the experiment, the color of the AAO templates had changed to black, indicating the presence of deposits. Low magnification SEM shows that nanowires were deposited on the whole top surface of the AAO template (Figure 9a). HRSEM observations reveal that these nanowires have the same shape and length (10-20 μm) as those grown on flat quartz slides, but have much smaller, uniform outer diameters of less than 100 nm (Figure 9c). According to SEM, the AAO template surface is coated with an amorphous-like layer, on top of which Ge nanowires have been deposited. After intentionally removing this surface coating with a blade, the AAO internal surface can be observed and it is apparent that some of the pores have been filled with nanowires (with closed cap ends, as indicated by the arrows in Figure 9d) or nanotubes (with open tube cores, as indicated by the circles in Figure 9d).

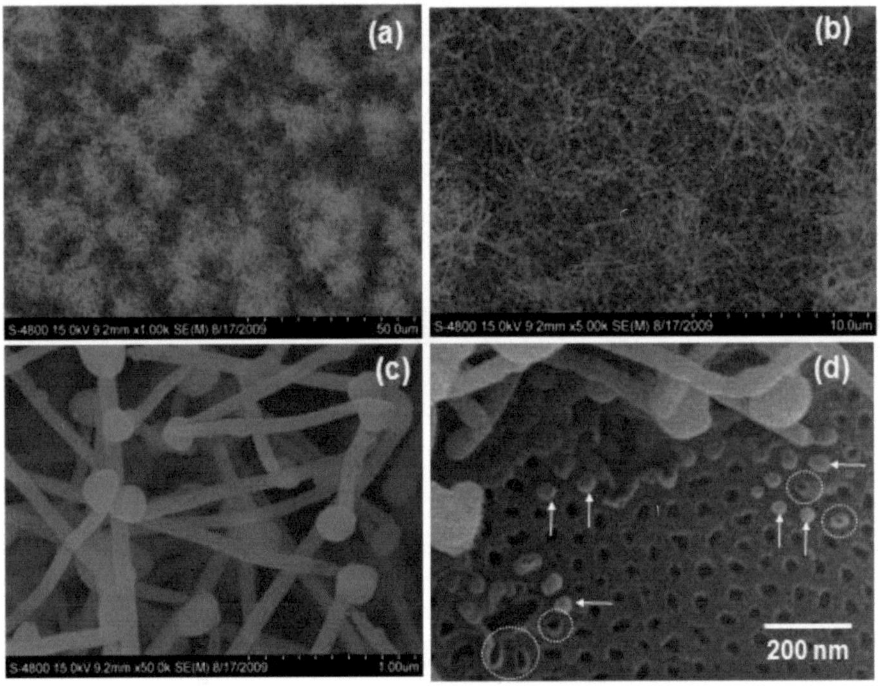

Fig. 9. Typical SEM images of nanowires by CVD of PTMG at 800 °C; deposited on AAO template surface (a, b and c); partly deposited within AAO channels (d).

From the cross-sectional view, it is clear that nanowires and nanotubes have grown within the AAO channels, although not along their whole length (Figure 10a and 10b). Energy dispersive spectra contain obvious Ge and C peaks (Figure 10c and 10d), consistent with the presence of MWCNT-encapsulated Ge nanowires. The Al and O peaks in the spectra derive from the AAO template, while Au derives from the SEM sample coating (applied to improve the image quality). These Ge@MWCNT nanowires present within the AAO channels have uniform outer diameter, close to the original AAO pore size of ~45nm.

Efforts to dissolve the AAO template using H_3PO_4 or KOH, as done previously for MWCNT/AAO arrays produced from the pyrolysis of xylene (Rajaputra et al., 2008), proved unsatisfactory. The amorphous-like coating covering the AAO surface after CVD may be responsible for slowing the acid dissolution process, which in turn suggests that reactions between the Ge clusters and Al(O) surface may have taken place during the CVD process. In this context it is pertinent to note that Al can catalyze Si nanowire growth via a vapor–solid–solid (VSS) rather than a VLS mechanism (Wang, et al., 2006). Hence, a similar process, involving a Ge-Al(O) phase, may be responsible for the growth of the amorphous-like coating. It is clear, however, that during the CVD process Ge clusters, formed from the PTMG, deposited on both the AAO top surface and the internal channels and catalyzed MWCNT nucleation and growth with incorporation of the Ge as nanowires.

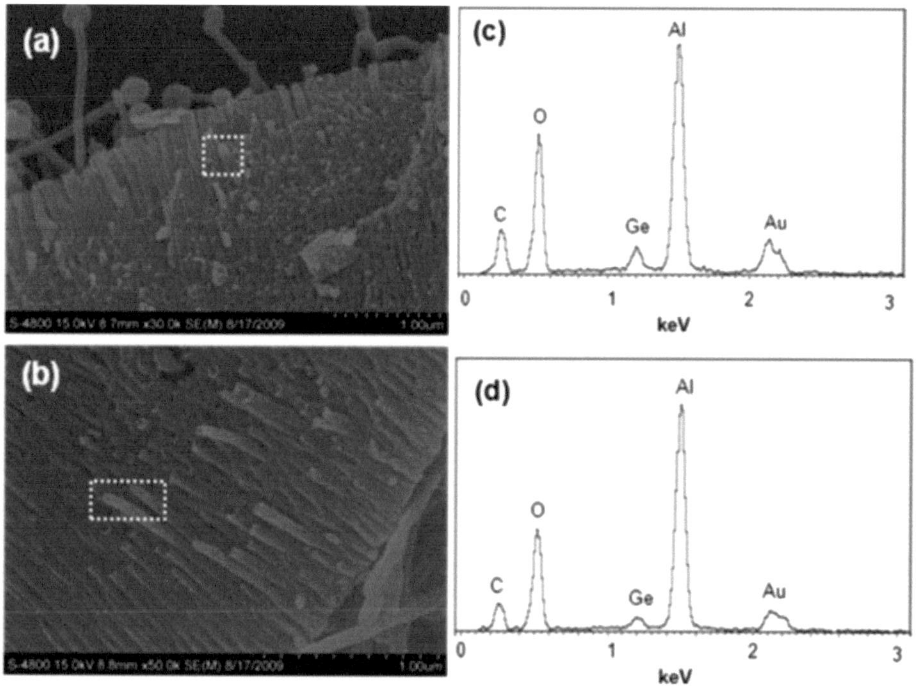

Fig. 10. Typical SEM images of Ge nanowires deposited within AAO channels after CVD of PTMG at 800 °C (a and b), and EDS collected from respective boxed regions (c and d).

5. Concluding remarks

CVD using a floating Ge catalyst represents a simple method for the preparation of carbon-encapsulated Ge nanowires. The process temperature and the partial pressure of carbon in the reactor are indicated as being critical factors for successful nanowire growth. By optimization of appropriate CVD parameters, such as the C:Ge ratio, reaction time and temperature, nanowires can be reproducibly synthesized with dimensions of 200-300 nm in diameter and 6-10 μm in length. Furthermore, by using thin AAO plates (i.e., 10-20 μm) as a template, the growth of MWCNT-encapsulated Ge nanowires with controllable diameter and orientation within AAO can be realized. Although there is a tendency for amorphous deposits to cover the AAO template, the coating can be mechanically removed by polishing to expose MWCNT-encapsulated Ge nanowire arrays which protrude out of the AAO matrix. Past work suggests that it should be possible to exploit these arrays to fabricate gas sensors and optoelectronic devices (Baughman et al., 2002; Agarwal & Lieber, 2006; Polyakov et al., 2006; Erts et al., 2006).

6. Acknowledgement

The authors would like to thank Drs. S. Rajaputra, R. Mangu and V.P. Singh for kindly supplying high quality AAO templates for this study. This work was supported in part by the Department of Science and Technology, Ministry of Science and Technology, Govt. of India, under young scientist programme award no. SR/BY/C-11/05. A.P. thanks the DST for the award of a BOYSCAST fellowship.

7. References

Agarwal, R. & Lieber, C.M. (2006). Semiconductor nanowires: optics and optoelectronics. *Appl. Phys. A*, 85, 209–15.

Andrews, R.; Jacques, D.; Rao, A.M.; Derbyshire, F.; Qian, D.; Fan, X.; Dickey, E.C. & Chen J. (1999). Continuous production of aligned carbon nanotubes: a step closer to commercial realization. *Chem. Phys. Lett.*, 303, 467-74.

Baughman, R.H.; Zakhidov, A.A. & de Heer, W.A. (2002). Carbon nanotubes - the route toward applications. *Science*, 297, 787-92.

Bom, D.; Andrews, R.; Jacques, D.; Anthony, J.; Chen, B.; Meier, M.S. & Selegue, J.P. (2002). Thermogravimetric analysis of the oxidation of multiwalled carbon nanotubes: evidence for the role of defect sites in carbon nanotube chemistry. Nano. Lett., 2, 615-9.

Bootsma, G.A. & Gassen, H.J. (1971). A quantitative study on the growth of silicon whiskers from silane and germanium whiskers from germane. *J. Cryst. Growth*, 10, 223-34.

Buffat, Ph. & Borel, J.-P. (1976). Size effect on the melting temperature of gold particles. *Phys Rev A*, 13, 2287-98.

Chen, W.W.; Sun, X.H.; Wang, S.D.; Lee, S.T. & Teo, B.K. (2005). Etching behavior of silicon nanowires with HF and NH_4F and surface characterization by attenuated total reflection fourier transform infrared spectroscopy: similarities and differences between one-dimensional and two-dimensional silicon surfaces. *J. Phys. Chem. B*, 109, 10871–9.

Cho, S.; Choi, S.; Hong, S.C.; Kim, Y.; Ketterson, J.B.; Kim, B.-J.; Kim, Y.C.; Jung, J.-H. (2002). Ferromagnetism in Mn-doped Ge. Phys. Rev. B 66, 033303.

Coleman, N.R.B.; Morris, M.A.; Spalding, T.R. & Holmes, J.D. (2001). The formation of dimensionally ordered silicon nanowires within mesoporous silica. *J. Am. Chem. Soc.*, 123, 187-8.

Couchman, P.R. & Jesser, W.A. (1977). Thermodynamic theory of size dependence of melting temperature in metals. *Nature*, 269, 481-3.

Cui, Y. & Lieber, C.M. (2001). Functional nanoscale electronic devices assembled using silicon nanowire building blocks. *Science*, 291, 851–3.

Cui, Y.; Wei, Q.; Park, H. & Lieber, C.M. (2001). Nanowire nanosensors for highly sensitive and selective detection of biological and chemical species. *Science*, 293 1289–92.

Cullis, A.G.; Canham, L.T.& Calcott, P.D.J. (1997). The structural and luminescence properties of porous silicon. *J. Appl. Phys.*, 82, 909-65.

Dai, J.Y.; Lauerhaas, J.M.; Setlur, A.A. & Chang, R.P.H. (1996). Synthesis of carbon-encapsulated nanowires using polycyclic aromatic hydrocarbon precursors. *Chem. Phys. Lett.*, 258, 547-53.

Das, K.; Chakraborty, A.K.; NandaGoswami, M.L.; Shinga, R.K.; Dhar, A.; Coleman, K.S. & Ray, S.K. (2007). Temperature dependent shape transformation of Ge nanostructures by the vapor-liquid-solid method. *J. Appl. Phys.*, 101, 074307.

Dujardin, E.; Ebbesen, T.W.; Hiura, H. & Tanigaki, K. (1994). Capillarity and wetting of carbon nanotubes. *Science*, 265, 1850-2.

Erts, D.; Polyakov, B.; Daly, B.; Morris, M.A.; Ellingboe, S.; Boland, J. & Holmes, J.D. (2006). High density germanium nanowire assemblies: contact challenges and electrical characterization". *J. Phys. Chem. B*, 110, 820-6.

Gu, G.; Burghard, M.; Kim, G.T.; Dusberg, D.S.; Chiu, P.W.; Krstic, V.; Roth, S. & Han, W.Q. (2001). Growth and electrical transport of germanium nanowires. *J. Appl. Phys.*, 90, 5747-51.

Hanrath, T. & Korgel, B.A. (2002). Nucleation and growth of germanium nanowires seeded by organic monolayer-coated gold nanocrystals. *J. Am. Chem. Soc.*, 124, 1424-9.

Hanrath, T. & Korgel, B.A. (2004). Chemical surface passivation of Ge nanowires. *J. Am. Chem. Soc.*, 126, 15466-72.

Heath, J.R. & LeGoues, F.K. (1993). A liquid solution synthesis of single crystal germanium quantum wires. *Chem. Phys. Lett.*, 208, 263-8.

Huang, Y.; Duan, X.; Cui, Y.; Lauhon, L.J.; Kim, K. & Lieber, C.M. (2001). Logic gates and computation from assembled nanowire building blocks. *Science*, 294, 1313.

Huang, Y.; Lin, J.; Zhang, J.; Ding, X.X.; Qi, S.R. & Tang, C.C. (2005). A novel method for preparing carbon-coated germanium nanowires. *Nanotech.*, 16, 1369-71.

Jin, C.-B.; Yang, J.-E. & Jo, M.-H. (2006). Shape-controlled growth of single-crystalline Ge nanostructures. *Appl. Phys. Lett.*, 88, 193105.

Kamins, T.I.; Williams, R.S.; Basile, D.P.; Hesjedal, T.; & Harris, J.S. (2001). Ti-catalyzed Si nanowires by chemical vapor deposition: Microscopy and growth mechanisms. *J. Appl. Phys.*, 89, 1008-16.

Kamins, T.I.; Li X.; Williams, R.S. & Liu, X. (2004). Growth and structure of chemically vapor deposited Ge nanowires on Si substrates. *Nano. Lett.*, 4, 503-6.

Kang, K.; Gu, G.H.; Kim, D.A.; Park, C.G. & Jo, M.-H. (2008). Self-organized growth of Ge nanowires from Ni-Cu bulk alloys. *Chem. Mater.*, 20, 6577-9.

Kodambaka, S.; Tersoff, J.; Reuter, M.C. & Ross, F.M. (2007). Germanium nanowire growth below the eutectic temperature. *Science*, 316, 729-32.

Lieber, C.M. (2003). Technical feature - nanoscale science and technology: building a big future from small things 2002 (MRS medalist presentation). *MRS Bull.*, 28, 486–91.

Loiseau, A.; Demoncy, N.; Stéphan, O.; Colliex, C. & Pascard, H. (2000). Filling carbon nanotubes using an arc discharge. In: Tománek, D; Enbody, RJ, eds. *Science and Application of Nanotubes*, New York: Kluwer Academic/Plenum Publishers; pp. 1-16.

Ma, D.D.D.; Lee, C.S.; Au, F.C.K.; Tong, S.Y. & Lee, S.T. (2003). Small-diameter silicon nanowire surfaces. *Science*, 299 1874–7.

Maeda, Y.; Tsukamoto, N.; Yazawa, Y.; Kanemitsu, Y. & Masumoto, Y. (1991). Visible photoluminescence of Ge microcrystals embedded in SiO_2 glassy matrices. *Appl. Phys. Lett.*, 59, 3168-70.

Mathur, S. & Barth, S. (2008). One-dimensional semiconductor nanostructures: growth, characterization and device applications". *Z. Phys. Chem.*, 222, 307–17.

Mathur, S.; Shen, H.; Sivakov, V. & Werner, U. (2004). Germanium nanowires and core-shell nanostructures by chemical vapor deposition of $[Ge(C_5H_5)_2]$. *Chem. Mater.*, 16, 2449-56.

Miikin, L.I. (1961). *Hand Book on X-ray Structural Analysis of Polycrystals*, Uspenskii, Ya.-S. (Ed.), Fizmatgiz, Moscow.

Moffatt, W.G. (1976). *The Handbook of binary phase diagrams*, Genium, Schenectady, NY.

Morales, A.M. & Lieber, C.M. (1998). A laser ablation method for the synthesis of crystalline semiconductor wires. *Science*, 279, 208-11.

Nguyen, P.; Ng, H.T. & Meyyappan, M. (2005). Growth of individual vertical germanium nanowires. *Adv. Mater.*, 17, 549-53.

Omi, H. & Ogino, T. (1997). Self-assembled Ge nanowires grown on Si(113). *Appl. Phys. Lett.*, 71, 2163-5.

Önce, C. & Yürüm, Y. (2006). Carbon nanotube synthesis via the catalytic CVD method: a review on the effect of reaction parameters. *Fullerenes, Nanotubes, and Carbon Nanostructures*, 14, 17-37, and references therein.

Pandurangan, A.; Morin, C.; Qian, D.; Andrews, R. & Crocker, M. (2009). Single-step synthesis of germanium nanowires encapsulated within multi-walled carbon nanotubes. *Carbon*, 47, 1708-14.

Park, Y.D.; Hanbicki, A.T.; Erwin, S.C.; Hellberg, C.S.; Sullivan, J.M.; Mattson, J.E.; Ambrose T.F.; Wilson, A.; Spanos, G.; Jonker, B.T. (2002). A group-IV ferromagnetic semiconductor: Mn_xGe_{1-x}. *Science*, 295, 651-654.

Pokropivnyi, V.V. (2001). Non-carbon nanotubes (review). II. Types and structure. *Powder Met. and Met. Ceramics*, 40, 582-94.

Polyakov, B.; Daly, B.; Prikulis, J.; Lisauskas, V.; Vengalis, B.; Morris, M.A.; Holmes, J.D. & Erts, D. (2006). High-density arrays of germanium nanowire photoresistors". *Adv. Mater*, 18, 1812-6.

Qian, D.; Andrews, R.; Jacques, D.; Kichambare, P.; Lian, G. & Dickey, E. C. (2003). Low-temperature synthesis of large-area CNx nanotube arrays. *J. Nanosci. Nanotech.* 3, 93-7.

Rajaputra, S.; Mangu, R.; Clore, P.; Qian, D.; Andrews, R. & Singh, V.P. (2008). Multiwalled carbon nanotube arrays for gas sensing applications. *Nanotechnology*, 19, 345502.

Ryan K.M.; Erts D.; Olin H.; Morris M.A. & Holmes J.D. (2003). Three dimensional architectures of ultra-high density semiconducting nanowires deposited on chip. *J. Am. Chem. Soc.*, 125, 6284-8.

Singh, A.K.; Kumar, V. & Kawazoe, Y. (2004). Ferromagnetism and piezomagnetic behavior in Mn-doped germanium nanotubes. *Phy. Rev. B.*, 69, 233406, 1-4.

Song, H.J.; Yoon, H.J.; Shin, H.-J.; Lim, H.; Park, C. & Choi, H.C. (2009). Growth of germanium nanowires using liquid $GeCl_4$ as a precursor: the critical role of Si impurities. *Chem. Commun.*, 2009, 5124–6.

Sun, X.H.; Didychuk, C.; Sham, T.K. & Wong, N.B. (2006). Germanium nanowires: synthesis, morphology and local structure studies. *Nanotechnology*, 17, 2925-30.

Sun, X.H.; Tang, Y.H.; Zhang, P.; Naftel, S.; Sammynaiken, R.; Sham, T.K.; Zhang, Y.F.; Peng, H.Y.; Wong, N.B. & Lee, S.T. (2001). X-ray absorption fine structure and electron energy loss spectroscopy study of silicon nanowires at the SiL3,2 edge. *J. Appl. Phys.*, 90, 6379–83.

Sun, X.H.; Wang, S.D.; Wong, N.B.; Ma, D.D.D.; Lee, S.T. & Teo, B.K. (2003). FTIR spectroscopic studies of the stabilities and reactivities of hydrogen-terminated surfaces of silicon nanowires. *Inorg. Chem.*, 42 2398–404.

Sutter, E.; Ozturk, B. & Sutter, P. (2008) Selective growth of Ge nanowires by low-temperature thermal evaporation. *Nanotechnology*, 19, 435607.

Sutter, E. & Sutter, P. (2006). Au-induced encapsulation of Ge nanowires in protective C shells. *Adv. Mater.*, 18, 2583-8.

Sze, S.M. (1981). *Physics of semiconductor devices* , Wiley, New York.

Takagi, D.; Hibino, H.; Suzuki, S.; Kobayashi, Y. & Homma, Y. (2007). Carbon nanotube growth from semiconductor particles. *Nano Lett.*, 7, 2272-5.

Wagner, R.S. & Ellis, W.C. (1964). Vapour–liquid–solid (VLS) mechanism of single crystal growth. *Appl. Phys. Lett.*, 4, 89-90.

Wang, D. & Dai, H. (2002). Low-temperature synthesis of single-crystal germanium nanowires by chemical vapor deposition. *Angew. Chem. Int. Ed.*, 41, 4783-6.

Wang, D. & Dai, H. (2006). Germanium nanowires: from synthesis, surface chemistry, and assembly to devices. *Appl. Phys. A*, 85, 217-25.

Wang, D., Chang, Y.-L., Liu, Z. & Dai, H. (2005). Oxidation resistant germanium nanowires: bulk synthesis, long chain alkanethiol functionalization, and Langmuir-Blodgett assembly. *J. Am. Chem. Soc.*, 127, 11871-5.

Wang, Y.; Schmidt, V.; Senz, S. & Gösele, U. (2006). Epitaxial growth of silicon nanowires using an aluminium catalyst. *Nature Nanotechnol.*, 1, 186-9.

Wu, Y. & Yang, P. (2000). Germanium nanowire growth via simple vapor transport. *Chem. Mater.*, 12, 605-7.

Wu, Y. & Yang, P. (2000). Germanium/carbon core-sheath nanostructures. *Appl. Phys. Lett.*, 77, 43-5.

Yao, Y. & Fan, S. (2007). Si nanowires synthesized with Cu catalyst. *Mater. Lett.*, 61, 177-81.

Yin, L.-W.; Li, M.-S.; Bando, Y.; Golberg, D.; Yuan, X. & Sekiguchi, T. (2007). Tailoring the optical properties of epitaxially grown biaxial ZnO/Ge, and coaxial ZnO/Ge/ZnO and Ge/ZnO/Ge heterostructures. *Adv. Funct. Mater.*, 17, 270–6.

Zhang, X.Q.; Li, H. & Liew, K.M. (2007). The structures and electrical transport properties of germanium nanowires encapsulated in carbon nanotubes. *J. Appl. Phys.*, 102, 073709.

Zhang, Y.F.; Tang, Y.H.; Wang, N.; Lee, C.S.; Bello, I. & Lee, S.T. (2000). Germanium nanowires sheathed with an oxide layer. *Phys. Rev. B*, 61, 4518-21.

Zhang, Y.; Suenaga, K.; Colliex, C. & Iijima, S. (1998). Coaxial nanocable: silicon carbide and silicon oxide sheated with boron nitride and carbide. *Science*, 281, 973-5.

Ziegler, K.J.; Polyakov, B.; Kulkarni, J.S.; Crowley, T.A.; Ryan, K.M.; Morris, M.A.; Erts, D. & Holmes, J.D. (2004). Conductive films of ordered nanowire arrays. *J. Mater. Chem.*, 14, 585-9.

7

Advances of SiO_x and Si/SiO_x Core-Shell Nanowires

Kuan Yew Cheong and Yi Ling Chiew
School of Materials & Mineral Resources Engineering, Engineering Campus, Universiti Sains Malaysia, 14300 Nibong Tebal, Seberang Perai Selatan, Penang, Malaysia

1. Introduction

Nanotechnology can be defined as the design, construction and utilization of functional materials with at least one of the dimension measured is in nano-scale to exploit the new properties and phenomena developed at that scale. These nanomaterials exhibit new and improved physical, chemical and biological properties, phenomena and processes due to the variation of wavelike properties of electrons inside matter as the size of matter is reduced and the electrons are confined.

Nanowires, one of the one-dimensional nanostructures, are wire-like structure that has diameter less than 100 nm. Single crystal nanowires grow along a specific axial direction though their side surfaces may not be well defined. The cross-section of the nanowires may either be round, hexagonal or polyhedron depending on the crystallography of the material (Lieber & Wang, 2007). The length of nanowires can vary from few hundred of nanometers to micrometer and even millimeters.

In this chapter, we review advances of SiO_x nanowires and Si/SiO_x core-shell nanowires. Nanowires, which are one form of one-dimensional nanostructures, have been used as versatile building blocks in the miniaturization of electronic and optoelectronic devices (Li et al., 2005). Various materials have been synthesized in the form of nanowires, such as silicon, germanium, gallium nitride, gallium arsenide, and silicon carbide (Spanier, 2006). Thus, it is important to understand properties, applications and synthesis methods of the nanowires so as to be able to produce nanowires with well-controlled properties and dimensions to be incorporated into electronic devices. Here in this review, properties, applications and how SiO_x nanowires and Si/SiO_x core-shell nanowires are made are discussed in the following sections. Section 2 gives a brief discussion on the SiO_x nanowires. Section 3 discusses the properties that enable SiO_x nanowires to be successfully used in different applications. Section 4 gives a short review on the potential applications of SiO_x nanowires. Section 5 reviews the different methods used to synthesize the nanowires, which also includes their phenomena, techniques and mechanism. Section 6 gives a brief discussion on Si/SiO_x core-shell nanowires. Section 7 discusses the important properties of Si/SiO_x core-shell nanowires that are linked to their potential applications. Section 8 reviews the different methods used to synthesize the core-shell nanowires. Finally, section 9 concludes by a short summary.

2. SiO_x nanowires

Silicon oxide (SiO_x) nanowires have been reported to be amorphous, where there is no long-range order of positions of the atoms (Sood et al., 2006). SiO_x may exist in different ratio of silicon and oxygen. The atomic ratio of silicon and oxygen in SiO_x has the x value between 1 and 2 (Zhang et al., 1999). Different methods had been used to synthesize SiO_x nanowires though the ratio of Si:O differed from one another but they remained in the range of $1 \leq x \leq 2$. Table 1 lists down the values of x of the SiO_x nanowires produced using different methods.

Methods	*References*	*x value in* SiO_x
Laser Ablation	Aharonovich et al. (2008)	1<x<2
Chemical Vapor Deposition	Ni et al. (2006)	2
	Zheng et al. (2002)	2
	Zhang et al. (2006)	1.3
	Jiang et al. (2005)	1.2
	Yang et al. (2007)	1.8 - 2.1
Sol-Gel Processing	Liang et al. (2000)	1.4
Rapid Thermal Annealing	Lai et al. (2008)	1:2
Ion Implantation	Sood et al. (2006)	1<x<2
Carbon-Assisted Growth	Zhu et al. (1998)	1<x<2
	Saulig-Wenger (2003)	2
Thermal Oxidation Route	Hu et al. (2003)	2

Table 1. List of Si:O ratio of the silicon oxide nanowires produced using different methods.

3. Properties of SiO_x nanowires

Since dimensionality alters the properties of a structure as the dimension decreases, it is then important to understand the properties of SiO_x nanostructures so as to enable the exploitation of these materials for new applications. Some of these properties are summarized in the Table 2.

Properties	*Value*
Young's Modulus	57 - 93 GPa
Electrical	Varies according to number of Si-O bonds (changes from metallic to insulating as number of bonds increase)
Fermi level	-3 eV to -7 eV (more positive as number of Si-O chain increases)
PL band energies	1.9 - 4.3 eV

Table 2. Properties of SiO_x nanowires (Wei et al., 2006, Jin et al., 2008, Ni et al., 2006 and Bilalbegović, 2006).

3.1 Optical properties of SiO_x nanowires

Various SiO_2 glasses and nanowires have different photoluminesence (PL) energy peaks ranging from 1.9 to 4.3 eV according to Wei et al. (2006). From the study performed by Jin et al. (2008), a strong blue-green emission was observed (Figure 1) with the peak centered at 2.5 eV (about 500 nm).

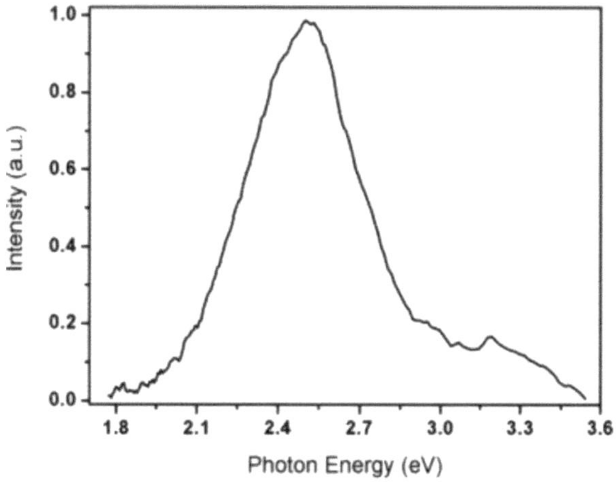

Fig. 1. Room temperature PL spectrum of the as-fabricated nanostructures showing a strong emission peak centered at 2.5 eV (Jin et al., 2008).

Similarly, Wu et al. (2001) reported that a stable and strong blue emission was found at 2.85 eV (435 nm) at room temperature under excitation at 260 nm while ultraviolet and blue light emission at 3.54 eV (350 nm), 3.0 eV (420 nm) and 2.7 eV (465 nm) could also be observed (Figure 2). The different values of energy bands had been attributed to certain phenomenon. The 3.0 eV band (415 nm) could be attributed to two-fold coordinated silicon lone-pair centers (Itoh et al., 1989). The 2.7 eV band (460 nm) was due to neutral oxygen vacancy. These defects are due to oxygen deficiency in the obtained products (Meng et al., 2003). These oxygen deficiencies in the SiO_x nanowires account for the blue-green emission around 500 nm. The 2.2 eV band (566 nm) was due to the self-trapped excitons that were confined to a SiO_4 tetrahedron (Lee et al., 2004). With the decrease of oxygen vacancies, the intensity of the green emissions would increase.

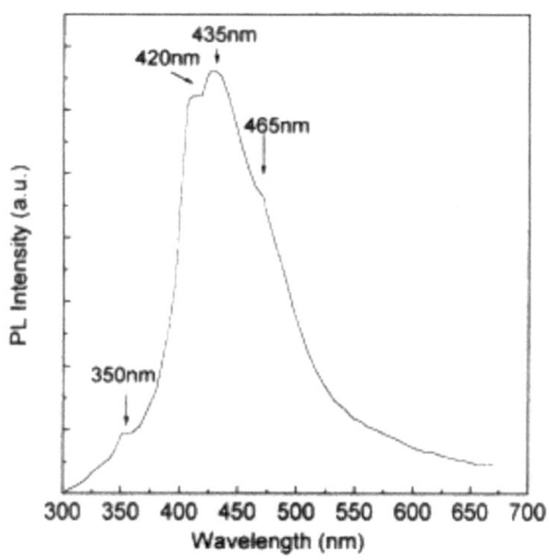

Fig. 2. PL spectrum of SiO_x nanowires under excitation at 260 nm (Wu et al., 2001).

3.2 Elastic modulus of SiO_x nanowires

Measurement of mechanical properties of the SiO_x nanowires is very important since these nanowires can be integrated into functional nanodevices without leading to malfunction or failure of the entire device. This is because the devices are always subjected to some form of mechanical forces during usage. Thus, if the nanowires used to produce the devices have good mechanical properties, then the devices can function effectively. To determine the elastic modulus of the amorphous SiO_x nanowires, nano-scale three-point bending tests can be performed directly on individual amorphous SiO_x nanowires using an atomic force microscope (AFM). Ni and Gao (2006) found that the elastic modulus of the amorphous SiO_2 nanowires was 76.6 ± 7.2 GPa. The amorphous SiO_2 nanowires also exhibited brittle fracture failure in bending.

Based on the assumption that the nanowire follows linear elastic theory of an isotropic material, the elastic modulus of the SiO_2 nanowire, E_n, can be calculated from the following equation (Ni and Gao, 2006).

$$E_n = \frac{FL^3}{192 d_n I} = \frac{k_n L^3}{192 I} \tag{1}$$

where

I is the moment of inertia and for a round-shaped nanowire, $I = \pi r4/4$,
r is the radius of nanowire,
L is the suspended length of nanowire,
F is the applied load at its midpoint position,
k_n is the spring constant of nanowire ($k_n = F/d_n$).

For all tested nanowires with a diameter ranging from 50 nm to 100 nm, the calculated elastic modulus ranged from 57 to 93 GPa and the average elastic modulus is 76.6 ± 7.2 GPa, which is close to the reported value of 73 GPa of thermally grown SiO_2 thin films and bulk SiO_2, but lower than that of plasma-enhanced CVD (PECVD) SiO_2 thin films (Ni and Gao, 2006). Normally, PECVD SiO_2 thin films usually exhibit a higher elastic modulus than thermally grown ones. It can be seen from Figure 3 that the elastic modulus of the amorphous SiO_2 nanowires was independent of the wire diameter in the range of 50–100 nm.

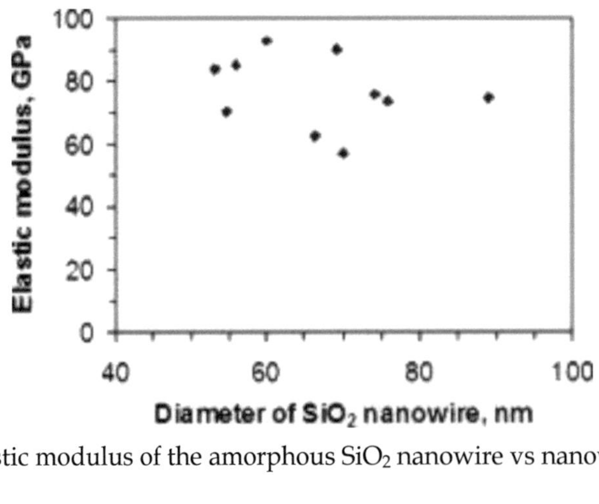

Fig. 3. Bending elastic modulus of the amorphous SiO_2 nanowire vs nanowire diameters (Ni et al., 2006).

3.3 Electronic properties of SiO_x nanowires

To integrate SiO_x nanowires into electronic components, it is important to understand their electronic properties. Bilalbegović (2006) used a computational method to investigate the electronic properties of the different structures of SiO_x nanowires consisting of linear chain, zigzag chain and long Si_4O_8 nanowires. The plot of the electronic structure of a linear, zigzag chain and Si_4O_8 nanowires are shown in Figure 4. The electronic structure of the linear chain in Figure 4(a) showed that one band crossed the Fermi level, and thus the system is metallic. For both zigzag and the long Si_4O_8 nanowires, the wires were insulator since there was no crossing of bands through the Fermi level. This indicated that as the number of neighbors in Si-O nanowires increased or changed from nanoscale to bulk size, the electronic behavior went from metallic to insulating.
The distances between the Si-O and Si-Si bonds in zigzag chains were computed to be smaller than that of linear chains simulated by Bilalbegović (2006). The Si–O distances in the three types of nanowires are larger than in the majority of silica bulk phases. It was due to rearrangement of the atoms that led to compression of bonds, and thereby removed a crossing band from the Fermi level in the zigzag chains and resulted in an insulating behavior in the structure. For the case of linear chains, the weak metallic behavior was due to weaker bonding and small coordination. The existence of a metallic state offered the possibility to use these nanowires in conducting nanodevices without doping.

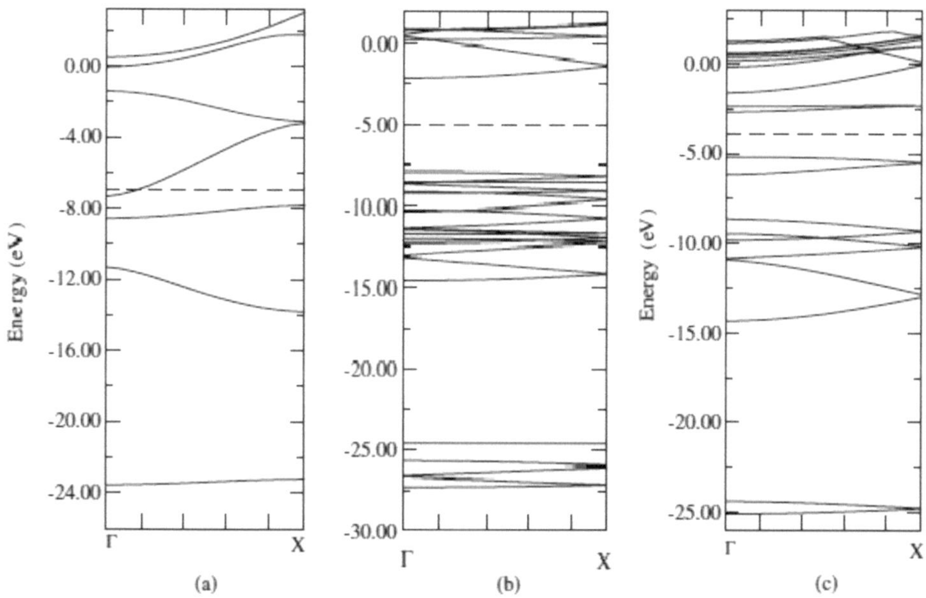

Fig. 4. Energy band structures of a) linear chain, b) zigzag chain and c) Si_4O_8 nanowires where the dashed lines denoted the Fermi levels (Bilalbegović, 2006).

4. Applications of SiO_x nanowires

SiO_x-based nanostructures have gained much interest due to their interesting properties as discussed in the previous section. Their photoluminescence properties in the range of 1.9 to 4.3 eV, emitting blue-green light makes them suitable to be made into photoluminescence materials such as intensive blue light emitters by Yu et al. (1998), integrated optical nanodevices and high resolution optical heads scanning near-field optical microscope (Cai et al., 2005). Furthermore, they are also being investigated for application as low-loss optical wave guiding (Tong et al., 2003). Their excellent bio-compatibility also makes these nanostructures suitable to be used in biomedical applications. For example, poly-L-lysine modified SiO_2 nanoparticles were developed by Li et al. (2005) as a nonviral vector for gene delivery.

5. Synthesis methods of SiO_x nanowires

The development of SiO_x nanowires have been reported since its discovery by Yu et al. (1998). Basically there are two different fabrication approaches, non-catalyst-based and catalyst-based methods. The non-catalyst-based methods usually involve silicon wafers (Dai et al., 2003), SiO_2 nanoparticles (Fang et al., 2005), or a mixture of Si/SiO_2 powder (Zhang et al., 2000) as source materials with the oxygen from either the oxygen flow, silica powder or residue O_2 gas in the chamber or carrier gas. The temperature involved in these methods is usually very high, reaching temperatures over 1000 °C as well as a long growth time. The catalyst-based methods, in the mean time, require a much lower temperature and shorter time. Different kinds of catalysts have been used to induce the growth of SiO_x nanowires such as TiN (Lee et al., 2004), Ga (Dai et al., 2005), In2O3 (Wang et al., 2003), Fe (Lee et al., 2003) and Pt (Lai et al., 2008). These methods are normally associated with vapor-liquid-solid mechanism. In the following sections, the different types of synthesis methods will be discussed briefly.

5.1 Excimer laser ablation

SiO_x or SiO_2 nanowires have been synthesized by this technique (Yu et al., 1998). The typical experiment was carried out by using an excimer laser to ablate the target in an evacuated quartz tube filled with Ar gas. The solid target could be highly pure SiO_2 powder mixed with metals (Fe or Ni). In a study performed by Yu et al. (1998), highly precise form of SiO_x nanowires were obtained by using ablation from an excimer laser to mill the tip of a SiO_2 material. The ablation process had been used to produce nanowires of diameter of 15 nm and hundreds of microns long and emit blue light from optical pumping.

Aharonovich, Tamir and Lifshitz (2008) also used laser ablation to produce SiO_x nanowires. The laser targets were pure Si and mixed Si–5%Au, where the Au is used a catalyst to induce the growth of nanowires. A pulsed neodymium doped yttrium aluminum garnet laser with wavelength of 266 nm was used to ablate the different targets in a heated evacuated tube, with the Au-coated or Ni-coated substrates on which the nanowires were supposed to grow placed along the tube. The products formed are as shown in Figure 5 that showed that the nanowires grown on sapphire substrates using different catalyst and target. The nanowires grown by Bi are longer than those of Au.

5.2 Chemical Vapor Deposition (CVD)

Chemical vapor deposition is a process which involves a chemical change occurring in the vapor phase, whether through rearrangement of the source materials' elements, made possible by excess energy provided to the system, or through reactions of the source vapors with the gasses introduced into the system. This process is normally associated with the vapor-liquid-solid (VLS) growth mechanism (Ni et al., 2006 and Zhang et al., 2006).

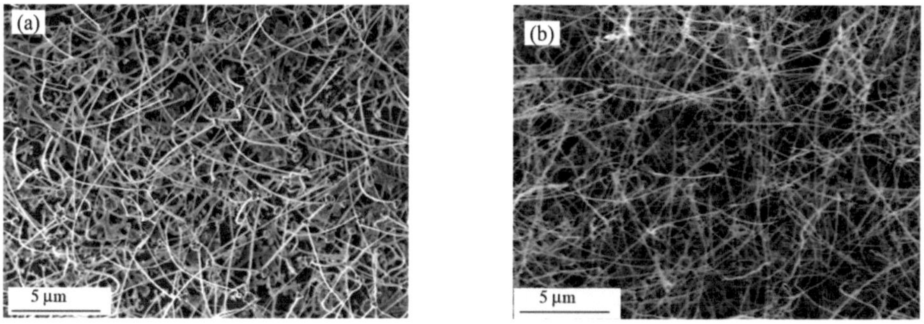

Fig. 5. HRSEM images of SiO_x nanowires on sapphire substrates after growth at (a) 1050 °C with Au catalyst, use of mixed Si–5%Au target and (b) 1150 °C with Ni catalyst, use of Si target (Aharonovich, Tamir and Lifshitz, 2008).

This mechanism was proposed by Wagner and Ellis in 1964 for silicon whisker growth and has been widely used to guide the growth of various kinds of one-dimensional nanostructures. This mechanism first involves the formation of a liquid eutectic alloy droplet composed of metal catalyst component such as Au and Ga under the reaction conditions that serves as a preferential site for nucleation site for crystallization and growth. During growth, the catalyst droplet alloy directs the growth direction of nanowires and defines the diameter of the crystalline nanowire. The nanowire stops growing when the temperature drops below the eutectic temperature of the catalyst alloy or the reactant is not available anymore. As a result, the nanowires obtained through this mechanism typically have a solid catalyst nanoparticle with diameter comparable to that of the connected nanowire.

Jiang et al. (2005) have reported the synthesis of SiO_2 nanowires and nanotubes through a simple CVD system using alumina wafers and silicon powder with Fe-Co-Ni alloy nanoparticles as catalyst. The nanowires formed had diameters of around 100 nm and length up to 100 μm. The nanowires formed had nanoparticles attached at each end of the nanowires [Figure 6(a)], which is a characteristic of product formed from the VLS growth mechanism and CVD.

Another study performed by Yang et al. (2007) using CVD produced amorphous SiO_2 nanowires of diameter 30 - 80 nm and length of several tens of micrometers. The source material used was pre-oxidized substrates catalyzed by Ni-based catalyst under ambient

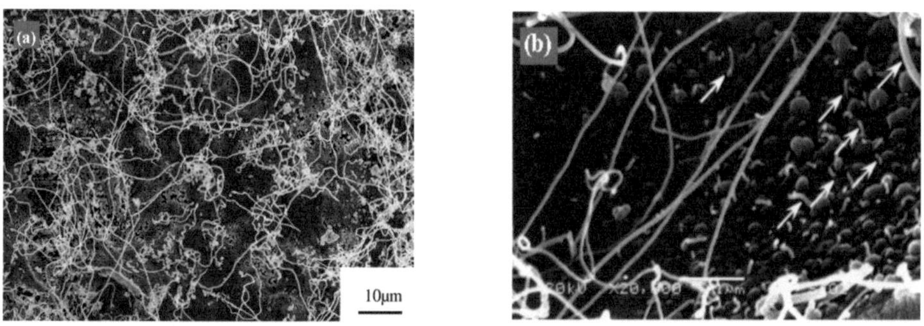

Fig. 6. SEM image of silica nanowires a) grown on alumina substrate (Jiang et al., 2005) and b) grown on pre-oxidised silicon substrate with some catalyst particles as indicated by the arrows (Yang et al., 2007).

pressure. Some catalyst particles were also observed on the SiO_2 nanowires [Figure 6(b)]. The growth was found to be non-uniform due to non-uniform decomposition of the catalyst $Ni(NO_3)_2$.

5.3 Sol-Gel template

SiO_2 nanowires can also be produced by this method. According to the study by Liang et al. (2000), mesoporous SiO_2 xerogel containing Fe nanoparticles, prepared by a sol–gel process from tetraethoxysilane (TEOS) hydrolysis in iron nitrate aqueous solution, was crushed into powder and mixed with silicon powder ground from a single crystal silicon plate with a molar ratio of SiO_2 to Si being slightly greater than 1. The source materials were then heated in a flowing Ar atmosphere. Most of the nanowires formed were straight and smoothly curved [Figure 7(a) and (b)], while others are helical-like [Figure 7(c)] or braided-like wires [Figure 7(d)]. The diameters of the braided and helical-like nanowires were about 10–25 nm, which were smaller than those of the straight and smoothly curved nanowires of about 40–50 nm. The lengths of all these nanowires were up to several tens of micrometers.

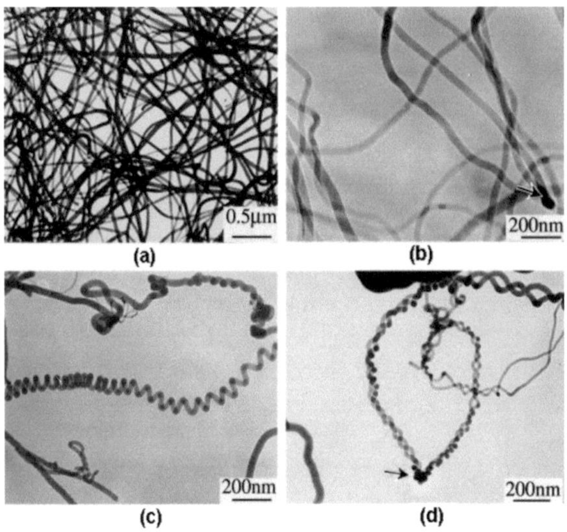

Fig. 7. TEM images of (a) straight, (b) smoothly curved, (c) helical-like and (d) braided-like SiO nanowires (Liang et al., 2000).

Zhang et al. (1999) also use a sol-gel method to prepare aligned SiO_x nanostructures on an anodic alumina as shown in Figure 8. The SiO_2 sol was prepared from mixing of TEOS solution, HCl and ethanol at room temperature and was left to age for several days at room temperature. The highly ordered nanochannel-array of anodic alumina was then dipped into the sol, removed to dried and finally heated in air at 200 °C for two day.

5.4 Rapid thermal annealing (RTA)

RTA refers to a semiconductor manufacturing process which heats the substrates to high temperatures (up to 1200 °C or greater) in a short time of several seconds but this rate does not apply to cooling. Slow cooling is performed to avoid the failure of substrates due to thermal shock. In addition, the soak time for RTA is much shorter than that for a furnace. These processes are used for a wide variety of applications in semiconductor manufacturing

Fig. 8. SEM image of silicon oxide nanowires obtained by immersing the anodic alumina in the sol which is aged at room temperature for 2 days (Zhang et al., 1999).

including dopant activation, thermal oxidation, metal reflow and synthesis of nanostructures (Nishi & Doering, 2000).

Lai et al. (2008) used RTA to synthesize SiO_x nanowires on a Pt-coated pre-oxidized p-type silicon (100) substrates. The oxide layer thickness was 11 nm whereas the Pt layer ranged from 15, 30 and 50 nm. The $Pt/SiO_2/Si$ structure was subjected to RTA in nitrogen atmosphere at 900 °C for 60 s. No nanowires were observed at the annealing temperature of 800 °C since the growth temperature of the SiO2 nanowires is above the Pt-Si eutectic temperature of 847 °C. Nanoparticles of Pt were found at the tip of the nanowires as shown in Figure 9, corresponding to the VLS growth mechanism. The diameter of the nanowires ranges from 30 nm to 150 nm with length up to 1 μm. The diameter was believed to be related to the size of the seed particles.

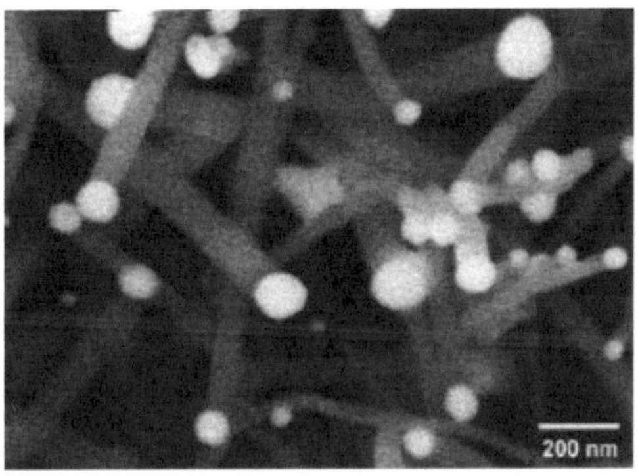

Fig. 9. SEM image of Pt-assisted SiO_2 nanowires grown after rapid thermal annealing at 900°C for 60 s (Lai et al., 2008).

5.5 Ion implantation

Ion implantation is a process in which ions of a material are implanted into another solid. The ions cause a change in the physical properties of the solid as well as in the chemical properties of the solid since the ions can be of different element than the solid (Nishi & Doering, 2000).

In the case of producing SiO_x nanowires using ion implantation as performed by Sood et al. (2006), Pd ions were first implanted on the prime grade Si (100) wafers in certain regions using an Al shadow mask. The ion dose varied from 5×10^{12} to 3×10^{6} Pd ions/cm^2. The ion-implanted wafers were then annealed at 1100 °C in argon ambient to grow the nanowires. The nanowires grown had diameters ranging from about 15 to 90 nm with length varied up to about 50 μm. It was also found that the nanowires' size and distribution could be controlled by varying the ion dose, as shown in Figure 10. It was found that an ion dose of about 10^{16} ions/cm^2 was required to produce the nanowires.

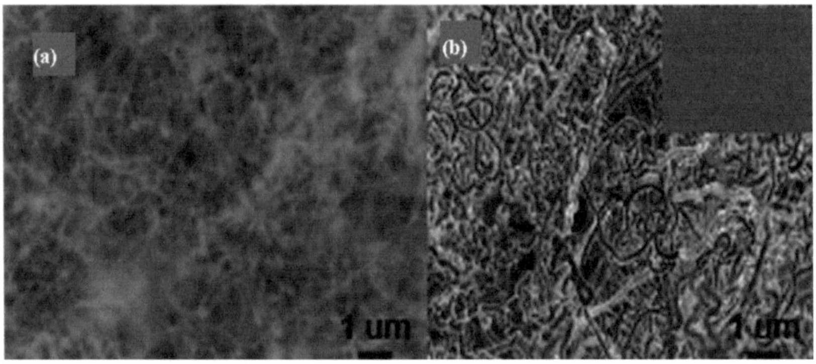

Fig. 10. Effect of implanted dose on the growth of SiO_2 nanowires heated at 1100 °C for 60 min where. (a) 3 x 10^{16} ions/cm^2—high density thinner nanowires and (b) 1 x 10^{16} ions/cm^2—thicker coiled nanowires of reduced density. Inset: 5 x 10^{12} ions/cm^2—absence of nanowires (Sood et al., 2006).

5.6 Carbon assisted growth

Carbon assisted growth is a process in which carbon plays an important role. Carbon can assist the growth of SiO_x nanowires but the opinions on the nanowire formation mechanism differ from one report to another report (Li et al., 2004). From information obtained from the Li et al. (2004), it was found that CO or CO_2 component played an important role in the synthesis of SiO_x nanowires depending on the setup used.

Saulig-Wenger et al. (2003) have reported the synthesis of amorphous SiO_2 nanowires with only silicon powder in the presence of graphite in a furnace without any catalyst. The SiO_2 nanowires had lengths up to 500 mm for diameters in the range of 10–300 nm as shown in Figure 11. The Ar : O_2 carrier gas ratio was 99.2 mol% : 0.8 mol%, and the growth

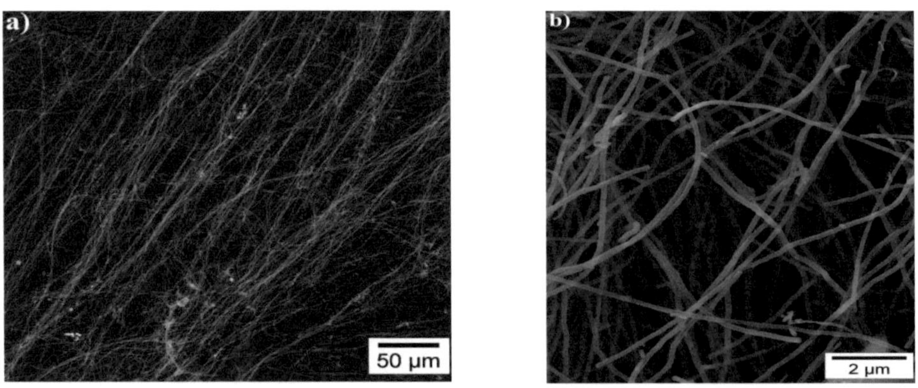

Fig. 11. SEM images of the SiO_2 nanowires formed through carbon-assisted growth (Saulig-Wenger et al., 2003).

temperature was 1200 °C. They found the relative O_2 flow rate played a very important role in the formation of the nanowires. Based on their study, it was found that having either no O_2 or a large amount of O_2 hindered the nanowire growth and that CO_2, not CO was the decisive factor in the growth of nanowires.
Li et al. (2004) also used carbon-assisted growth to synthesize SiO_x nanowires with a short growth time (30 min) on Au-coated silicon substrates. The formation of the SiO_x nanowires depends on the experimental conditions, such as substrate coating (Au), the presence of graphite powder, the substrate temperature, the oxygen flow, and the growth time. The optimum growth conditions in this study are a substrate temperature of 1000 °C, an Ar flow of 250 sccm, and an O_2 flow of 5 sccm. It was also demonstrated that the formation of the SiO_x nanowires was due to a solid-liquid-solid (SLS) mechanism, and the locally catalytic oxidation of CO by Au nanoclusters may play a role in accelerating nanowire formation.

6. Si/SiO_x core-shell nanowires

SiO_x nanowires sometimes also form a core-shell structure with the crystalline core and amorphous shell. From previous studies, the core can be either the combination of Ga, Ni, Si and O (Cai et al., 2005), Ge (Arnold et al., 2009), SiC (Liu and Yao, 2005) and Si (Jia et al. 2007, Pan et al., 2001, Kolb et al., 2004, King et al., 2008, Park and Yong, 2004, Yao et al., 2005). There was also report on SiO_x/ZnO core-shell nanowires, where the core consisted of amorphous SiO_x while the shell consisted of crystalline ZnO (Kim et al., 2007). Here, we will only focus on Si/SiO_x core-shell nanowires whereby the core consisted of crystalline silicon and the shell of amorphous SiO_x. Silicon nanowires can also be obtained by etching the silicon oxide off the core-shell nanowires using HF solution. The applications of Si/SiO_x core-shell nanowires are similar to those of silicon nanowires.

7. Properties of Si/SiO_x core-shell nanowire

The properties of these core-shell structures are similar to those of silicon nanowires. The amorphous shell helps to prevent the mechanical or radiation damage and suppress chemical reactivity, which could lead to oxidation and contamination in the silicon nanowire. However, there are only a few studies that report on the properties on the core-shell nanowires (Jia et al., 2007, King et al., 2008). Below is compilation of the properties being reported.

7.1 Optical properties of Si/SiO_x core-shell nanowires

Research has been carried out on the origin of light emission in the nanowires. Several origins have been proposed, including the existence of excess silicon atoms in the silicon nanostructure, defect centers in the SiO_x layer that surrounds the nanowires defects in the SiO_x and the interface between the SiO_x and nanoparticles in the case of chainlike nanowires incorporating crystalline silicon spheres (King et al., 2008). The Si/SiO_2 core-shell nanowires formed had different morphologies ranging from chain-like structure [Figure 12(a)], tadpole-like structure [Figure 12(b)] and straight core-shell nanowires [Figure 12(c) and (d)], with each synthesized using different methods. The chain-like structure and tadpole-like structure were produced using oxide-assisted growth while the straight core-shell nanowires were produced using thermal evaporation and CVD. PL occurred across the

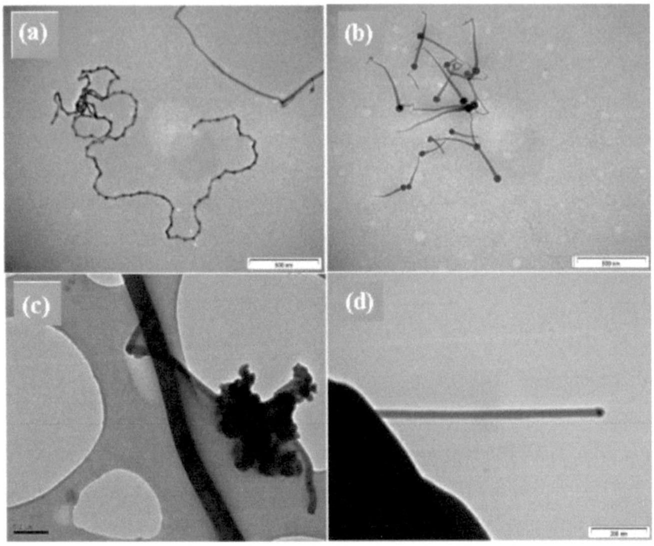

Fig. 12. TEM study of nanostructures: (a) Chain-like structure of crystalline silicon spheres encapsulated by silicon oxide, (b) Tadpole-like structure containing crystalline silicon spheres encapsulated by silicon oxide, (c) Straight nanowires of crystalline silicon encapsulated by SiO_x produced using thermal evaporation and (d) Straight nanowires of crystalline silicon nanowires encapsulated by SiO_x produced using CVD (King et al., 2008).

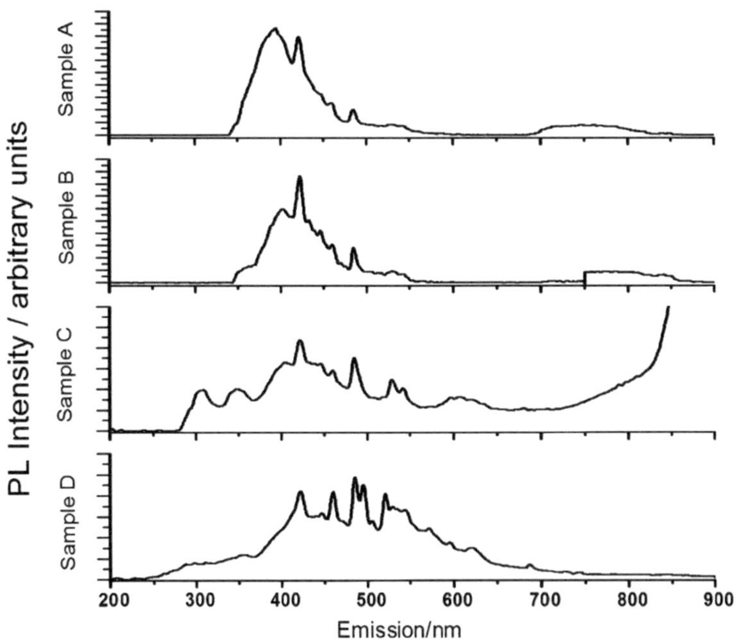

Fig. 13. PL spectrum of each sample when excited at 200 nm (King et al., 2008).

visible region for all the samples. PL in the red region (700 - 800 nm) was due to the radiative recombination of carriers in silicon nanocrystallites formed at the Si/SiO_x interface

(Figure 13). The radiative decay of self trapped excitons at the Si/SiO_x interface resulted in the PL characteristics in the yellow region, which is the main emission in the region for the straight core-shell nanowires, whereas for both chain-like and tadpole-like structures, the main emission region occurred in the ultra-violet to violet region (350 - 450 nm), which was attributed to the direct recombination of carriers at the interface of crystalline silicon nanoclusters and oxide shell.

A study from Jia et al. (2007) on the cathodoluminescence (CL) properties of Si/SiO_x core-shell nanowires showed three main bands at 77 K for all samples [Figure 14(a)], with their maxima at about 620 - 650 nm with wavelength of 2.0 - 1.91 eV (peak 1), 920 nm with wavelength of 1.35 eV (peak 2) and 1270 nm with wavelength of 0.97 eV (peak 3) as well as a shoulder at about 500 nm (peak 0). At room temperature as shown in Figure 14(b), an additional peak around 1570 nm of wavelength 0.79 eV was observed. Peak 1 was attributed to the oxygen deficient centers (ODCs) in the silicon oxide matrix. Peak 2, which was the second order diffraction of peak 0, was attributed to the band-band recombination of crystalline silicon bulk materials and the crystalline core of the silicon nanowires. Peak 3, which was the second order diffraction of peak 1, was attributed to the G-centers near certain twin boundaries. The additional peak at higher temperatures, peak 4 was attributed to the disruption of the efficiency of the recombination channels due to confinement of excess carriers in the silicon nanowires. Infrared emission was also reported at around 1550 nm due to high density of extended defects and oxygen accommodation within the extended defects.

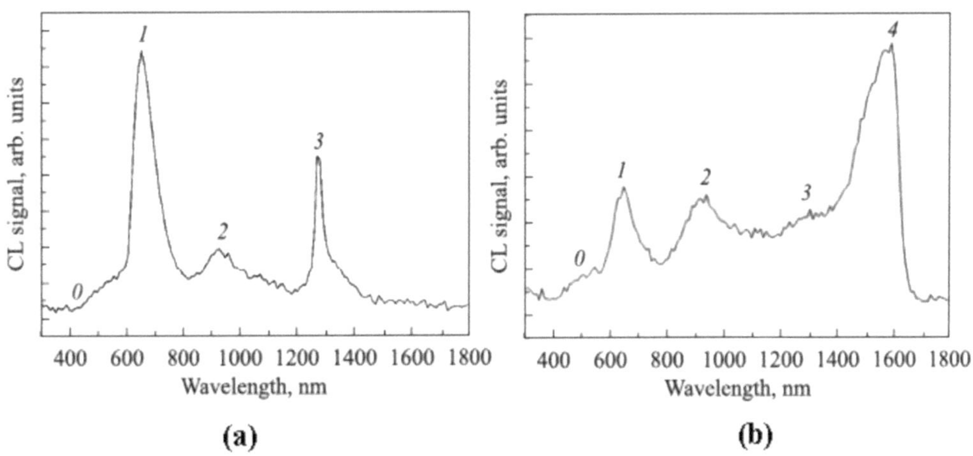

Fig. 14. CL spectrum at (a) liquid nitrogen temperature and (b) at room temperature (Jia et al., 2007).

7.2 Absorption characteristics of Si/SiO_x core-shell nanowires

An understanding of the absorption characteristics of the nanowires will allow the researchers to determine the quantum confinement effects of nanowires. Quantum confinement is said to correspond to the red-shift Γ25 - Γ15 (3.4 eV) direct transition. King et al. (2008) had performed the absorption test besides the PL test. The four samples, chain-like structure [Figure 12(a)], tadpole-like structure [Figure 12(b)], straight Si/SiO_x core-shell nanowires [Figure 12(c) and (d)] showed different absorption spectrum (Figure 15). Both chain-like structure and tadpole-like structure exhibited a strong increase in the absorption above the X4 - X1 (4.3 eV) direct transition. The straight core-shell nanowires produced using thermal evaporation exhibited a broad shoulder in the X4 - X1 and a weak absorption compared to that in the Γ25 - Γ15 region while the straight core-shell nanowires produced using CVD exhibited an absorption shoulder at 3.8 eV and a strong absorption at 4.3 eV. The quantum confinement phenomenon could only be observed when the nanowires contained crystalline cores due to nucleation kinetics at the Si/SiO_x interface.

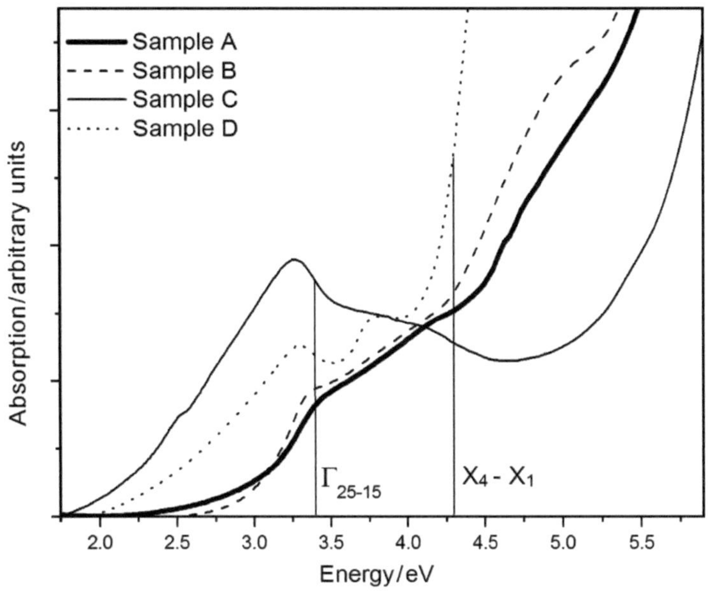

Fig. 15. Optical absorption spectrum of nanowires of different morphologies (King et al. 2008).

8. Synthesis methods of Si/SiO_x core-shell nanowires

Si/SiO_x core-shell nanowires have been synthesized as by-products in the process of obtaining Si nanowires through different methods such as oxide-assisted growth (King et al., 2008, Park & Yong, 2004, Yao, Li & Lee, 2005) and thermal evaporation of silicon monoxide (Jia et al. 2007, Pan et al., 2001, Kolb et al., 2004).

8.1 Oxide-Assisted Growth (OAG)

Oxide-Assisted Growth was first proposed by Zhang et al. (2000) to synthesize these core-shell nanowires in the absence of metal catalyst by laser ablating a mixture of Si and SiO_2 powder. The vapor phase Si_yO generated was the main component in the oxide-assisted process. The formation of the nanowires was believed to occur through two steps:

$$Si_yO \rightarrow Si_{y-1} + SiO, y > 1 \quad (2)$$

$$2SiO \rightarrow Si + SiO_2 \quad (3)$$

It was suggested that these decomposition reactions first led to precipitation of Si/SiO_x core-shell nanoparticles. Some of these particles might be piled up and served as seeds for the growth of nanowires in the following steps. Si_yO (y>1) layer at the tip of each nanowire seemed to have a catalytic effect. This layer might be in or near a molten state and thus capable of enhancing atomic absorption, diffusion and deposition. The SiO_2 component in the shell might help to retard the lateral growth of each nanowire. The precipitation, nucleation and growth of Si nanowires always occurred in the region closest to the cold region indicating that the temperature gradient was the driving force for the nanowire growth (Li et al., 2002, Lee, Wang & Lee, 2000). A TEM image of the core-shell nanowires formed through oxide-assisted growth using zeolite and SiO as starting materials by Li et al. (2002) is as shown in Figure 16. The SiO nanoclusters deposited onto the surface of zeolites and disproportionated to form Si and SiO_2 similar to the reactions as explained above.
The OAG method can also be used to synthesize the core-shell nanowires in the presence of Au catalyst. This route was found to have better dimensional control on the nanowires and required a lower growth temperature. Yao et al. (2005), using Au catalyst, successfully produced oriented Si/SiO_x core-shell nanowires through this fabrication route. This route is just the combination of the VLS and OAG methods. The reduced growth temperature relative to that of catalyst-free OAG might be due to the Au catalytic effect in lowering the SiO decomposition temperature. The Au-SiO approach offered some advantages over VLS growth and OAG, such as the absence of toxic and flammable gases and the control of size and epitaxial growth of silicon nanowires. The nanowires generally grow along the [112] and [110] directions, similar to those grown by catalyst-free OAG method and the product in Figure 17.

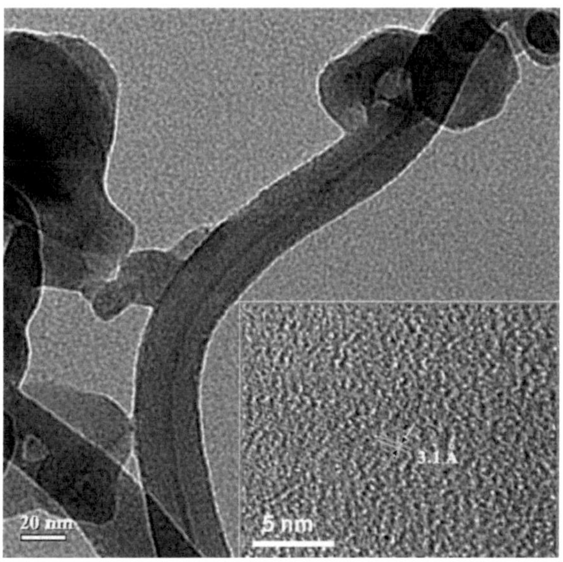

Fig. 16. TEM image of a Si/SiO_2 core-shell nanowires and the inset showed HRTEM image of the same silicon nanowire (Li et al., 2002).

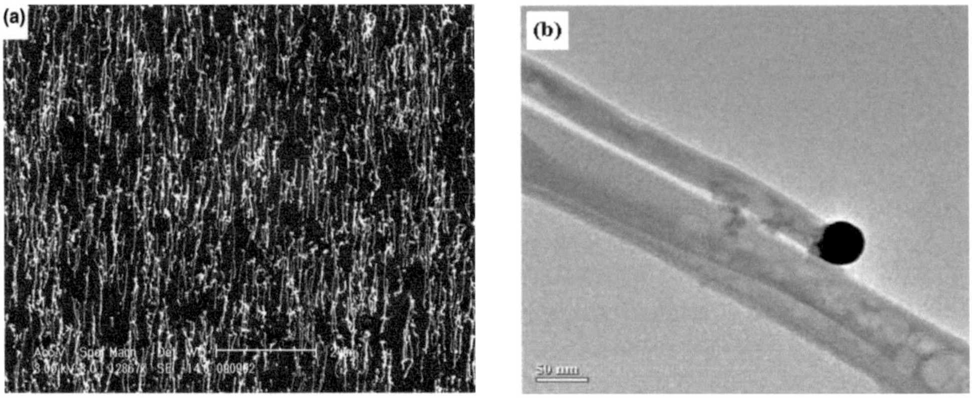

Fig. 17. (a) SEM image of oriented Si/SiO_x core-shell nanowires synthesized at 800 °C and (b) TEM image of a nanowire with a metal particle at the tip (Yao, Li and Lee, 2005).

8.2 Thermal evaporation of SiO powder

Thermal evaporation of SiO powder is a process in which the oxides were found to play an important role in the nucleation and growth of nanowires (Niu et al, 2004, Pan et al., 2001, and Zhang et al., 2000). During annealing, SiO evaporates and transported by the carrier gas to the lower-temperature region to decompose on substrates. It was found that the morphologies of the nanostructures differed as the distance from the SiO source increased when the SiO was transported to lower-temperature region as shown in Figure 18 in a study by Pan et al. (2001). The reaction involved in thermal evaporation of SiO is as follow:

$$2SiO \rightarrow Si + SiO_2 \tag{4}$$

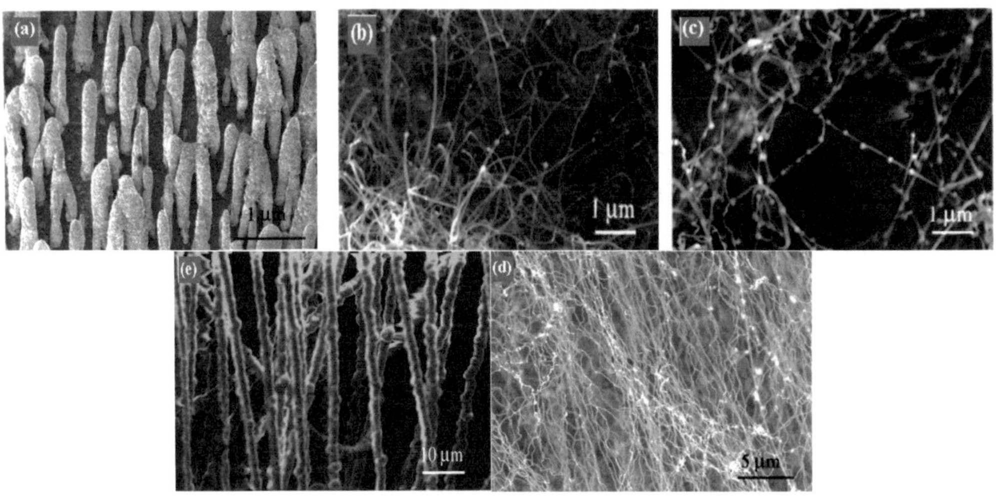

Fig. 18. SEM images of Si/SiO_x core-shell nanowires formed as distance increased from the silicon source (a) Aligned round-tip rods in Zone 1, (b) Yellow circular hard shell of numerous pin-like nanowires in Zone 2, (c) Very fine pinlike nanowires in Zone 3, (d) Si nanowires, which grow along the flowing direction of the carrier gas and form quasi-aligned structures in Zone 4 and (e) Amorphous SiO wires in Zone 5 (Pan et al., 2001).

When the silicon atoms precipitate, the atoms centralize to form the nuclei on some locations on the substrate at which the energy is low such as defects. As the silicon atoms increase, the nuclei will grow up into wires. In this growth route, some of the growth directions of the nanowires will be limited to <111> and <112> orientations due to oxide reaction and growth energy (Niu et al., 2004, Pan et al., 2001).

9. Conclusion

In this review, we have introduced SiO_x nanowires and Si/SiO_x core-shell nanowires. Nanowires have achieved recognition within a very short period of time. The properties of SiO_x and Si/SiO_x core-shell nanowires were reviewed and linked to their relevant applications. Wide range of synthesis methods used to produce the nanowires have also been presented. The nanowires have excellent photoluminescence and mechanical properties as well as biocompatibility, making them potential in the fields of nano-optoelectronics and nono-biotechnology.

10. Acknowledgements

This work was supported by Research University Grant provided by Universiti Sains Malaysia.

11. References

Aharonovich, I., Tamir, S. and Lifshitz, Y., (2008). Growth of SiO_x nanowires by laser ablation, *Nanotechnology*, Vol., 19, 065608 (8 pp), 0957-4484.

Arnold, D.C., Hobbs, R.G., Zirngast, M., Marschner, C., Hill, J.J., Ziegler, K.J., Morris, M.A. and Holmes, J.D., (2009). Single step synthesis of Ge-SiO_x core-shell heterostructured nanowires, *Journal of Materials Chemistry*, Vol., 19, 954-961, 0959-9428.

Bilalbegović, G., (2006). Electronic properties of silica nanowires, *Journal of Physics: Condensed Matter*, Vol., 18, 3829-3836, 0953-8984.

Cai, X.M., Djurišić, A.B. and Xie, M.H., (2005). Growth of SiO_x nanowires, *Proceedings of 2005 5th IEEE Conference on Nanotechnology*, Nagoya, Japan.

Dai, L., Chen, X.L., Jian, J.K., Wang, W.J., Zhou, T. and Hu, B.Q., (2003). Strong blue photoluminescence from aligned silica nanofibers, *Applied Physics A: Materials Science and Processing*, Vol., 76, 625-627, 0947-8396.

Dai, L., You, L.P., Duan, X.F., Lian, W.C. and Qin, G.G., (2005). Growth of silica nanowire arrays by reaction of Si substrate with oxygen using Ga as catalyst, *Physics Letters A*, Vol., 335, 304-309, 0375-9601.

Fang, X.S., Ye, C.H. Xie, T., He, G., Wang, Y.H. and Zhang, L.D., (2005). Synthesis and characterization of ultra-long silica nanowires, *Applied Physics A: Materials Science and Processing*, Vol., 80, 423-425, 0947-8396.

Itoh, C., Suzuki, T. and Itoh, N., (1989). Luminescence and defect formation in undensified and densified amorphous SiO_2, *Physical Review B*, Vol., 41, 3794-3799, 1098-0121.

Jia, G., Arguirov, T., Kittler, M., Su, Z., Yang, D. and Sha, J., (2007). Cathodoluminescence investigation of silicon nanowires fabricated by thermal evaporation of SiO, *Semiconductors*, Vol., 41, No. 4, 391-394, 1063-7826.

Jiang, Z., Xie, T., Yuan, X.Y., Geng, B.Y., Wu, G.S., Wang, G.Z., Meng, G.W., and Zhang, L.D., (2005). Catalytic synthesis and PL of silicon oxide nanowires and nanotubes, *Applied Physics A: Materials Science and Processing*, Vol., 81, 477-479, 0947-8396.

Jin, L., Wang, J.B., Cao, G.Y. and Choy, W.C.H., (2008). Fabrication and characterization of amorphous silica nanostructures, *Physics Letters A*, Vol., 372, Issue 25, 4622-4626, 0375-9601.

Kim, H.W., Shim, S.H., Lee, J.W., Lee, C.M. and Jeoung, S.C., (2008). ZnO-sheathed SiO_x nanowires: Annealing effect, *Optical Materials*, Vol., 30, Issue 7, 1221-1224, 0925-3467.

King, S.M, Chaure, S., Krishnamurthy, S., Blau, W.J., Colli, A. and Ferrari, A.C., (2008). Optical characterization of oxide encapsulated silicon nanowires of various morphologies, *Journal of Nanoscience and Nanotechnology*, Vol., 8, 4204-4206, 1533-4880.

Kolb, F.M., Hofmeister, H., Scholz, R., Zacharias, M., Gösele, U., Ma, D.D. and Lee, S.T., (2004). Analysis of silicon nanowires grown by combining SiO evaporation with the VLS mechanism, *Journal of the Electrochemical Society*, Vol., 151. No. 7, G472-G475, 0013-4651.

Lai, Y.S., Wang, J.L., Liou, S.C. and Tu, C.H., (2008). Tailoring of amorphous SiO_x nanowires grown by rapid thermal annealing, *Chemical Physics Letters*, Vol., 453, Issue 1-3, 97-100, 0009-2614.

Lee, K.H., Lee, S.W., Vanfleet, R.R. and Sigmund, W., (2003). Amorphous silica nanowires grown by the VS mechanism, *Chemical Physics Letters*, Vol., 376, 498-503, 0009-2614.

Lee, K.H., Loftona, C., Kim, K.H., Seo, W.S., Lee, Y.H., Lee, M.H. and Sigmund, W., (2004). Photoluminescence from amorphous silica nanowires synthesized using TiN/Ni/SiO_2/Si and TiN/Ni/Si substrates, *Solid State Communications*, Vol., 131, 687-692.

Lee, S.T., Wang, N. and Lee, C.S., (2000). Semiconductor nanowires:synthesis, structure and properties, *Materials Science and Engineering A*, Vol., 286, Issue 1, 16-23.

Li, C.P., Sun, X.H., Wong, N.B., Lee, C.S., Lee, S.T. and Teo, B.K., (2002). Ultrafine and uniform silicon nanowires grown with zeolites, *Chemical Physics Letters*, Vol., 365, 22-26, 0009-2614.

Li, S.H., Zhu, X.F. and Zhao, Y.P. (2004). Carbon-assisted growth of SiO_x nanowires, *Journal of Physics Chemistry B*, Vol., 108, 17032-17041.

Li, Z., Zhu, S.G., Gan, K., Zhang, Q.H., Zeng, Z.Y., Zhou, Y.H., Liu, H.Y., Xiong, W., Li, X.L. and Li, G.Y., (2005). Poly-L-lysine-Modified Silica Nanoparticles: A Potential Oral Gene Delivery System, *Journal of Nanoscience and Nanotechnology*, Vol., 5, No. 8, 1199-1203.

Liang, C.H., Zhang, L.D., Meng, G.W., Wang, Y.W. and Chu, Z.Q., (2000). Preparation and characterization of amorphous SiO_x nanowires, *Journal of Non-Crystalline Solids*, Vol., 277, Issue 1, 63-67, 0022-3093.

Lieber, C.M. and Wang, Z.L., (2007). Functional Nanowires, *MRS Bulletin*, Vol., 32, 0883-7694.

Liu, X.M. and Yao, K.F., (2005). Large -scale synthesis and photoluminescence properties of SiC/SiO_x nanocables, *Nanotechnology*, Vol., 16, 2932-2935, 0957-4484.

Meng, G.W., Peng, X.S. Wang, Y.W., Wang, C.Z., Wang, X.F. and Zhang, L.D., (2003). Synthesis and photoluminescence of aligned SiO_x nanowire arrays, *Applied Physics A: Materials Science & Processing*, Vol., 76, No. 1, 119-121.

Ni, H., Li, X.D. and Gao, H.S., (2006). Elastic modulus of amorphous SiO_2 nanowires, *Applied Physics Letter*, Vol., 88, 043108, 0003-6951.

Nishi, Y. and Doering, R., (2000). *Handbook of semiconductor manufacturing technology*, CRC Press.

Niu, J.J., Sha, J. and Yang, D., (2004). Silicon nanowires fabricated by thermal evaporation of silicon monoxide, *Physica E: Low-dimensional Systems and Nanostructures*, Vol., 23, Issues 1 - 2, 131-134, 1386-9477.

Pan, Z.W., Dai, Z.R., Xu, L., Lee, S.T. and Wang, Z.L., (2001). Temperature-controlled growth of silicon-based nanostructures by thermal evaporation of SiO powders, *Journal of Physical Chemistry B*, Vol., 105, 2507-2514, 1520-6106.

Park, B.T. and Yong, K.J., (2004). Controlled growth of core-shell Si-SiO_x and amorphous SiO_2 nanowires directly from NiO/Si, *Nanotechnology*, Vol., 15, S365-S370, 0957-4484.

Saulig-Wenger, K., Cornu, D., Chassagneux, F., Epicier, T. and Miele, P., (2003). Direct synthesis of amorphous silicon dioxide nanowires and helical self-assembled nanostructures derived therefrom, *Journal of Materials Chemistry*, Vol., 13, 3058-3061, 1520-6106.

Sood, D.K., Sekhar, P.K. and Bhansali, S., (2006). Ion implantation based selective synthesis of SiO_2 nanowires on silicon wafers, *Applied Physics Letters*, Vol., 88, 143110, 0003-6951.

Spanier, J.E., (2006). One-dimensional semiconductors and oxide nanostructures, In: *Nanomaterials Handbook*, 1st edition Yuri Gogotsi, CRS Press, Taylor & Francis Group, United States of America.

Tong, L.M., Gattass, R.R., Ashcom, J.B., He, S.L., Lou, J.Y., Shen, M.Y., Maxwell, I. and Mazur, E., (2003). Subwavelength-diameter silica wires for low-loss optical wave guiding, *Nature*, Vol., 426, 816-819, 0028-0836.

Wang, J.C., Zhan, C.Z. and Li, F.G., (2003). The synthesis of silica nanowire arrays, *Solid State Communications*, Vol., 125, 629-631, 0038-1098.

Wei, Q., Meng, G.W., An, X.H., Hao, Y.F. and Zhang, L.D., (2006). Synthesis and photoluminescence of aligned straight silica nanowires on silicon substrate, *Solid State Communications*, Vol., 138, Issue 7, 325-330, 0038-1098.

Wu, X.C., Song, W.H., Wang, K.Y., Hu, T., Zhao, B. Sun, Y.P. and Du, J.J., (2001). Preparation and photoluminescence properties of amorphous silica nanowires, *Chemical Physics Letters*, Vol., 336, pp 53-56, 0009-2614.

Yang, Z.M., Yu, Z., Chen, H., Jiao, Z.F., Jin, Y., He, Y., Gong, M. and Sun, X.S., (2007). Growth of amorphous SiO_2 nanowires on pre-oxidised silicon substrate via CVD, *Journal of Non-Crystalline Solids*, Vol., 354, Issue 15-16, 1731-1735, 0022-3093.

Yao, Y., Li, F.H. and Lee, S.T., (2005). Oriented silicon nanowires on silicon substrates from oxide-assisted growth and gold catalysts, *Chemical Physics Letters*, Vol., 406, 381-385, 0009-2614.

Yu, D.P., Hang, Q.L., Ding, Y., Zhang, H.Z., Bai, Z.G., Wang, J.J., Zou, Y.H., Qian, W., Xiong, G.C. and Feng, S.Q., (1998). Amorphous silica nanowires: Intensive blue light emitter, *Applied Physics Letters*, Vol., 73, 3076-3079, 0003-6951.

Zhang, J., Xu, B.L., Yang, Y.D., Jiang, F.H., Li, J.P., Wang, X.C. and Wang, S.M., (2006). Catalyzed-assisted growth of well-aligned silicon oxide nanowires, *Journal of Non-Crystalline Solids*, Vol., 352, Issues 26-27, pp 2859-2862, 0022-3093.

Zhang, M., Bando, Y., Wada, K. and Kurashima, K., (1999). Synthesis of nanotubes and nanowires of silicon oxide, *Journal of Materials Science Letters*, Vol., 18, 1911-1913.

Zhang, Y.F., Tang, Y.H., Lam, C., Wang, N., Lee, C.S., Bello, I. and Lee, S.T., (2000). Bulk-quantity Si nanowires synthesized by SiO sublimation, *Journal of Crystal Growth*, Vol., 212, 115-118, 0022-0248.

8

Yttrium Oxide Nanowires

Nan Li[1] and Kazumichi Yanagisawa[2]
[1]*School of Materials Science and Engineering, Jilin University*
[2]*Research Laboratory of Hydrothermal Chemistry, Faculty of Science, Kochi University*
[1]*China*
[2]*Japan*

1. Introduction

Yttrium is one of the most important rare-earth elements. Yttrium compounds show promise for applications such as high quality phosphors, catalysts, up-conversion materials and fine ceramics because of their optical, catalytic properties and high-temperature stability. Yttrium oxide is the most familiar yttrium compound, which is popularly known as host for ion doping of other rare earth elements. For example, europium doped yttrium oxide, Y_2O_3:Eu^{3+}, is a superior red phosphor with a quantum efficiency of nearly 100%. It presents good luminescent characteristic, acceptable atmospheric stability, reduced degradation under applied voltages, and the lack of hazardous constituents as opposed to sulphide phosphors, thus it has commercial applications in fluorescent lamps, projection televisions, and field emission displays (FEDs). With respect to the physical properties, yttrium oxide has a high melting point (T_m = 2430 °C), which is higher than that of a number of other well-known oxides, such as alumina, zirconia, yttrium aluminum garnet (YAG), and spinel. Yttrium oxide exists as a cubic phase and is stable up to melting point without any phase transformations. Furthermore, it has a very large unit cell, which results in large unit slip distances. Hence, it is expected that plastic deformation in yttrium oxide by dislocation motion would be difficult. These properties endow it with usefulness as bulk ceramics for refractory applications. In addition, yttrium oxide has found applications in a wide variety of catalytic reactions owing to its basic nature. Therefore, the preparation of yttrium oxide, as well as its precursors, has attracted much academic attention. Yttrium oxide has been prepared in many shapes, like spherical particle (Tomaszewski et al., 1997; Sharma et al., 1998; Yang et al., 2007), rod (Wan et al., 2005), tube (Li et al., 2004; Fang et al., 2003; Tang et al., 2003; Wang et al., 2003; Zhang et al., 2008), prism (Zhang et al., 2005), plate (Wan et al., 2005) or sheet (Zhang et al., 2008), and wire. Our group has systematically investigated the phase distribution and morphology of products synthesized under hydrothermal conditions (Li & Yanagisawa, 2008). It was found that by simply adjusting the hydrothermal temperature and initial pH value of the starting solution, yttrium oxide with a diversity of well-defined morphologies like sheet, rod, needle and tube were successfully fabricated from different precursors. At the same time, the particle size of products could be controlled in a wide range.

Hitherto, the reported synthetic pathways to yttrium oxide nanowire were unlimitedly accomplished by the preparation of its precursors through hydrothermal reaction followed by decomposing these precursors into oxide. Hydrothermal reaction will not result in yttrium oxide directly, because yttrium oxide is not stable under hydrothermal conditions at temperature lower than 550 °C, according to the phase diagram of Y_2O_3-H_2O system (Shafer & Roy, 1959).

Reactant	pH	Hydrothermal T.	Size	Reference
Y_2O_3 + 10% HNO_3 + 10% NaOH/KOH	13	180 °C for 12-24 h	Did not mention	Wang & Li, 2003
$Y(NO_3)_3 \cdot 6H_2O$ + 10% KOH	10	180 °C for 10 h	D: 30-40 nm; L: 1.4 μm	Li et al., 2004
Y_2O_3 + concentrated HNO_3 + 0.2M NaOH	12-13	170-180 °C for 12 h	D:100-300 nm; L: 1-2 μm	Bai et al., 2005
Y_2O_3 + concentrated HNO_3 + NaOH + PEG 2000	13	170 °C for 24 h	D: 85-95 nm; L: several tens of microns	Wu et al., 2005
Y_2O_3 + 3.0M HNO_3 + ammoina solution	12	200 °C for 24 h	D: 40-100 nm; L: ~1.4 μm	Li & Yanagisawa, 2008
Y(NO3)3 ·6 H_2O (5mol% Sm) +15% TEAH	13	180 °C for 10 h	D: 50 ± 3 nm; L: several tens of microns	Zahir et al., 2009

Table 1. Summarized literatures on $Y(OH)_3$ nanowires synthesized by precipitation-hydrothermal synthetic method

Wirelike yttrium hydroxide with hexagonal structure has been used as the precursor of yttrium oxide nanowires. Generally, the reported synthesis of yttrium hydroxide nanowires is accomplished through a two-step route, which is also called precipitation-hydrothermal synthetic method. Firstly, the precipitation of precursor colloids is obtained by adjusting the pH value of $Y(NO_3)_3$ aqueous solution, sometimes produced by dissolving Y_2O_3 in nitric acid, to 10-14 through the addition of NaOH/KOH solution. Then this precipitation is hydrothermally treated at 120-180 °C. The products are usually composed of discrete, single-crystalline hydroxide nanowires with diameter of 30-300 nm. Corresponding oxide nanowires were received after thermal treating the hydroxide at 450-700 °C in air. For instance, Wang et al. reported the synthesis of hydroxide nanowires of yttrium and of other rare-earth elements through this synthetic method at 180 °C and pH around 13 (Wang & Li, 2003). The driving force for the growth was attributed to the crystal structure of yttrium hydroxide. Li et al. explained the formation mechanism of yttrium hydroxide nanowires by the complex interaction and balance between the chemical potential and the rate of ionic motion (Li et al., 2004). Europium doped yttrium oxide nanowires were also prepared by this method by mixing appropriate amount of europium oxide and yttrium oxide. The obtained Y_2O_3:Eu^{3+} nanowires exhibited strong red $^5D_0$-$^7F_2$ transitions in fluorescent spectra. Our group found that using ammonia solution replacing NaOH/KOH as precipitation reagent, yttrium hydroxide nanowires could be obtained at pH 12.5 via hydrothermal reaction at 200 °C (Li & Yanagisawa, 2008). Several other researches also reported similar synthetic routes to yttrium hydroxide and yttrium oxide nanowires. Table 1 summarized the synthesis conditions and product information obtained in literatures. However, the hydroxide and oxide nanowires prepared through this synthetic method usually exhibit low aspect ratio, at most 50, which makes it look more like needles than wires, as shown in Fig. 1. Yttrium hydroxide nanowires with higher aspect ratio could be prepared by a polymer-assisted hydrothermal method (Wu et al., 2005). It is presumed that polyethylene glycol (PEG) played an important role on the formation of hydroxide nanowires. However, the final products coexisted with a mixture of nanorods and nanobelts. Recently, it is reported that after the doping of 5 mol% Sm^{3+} into yttrium and using tetraethylammonium hydroxide (TEAH) as precipitator, very well separated yttrium hydroxide nanowires with high aspect ratio were prepared (Zahir et al., 2009). The authors attributed the growth of yttrium hydroxide nanowires to the presence of Sm^{3+} and tetraethylammonium ion.

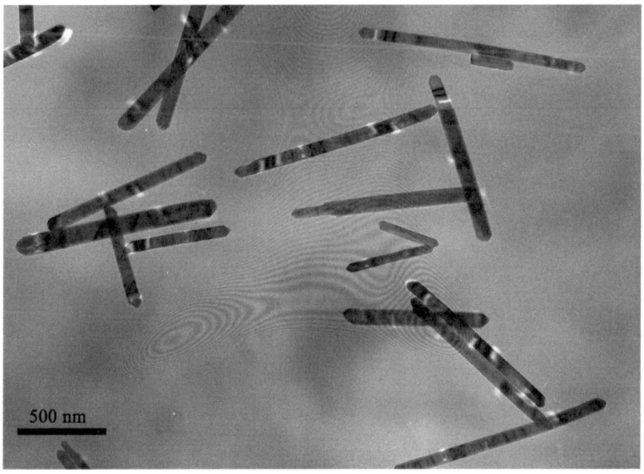

Fig. 1. TEM image of hexagonal $Y(OH)_3$ needles prepared by precipitation-hydrothermal synthetic method. The reaction was conducted at pH 12.5, 160 °C for 24h, with ammonia solution as precipitator

Our group has recently exploited a simpler one-step hydrothermal synthetic method to prepare yttrium hydroxide nanowires, where yttrium oxide powder instead of soluble yttrium salts was directly used as a starting material, and the reaction was conducted under near neutral conditions. Unlike yttrium hydroxide nanowires obtained by precipitation-hydrothermal synthetic method, the nanowires synthesized via this one-step method showed unusual, bundle-like morphology with much higher aspect ratio (Li et al., 2009). Furthermore, we found that two other yttrium compounds, yttrium oxide nitrate hydroxide (Li & Yanagisawa, 2008) and yttrium chloride hydroxide could also be prepared in wire shape in nanometer size. Both of them were synthesized through precipitation-hydrothermal synthetic method, using yttrium nitrate and yttrium chloride as starting material respectively. In this chapter, we will introduce the preparation and characterization of the three wirelike yttrium compounds, as well as their conversion to yttrium oxide nanowires.

2. Yttrium oxide nanowires prepared from yttrium hydroxide

As has introduced above, the yttrium hydroxide nanowires prepared from the precipitation-hydrothermal synthetic method usually exhibit low aspect ratio. Recently, we synthesized yttrium hydroxide nanowires by an acetic acid (HAc) assisted hydrothermal method under near neutral conditions. The as-synthesized nanowires are of higher aspect ratio up to 2000. The preparation was effected through a one-step hydrothermal procedure, simpler than the precipitation-hydrothermal synthetic method. Typically, the appropriate quantities of Y_2O_3 powder and 0.067 mol/L acetic acid aqueous solution were mixed at room temperature and hydrothermally treated at 200 °C for 24 h under agitation. Powder X-ray diffraction (XRD) characterization indicated that the product could be indexed to a pure hexagonal phase of $Y(OH)_3$, with lattice cell constants of a = 6.2610Å and c = 3.5440 Å, identical to the reported data in JCPDS cards (JCPDS 83-2042), as shown in Fig. 2a. The (110) reflection around 2θ of 28° is much more pronounced compared with the literature data, indicating the preferred growth direction of these nanowires is along c-axis. Scanning electron microscopy (SEM) and transmission electron microscopy (TEM) observations indicated that the product was composed of bundles of nanowires, as shown in Fig. 3a-c. These bundles were relatively uniform in size with length of 30-40 µm. Each bundles consisted of numerous of nanowires with diameter of 20-50 nm. Assuming each nanowire spans the whole length of the bundle, the nanowires had an aspect ratio of 600-2000, which was much higher than that of the nanowires fabricated from the precipitation-hydrothermal synthetic method. High resolution transmission electron microscopic (HRTEM) image and selective area electron diffraction (SAED) pattern taken from an individual nanowire revealed its single-crystalline nature, as shown in Fig. 3d, e. After calcinations at 500 °C, the obtained yttrium oxide nanowires almost remained unchanged in morphology except that the wire diameter decreased slightly, owing to the higher density of oxide compared with that of hydroxide, as shown in Fig. 3f. The decomposition behaviours of yttrium hydroxide and other precursors will be summarized later.

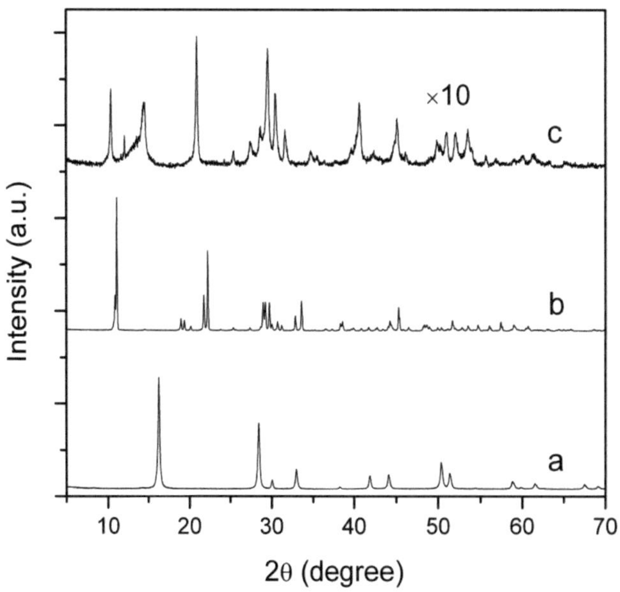

Fig. 2. Typical XRD patterns of three precursors of yttrium oxide. (a) hexagonal $Y(OH)_3$, (b) $Y_4O(OH)_9(NO_3)$ and (c) yttrium chloride hydroxide hydrate

Fig. 3. SEM (a, b), TEM (c), HRTEM (d) and SAED (e) images of bundles of $Y(OH)_3$ nanowires. SAED pattern was taken from the region indicated by white circle in (c). (f) TEM image of corresponding Y_2O_3 nanowires.

It was found that the formation of yttrium hydroxide nanowires was closely correlated with the presence of HAc and its concentration. Hydrothermal reaction conducted in pure water suggested that a majority of yttrium oxide could be converted into hydroxide after 72 hours of hydrothermal treatment, indicating that HAc was not necessary for this transformation. This result was consistent with the Y_2O_3-H_2O phase diagram. However, the conversion process without the participation of HAc was very slow, and the product morphology totally differed from that of yttrium hydroxide prepared with the assistance of HAc. Our experiment results proved that in the presence of HAc the whole reaction was completed within several hours. Moreover, it was also found that the product morphology was sensitive to HAc concentration. Yttrium hydroxide could be prepared at concentration between 0.001 and 0.01 mol/L. The optimum HAc concentration for the growth of uniform nanowires was 0.006-0.007 mol/L. Lower concentration would result in irregular rods in a wide scale of size, while higher concentration would lead to the formation of an unknown phase, probably being yttrium acetate compound.

The investigation on the intermediate products obtained during different growth stages suggested that with the process of time, yttrium oxide disappeared step by step, while the appearance of hydroxide was observed at the same time. The transformation from oxide into hydroxide was completed within two hours. SEM images (Fig. 4) indicated that at the early stage of hydrothermal reaction, the sample was composed of irregular blocks and a small quantity of microrods with quasi-hexagonal geometry, which was unreacted yttrium oxide and the formed hydroxide, respectively. The individual microrod was constructed by numerous of nanorods fused in the same orientation. Along with the proceeding of hydrothermal reaction, these rods became less faceted, and the diameter is not uniform along the long axis, forming a spindle-like structure. The tips of the rod looked like tubes. As the reaction proceeded, the size of these microrods, as well as the aspect ratio increased, suggesting that the growth along the long-axis was faster than that along the short-axis. This result was supported by XRD results that the intensity of (110) diffraction of yttrium hydroxide increased remarkably in comparison with that of (101) diffraction. Subsequently, the tips of these rods began to split into wires and the whole rod developed into a bundle of nanowires eventually.

It is likely that the growth is accomplished through an oriented attachment and the subsequent dissolution process from the defect sites. Firstly, yttrium oxide gradually dissolves into the solution, and yttrium hydroxide nucleates into 1D nanocrystals owing to the high anisotropic structure along *c*-axis of hexagonal phase. These nanocrystals then align side by side and fuse together spontaneously by oriented attachment (Penn & Banfield, 1999; Banfield et al., 2000; Penn et al., 2001), giving rise to the formation of quasi-single crystals, where the thermodynamic driving force is the substantial reduction of surface energy contributed by the interface elimination. During that procedure, some defects like dislocation are introduced into the microrods as a result of imperfect attachment (Penn & Banfield, 1998), which accumulates during the crystal growth. Therefore, the microrod gradually loses its anisotropy. When the starting yttrium oxide is depleted, the reactant concentration in the solution begins to drop. As the concentration drops to a certain level, the particles stop its growth on the whole. However, because the site where defect exists has a higher solubility than those free of defects, there is an internal concentration gradient. Consequently, the microrods dissolve from the defects and grow onto other facets to maintain the solubility equilibrium, resulting in bundles of nanowires. This process is similar as the intraparticle ripening mechanism proposed by Peng et al (Peng & Peng, 2001). The formation mechanism of the nanowires is schematically shown in Fig. 5. The growth of the yttrium hydroxide nanowires begins from 1D nanocrystals, which align in the same direction and fuse together via oriented attachment process, leading to microrods. These microrods grow into tube-like structures by loss of anisotropy and growing preferable along circumference. Finally, the microrods are dissolved from the defect sites and split into nanowires.

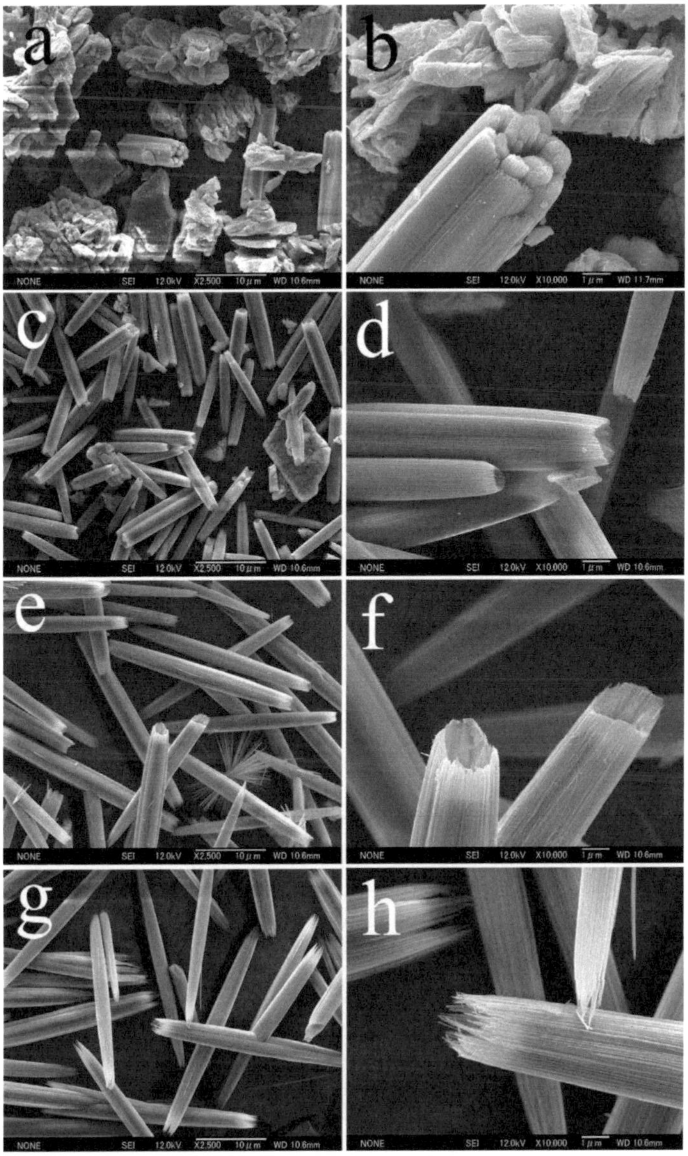

Fig. 4. SEM images of products synthesized at 200 °C in 0.067 mol/L HAc for 0h (a, b), 0.5h (c, d), 1.0 h (e, f) and 2.0 h (g, h)

HAc plays a key role in the formation of the unusual nanowires, probably acting in two ways. Firstly, the addition of HAc changes the pH value of solution, thus accelerates the conversion from yttrium oxide to hydroxide. Secondly, acetic acid molecules or ions adsorb to the surface of yttrium hydroxide crystallites, introducing defects and blocking the attachment mechanism.

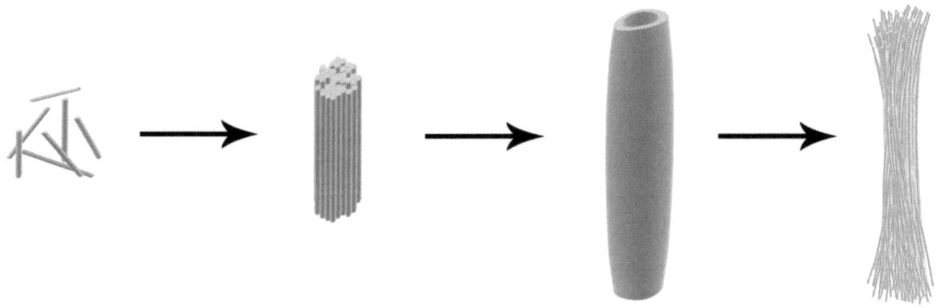

Fig. 5. Sketch for the growth mechanism of $Y(OH)_3$ nanowires. Reprinted with permission from Cryst. Growth Des, 9 (2):978-981, Copyright 2009 American Chemical Society.

3. Yttrium oxide nanowires prepared from yttrium oxide nitrate hydroxide

As we have mentioned in the introduction of this chapter, yttrium hydroxide nanowires could be prepared through a precipitation-hydrothermal synthetic method. Our group have systematically investigated the precipitation-hydrothermal reaction conducted in a wide range of temperature and pH value. It was found that three compounds, hydrated yttrium nitrate hydroxide $Y_2(OH)_{5.14}(NO_3)_{0.86}$, yttrium oxide nitrate hydroxide $Y_4O(OH)_9(NO_3)$, and hexagonal yttrium hydroxide $Y(OH)_3$ were obtained when hydrothermal reaction were conducted at 80-220 °C and pH 6.0-13.5. $Y_2(OH)_{5.14}(NO_3)_{0.86}$ and $Y_4O(OH)_9(NO_3)$ occurred under near neutral to weak basic conditions, and low hydrothermal temperature would result in $Y_2(OH)_{5.14}(NO_3)_{0.86}$, while $Y_4O(OH)_9(NO_3)$ was received at higher temperature. If the hydrothermal reaction was conducted under strong basic condition, $Y(OH)_3$ was obtained. The pH boundary where $Y(OH)_3$ was formed was 11.25 and 13.0 when ammonia and NaOH solution was used as precipitator respectively. Among the three yttrium compounds, yttrium oxide nitrate hydroxide $Y_4O(OH)_9(NO_3)$ could also be synthesized in wire shape, which made it a alternative precursor for the preparation of yttrium oxide nanowires.

The synthesis of $Y_4O(OH)_9(NO_3)$ was almost identical with that of yttrium hydroxide with the exception that the reaction was conducted at a slightly lower pH value. In a typical synthesis process, an appropriate quantity of yttrium oxide was dissolved in nitric acid. Then ammonia/NaOH solution was added to adjust the solution to a designated pH value. Products were received after hydrothermal treatment at 80-220 °C for 24 h and then investigated by XRD and TEM characterizations. As the XRD pattern shown in Fig. 2b, the product could be attributed to a pure monoclinic phase. TEM images (Fig. 6a, b) illustrated that these nanowires were straight and well-dispersed with uniform diameter along the longitudinal axis. The diameter ranged from 30 to 50 nm, while the length was up to tens of micrometers. Some nanowires were as long as several millimetres, spanning the whole visual field under the TEM observation at low magnification (×4). That means the nanowires could be prepared with extraordinary high aspect ratio. Some of the shorter nanowires probably were broken fragments from the longer ones by the ultrasonic during the TEM pre-treatment. HRTEM and SAED characterization revealed its single crystalline nature. It seems that the driving force for the spontaneous growth of nanowires is the anisotropic characteristic of monoclinic structure of yttrium oxide nitrate hydroxide. The yttrium oxide obtained through calcinations maintained the wire shape. However, it became porous in contrast with its precursor, as shown in Fig. 6c.

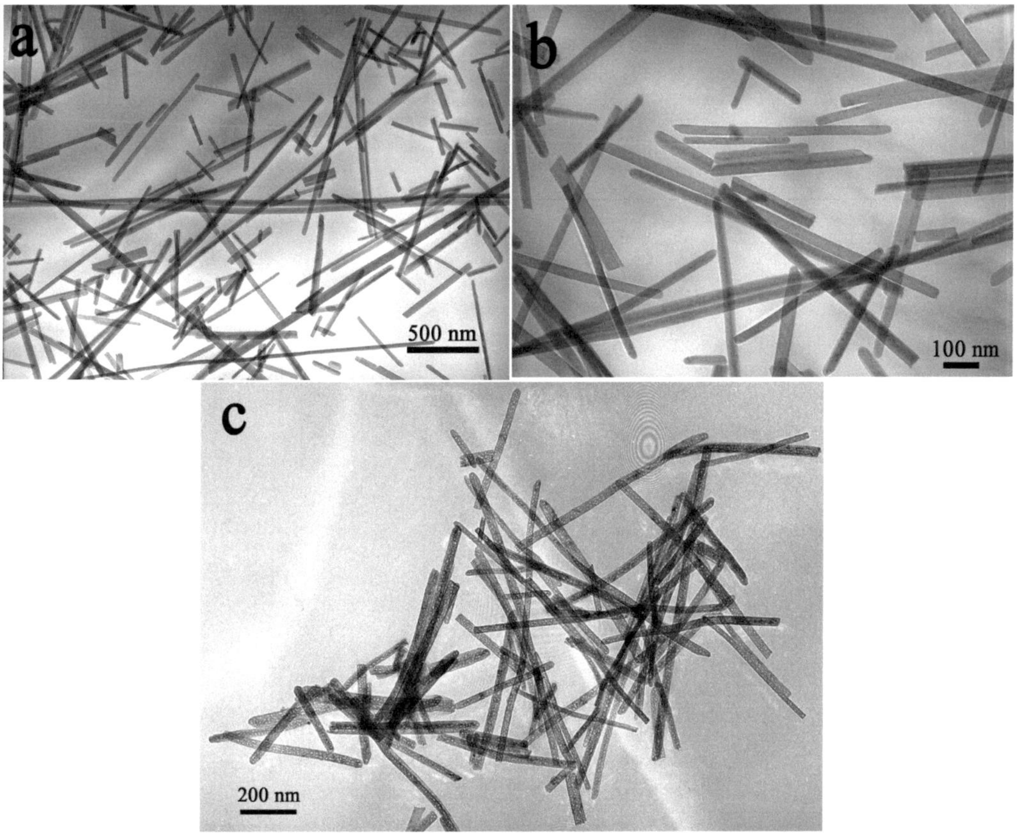

Fig. 6. TEM images of $Y_4O(OH)_9(NO_3)$ nanowires hydrothermally synthesized at 200 °C, pH 12.5 (a, b), and of Y_2O_3 nanowires prepared thereby (c)

This yttrium compound could be prepared in a wide pH range, from 6.0 to 13.0. However, nanometer sized product was only obtained at relatively high pH value, for example, pH 12.5, while low pH value would lead to the formation of microrods with hexagonal cross-section. Along with the increasing pH value, the size of the product decreased. This is because that high pH value leads to the formation of large amount of nuclei and small particles is favoured due to the suppressed growth of each nuclear. It should be noted that further increasing the pH value would give rise to the formation of yttrium hydroxide, which contains more hydroxyl in its structure and thus is more stable than yttrium oxide nitrate hydroxide under stronger basic conditions. Therefore, by carefully controlling the pH value of the colloidal solution and its concentration, nanosized products could be prepared. The effects of hydrothermal temperature on products were also studied by fixing the pH value and changing the reaction temperature from 140 °C to 220 °C. The products did not show much difference in morphology except for the size distribution narrowed down with increasing temperature, indicating that pH value was the predominant factor in strong basic media, and the morphologies were less dependent on reaction temperature as compared with pH value.

4. Yttrium oxide nanowires prepared from yttrium chloride hydroxide

In the aforementioned precipitation-hydrothermal synthesis, it is yttrium nitrate that was used as starting material. What will happen if yttrium chloride took the place of yttrium nitrate? Our study revealed that different products were received in this case, and Cl element was usually involved in the final products. Among them, an yttrium chloride hydroxide with indefinite composition could be prepared in wire shape.

Fig. 7. SEM image of yttrium chloride hydroxide nanowires

Similar as that of $Y_4O(OH)_9(NO_3)$, the synthesis was started by dissolving yttrium oxide in hydrochloric acid. Ammonia solution was then added to adjust the solution to designated pH value. Flocculent product was received after hydrothermal treatment at 200 °C for 12 h. The crystal structure and morphology of the products were studied by XRD and SEM, respectively. Fig. 2c shows the XRD pattern of the product hydrothermally synthesized at pH 9.5. It was poorly crystallized and could not be identified to any known phase. SEM observation revealed that the product consisted of sub-micrometer wires. The wires were of uniform diameter ranging from 100 to 300 nm, while the length was up to tens of microns, as shown in Fig. 7. It is possible that the wirelike product synthesized at pH 9.5 belongs to chloride hydroxide hydrate, represented by $Y(OH)_xCl_{3-x} \cdot yH_2O$. These wirelike products could be prepared within the pH range from 9.50 to 10.25.

5. Decomposition behaviours of yttrium oxide precursors

All of the three above-mentioned yttrium compounds could be converted into yttrium oxide through calcination in air. Their wirelike morphologies were maintained except for the slight shrinkage in size, arising from the higher density of yttrium oxide compared with their precursors. Because of the difference in structural compositions, these compounds exhibited different characteristics during calcination. Their decomposition behaviours were illustrated in Fig. 8.

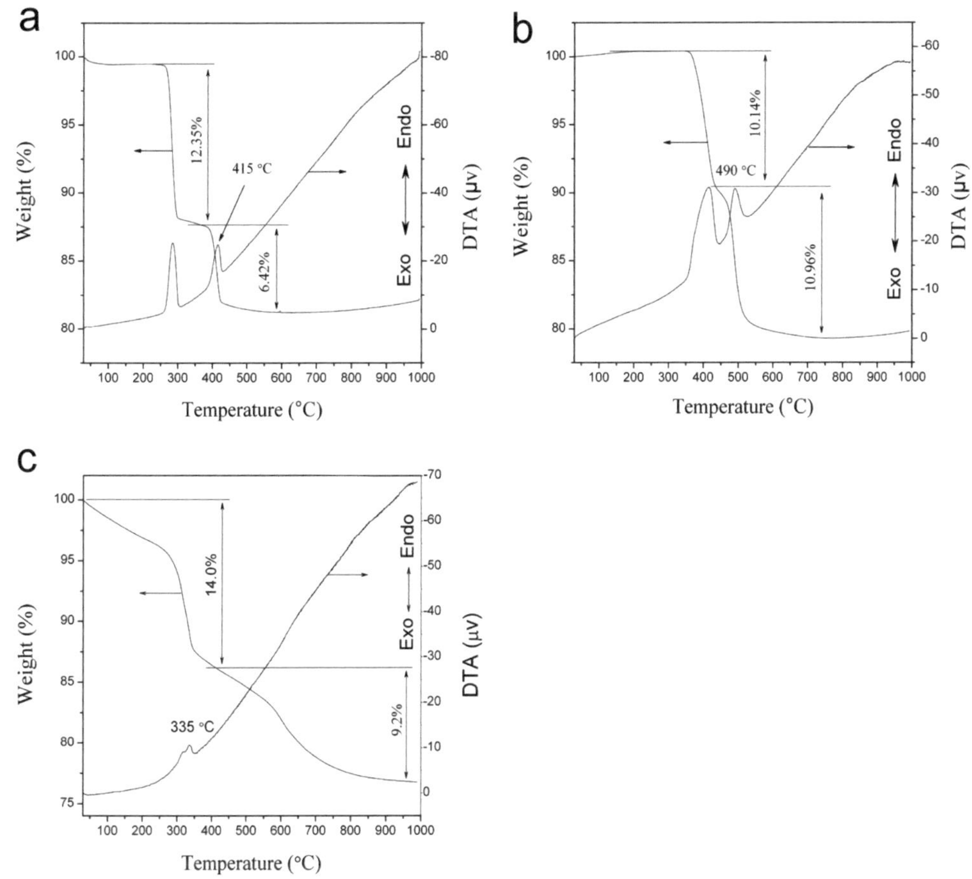

Fig. 8. Typical TG-DTA curves of (a) hexagonal $Y(OH)_3$, (b) $Y_4O(OH)_9(NO_3)$ and (c) yttrium chloride hydroxide hydrate. Reproduced in part with permission from J. Solid State Chem. 181 (8):1738-1743, Copyright 2008 Elsevier.

During heat treatment, yttrium hydroxide nanowires synthesized by the HAc-assisted hydrothermal method underwent two step-wised decomposition procedures, where yttrium hydroxide was firstly transformed into an intermediate oxyhydroxide, YOOH, and then it converted into oxide at elevated temperature. The conversions occurred at around 280 and 415 °C, respectively, as indicated by two DTA endothermic peaks in Fig. 8a.

For $Y_4O(OH)_9(NO_3)$ nanowires, there were two decomposition procedures during the heat treatment, as shown in Fig. 8b. Ion chromatography analysis suggested that the NO_3^- content of intermediate products did not decrease until temperature increased to 460 °C, indicating that N-O species were released in the second step. The weight loss of the first procedure between 330 °C to 450 °C was 10.14%, lower than the theoretical value associated with the total release of water (13.8%). It suggests that some hydroxyl was stilled remained in the structure. Following this step, sample transformed into oxide at around 490°C, accompanied by the release of N-O species and water.

With respect to yttrium chloride hydroxide, it also underwent two decomposition procedures during calcinations, as shown in Fig. 8c. The first one below 400 °C gave weight loss of 14.0%, which was consistent with the release of water, and the weight loss of 9.2% during the second step corresponded to the release of HCl. Its calcination behaviour exhibited an overall weight loss of 23.2%.

6. Conclusion

In summary, we have prepared three wirelike yttrium compounds by hydrothermal method, which were hexagonal yttrium hydroxide, yttrium oxide nitrate hydroxide, and yttrium chloride hydroxide. Hexagonal yttrium hydroxide nanowires were synthesized as single crystals from yttrium oxide powder by a simple acetic acid-assisted hydrothermal method under near neutral conditions. The nanowires show unusual, bundle-like morphology with diameter of 20-50 nm and aspect ratio of 600-2000. The growth of nanowires involved the oriented attachment of 1D nanocrystals to form microrods and selective dissolution from defect sites to form bundles of nanowires. Another wirelike yttrium compound, $Y_4O(OH)_9(NO_3)$ were obtained through conventional precipitation-hydrothermal synthetic route under basic condition. These nanowires were straight and well-dispersed single crystals, with extraordinary high aspect ratio and uniform diameter ranging from 30 to 50 nm. By substituting yttrium nitrate with yttrium chloride, yttrium chloride hydroxide nanowires with diameter of 100-300 nm and length up to tens of microns were synthesized at pH 9.50-10.25 via hydrothermal reaction.

All of the three compounds could be utilized to fabricate yttrium oxide nanowires through thermal treatment. This one dimensional (1D) structure may lead to new opportunities in yttrium chemistry. For example, the yttrium oxide nanowires showed intriguing high-temperature stability. Our experiments suggested that after calcination at 1400 °C, the yttrium oxide nanowires obtained from yttrium chloride hydroxide remained their wire shape, which makes it promising material for refractory applications, such as refractory insulation and high temperature gas filtration. Furthermore, the formation of the three compounds all went through a dissolution-crystallization procedure. Therefore, homogeneous rare earth doped yttrium oxide nanowires could be prepared by using mixture of yttrium and other rare earth oxide as starting material. By this means, 1D luminescent materials like Y_2O_3:Eu^{3+} nanowires can be fabricated, which may extend its application and benefit the understanding of the luminescent mechanism in low dimensional materials.

7. References

Bai, X., Song, H.; Yu, L.; Yang, L.; Liu, Z.; Pan, G.; Lu, S.; Ren, X.; Lei, Y. & Fan, L. (2005). Luminescent properties of pure cubic phase Y_2O_3/Eu^{3+} nanotubes/nanowires prepared by a hydrothermal method. *The Journal of Physical Chemistry B*, Vol. 109, No. 32, (Aug 2005) 15236-15242, ISSN 1520-6106

Banfield, J. F.; Welch, S. A.; Zhang, H.; Ebert, T. T. & Penn, R. L. (2000). Aggregation-based crystal growth and microstructure development in natural iron oxyhydroxide biomineralization products. *Science,* Vol. 289, No. 5480, (Aug 2000) 751-754, ISSN 0036-8075

Fang, Y. P., Xu, A. W., You, L. P., Song, R. Q., Yu, J. C. Zhang, H. X., Li, Q. & Liu, H. Q. (2003). Hydrothermal synthesis of rare earth (Tb, Y) hydroxide and oxide nanotubes. *Advanced Functional Materials,* Vol. 13, No. 12, (Dec 2003) 955-960, ISSN 1616-3028

Jadhav, A. P., Kim, C. W., Cha, H. G., Pawar, A. U., Jadhav, N. A., Pal, U. & Kang, Y. S. (2009). Effect of different surfactants on the size control and optical properties of Y_2O_3:Eu^{3+} nanoparticles prepared by coprecipitation method. *Journal of Physical Chemistry C,* Vol. 113, No. 31, (Aug 2009) 13600-13604, ISSN 1932-7447

Li, N. & Yanagisawa, K. (2008). Controlling the morphology of yttrium oxide through different precursors synthesized by hydrothermal method. *Journal of Solid State Chemistry,* Vol. 181, No. 8, (Aug 2008) 1738-1743, ISSN 0022-4596

Li, N., Yanagisawa, K. & Kumada, N. (2009). Facile hydrothermal synthesis of yttrium hydroxide nanowires. *Crystal Growth & Design,* Vol. 9, No. 2, (Feb 2009) 978-981, ISSN 1528-7483

Li, Q., Feng, C., Jiao, Q., Guo, L., Liu, C. & Xu, H. B. (2004). Shape-controlled synthesis of yttria nanocrystals under hydrothermal conditions. *Physica Status Solidi (a),* Vol. 201, No. 14, (Nov 2004) 3055-3059, ISSN 0031-8965

Li, W., Wang, X. & Li, Y. (2004). Single-step in situ synthesis of double bond-grafted yttrium-hydroxide nanotube core-shell structures. *Chemical Communications,* No. 2, (Jan 2004) 164-165, ISSN 1359-7345

Mao, Y. B., Bargar, J., Toney, M. & Chang, J. P. (2008). Correlation between luminescent properties and local coordination environment for erbium dopant in yttrium oxide nanotubes. *Journal of Applied Physics,* Vol. 103, No. 9, (May, 2008) 094316 :1-8 ISSN 0021-8979

Mao, Y. B., Guo, X., Tran, T., Wang, K. L., Shih, C. K. & Chang, J. P. (2009). Luminescent properties of ensemble and individual erbium-doped yttrium oxide nanotubes. *Journal of Applied Physics,* Vol. 105, No. 9, (May 2009) 094329 :1-4 ISSN 0021-8979

Peng, L. X., Hong, Z. L., Li, G. H. & Wang, M. Q. (2007). Synthesis of yttrium oxide hydroxide nitrate nanorods by hydrothermal method. *Proceeding of 4th China International Conference on High-Performance Ceramics,* Chengdu, Oct 2005, *High-Performance Ceramics IV, Pts 1-3,* Vol. 336-338, 2151-2153, Trans tech publications LTD, ISBN 1013-9826, Switzerland

Peng, Z. A. & Peng X. G. (2001). Mechanisms of the shape evolution of CdSe nanocrystals. *Journal of the American Chemical Society,* Vol. 123, No. 7, (Feb 2001) 1389-1395, ISSN 0002-7863

Penn, R. L. & Banfield, J. F. (1998). Imperfect oriented attachment: dislocation generation in defect-free nanocrystals. *Science,* Vol. 281, No. 5379 (Aug 1998) 969-971, ISSN 0036-8075

Penn, R. L. & Banfield, J. F. (1999). Morphology development and crystal growth in nanocrystalline aggregates under hydrothermal conditions: Insights from titania. *Geochimica Et Cosmochimica Acta,* Vol. 63, No. 10, (May 1999) 1549-1557, ISSN 0016-7037

Penn, R. L.; Oskam, G.; Strathmann, T. J.; Searson, P. C.; Stone, A. T. & Veblen, D. R. (2001). Epitaxial assembly in aged colloids. *Journal of Physical Chemistry B*, Vol. 105, No. 11, (Mar 2001) 2177-2182, ISSN 1089-5647

Shafer, M. W. , & Rustum, R. (1959). Rare-earth polymorphism and phase equilibria in rare-earth oxide-water systems. *Journal of the American Ceramic Society*, Vol. 42 No. 11 (Nov 1959) 563-570, ISSN 1551-2916

Tang, Q., Liu, Z., Li, S., Zhang, S., Liu, X. & Qian, Y. (2003). Synthesis of yttrium hydroxide and oxide nanotubes. *Journal of Crystal Growth*, Vol. 259, No. 1-2, (Nov 2003) 208-214, ISSN 0022-0248

Tanner, P. A. & Fu, L. S. (2009). Morphology of Y_2O_3:Eu^{3+} prepared by hydrothermal synthesis. *Chemical Physics Letters*, Vol. 470, No. 1-3, (Feb 2009) 75-79, ISSN 0009-2614

Tomaszewski, H., Weglarz, H. & Roger, D. G. (1997). Crystallization of yttria under hydrothermal conditions. *Journal of the European Ceramic Society*, Vol. 17, No. 2-3, 403-406, ISSN 0955-2219

Wan, J., Wang, Z., Chen, X., Mu, L. & Qian, Y. (2005). Shape-tailored photoluminescent intensity of red phosphor Y_2O_3:Eu^{3+}. *Journal of Crystal Growth*, Vol. 284, No. 3-4, (Nov 2005) 538-543, ISSN 0022-0248

Wang, X. & Li, Y. D. (2003). Rare-earth-compound nanowires, nanotubes, and fullerene-like nanoparticles: Synthesis, characterization, and properties. *Chemistry-a European Journal*, Vol. 9, No. 22, (Nov 2003) 5627-5635, ISSN 0947-6539

Wu, X. C., Tao, Y. R., Gao, F., Dong, L. & Hu, Z. (2005). Preparation and photoluminescence of yttrium hydroxide and yttrium oxide doped with europium nanowires. *Journal of Crystal Growth*, Vol. 277, No. 1-4, (Apr 2005) 643-649, ISSN 0022-0248

Xu, Z. X., Hong, Z. L., Zhao, Q. C., Peng, L. X. & Zhang, P. Y. (2006). Preparation and luminescence properties of Y_2O_3:Eu^{3+} nanorods via post annealing process. *Journal of Rare Earths*, Vol. 24, (Dec 2006) 111-114, ISSN 1002-0721

Yang, J., Quan, Z. W., Kong, D. Y., Liu, X. M. & Lin, J. (2007). Y_2O_3:Eu^{3+} microspheres: Solvothermal synthesis and luminescence properties. *Crystal Growth & Design*, Vol. 7, No. 4, (Apr 2007) 730-735, ISSN 1528-7483

Zahir, M. H., Suzuki, T., Fujishiro, Y. & Awano, M. (2009). Synthesis and characterization of Sm^{3+}-doped $Y(OH)_3$ and Y_2O_3 nanowires and their NO reduction activity. *Journal of Alloys and Compounds*, Vol. 476, No. 1-2, (May 2009) 335-340, ISSN 0925-8388

Zhang, J., Liu, Z., Lin, J. & Fang, J. (2005). Y_2O_3 Microprisms with trilobal cross section. *Crystal Growth & Design*, Vol. 5, No. 4, (Jul-Aug 2005) 1527-1530, ISSN 1528-7483

Zhang, N., Liu, X. H., Yi, R., Shi, R. R., Gao, G. H. & Qiu, G. Z. (2008). Selective and controlled synthesis of single-crystalline yttrium hydroxide/oxide nanosheets and nanotubes. *Journal of Physical Chemistry C*, Vol. 112, No. 46, (Nov 2008) 17788-17795, ISSN 1932-7447

Zhong, S. L., Wang, S. J., Xu, H. P., Hou, H. Q., Wen, Z. B., Li, P., Wang, S. P. & Xu, R. (2009). Spindlelike Y_2O_3:Eu^{3+} nanorod bundles: hydrothermal synthesis and photoluminescence properties. *Journal of Materials Science*, Vol. 44, No. 14, (Jul 2009) 3687-3693, ISSN 0022-2461

9

Polymer Nanowires

Baojun Li and Xiaobo Xing
State Key Laboratory of Optoelectronic Materials and Technologies, School of Physics and Engineering, Sun Yat-Sen University, Guangzhou 510275, China

1. Introduction

The development of nanotechnology in photonics offers significant scientific and technological potentials [1,2]. It fosters the substantial efforts for exploring novel materials, developing easy fabrication techniques, reducing the size of photonic components, improving device integration density, and fabricating low-cost nanodevices. Since nanometer-scale photonic wires are highly desirable for applications in high density and miniaturized photonic integrated circuits (PICs), subwavelength-diameter wires have been drawn and demonstrated by flame-heated silica fiber [3] and bulk glasses [4] methods. The method provides an easy and cheap photonic wires manufacturing technique, but a steady temperature distribution is required in the drawing region and the lengths of the fabricated wires are limited to 4 mm and tens of millimeters. Later, flame-brushing and microheater-brushing techniques were proposed to fabricate nanowires from silica fibers [5−7] and compound-glass fibers [8], respectively. The length of the fabricated nanowire is extended to 110 mm, but this technique requires extremely good control of processing temperature and airflow around the nanowires. On the other hand, due to relatively low flexibility of silica and glasses, only wire-based simple devices such as 2×2 branch coupler [4,9], single-ring resonantor [3,9], and single Mach-Zehnder (MZ) interferometer [10] were assembled and demonstrated. Compared with silica and glasses, polymers have high flexibility, and thus can be randomly bent/squeezed and molded to have variety of shapes [11,12].

In this chapter, we will introduce a novel polymer material of poly(trimethylene terephthalate) (PTT) which was used for the first time in nanophotonics. Then we will introduce a simple PTT polymer fabrication method, i.e. one-step drawing process. Followed by a description of a series of ultracompact devices and nanophotonic device arrays assembled by the PTT nanowires. Final is a perspective.

2. PTT polymer

PTT polymer can be extracted from petroleum or refined from corn and was original being used in carpet and textile. It was first synthesized by Whinfield and Dickson [13]. It is a semicrystalline aromatic polyester synthesized by the polycondensation of terephthalic acid and 1,3-propanediol (PDO) [14]. Mainly because of its raw materials, PDO, was very expensive, so the PTT with three methylene groups in the polymer's repeating unit remained an obscure polymer and never went beyond laboratory status despite having some

outstanding properties as a textile fiber [15]. Fortunately, recent breakthroughs in PDO synthesis via hydroformylation of ethylene oxide [16], process improvements in the traditional synthetic route through acrolein and the promising bio-engineering route using glycerol as starting material for PDO have reduced the price of PDO by at least an order of magnitude.
PTT is made by the melt polycondensation of PDO with either terephthalic acid or dimethyl terephthalate. The chemical structure is shown as follows:

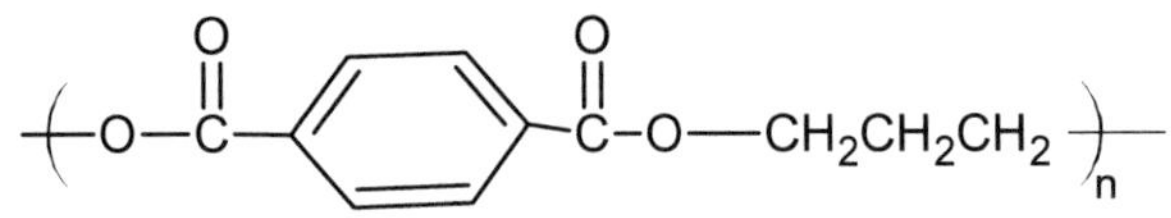

2.1 Thermal properties

PTT is a semicrystalline polymer with a differential scanning calorimetry (DSC) peak melting point of 228 °C. The equilibrium melting points, T_m, are 238 °C [17, 18], 244 °C [19] and 248 °C [20, 21]. Since the T_m values of semicrystalline polymers are usually 15–25 °C higher than their DSC T_m values, and also because the lower-melting poly(butylenes terephthalate) (PBT) has a T_m of 245 °C, it is not unreasonable to assume that 248 °C might be a more appropriate T_m for PTT.
Double DSC melting peaks are frequently observed in PTT, especially when scanning at a low heating rate of < 5 °C/min. Huang *et al.* [21] studied the effect of crystallization temperature, time and cooling rate on these PTT double melting peaks. PTT had two melting peaks at about 222 and 228 °C when it was crystallized at 210 °C between 10 and 60 min. With prolonged crystallization to 360 min, the two peaks merged into one with a peak temperature of 225.5 °C. However, when it was crystallized at a lower temperature of 180 °C, instead of having two melting peaks, PTT had a main 228 °C endotherm with a shoulder at about 219 °C. Unlike the 210 °C crystallized samples, prolonging the crystallization time to 360 min did not change the overall shape of the DSC curves. The lower-melting shoulder persisted and moved slightly to a higher temperature. The origin of the double melting peaks was attributed to the lower-melting crystals being recrystallized and melted at a higher temperature during the heating scan, or to the polymer having two populations of crystals of substantially different sizes.

2.2 Mechanical and physical properties

The mechanical and physical properties of PTT [22], measured from injection molded American Society of Testing Methods (ASTM) Type II samples, is given in Table 1.

2.3 Intrinsic birefringence

Birefringence is the difference between the refractive index parallel, $n_{//}$, and perpendicular, $n_{\perp}$, to the draw direction. When the polymer is perfectly oriented, the birefringence is called intrinsic birefringence. For PTT crystal, the average refractive index, $n_{av} = (n_{xx} + n_{yy} + n_{zz})/3$, was found 1.638 [24].
For uniaxial orientation, the transverse refractive index $n_{\perp}$ of PTT, averaged from $(n_{xx} + n_{yy})/2$, was 1.636. The intrinsic birefringence of PTT crystal, $\Delta n_c^o = n_{//} - n_{\perp}$ was 0.029. The two-phase model explains why highly oriented PTT has such a low birefringence; the contribution of the crystalline-phase orientation to the overall birefringence is very small, although the polymer crystallinity and degree of crystal orientation are very high.

Property	PTT
Tensile strength (MPa)	67.6
Flexural modulus (GPa)	2.76
Heat distortion temperature, at 1.8 MPa (°C)	59
Notched Izod impact (J/m)	48
Specific gravity	1.35
Mold shrinkage (m/m)	0.02
Dielectric strength (V/mil)	530
Dielectric constant, at 1 MHz	3.0
Dissipation factor, at 1 MHz	0.015
Volume resistivity (ohm cm)s	1.00×10^{16}

Table 1. The mechanical and physical properties of PTT [22, 23]

2.4 Transmittance

PTT (M_n =17,300, M_w =35,200, Shell Chemical Company) was compression molded at 250 °C into 25-μm-thick film. It was quenched in water. Wide-angle X-ray diffraction data (WAXD) were taken on a Bruker D8 Advance Diffractometer (Germany). Figure 1 shows the WAXD spectra of PTT film. There is no apparent diffraction peak in the WAXD pattern, so the PTT film is amorphous pattern. Figure 2 shows that the transmittance of the PTT film (25-μm amorphous film) is about 90% in the wavelength region of 400 to 2000 nm [25]. Its good transparency from visible to near-infrared together with its relatively large refractive index (1.638) [24] can provide fine optical confinement for advanced nanowires and nanophotonic devices.

2.5 Elastic recovery

The fiber industry has long been aware of PTT's good tensile elastic recovery. Ward *et al.* [15] studied the deformation behavior of PTT fiber. Figure 3 shows that the PTT elastic recovery and permanent set nearly tracked that of nylon 66 up to 30% strain.

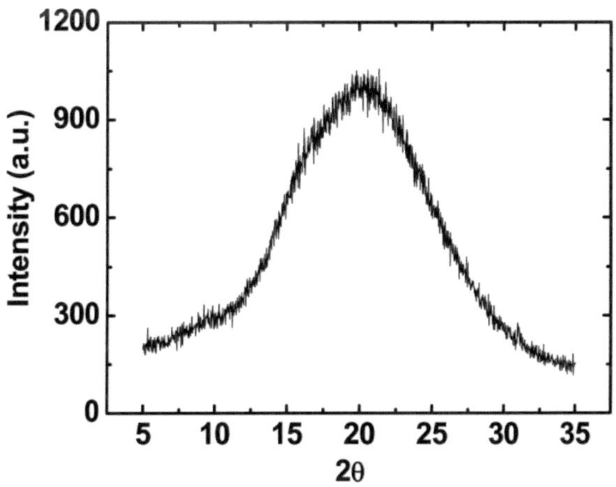

Fig. 1. The WAXD spectra of a 25-μm-thick PTT film.

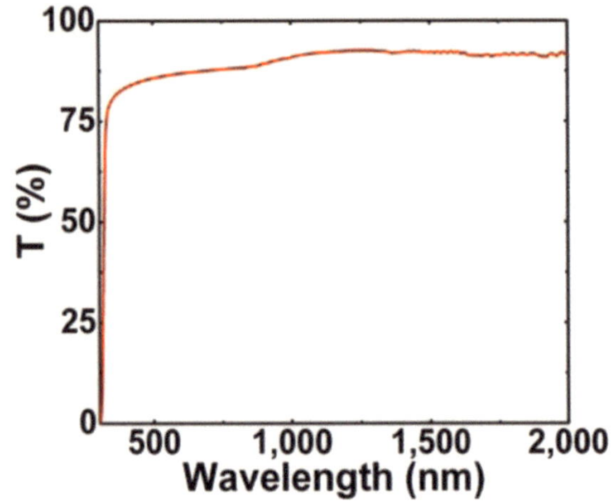

Fig. 2. Room-temperature transmission spectrum of a 25-μm-thick amorphous PTT film. It shows the transmittance of about 90% in the wavelength region of 400 to 2000 nm.

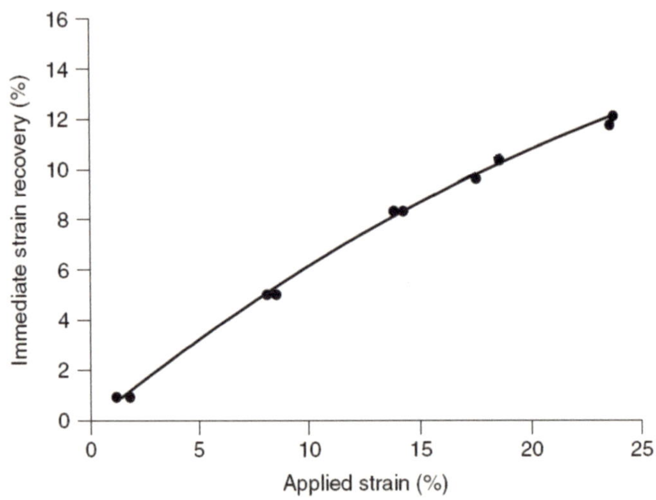

Fig. 3. PTT elastic recovery at various strains [26]

The unusually good PTT elastic recovery property was thought to relate to the plateau region of its stress–strain curve. Jakeways *et al.* [27] deformed PTT fiber *in situ* in a wide-angle X-ray diffractometer, and measured the changes in the fiber period *d*-spacing along the *c*-axis as a function of strain. The crystalline chain responded and deformed immediately to the applied strain. It increased in direct proportion to the applied strain up to 4% before deviating from affine deformation (Figure 4). Furthermore, the deformation below this critical strain was reversible. This microscopic reversible crystal deformation was tied to PTT chain conformation. Since initial deformation involves torsional rotation of the *gauche* methylene C–C bonds, the force is only a fraction of the bond stretching force. Thus, polymer with a helical chain conformation tends to have a low crystal modulus, about 20% of the predicted modulus if the chains were in all-*trans* conformations [28].

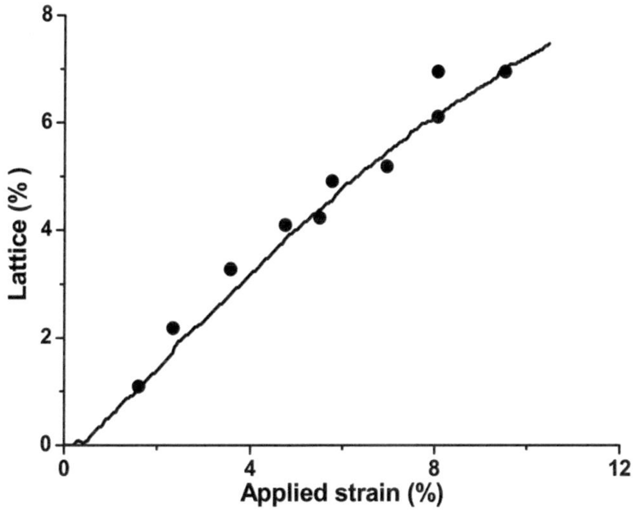

Fig. 4. Changes of PTT fiber *c*-axis lattice strains measured from X-ray diffraction spacings as a function of applied external strains [26].

PTT indeed has a very low X-ray crystal modulus of 2.59 GPa [29]. This value is probably too low because a highly oriented PTT fiber with about 50% crystallinity already had a 2.5 GPa modulus [15]. Using the CERRIUS II molecular simulation program, Jo [30] calculated a 12.2 GPa crystal modulus. Because of the low crystal modulus, the PTT crystalline chain responded and deformed immediately with applied macroscopic strain. The crystalline chain deformation is reversible, and is the driving force for the good elastic recovery. Jakeways *et al.* [27] addressed only the crystalline chain deformation to explain PTT's elastic recovery. The macroscopic deformation must also simultaneously involve the partially irreversible amorphous chain deformation. The higher the applied strain, then the more dominant was the irreversible amorphous deformation with deviation from affine deformation.

2.6 Drawing behavior

Figure 5 shows the tensile stress–strain curves of PTT at various temperatures [31]. At room temperature, PTT is ductile. It yields at 5.4% strain, cold draws with a natural draw ratio of about 3.2, strain-hardens and breaks at 360% strain. With increasing draw temperature, the yield stress decreases and the elongation at break increases. At 50 °C, just above the glass transition temperature, T_g, PTT becomes rubbery. The Young's modulus decreases by about two orders of magnitude from 1140 to 12.9 MPa, and the overall drawability increases with a strain at break of nearly 600%. However, when the draw temperature was increased to 75 °C, 30 °C above the T_g, instead of becoming more rubbery and capable of higher draw, PTT became ductile again. The modulus unexpectedly increased by more than tenfold to 189 MPa. The overall drawability decreased with a drop in breaking strain to 390%. In fact, the 75 °C stress–strain curve looked similar to that of the one room temperature. Instead of the conventional experience of increasing drawability with increasing temperature, the PTT draw first increased, went through a maximum and decreased, all happening over a narrow range of temperature from room temperature to T_g +30 °C.

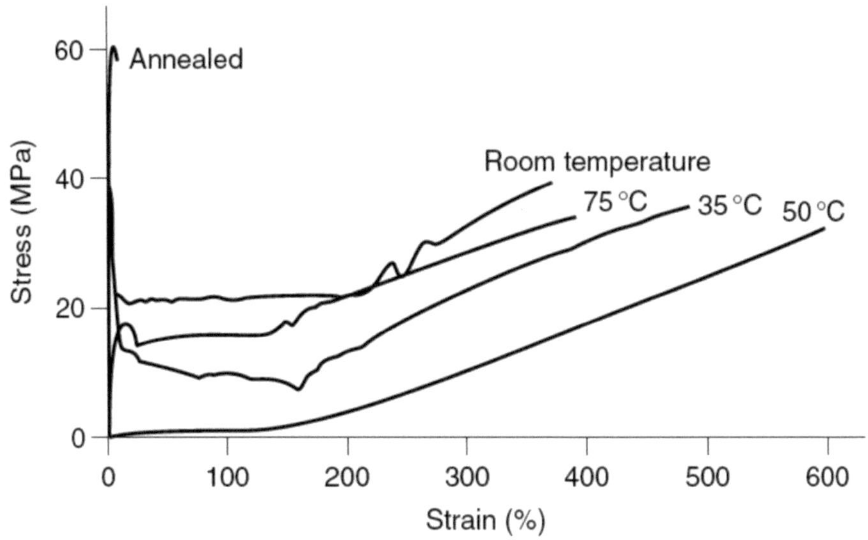

Fig. 5. PTT stress–strain curves at draw temperatures below and above the glass transition temperature [31].

This unexpected drawing behavior was due to the onset of cold crystallization competing with drawing. To draw a polymer, it is usually heated to temperatures above its T_g so that the polymer became soft to facilitate draw. However, when the polymer cold-crystallized during hot drawing, the increase in crystallinity increased the polymer's modulus and had an opposing effect to hot drawing, and therefore reduced the drawability. When cold-crystallization proceeded at a fast rate, PTT transitioned *in situ* from rubbery to ductile, such as the 75 °C draw shown in Figure 5. At higher temperatures, the polymer could become brittle and cause draw failure. Thus, PTT drawability depends on its initial thermal history and morphology, and whether it can cold-crystallize or not during hot drawing. This behavior must be taken into account in PTT fiber spinning and drawing.

3. PTT nanowire drawing

The PTT nanowire can be fabricated by using the electrospinning method reported in Ref. 32. However, the nanowire fabricated by electrospinning with large surface roughness and length inhomogeneity induces high optical loss while the PTT nanofibrous mats with diameters only 200−600 nm. The simple way to fabricate PTT wire would be direct drawing technique, which is a one-step tip-drawing process.

Figure 6 shows the schematic illustration of the drawing process. Figure 6a shows a vertical direction tip-drawing process while Figure 6b shows a random direction tip-drawing process. PTT pellets (melt temperature T_m = 225°C) was melt by a heating plate and the temperature was kept at around 250°C during the wire drawing. First, an iron or silica rod/tip with radius of about 125 μm is being approached and its tip is immerged into the molten PTT. Then the rod tip is retracted from the molten PTT with a speed of 0.1−1 m/s, leaving a PTT wire extending between the molten PTT and the tip. The extended PTT wire is quickly quenched in air and finally, a naked amorphous PTT nanowire is formed.

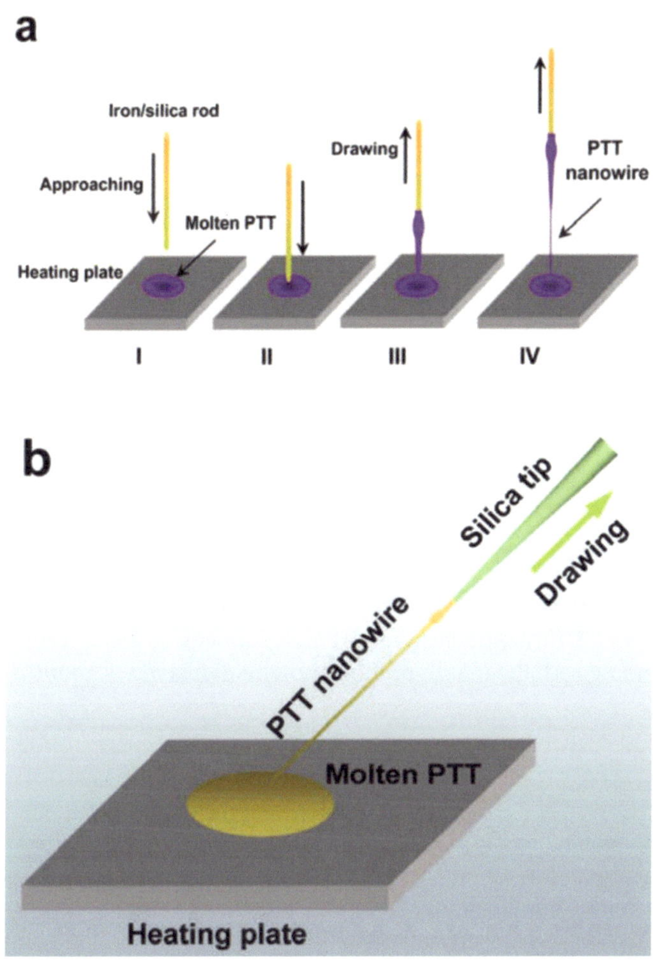

Fig. 6. Schematic illustration of nanowires fabrication by direct drawing process from PTT melt. (a) vertical direction drawing. I, An iron or silica rod is approaching the molten PTT. II, The rod end is immersed into the molten PTT. III, The rod conglutinated PTT is being drawn out. IV, A PTT nanowire is formed. (b) random direction drawing. The arrow shows the drawing direction.

4. PTT nanowire characterization

In this section, we will show the PTT nanowires by SEM/TEM images and introduce a systematic study on the dependence of nanowire evanescent wave coupling efficiency on wavelength, cross-angle, and core-diameter. Finally, optical losses of the PTT nanowires will be given, which were measured by evanescent coupling.

4.1 SEM and TEM images

To display the fabricated PTT nanowire, a 250-mm-long PTT nanowire was coiled on a 12-μm-diameter PTT bending rod. A scanning electron microscope (SEM) image (Figure 7) shows part of the coiled nanowire with a length of about 200 mm and an average diameter of 280 nm. The diameter variation ratio is about 8.4×10^{-8}. Figure 8 show a nanowire with diameter of 105 nm.

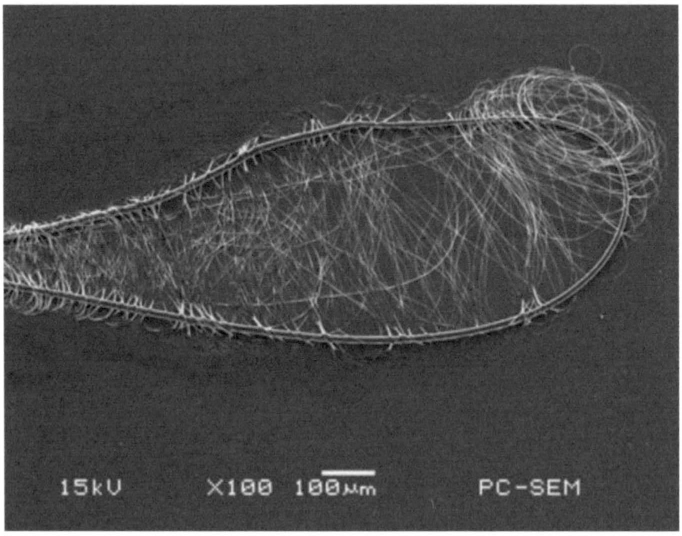

Fig. 7. SEM image of a PTT nanowire with average diameter of 280 nm coiled on a 12-μm-diameter PTT bending rod, the length of the nanowire displayed is about 200 mm.

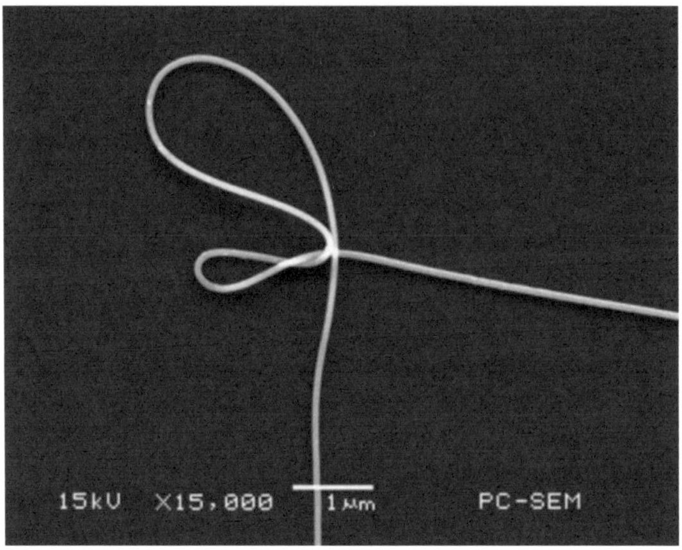

Fig. 8. 105-nm-diameter flexible nanowire rings.

For comparison, a scanning electron microscope (SEM) image (Figure 9a) shows that a 167-nm-diameter PNW with length of about 155 μm is positioned together with a 1.26-μm-diameter straight PTT rod. Figure 9b shows that four 110-nm-diameter PNWs are bent and positioned together. Figure 9c demonstrates flexible and elastic connection by pulling the PNWs with diameters of 140 and 170 nm. To examine surface roughness of the PNWs, high-magnification transmission electron microscope (TEM) was done. Figure 9d shows a TEM image of a 190-nm-diameter nanowire, indicating no visible defect and irregularity on the surface of the PNWs. Typical average sidewall root-mean-square roughness of the PNW is 0.28 nm. The electron diffraction pattern (inset of Figure 9d) demonstrates that the obtained PNW is amorphous. The results demonstrate that the obtained PNWs exhibit high surface smoothness, length uniformity, high mechanical properties, and excellent flexibility.

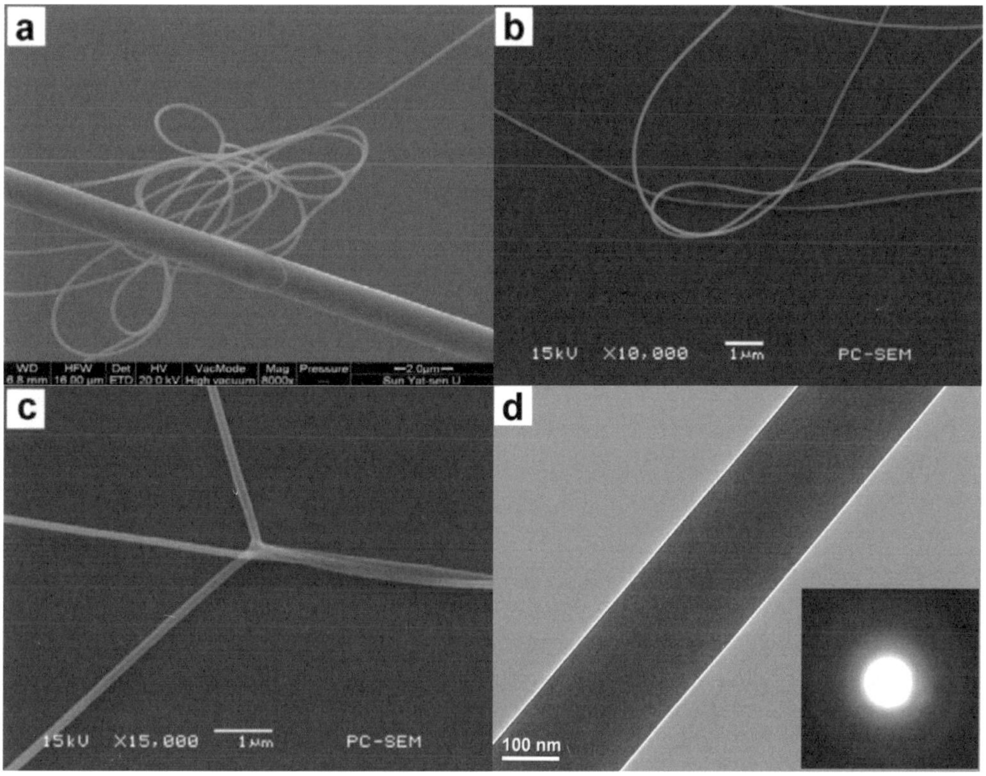

Fig. 9. Electron micrographs of PTT nanowires and nanowire structures. a–c, SEM images: (a) Comparison of a 167-nm-diameter PNW (about 155 μm long) with a 1.26-μm-diameter straight PTT rod. (b) Four 110-nm-diameter PNWs are positioned together. (c) Flexible and elastic enough PNWs connection with diameters of 140 and 170 nm. (d) TEM image of a 190-nm-diameter PNW. The inset shows its electron diffraction pattern.

4.2 Evanescent coupling [33]

In the characterization of optical nanowires, subwavelength optical wires and miniaturized photonic devices, an unavoidable issue is to launch lights of different wavelengths into them efficiently. One can use end-to-end direct optical coupling, grating coupling or evanescent wave coupling methods. The most efficient way to launch light into nanowires and nanowire-based photonic devices is evanescent wave coupling using a silica tapered fiber. However, the coupling efficiency depends strongly on the launched wavelength, the cross-angle and the core-diameter of the silica taper and the measured nanowire. In this section, we will introduce a systematic experimental study on the wavelength, cross-angle, and core-diameter dependence of coupling efficiency in nanowire evanescent wave coupling between silica tapered fiber and nanowires.

In the measurement, the PTT nanowire (PNW) was fixed by two micro-stages supports on a rotation disk by using a micromanipulator under an optical microscope. An 800-nm-diameter silica tapered fiber, which was fabricated by a direct drawing method, was fixed by a precision manipulator under an optical microscope and was used for light launching. Figure 10a shows the schematic measurement setup. When the silica tapered fiber and the PNW are close enough, they will contact each other through van der Waals and electrostatic attractive forces. In the experiment, when we rotate the disk slowly, different cross-angles between the silica tapered fiber and the PNW can be formed. The output optical power from the PNW was measured using an optical power meter. Figure 10b shows schematically the measurement of the output power at the end of the PNW.

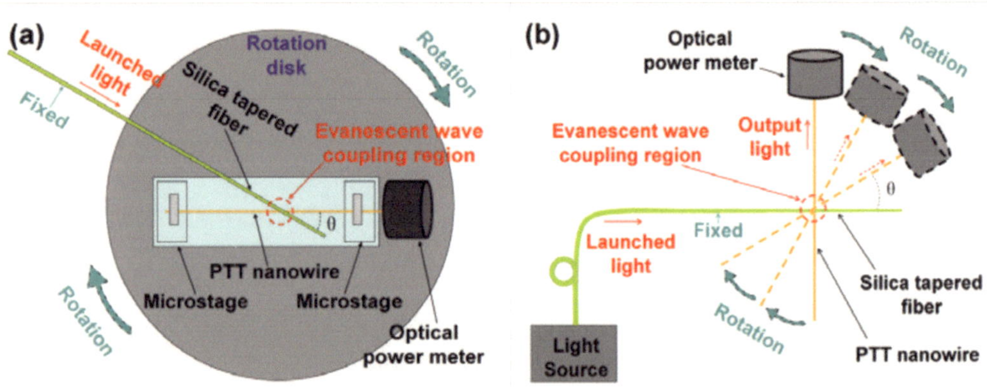

Fig. 10. Schematic diagram of the experimental setup. (a) Top view of the setup, (b) illustration of the measurement with different cross angles.

In the experiment, we first launched green (532 nm) light into a 640-nm-diameter PNW through the 800-nm-diameter silica tapered fiber, and measured the output optical power coupled into the PNW at different cross angles by rotating the rotation disk. Figure 11a shows the optical microscope images of the coupling of green light at different cross angle from 90° to 0° with a 5° changing step. It can be seen that when the silica tapered fiber (horizontal one) and the PNW crosses perpendicularly there is negligible coupling to the PNW. As a result, a bright spot is observed at the end of the silica taper with a small weak spot occurring at the cross junction. Since the cross junction under the perpendicular situation is very small, the junction can be considered as a scattering point (about 640 to 1440 nm in diameter) and there is no optical coupling. With the cross angle is decreased the measured optical power from the end of the PNW increases; the spot at the end of the silica tapered fiber becomes dark and the spot at the junction disappears gradually. This is because the cross angle is decreased the coupling strength between the silica tapered fiber and the PNW increases, and more power is coupled into the PNW through evanescent wave coupling. As a result, the bright spot at the junction gradually disappears and the optical power coupled into the PNW reaches a maximum. Similar phenomena were observed when blue (473 nm), red (650 nm), and near-infrared (1310 and 1550 nm) lights were launched.

Figure 11b shows the measured coupling efficiency versus cross-angle (θ) at different wavelengths (473, 532, 650, 1310 and 1550 nm). From Figure 11b, it can be seen that the measured coupling efficiency decreases with increasing cross-angle (θ). This is because a larger cross-angle will induce a smaller overlap length between the silica tapered fiber and the PNW, and the coupling efficiency decreases with increasing cross-angle (θ). For example, for green (532 nm) light, at $\theta = 0°$, 63.4% optical power was coupled into the PNW from the silica taper. When $\theta = 45°$, the coupling efficiency was only 1.65%. Also, the measured coupling efficiency increases with increasing launched wavelength. This is because more optical power leaks out of the nanowire core when the launched light wavelength increases, and the leakage of the optical power enhances energy exchange between the nanowires within a short interaction length. For example, about 80% coupling

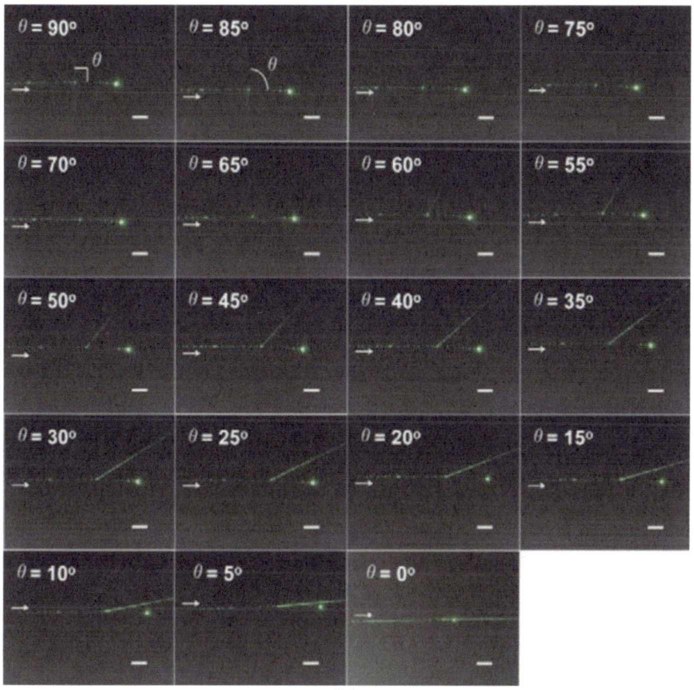

(a)

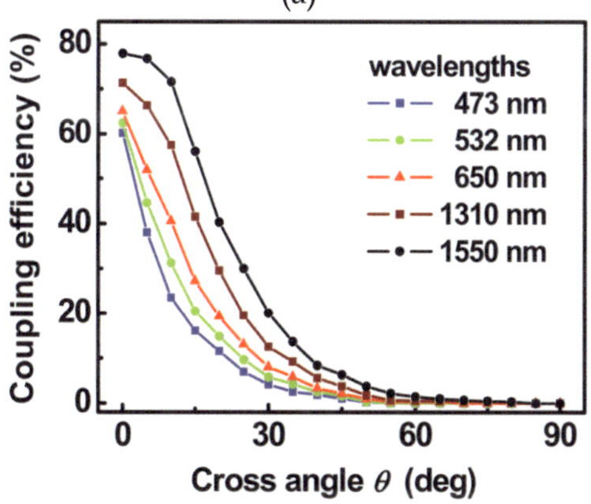

(b)

Fig. 11. (a) Optical microscope images of the evanescent wave coupling. Green (532 nm) light was launched from an 800-nm-diameter silica tapered fiber into a 640-nm-diameter PNW at different cross-angles from 90° to 0° with a 5° changing step. The white arrows show the propagation directions of the launched lights. The scale bar represents 20 μm. (b) Measured coupling efficiency versus cross-angle (θ) at different wavelengths.

efficiency is obtained for 1550 nm near-infrared light at $\theta = 0°$, and for 473 nm blue light the coupling efficiency is about 60% at $\theta = 0°$.

Similar dependence behavior of the coupling efficiency is also observed in a 790-nm-diameter PNW. Figure 12a shows red (650 nm) light was coupled from the 800-nm-diameter silica tapered fiber (horizontal) into a 790-nm-diameter PNW. Figure 12b shows the measured coupling efficiency versus cross-angle (θ) when lights with different wavelengths

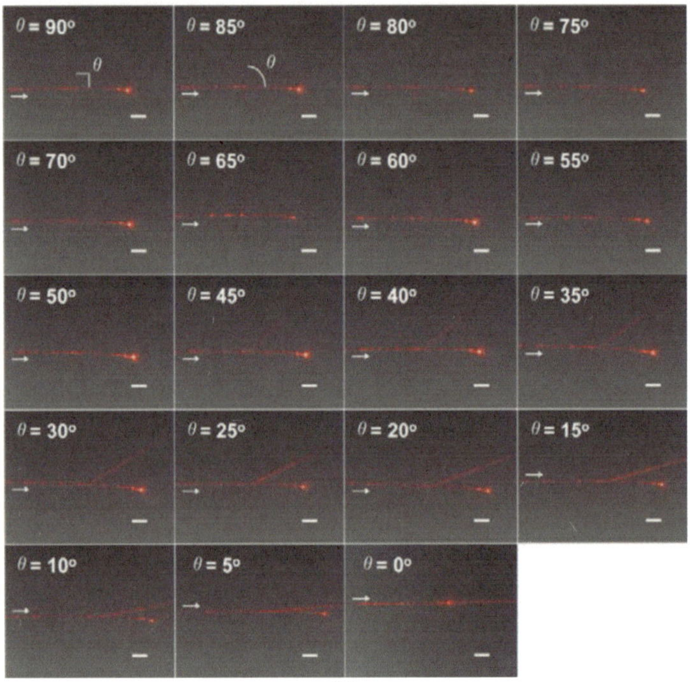

(a)

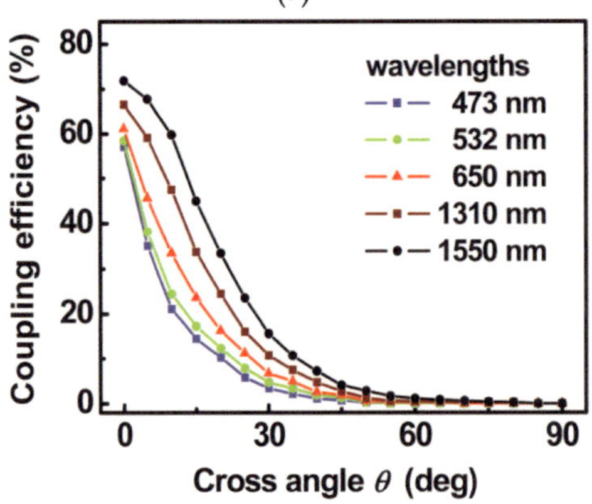

(b)

Fig. 12. (a) Optical microscope images of the evanescent wave coupling. Red (650 nm) light was launched from an 800-nm-diameter silica tapered fiber into a 790-nm-diameter PNW at different cross-angles from 90° to 0° with a 5° changing step. The white arrows show the propagation directions of the launched lights. The scale bar represents 20 μm. (b) Measured coupling efficiency versus the cross angle (θ) at different wavelengths.

(473, 532, 650, 1310, and 1550 nm) were individually launched into the 790-nm-diameter PNW. With decreasing cross-angle (θ) the output power increases. For example, taking the case if red (650 nm) light, when θ = 45°, only 1.88% optical power was coupled into the PNW. When θ = 15°, 24.1% optical power was coupled into the PNW. When θ = 0° the coupling efficiency reached a maximum value of 62.1%.

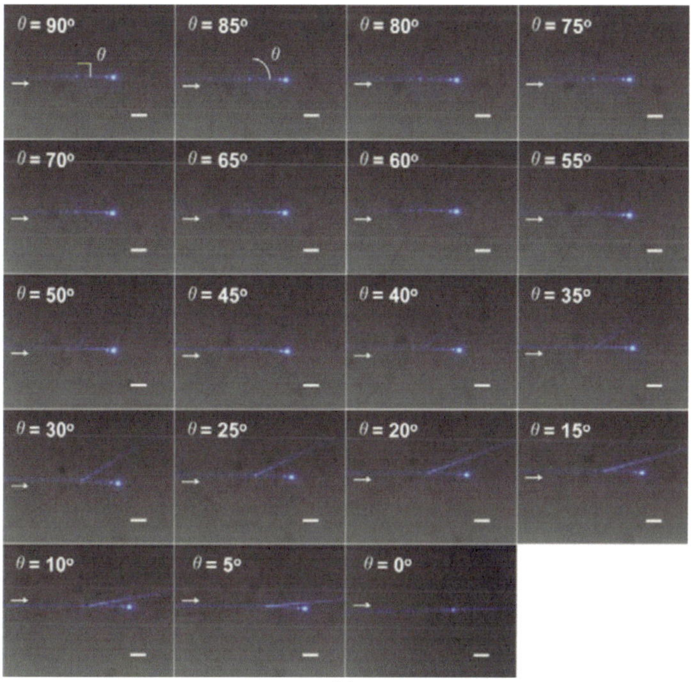

(a)

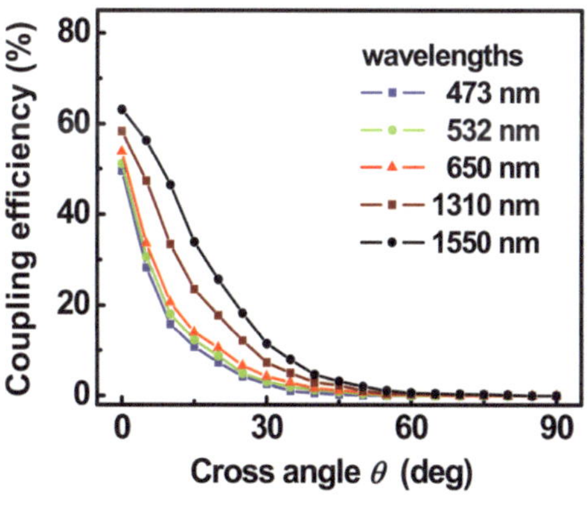

(b)

Fig. 13. (a) Optical microscope images of the evanescent wave coupling. Blue (473 nm) light was launched from an 800-nm-diameter silica tapered fiber into a 950-nm-diameter PNW at different cross-angles from 90° to 0° with a 5° changing step. The white arrows show the propagation directions of the launched lights. The scale bar represents 20 µm. (b) Measured coupling efficiency versus the cross angle (θ) at different wavelengths.

Figure 13a shows blue (473 nm) light coupled from the 800-nm-diameter silica tapered fiber (horizontal) to a 950-nm-diameter PNW. Figure 13b shows the measured coupling efficiency versus cross-angle (θ) when light at different wavelengths (473, 532, 650, 1310, and 1550 nm) were individually launched into the 950-nm-diameter PNW. For example, taking the case of blue (473 nm) light, when $\theta = 45°$, 21.9% power was coupled into the PNW. When $\theta = 15°$, about 10% power was coupled into the PNW. When $\theta = 0°$ the coupling efficiency reached a maximum value of 46.5%.

To investigate the core-diameter dependence, the coupling efficiency versus cross angle for different core-diameters at different wavelengths was plotted. Figure 14 shows the coupling efficiency of different nanowire diameters as a function of launched wavelength at $\theta = 0°$ and different nanowire diameters. It has been concluded that: 1) for the same cross angle and same nanowire diameter, the coupling efficiency will be higher for longer wavelength; 2) for the same wavelength and same cross angle, the coupling efficiency will be higher for smaller nanowire diameter; 3) for the same wavelength and same nanowire diameter, the coupling efficiency will be higher for smaller cross angle. As expected, to get the maximum coupling efficiency the silica tapered fiber used for light launching and the nanowire must be placed in parallel. Here it should be pointed out that, even for zero cross-angle (the two wires contacted in parallel), the coupling efficiency will be different for different overlap length.

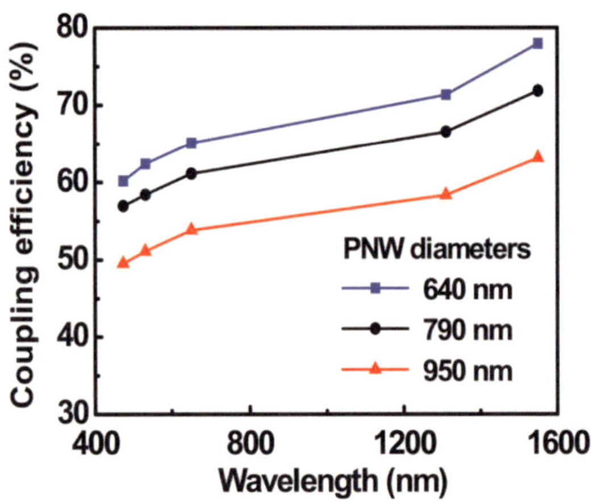

Fig. 14. Measured coupling efficiency versus different wavelengths of launched light at $\theta = 0°$ and different nanowire diameters.

4.3 Optical loss

Guided optical properties of the PTT nanowires were characterized by fixing the nanowires by two microstage supports, and launched lights of different wavelength into them by evanescent coupling through directional coupling as shown in Figure 15a. As an example, the figure shows green light (532 nm) was coupled into a 470-nm-diameter PTT nanowire bend from a submicro-taper silica fiber with a coupling length of 10.5 μm, where the upper red and yellow colors show the simulated evanescent coupling by the beam propagation method (BPM). It should be emphasized that some light scattering in the PTT nanowire was induced by surface contamination rather than surface roughness. The output powers from the PTT nanowires were measured by an optical power meter together with an optical spectrum analyzer. The optical losses of the PTT nanowires were measured by a cutback method. In the measurement, the original nanowire we used to evaluate the loss is 5 to 20 cm. In each cutback of the measurement, about 1-mm-long fiber was cut off from the output end of the nanowire. Figure 15b shows the plots of the PTT wire diameter versus measured optical loss at the wavelengths of 473, 532, 650, 1310, and 1550 nm. The measured rate of the PTT nanowires' loss over time is 0.4 dB/hour, which is smaller than that of the silica nanowires (about 1 dB/hour). The measured coupling efficiency is as high as 95% when a silica taper is parallel to the PTT wire.

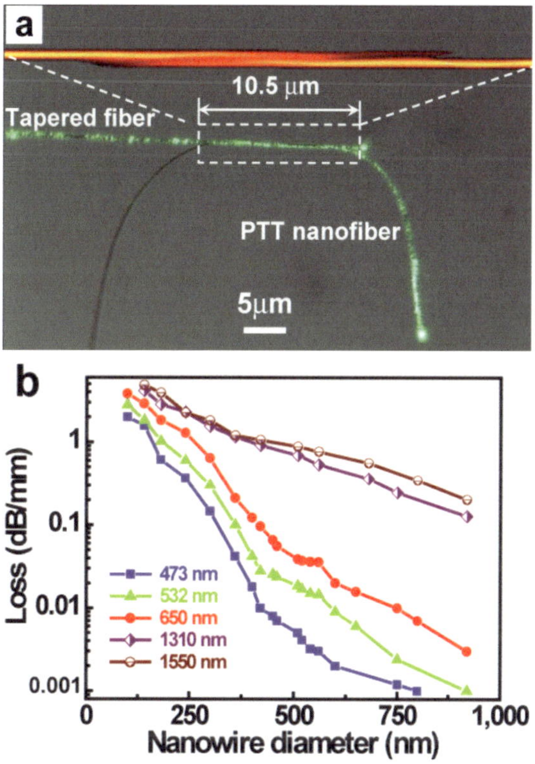

Fig. 15. Optical coupling method and the measured optical loss. (a) Optical microscope image of a tapered silica fiber launched 532 nm green light into a 470-nm-diameter PTT nanowire bend by evanescent coupling, where the coupling length is 10.5 μm. The upper red and yellow colors illustrate the coupling region simulated by the beam propagation method. (b) Optical loss of PTT wires versus different diameter at the wavelengths of 473, 532, 650, 1310, and 1550 nm.

5. Twisting method

There are several methods to assemble nanowire devices. Here we introduce a simple nanowire device assembly method, named "twisting method".

Figure 16 shows the twisting process. As an example, we use two nanowires to demonstrate the method: First, a nanowire was drawn from the PTT melt by a one-step direct drawing process and cut into two segments. Second, the two segment nanowires were placed in parallel on two microstage supports with their ends fixed (Figure 16a). Third, we rotated the right support in anti clockwise direction with a high precision while keeping left support fixed (Figure 16b). The rotation was stopped when desirable number of turns was obtained in the twisted region (*e.g.*, four twisted turns, Figure 16c). Finally, a twisted 2 × 2 coupling device was formed.

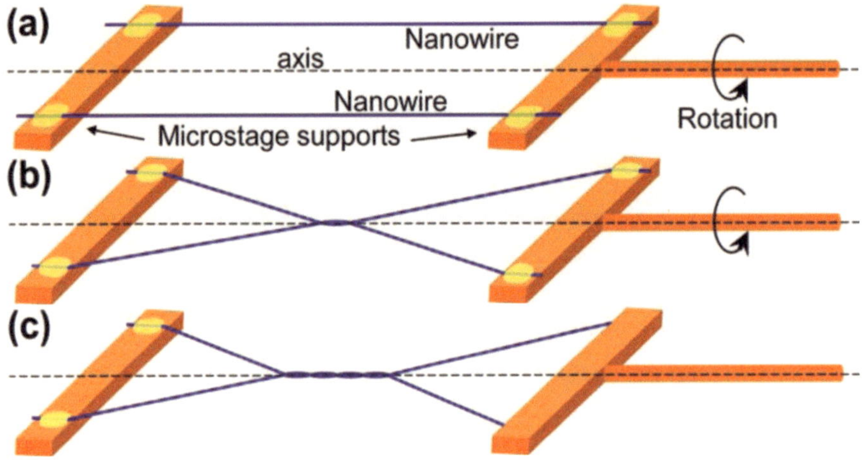

Fig. 16. (a) Two parallel PTT nanowires were fixed by two microstage supports. The left support is fixed and the right support can be rotated around the axis. (b) A twisted 2 × 2 structure with one twisted turn was formed by rotating the right support. (c) A twisted 2 × 2 structure with four twisted turns was formed by further rotating the right support.

By using the simple twisting method with microstage supports under a microscopy, any coupling device with multi-branches can be easily assembled depending on how many nanowires were used.

6. Nanodevices

6.1 SEM images of arbitrary nanodevices [34]

A series of nanodevices were assembled, as examples, Figure 17a shows a nano bird's nest (top right of the Figure). The nanowires can also be bent from 0 to 180°, as an example, Figure 17b shows a 45° bend with diameter of 280 nm while Figure 17c shows a 155° sharp bend with diameter of 160 nm. Figure 17d shows that a 340-nm-diameter nanowire was bent to a tweezer-shaped structure with a bending radius of 1.6 μm. Similarly, a 70-nm-diameter nanowire was first twisted, and then bent to form a scissor-shaped structure (Figure 17e). Figure 17f shows a 2×2 coupler with three twist turns in the coupling region by twisting 110-nm- and 150-nm-diameter nanowires.

Figure 18 shows the SEM images of some nanowire devices and device arrays. Figure 18a shows a directional coupler assembled by two parallel 750-nm-diameter nanowires with a 110 nm coupling gap. Figure 18b shows a Y-branching coupler assembled by two 360-nm-diameter nanowires without coupling gap. We further assembled a bending Y-branching coupler by a 210-nm-diameter 155° bend (inside up) and a 270-nm-diameter 120° bend (outside down) (Figure 18c), a 2×2 coupler by two 150-nm-diameter nanowires (Figure 18d), and a basic asymmetric MZ coupler by two 60-nm-diameter nanowires (Figure 18e).

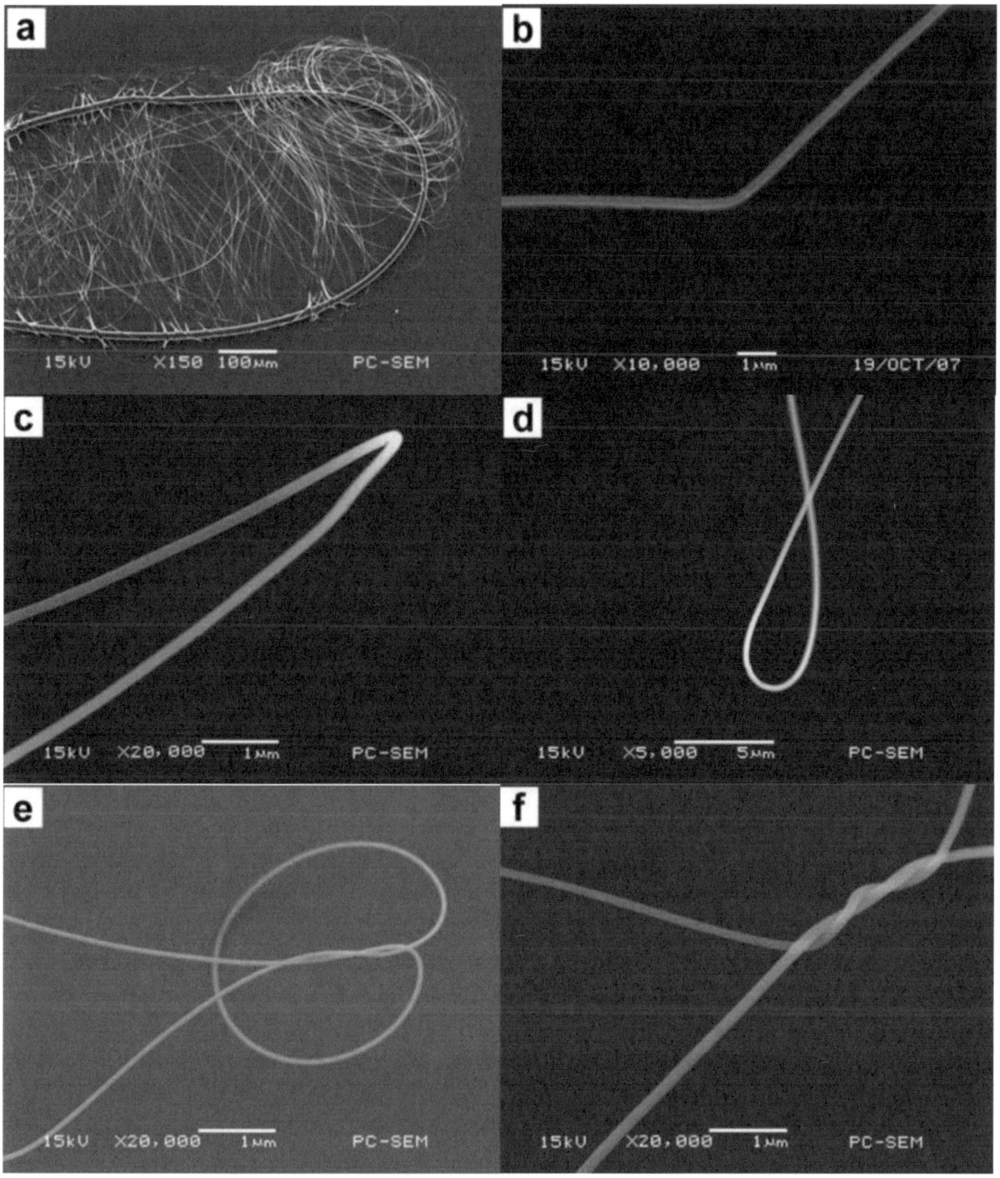

Fig. 17. Scanning electron microscope (SEM) images of nanodevices. (a) A nano bird's nest with average diameter of 280 nm on a 12-μm-diameter PTT bending rod (top right). (b) A 45° bend with a diameter of 280 nm. (c) A 155° sharp bend with a diameter of 160 nm. (d) A 340-nm-diameter tweezer-shaped nanowire. (e) A 70-nm-diameter scissor-shaped nanowire. (f) A twisted 2×2 coupler consists of 110-nm- and 150-nm-diameter nanowires.

Furthermore, integrated device arrays were constructed (Figures 18f–i), where the insets in yellow colour show their respective schematic structures. Figure 18(f) shows an integrated structure cascaded by two MZ couplers using 100-nm-diamter nanowires. Figure 18g shows an integrated device cascaded by a 2×2 coupler and a MZ coupler using 130-nm-diameter nanowires. Figure 18h shows a three inputs four outputs device array formed by two 300-

nm-diameter 2×2 couplers in parallel and, Figure 18i shows a three inputs three outputs device array formed by four X-crosses with nanowire diameter of 560 nm.

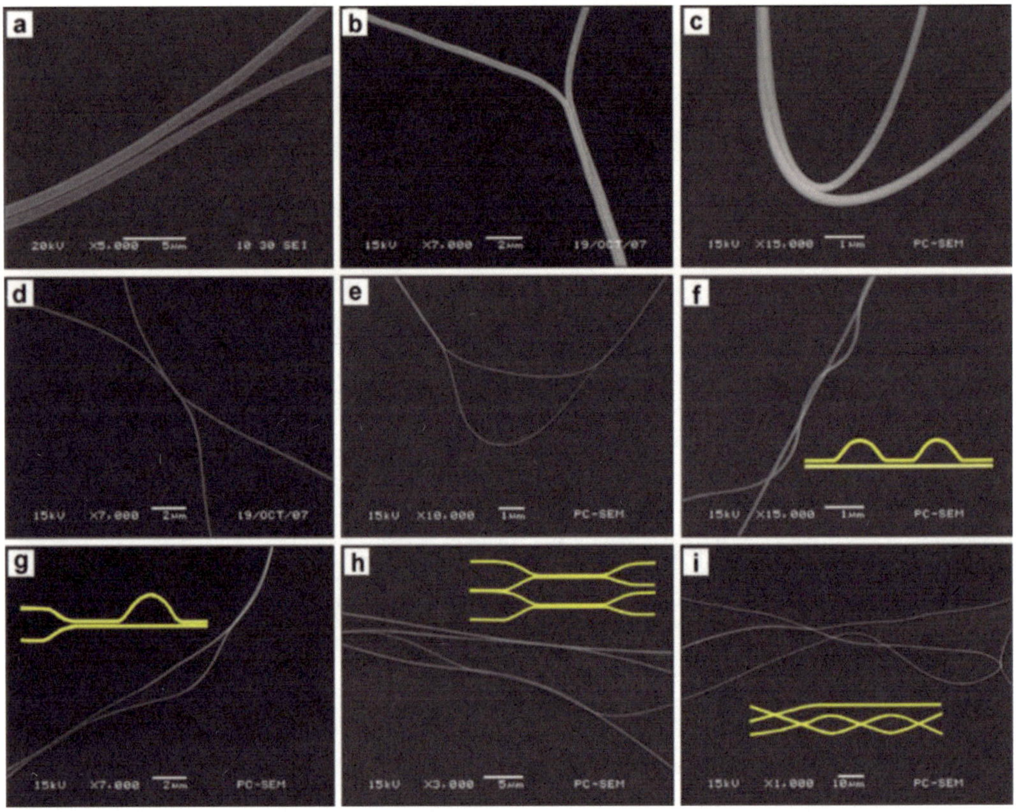

Fig. 18. SEM images of photonic devices and device arrays constructed by PTT nanowires. (a) A directional coupler with a 110 nm coupling gap between the two 750-nm-diameter nanowires. (b) A 360-nm-diameter Y-branching coupler without coupling gap. (c) A zero coupling gap coupler cascaded by a bending Y-branch (inside bend: 155°-bending-angle, 210-nm-diameter; outside bend: 120°-bending-angle, 270-nm-diameter). (d) A 2×2 coupler (150-nm-diameter). (e) A basic asymmetric MZ structure (60-nm-diameter). (f) An integrated coupler cascaded by two MZ structures (100-nm-diameter). (g) An integrated device cascaded by a 2×2 coupler and a MZ structure (130-nm-diameter). (h) A three inputs four outputs device array integrated by two 2×2 couplers (300-nm-diameter). (i) A three inputs three outputs device array integrated by four X-crosses (560-nm-diameter). The insets in yellow colour in (f)−(i) show respective schematic diagrams of the integrated structures.

6.2 Optical images of arbitrary nanodevices [34]

To analyze the optical characteristics at different wavelengths, we launched lights of different wavelengths into the assembled nanowire devices and device arrays. As an example, Figure 19a shows the launched blue light (473 nm) in the input port B is split into two parts by a 2×2 branching splitter, which was formed by twisting two 340-nm-diameter nanowires with one twist turn. The twisted region is shown in the inset of Figure 19a. The branching angles is 25° and the splitting ratio is 46:54 for the outputs C:D. The measured insertion loss is less than 0.3 dB. We have also assembled a 1×3 branching splitter by twisting three 290-nm-diameter nanowires (Figure 19b), and coupled blue light into the input port A with a splitting ratio of about 35:40:25 (from output ports B to D) and a total insertion loss of less than 0.3 dB. Figure 19c further shows red light (650 nm) in a 1×4 branching splitter with diameter of 350nm and splitting ratio of 23:20:28:29 (from output ports B to E). The measured total insertion loss is about 0.35 dB. From Figures 19a–c, we can see that the scattering lights are extremely weak in the branching areas, which is desirable for high performance optical beam splitters/couplers. In Figure 19d, we launched red light into a 500-nm-diameter twisted spiral nanowire with a 680-nm-radius ring at its end and found that that the optical power was well confined within the fiber core. BPM analysis shows that the bending loss is about 0.15 dB through the nanoring with radius as small as 680 nm. Figure 19e shows green light in an 850-nm-radius ring with nanowire diameter of

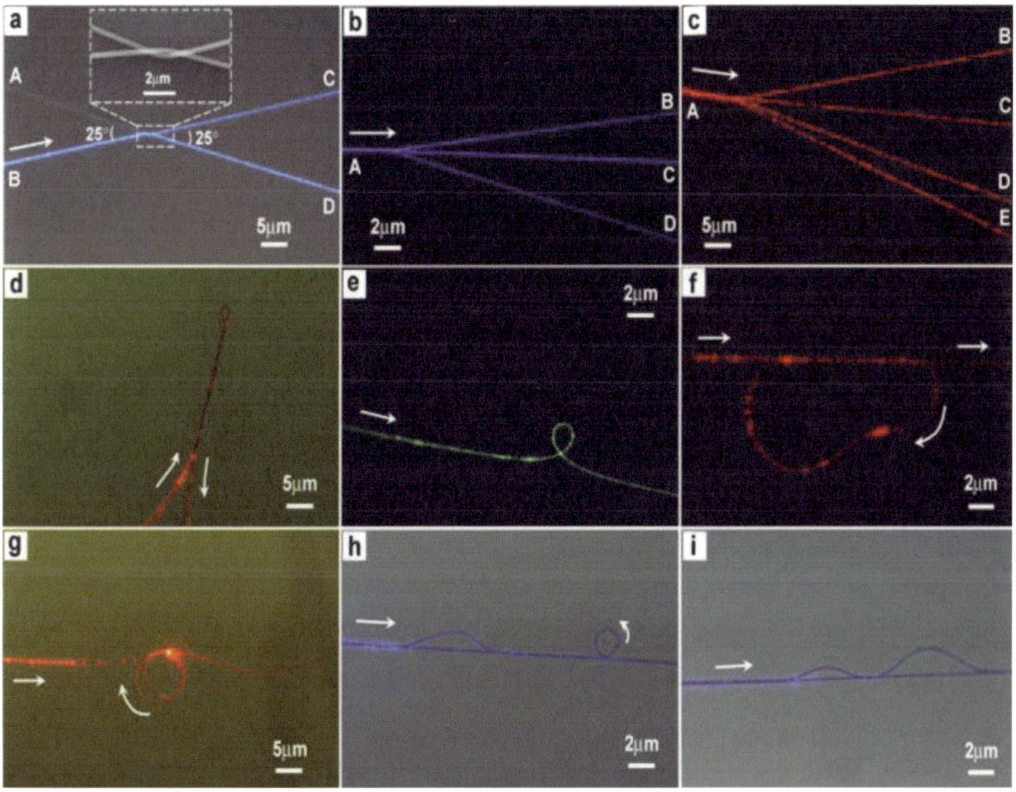

Fig. 19. Optical microscope images of the guided visible lights in different PTT nanowires and nanowire devices. (a) Blue light in a 2×2 (340-nm-diameter) branching splitter. (b) Blue light in a 1×3 (290-nm-diameter) branching splitter. (c) Red light in a 1×4 (350-nm-diameter) branching splitter. (d) Red light in a 500-nm-diameter twisted spiral nanowire with a 680-nm-radius ring. (e) Green light in an 850-nm-radius ring (120-nm-diameter). (f) Red light in a 230-nm-diameter racetrack-shaped resonator. (g) Red light in a 550-nm-diameter doubled-loop resonator (average radius about 3.8 μm). (h) Blue light in a 210-nm-diameter integrated device cascaded by a MZ structure and an 850-nm-radius ring. (i) Blue light in an integrated structure cascaded by two MZ structures (210-nm-diameter).

120 nm. The result shows that the optical field is confined in a small area of about 0.3 μm × 0.3 μm around the nanowire with good optical confinement. We have also characterized a 230-nm-diameter racetrack-shaped resonator (Figure 19f), which exhibits good transmission with an insertion loss of about 1.0 dB. In addition to above, optical microscope image of our double-loop resonator (550-nm-diameter) with an average radius of about 3.8 μm (Figure 19g) demonstrates that light can also be transmitted in multiple-rings (the measured total insertion loss is 2.4 dB). Therefore, a variety of ring-typed structures, ranging from simple fiber knot to complex coil/microcoil resonators can be achieved by proper assembling of the PTT nanowires. Figure 19h shows blue light in a 210-nm-diameter integrated structure cascaded by a basic asymmetric MZ structure and an 850-nm-radius ring. The total insertion loss is 0.38 dB. Moreover, Figure 19i shows blue light in a 210-nm-diameter integrated structure cascaded by two basic asymmetric MZ couplers with a total insertion loss of 0.33 dB. These structures demonstrate good feasibility of PTT nanowire-based integrated devices for miniaturized photonic integrated circuits.

6.3 Optical couplers and splitters [35]

A series of ultracompact photonic coupling splitters with multi-input/output ports were assembled by twisting flexible polymer nanowires. They are desirable for high density photonic integrated circuits (PICs) and nanonetworks.

Figure 20a shows SEM image of a 2×2 photonic coupling splitter with a branching angle of 25°, which was formed by twisting two polymer nanowires (PNWs) with diameters of 460 nm (branch A-1) and 548 nm (branch B-2). The inset of Figure 20a shows that there are three twisted turns in the coupling region (about 14.7 μm long and 1 μm wide). To demonstrate its optical coupling and splitting properties, we launched visible lights into different input branches by evanescent coupling. As examples, Figure 20b shows a 650 nm red light was launched into the branch A, coupled through the three-turn coupling region, and then divided into the output branches 1 and 2. Figure 20c shows a 532 nm green light was launched into the branch A, coupled through the three-turn coupling region, and then divided into the output branches 1 and 2 with a splitting ratio of about 54:46. The excess loss of the device, defined as -10log(sum[P_{output}]/P_{input}), is 0.65 dB, including 0.48 dB input and output coupling loss. The propagation loss is 0.007 dB and the scattering loss is 0.163 dB in the twisted region. Similarly, Figure 20c shows that a 473 nm blue light is divided by the device with a splitting ratio of about 60:40. The measured excess loss is 0.63 dB, which is composed of coupling loss (0.48 dB), propagation loss (0.001 dB), and scattering loss (0.149 dB). It is revealed that the device is very efficient in guiding and splitting lights. To investigate the influence of branching angle on the device performance, input/output branching angles of the splitter was changed from 8° to 90°. The calculated results show that the splitting ratios for branches 1 and 2 are ranging from 46:54 to 56:44 if a green light is launched into the branch A, and ranging from 55:45 to 62:38 if a blue light is launched into the branch B. In our experiment, we found that if the branching angles of the 2×2 splitter are 10° or 40°, a 3-dB splitter can be achieved in the case of the green light was launched into the branch A. The splitter with large branching angle operates at a large excess loss, because the bending loss and the scattering loss increase with the increase of the branching angle.

Figure 21 further shows the assembled 2 × 2 splitter (different branching angles) with visible lights in the nanowires. The diameter of the PTT nanowires is 580 nm and the branching angles of the splitter are 37° and 45°. There is one twisted turn (about 2 μm long) in the coupling region. The coupling region (one twisted turn) of the splitter is 1.16 μm wide

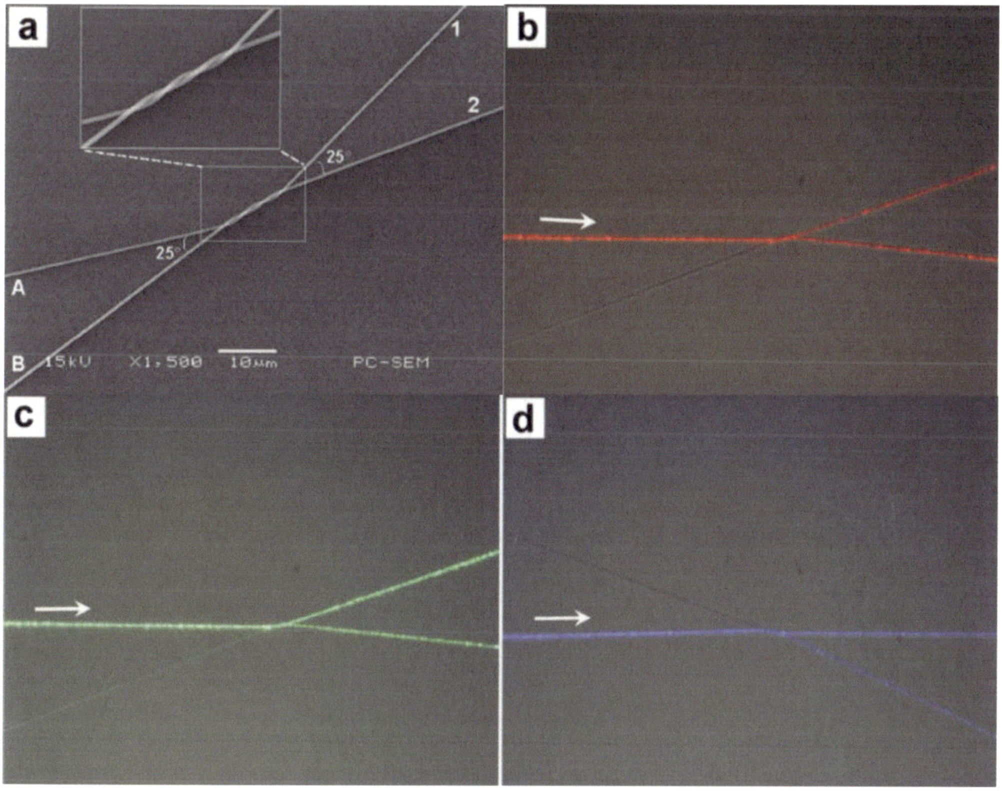

Fig. 20. A 2×2 photonic coupling splitter. (a) SEM image of the splitter with diameters of 460 and 548 nm for branches A-1 and B-2, respectively. (b)-(d) Optical microscope images of the guided red (650 nm), green (532 nm) and blue (473 nm) lights, respectively. The arrows show the propagation directions of the launched lights.

and about 2 µm long. Figure 21a shows a 650 nm red light launched into the input branch B, coupled through the one-turn coupling region, and then divided into the output branches C and D with a splitting ratio of about 40:60. Figure 21b shows that a 532 nm green light is divided by the device with a splitting ratio of about 45:55. Figure 21c shows a 473 nm blue light divided into branches C and D with a splitting ratio of about 40:60. The excess loss of the splitter is 0.63 dB. We also launched the visible lights into the input branch A of the splitter, the measured respective splitting ratios are 60:40 (red), 58:42 (green), and 50:50 (blue), the measured excess loss is 0.635 dB.

To observe the influence of the branching angles on the splitting properties of the device, the branching angles of the 2 × 2 splitter were changed to 44° and 38°. Figure 22 shows the guiding optical images of the device when red, green, and blue lights are coupled into the input branch A. The measured splitting ratios are listed in Table 2. The measured excess loss is 0.60 to 0.65 dB. By comparison the splitting properties of the 2 × 2 splitters, we found that the splitting ratios are different when the operating wavelengths or the branching angles are changed. This is because that of the coupling conditions is different in the coupling region when the operating wavelengths or the input/output branching angles are changed. For a fixed wavelength, desirable splitting ratios from 0 to 100% would be achieved by changing the branching angles.

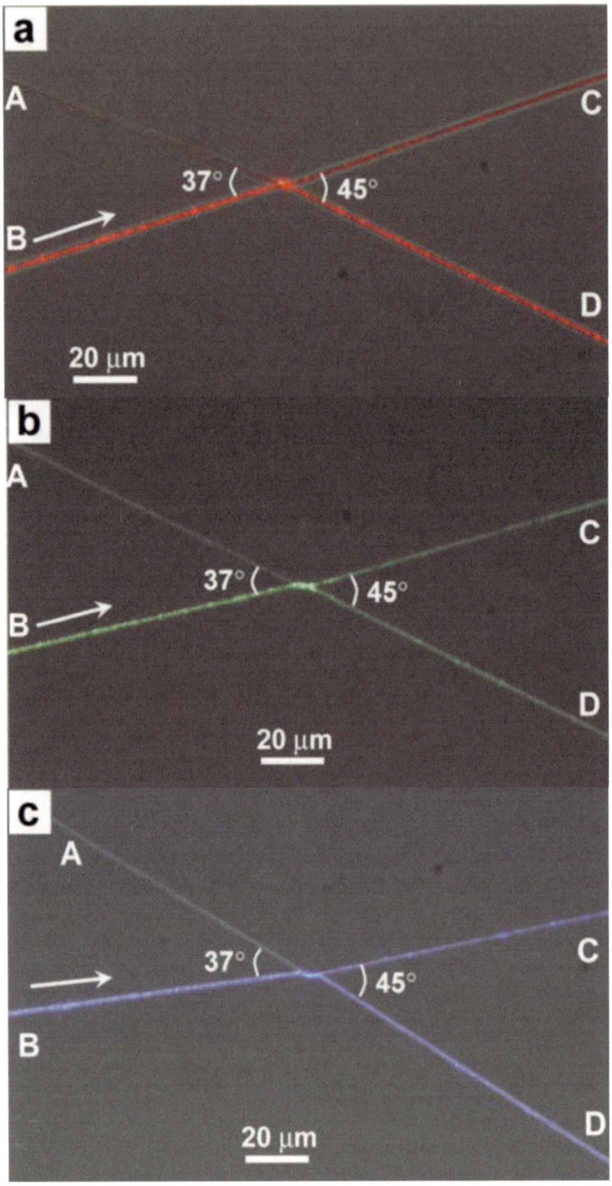

Fig. 21. Optical microscope images of the guided visible lights in the 2 × 2 optical beam splitters with branching angles of 37° and 45°. The diameter of the nanowires is 580 nm. The arrows show the propagation directions of the launched lights with (a) 650 nm red light, (b) 532 nm green light, and (c) 473 nm blue light.

Figure 23 shows a 3 × 3 optical beam splitter, which was assembled by twisting three 410-nm-diameter PTT nanowires. The length (*L*) of the coupling region is about 12-μm long and the width is 1.23 μm. Figure 23a shows blue light is launched into the branch A and divided into three parts, with a splitting ratio of 50:30:20 for the output branches 1 to 3. If blue light is launched into the branch B (Figure 23b), the splitting ratio is changed to be 30:40:30. When blue light is launched into the branch C (Figure 23c), the splitting ratio is 25:38:37. We also launched the blue light into the branches B and C. The measured excess losses are 0.72, 0.61, and 0.68 dB for the input branches A, B, and C, respectively. Table 3 lists the corresponding splitting ratios of the 3 × 3 optical beam splitters at different wavelengths.

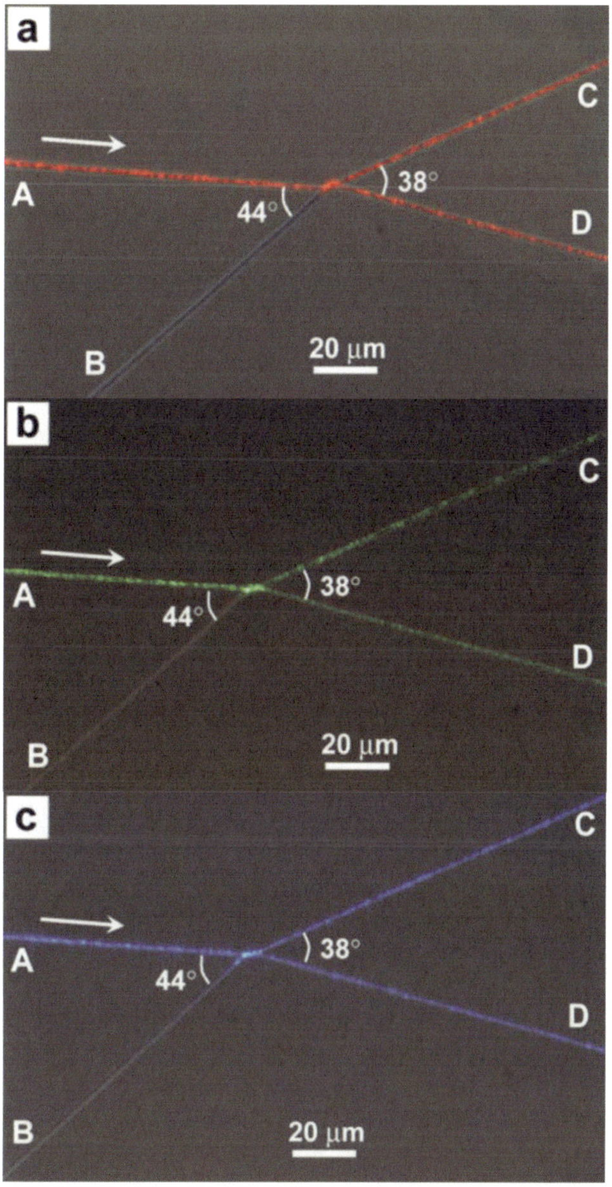

Fig. 22. Optical microscope images of the guided visible lights in the 2 × 2 optical beam splitters with branching angles of of 44° and 38°. The diameter of the nanowires is 580 nm. The arrows show the propagation directions of the launched lights with (a) 650 nm red light, (b) 532 nm green light, and (c) 473 nm blue lights.

Branching angles	Input branch	Splitting ratio (C : D)		
		Red light	Green light	Blue light
37° and 45°	A	60:40	58:42	50:50
	B	40:60	45:55	40:60
44° and 38°	A	50:50	55:45	52:48
	B	50:50	51:49	42:58

Table 2. Splitting ratios of the 2 × 2 optical beam splitters

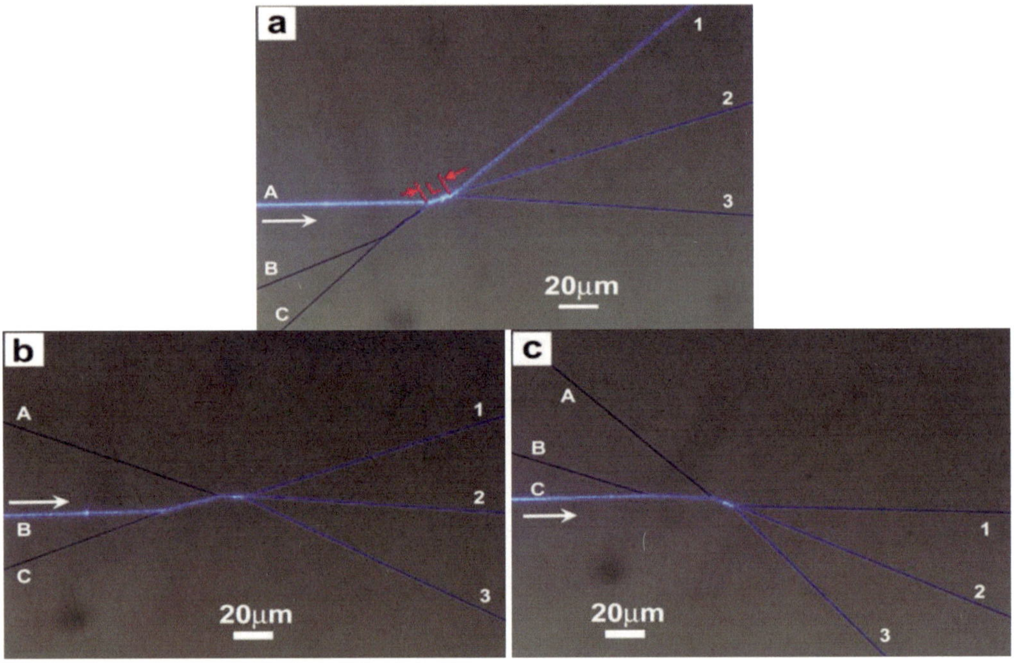

Fig. 23. Optical microscope images of the guided blue lights in a 3 × 3 optical beam splitter (410-nm-diameter). The arrows show the propagation directions of the launched blue light.

Input branch	Splitting ratio (1 : 2 : 3)		
	Red light	Green light	Blue light
A	50:25:25	55:25:20	50:30:20
B	31:38:31	32:36:32	30:40:30
C	35:33:32	32:34:34	25:38:37

Table 3. Splitting ratios of the 3 × 3 optical beam splitters

Figure 24a shows a 4×4 photonic coupling splitter assembled by twisting four PNWs with diameters of 450, 450, 510, and 570 nm for branches A to D. The inset of Figure 24a shows that the coupling section is composed of a 3×4 and a 1×4 couplers, where the total width of the coupling section is 1.98 μm. The maximum length of the coupling region (Figure 24a, inset) is about 16.1 μm, and that of the 1×4 splitter is about 8.5 μm. Figure 24b shows that a 650 nm red light is sent into the branch B and divided into the output branches 1 to 4 with a splitting ratio of about 24:25:32:19. Figures 24c and 24d show lights are simultaneously launched into the branches A and B with a power ratio of about 1:2.

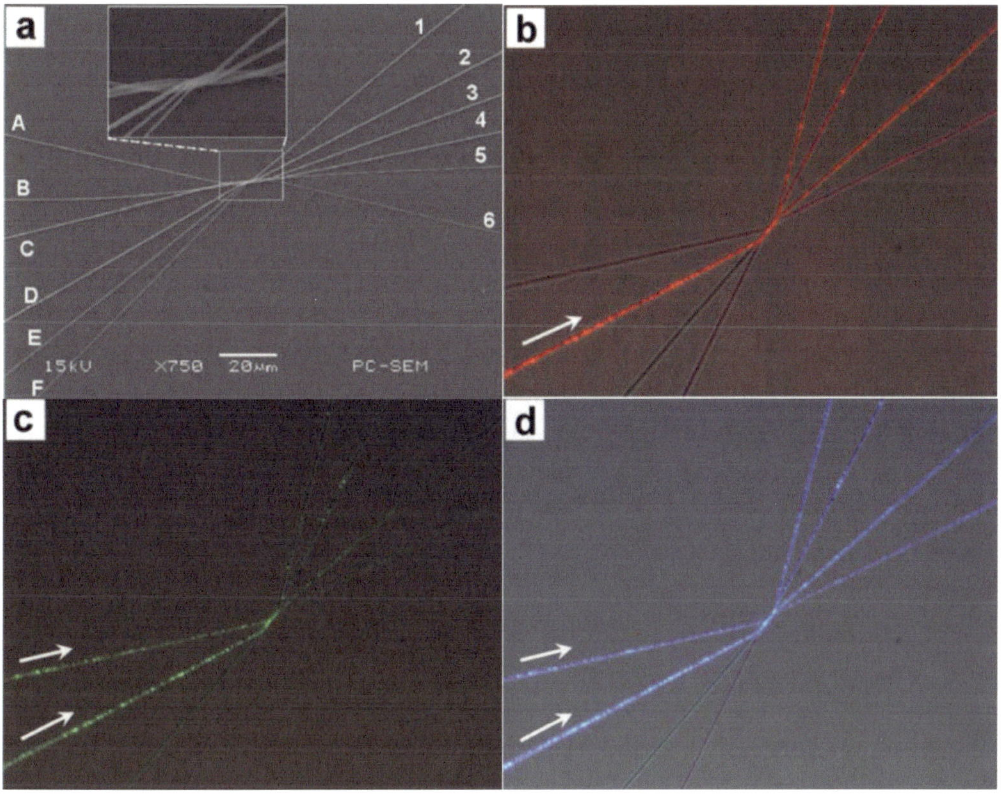

Fig. 24. A 4×4 photonic coupling splitter. (a) SEM image of the device with diameters of 450, 450, 510, and 570 nm for branches A to D. (b)-(d) Optical microscope images of the guided red, green and blue lights. The arrows show the propagation directions of the launched lights.

Figure 25a further shows a 6×6 photonic coupling splitter, which was formed by twisting six PNWs with diameters of 520, 540, 540, 540, 420, and 360 nm for branches A to F, respectively. The inset of Figure 25a shows the magnified twisted section with a coupling length of 11 to 20 μm and a coupling width of 2.92 μm. Figure 25b and 25c show red and green lights were launched into the branch C and divided into six parts, respectively. When blue light is launched into the branch D (Figure 25d), the device exhibits good power distribution uniformity and its power uniformity is about 0.03 dB. Different visible lights were also launched into the branches C and D.

An 8×8 photonic coupling splitter (Figure 26) with a longer coupling region was further assembled by twisting eight PNWs with diameters of 400, 400, 400, 400, 400, 750, 750, and 600 nm from branches A to H. The coupling section of the splitter (Figure 26a, inset) is 38-μm-long and 2.5-μm-wide. Here we launched visible lights into the device to observe the splitting phenomenon. Figures 26b and 26c show that red light is coupled into the branches E and G, respectively, and divided into branches 1 to 8. Figure 26d shows that blue light is coupled into the branch G and divided into branches 1 to 8.

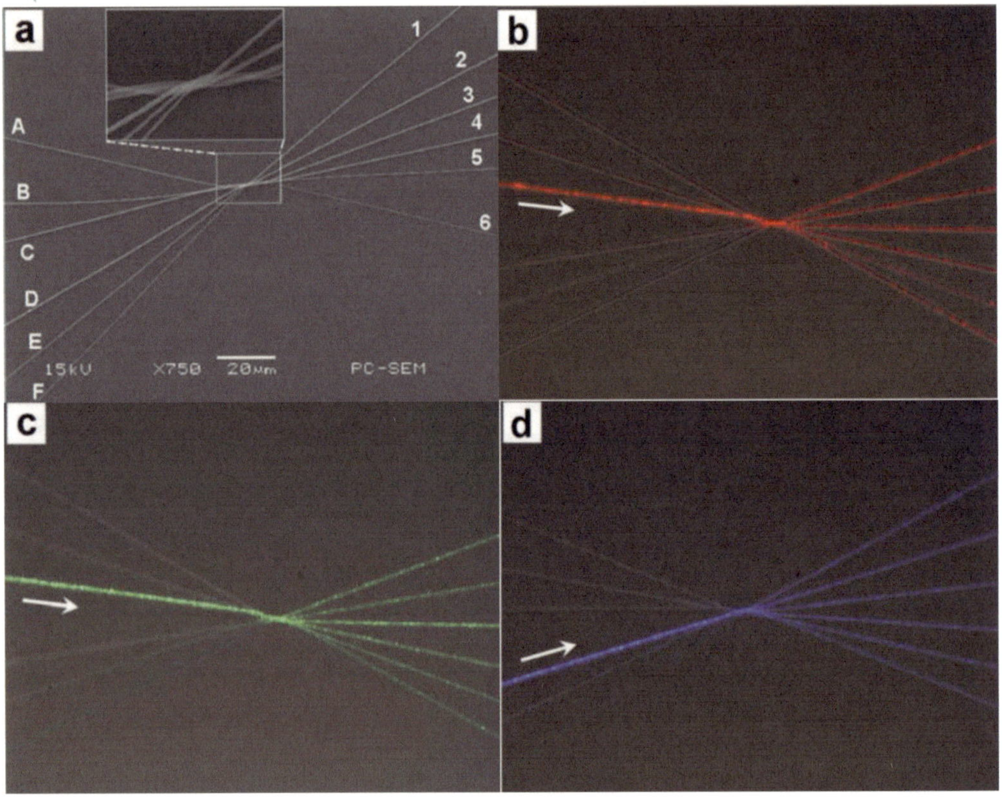

Fig. 25. A 6×6 photonic coupling splitter. (a) SEM image of the device with diameters of 520, 540, 540, 540, 420, and 360 nm for branches A to F. (b)-(d) Optical microscope images of the launched red and green lights from the input branch C, and blue light from the input branch D are split into six parts, respectively. The arrows show the propagation directions of the launched lights.

Using the facile twisting technique, M×N photonic coupling splitters can be fabricated. As examples, in the following, we demonstrate 2×4, 3×3, and 5×5 photonic coupling splitters. Figure 27 shows that visible lights are coupled into a 2×4 coupling splitter with diameters of 330, 330, 330 and 320 nm for the right branches 1 to 4. The length of the twisted section is about 30 μm. The input branch B was formed by twisting three 330-, 330-, and 320-nm-diameter PNWs. Figure 27a shows that the red light is coupled into the branch A and split into four parts through the coupling region, with a splitting ratio of 26:26:26:22 for output branches 1 to 4. Figures 27b and 27c show that the green and blue lights launched from the branch B are divided into the branches 1 to 4 with splitting ratios of 30:21:20:29 and 26:25:25:24, respectively.

Figure 28 further shows a 3×3 coupling splitter cascaded by a 1×3 and a 2×3 coupling splitters, with diameters of 550, 530, and 480 nm for branches 1 to 3. The length of the coupling region of the 1×3 splitter is about 23 μm, and that of the 2×3 splitter is about 95 μm. The measured splitting ratios are 42:30:28 and 30:33:37 for output branches 1 to 3 when red (Figure 28a) and blue (Figure 28b) lights are launched into the branch A, respectively. The excess losses are 0.725 and 0.662 dB.

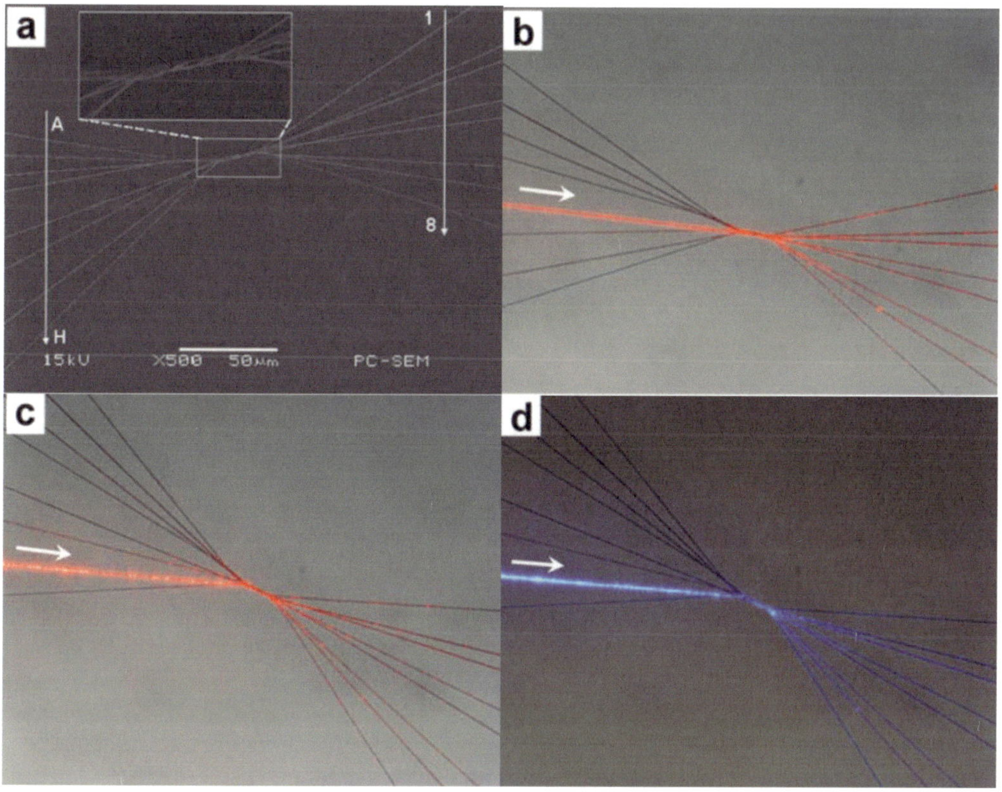

Fig. 26. An 8×8 photonic coupling splitter. (a) SEM image of the device with diameters of 400, 400, 400, 400, 400, 750, 750, and 600 nm for branches A to H. (b)-(d) Optical microscope images of the guided red and blue lights, respectively. The arrows show the propagation directions of the launched lights.

Figure 29a shows that red light is launched into the branch D of a 5×5 splitter, which was assembled by twisting five PNWs with diameters of 410, 450, 470, 470 and 470 nm for branches 1 to 5. The splitting ratio is 16:17:24:23:20 for the branches 1 to 5 and the excess loss is 0.775 dB. Figure 29b shows a 5×5 coupling splitter with diameters of 480, 450, 350, 380 and 400 nm for branches 1 to 5. Green light is launched into the branch B and split into five parts with a splitting ratio of 55:20:10:9:6 and an excess loss of 0.765 dB. Figure 29c shows that blue lights are simultaneously launched into the branches A, B, and C of a 5×5 coupling splitter, which was assembled by twisting five PNWs with diameters of 420, 540, 410, 550 and 570 nm for branches 1 to 5. The measured splitting ratio is 15:23:16:25:21 for branches 1 to 5 and the excess loss is 0.68 dB.

6.4 Interferometers [36]

Optical interferometers assembled is one of important applications of nanowires. To ensure the PTT wires used for assembly of interferometers have same diameter, a PTT wire was cut into two segments. In device assembly, the two segment wires were placed on a glass substrate, which was supported by a micromanipulator with a high precision. The first segment wire was formed to an S-shaped bend under an optical microscope. The second segment wire was pulled straight and moved to approach the S-shaped one. When the two

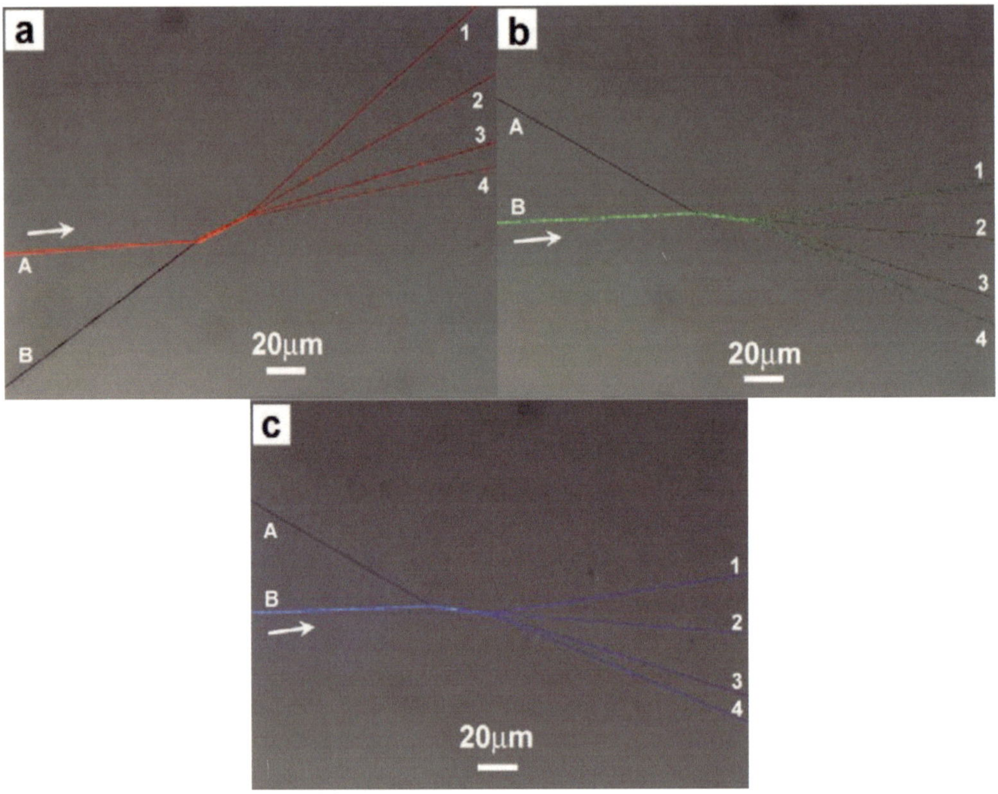

Fig. 27. A 2×4 photonic coupling splitter. The diameters of the PNWs are 330, 330, 330 and 320 nm for right branches 1 to 4. Three PNWs with diameters of 330, 330, and 320 nm are twisted together to form the left input branch B. (a)–(c) Optical microscope images of the guided red, green, and blue lights, respectively. The arrows show the propagation directions of the launched lights.

wires were close enough, they attracted each other and formed into couplers because of van der Waals force. Finally, by careful adjustment, a desired cascaded Mach–Zehnder interferometer (MZI) was assembled. The whole structure was placed on a glass substrate and the device shape is stable. Figure 30 shows a schematic diagram of the cascaded MZI. It consists of two bow-shaped MZIs (MZI 1 and MZI 2). By adjusting the contacting length between each bow-shaped bending wire and the straight wire, the lengths of the couplers (C_1, C_2, and C_3) and path-length difference (ΔL) of each bow-shaped MZI can be changed to the desired ones.

Figure 31 shows an optical microscope image of the assembled two-cascaded MZI (wire diameter, 900 nm). The inset (a) shows a scanning electron microscope image of MZI 1 and the inset (b) shows guided red light (650 nm) in the cascaded MZI. The measured insertion loss is about 0.94 dB for the red light. The total length of the cascaded MZI is 327 μm. The width and length of each bow-shaped MZI are 32 μm and 121 μm, respectively. According to the analysis, to get coupling ratios of 0.147, 0.501, and 0.147, the lengths of the couplers C_1, C_2, and C_3 are 27, 31, and 27 μm, respectively. The estimated total path-length difference is 40 μm. The bright spot in the inset (b) of Figure 31 is the scattering spot of the input light at the end of the tapered fiber I.

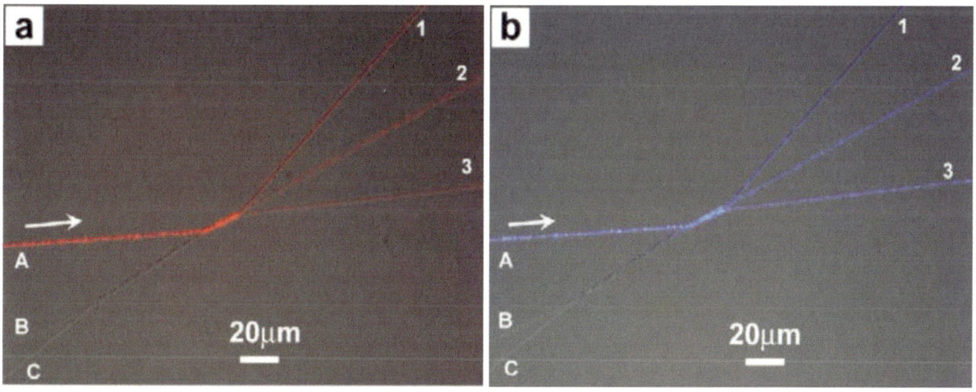

Fig. 28. Optical microscope images of the guided red and blue lights in a 3×3 photonic coupling splitter cascaded by a 1×3 and a 2×3 coupling splitters. The diameters of the PNWs are 550, 530, and 480 nm for branches 1 to 3. The arrows show the propagation directions o the launched lights.

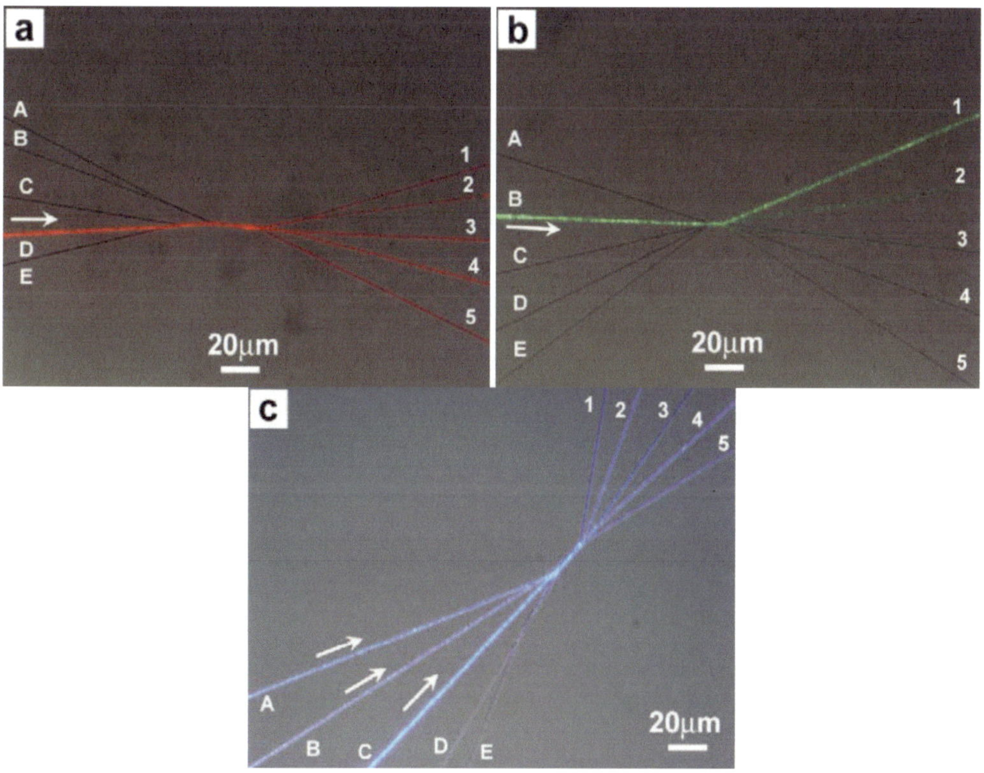

Fig. 29. Optical microscope images of three 5×5 photonic coupling splitters. (a) Red light is launched into the branch D of a 5×5 splitter with diameters of 410, 450, 470, 470 and 470 nm from branches 1 to 5. (b) Green light is launched into the branch B of a 5×5 splitter with diameters of 480, 450, 350, 380 and 400 nm from branches 1 to 5. (c) Blue lights are simultaneously launched into the branches of A, B, and C of a 5×5 splitter with diameters of 420, 540, 410, 550 and 570 nm from branches 1 to 5. The arrows show the propagation directions of the launched lights.

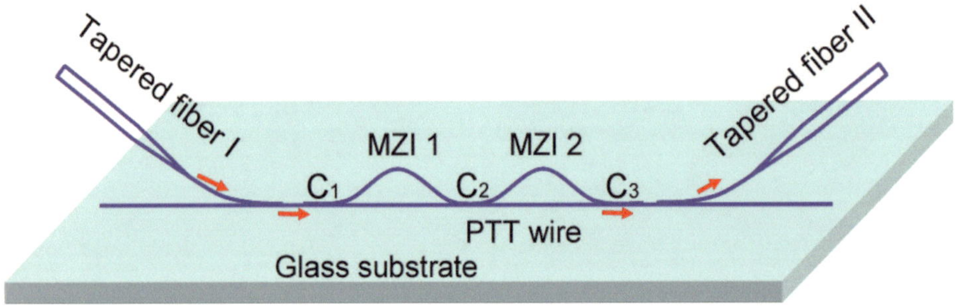

Fig. 30. Schematic diagram of a cascaded MZI. It is assembled by two PTT wires with same diameter and consists of bow-shaped MZI 1 and MZI 2. Tapered fibers I and II are used to launch and collect light by evanescent coupling. The red arrows indicate transmission direction of the light.

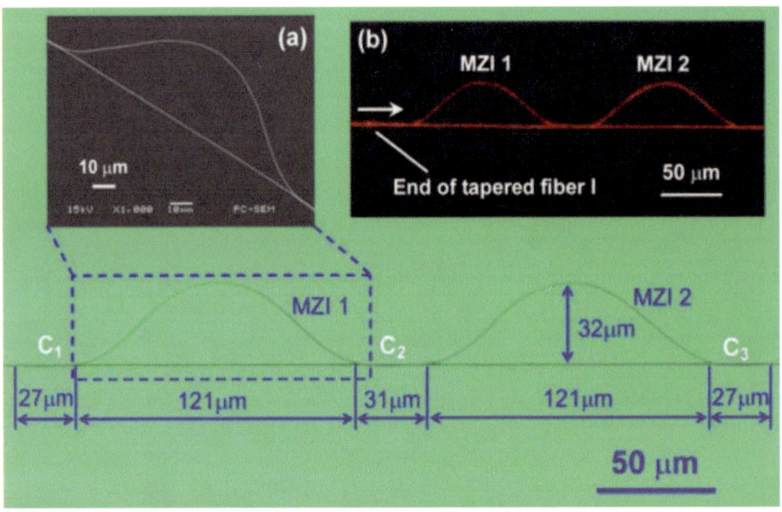

Fig. 31. Optical microscope image of the assembled cascaded MZI (wire diameter, 900 nm). Inset (a) shows the scanning electron microscope image of bow-shaped MZI 1 and inset (b) shows the optical microscope image of guided red light (650 nm) in the cascaded MZI. Estimated path-length difference is 40 μm. The white arrow in the inset (b) indicates propagation direction of the light.

Figure 32 shows the measured transmission spectrum (red dot-line) versus near-infrared wavelengths of the two bow-shaped MZI. For comparison, transmission spectrum of single bow-shaped MZI was also measured by repositioning the tapered fiber II to the straight branch of MZI 2, and depicted in Figure 32 (black dot-line). According to the curves of Figure 32, the calculated path-length differences are 21 and 42 μm for the single bow-shaped MZI and the cascaded MZI, respectively. This is well consistent with the estimated value of 40 μm for the cascaded MZI from Figure 31. The measured peak/valley ratio at the top region of the spectrum for the cascaded MZI is 1.3 to 1.7 dB. A broad pass-band appears at the top of the curve (red dot-line) and the band edge becomes sharper by using the cascaded MZI. Compared with the bandwidth (24 nm) of the single MZI, the average bandwidth of 3-

dB pass-band is improved to 33 nm for the cascaded MZI over the wavelengths of 1.3 to 1.6 μm. This is good for band-pass filter applications. The measured extinction ratio for the cascaded MZI is 16 to 19 dB, which has an improvement of 5 to 6 dB compared with the single MZI (11 to 13 dB). The measured optical insertion loss is 1.1 to 1.8 dB at wavelengths of 1.3 to 1.6 μm.

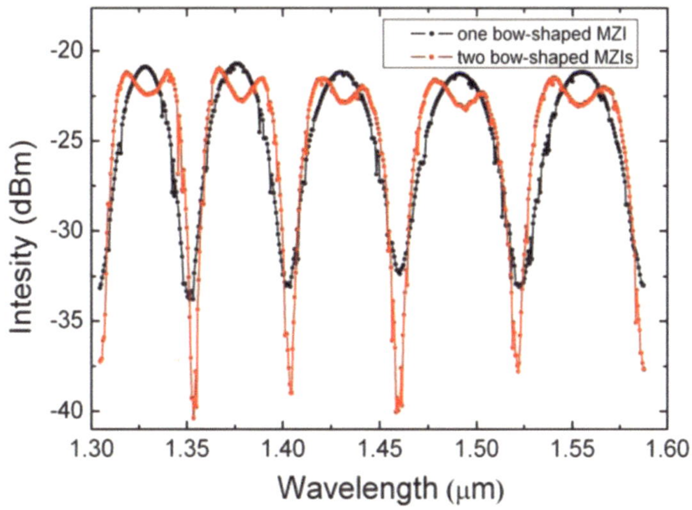

Fig. 32. Transmission spectrum versus near-infrared wavelengths. The black dot-line is the spectrum of the single bow-shaped MZI whereas the red dot-line is the spectrum of the cascaded MZI (two bow-shaped MZIs). The calculated path-length differences, from the curves, are 21 and 42 μm for the single MZI and the cascaded MZI, respectively.

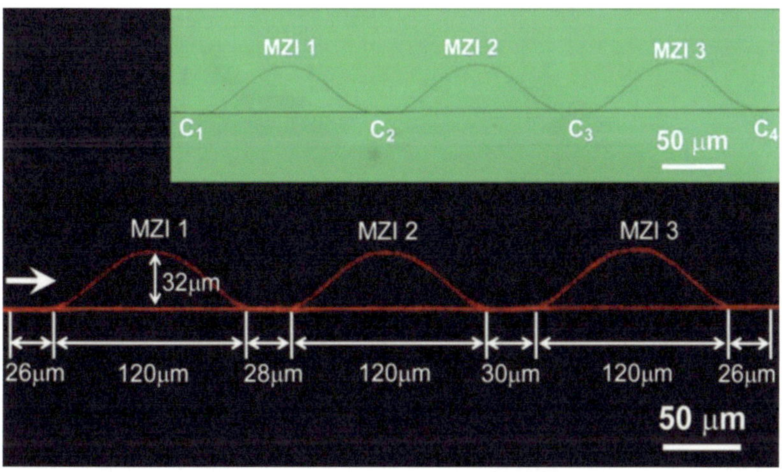

Fig. 33. Optical microscope image of the guided red light (650 nm). The inset shows the original optical microscope image of the MZI (900-nm-diameter). The estimated total path-length difference is about 61.5 μm. The white arrow indicates propagation direction of the light.

To further improve extinction ratio of the device, a cascaded MZI with three bow-shaped MZIs was assembled by two 900-nm-diameter PTT wires. Figure 33 shows an optical microscope image of the guided red light in the cascaded MZI. The measured insertion loss is 1.15 dB for the red light. The inset shows its original optical microscope image. According to the analysis, to get coupling ratios of 0.050, 0.188, 0.388, and 0.050, the lengths of the couplers C_1, C_2, C_3, and C_4 are 26, 28, 30, and 26 μm, respectively. The size of each bow-shaped MZI is 120 μm long and 32 μm wide. Total dimension of the cascaded MZI is 470 μm × 32 μm. Estimated total path-length difference for the cascaded MZI is 61.5 μm. Figure 34 shows its transmission spectrum (blue dot-line). The curve in black dot-line depicts the spectrum of the single bow-shaped MZI. From Figure 34, we calculated that the total path-length difference is 64 μm, which agreed with the estimated value of 61.5 μm. The peak/valley ratio of the cascaded MZI at the top region of the spectrum is 0.9 to 1.6 dB. The measured total insertion loss for the cascaded structure is 1.5 to 2.3 dB at wavelengths of 1.3 to 1.6 μm. The measured extinction ratio is 23 to 25 dB, which has an improvement of 12 to 13 dB compared with the single MZI (11 to 12 dB). The top region of its band-pass is wider and its band edge becomes sharper. The average bandwidth of the 3-dB band-pass for the three-cascaded MZI is 33 nm.

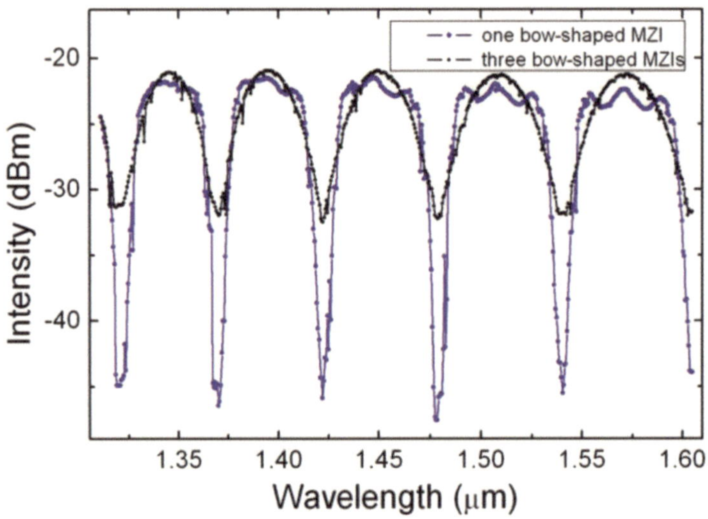

Fig. 34. Transmission spectrum versus near-infrared wavelengths. The black dot-line is the spectrum of the single bow-shaped MZI whereas the blue dot-line is the spectrum of the cascaded MZI (three bow-shaped MZIs). The total path-length difference of the cascaded MZI calculated from the spectrum is 64 μm.

In conclusion, the interferometers have a large extinction ratio with a flat-top transmission over a wide wavelength range of 1.3 to 1.6 μm. The average 3-dB bandwidth for the cascaded MZIs is broadened to 33 nm compared with the bandwidth (24 nm) of the single MZI. The extinction ratio and the measured insertion loss are 16 to 19 dB and 1.1 to 1.8 dB for the two-cascaded MZI, respectively. For the three-cascaded MZI, the measured insertion loss is 1.5 to 2.3 dB and the extinction ratio is improved to 23 to 25 dB. By carefully adjusting the bending radius of the bow-shaped MZIs (also coupling length of the PTT wires), the path-length difference could be easily controlled and the output light intensity could be tuned.

6.5 Nanosensors [37]

Nanosensor, as a promising application of nanowires, have been widely investigated. In the section, we report a tunable refractive index sensor with ultracompact structure in a 2 × 2 PTT nanowire coupling splitter assembled by twisting two flexible PTT nanowires. The sensor consists of two input branches, a twisted coupling region, and two output branches.
The sensor consists of two input branches, a twisted coupling region, and two output branches, which were assembled by a simple twisting method with microstage supports

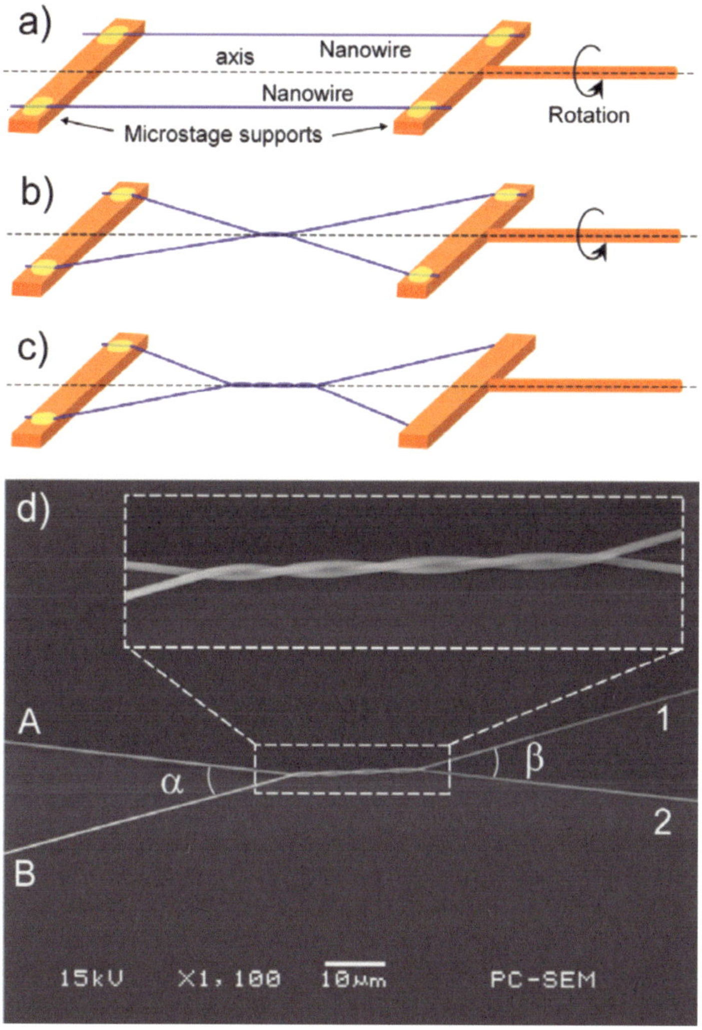

Fig. 35. Fabrication process of the PTT nanowire-based tunable refractive index sensor. (a) Two parallel PTT nanowires were fixed by two microstage supports. The left support was fixed and the right support can be rotated around the axis. (b) A twisted 2 × 2 coupling sensor with one twisted turn was formed by rotating the right support. (c) A twisted 2 × 2 coupling sensor with four twisted turns was formed by further rotating the right support. (d) SEM image of the 2 × 2 coupling sensor, which was assembled by twisting two 440-nm-diameter PTT nanowires. The inset shows the twisted region with four twisted turns. The branching angles α and β are 20°.

under a microscopy. Figure 35a–c shows assembly process of the sensor. Figure 35d shows a scanning electron microscope (SEM) image of the 2 × 2 coupling sensor with four twisted turns in the coupling region (twisted region). Diameter of the PTT nanowires used is 440 nm and coupling region (inset of Figure 35d) is about 23 μm long and 880 nm wide. Both the input and the output branching angles, α and β are 20°.

For sensing applications, the twisted region (used as the sensing area) was immersed in various sample solutions with differing refractive indices. Then the output optical signals delivered through tapered fibers get detected by optical detector and fed to power meter. In our characterization, sodium chloride aqueous solutions with different mass concentrations (different refractive indices of surrounding mediums) were chosen as sample solutions. Each solution droplet was dropped on the sensing area of the sensor using a micro-injector. After each measurement, the sensor was cleaned by purified water and dried for different concentration solution use.

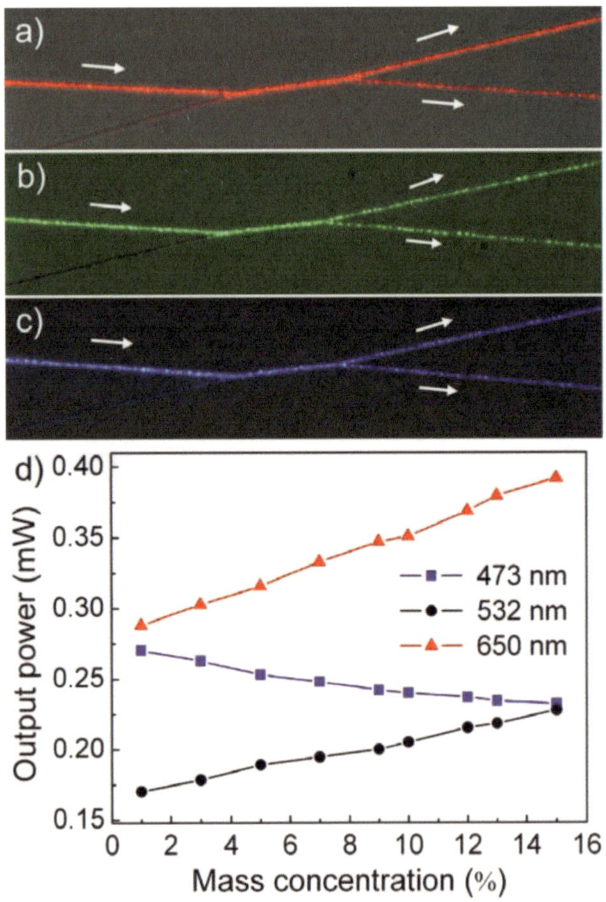

Fig. 36. Optical microscope images of the sensor with (a) red light (λ = 650 nm), (b) green light (λ = 532 nm), and (c) blue light (λ = 473 nm) in it (without sample solution). The white arrows show the propagation directions of the light. (d) Output power in the output branch 1 as a function of mass concentration of solution at branching angles of 20° and light of different wavelengths (473, 532, and 650 nm).

Figure 36a–c show optical microscope images of red, green, and blue lights propagated in the sensor (without sample solution), respectively. The three wavelengths were launched individually. The output power splitting ratios are 70:30, 50:50, and 55:45, for red, green, and blue lights, respectively. Figure 36d shows the measured power (P_1) in the output branch 1 as a function of the mass concentration of the solution at branching angle of 20° for wavelengths of 473 nm, 532 nm, and 650 nm. As evident, the output power is approximately linearly related to the mass concentration. Therefore, there is a direct relationship between output power and refractive index. In reverse, one can get to know the matter of surrounding according to the changes of the output power. The sensitivity S of the sensor is defined as $S = |\Delta P_1/\Delta n|$, where ΔP_1 is the change of the output power in the output branch 1 and Δn is the change of the refractive index of surrounding medium. From Figure 36d, we have calculated that the sensitivities for the blue (473 nm), green (532 nm), and red (650 nm) lights are 1.47, 2.19, and 4.25 mW/RIU (refractive index unit), respectively. Since the resolution of the optical power meter, which we have used is 5 nW, therefore, the detection limits on the refractive index change are about 3.40×10^{-6} (blue light), 2.28×10^{-6} (green light), and 1.18×10^{-6} (red light) for the sensor with four twisted turns in the sensing area.

The sensing mechanism of the twisted 2 × 2 coupling sensor can be explained as follows: Since the structure was assembled by the nanometer-scale PTT wires, launched light will be guided along the nanowires as strong evanescent waves. The change of the refractive index in the surrounding medium results in the change of the mode profile. As a result, the optical coupling property between the two nanowires in the sensing area (twisted region or coupling region) will be different and the optical power in the output branches will be changed.

To investigate influence of twisted turns on the sensing properties of the 2 × 2 coupling sensor, the number of the twisted turns was changed by rotating the right support. Figure 37a shows the measured output power change in the output branch 1 as a function of the refractive index at different twisted turns (2, 3, 4, 6, 8, 12, 16, 20, 23 and 24 turns). The change of the angles α and β caused by the change of the twisted turns was ignored here. In experiment, a longer coupling region causes a larger output power change for the same refractive index change. As a result, the sensitivity increases with the increase of the twisted turns. The sensitivity will reaches a maximum value when the twisted turns reach a particular value. In case of continuous increase in the twisted turns, the output power in the output branch will no longer be a monotone function of the refractive index and will not be able to decide the refractive index of the surrounding medium according to the output power. In experiment, the refractive index varies from 1.3321 to 1.3565 for red light, the sensitivity reaches a maximum when the twisted turns is 23. Figure 37b shows the sensitivity versus the twisted turns. The calculated highest sensitivity is about 26.96 mW/RIU. In this scenario, the calculated detection limit on the refractive index change is about 1.85×10^{-7} and the measured tenability is 1.2 mW per RIU per twisted turn.

To investigate tunability and/or branching angle dependence of the sensor, each branch end of the 2 × 2 sensor was removed from the microstage supports and refixed by the PTT melt on four tunable microstage supports as shown schematically in Figure 38. Since the sensitivity of the sensor increases with the twisted turns and reaches a maximum value at 23 twisted turns, we chose a twisted turn number less than 23 to investigate the branching angle dependence. As an example, we rotated the sensor to 14 twisted turns and fixed it on the tunable microstage supports as shown in Figure 38. It should be pointed out that the branching angles of the sensor are changeable by adjusting the tunable microstage supports.

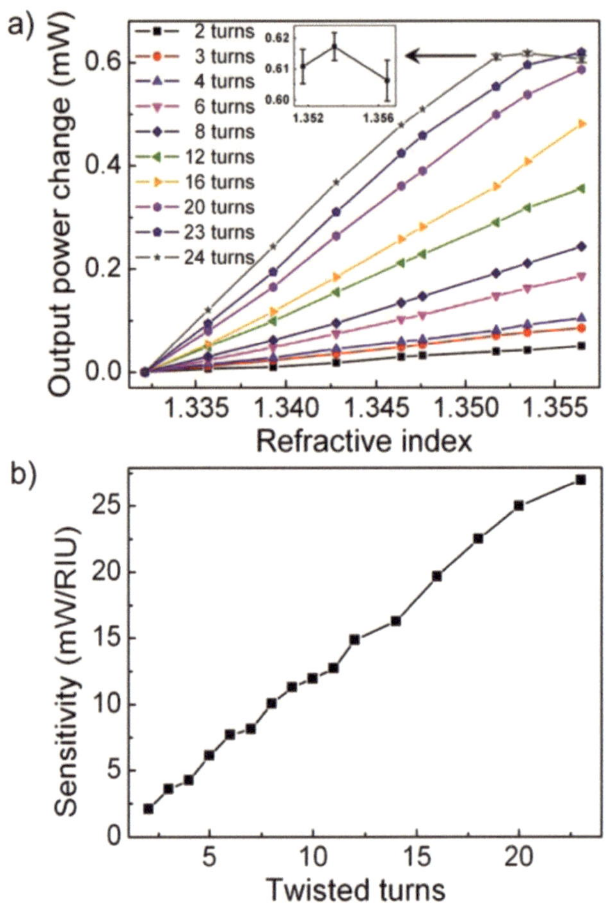

Fig. 37. Twisted turn dependence of the sensor. (a) Output power change in the output branch 1 as a function of refractive index at 650 nm wavelength with different twisted turns. The diameter of the nanowire is 440 nm and the input/output branching angles are around 20°. The inset shows error bars of the output power for 24 turns with refractive index larger than 1.35. (b) The sensitivity of the sensor *versus* the twisted turns.

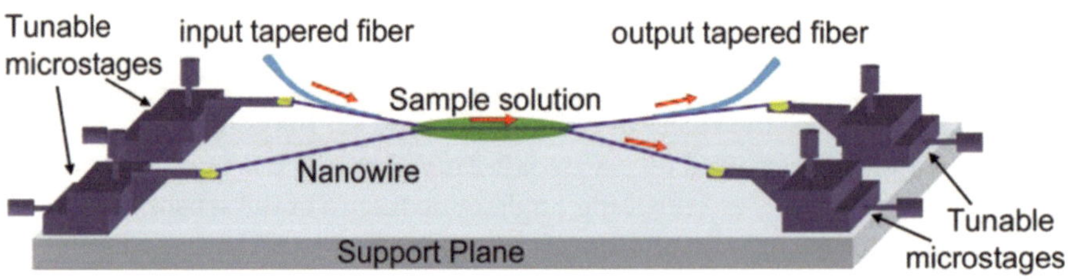

Fig. 38. Schematic diagram of the sensor with 14 twisted turns for tunable characterization. Each end of the sensor was fixed by a tunable microstage. The sensing area was immersed into different sample solutions. Red arrows show the propagation directions of the light. The light is launched into the sensor by the input tapered fiber and collected by the output tapered fiber.

The input/output branching angles α and β were changed from 5° to 35° with a step of 5° by moving the tunable microstages. Figure 39a shows the measured output power change in the output branch 1 with different branching angles at 650 nm wavelength. The measured tunability is 1.8 mW per RIU per 5° branching angle change. Experimental results show that the sensitivity becomes higher with smaller angles of α and β, which is shown in Figure 39b. This can be explained as that the effective coupling length decreases when the branching angle increases, which results in a decrease of the sensitivity. Another reason is that the bending loss in the branching region increase when the branching angle increases, leading to a decrease of the change amplitude of the output power for the same refractive index change, which will lower sensitivity.

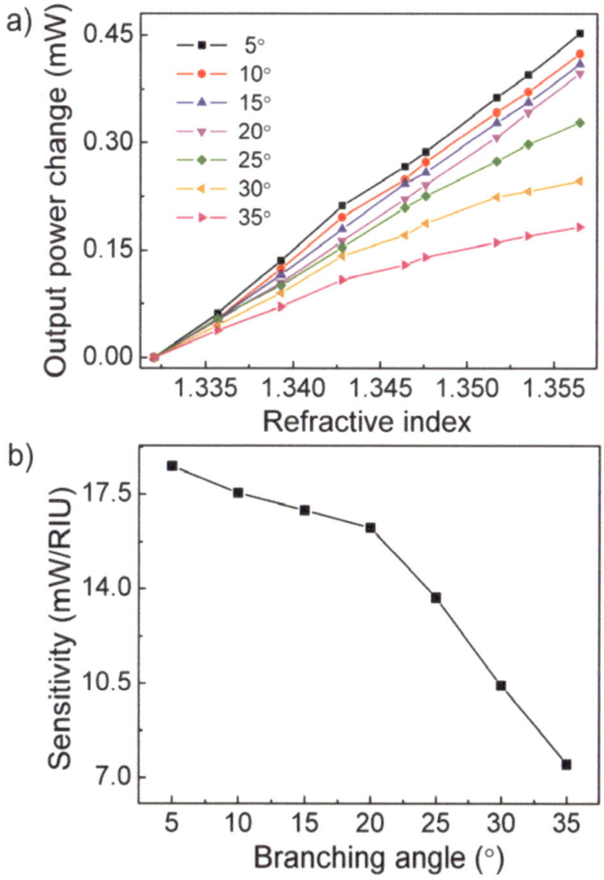

Fig. 39. Input/output branching angle dependence of the sensor. (a) Output power change in the output branch 1 as a function of refractive index at 650 nm red light with different branching angles and 14 twisted turns. Input/output branching angles change from 5° to 35° with a step of 5°. (b) The sensitivity of the sensor *versus* the branching angles.

The highest sensitivity of sensor is 26.96 mW/RIU with 23 twisted turns at 650 nm red light and the refractive index detection limit is 1.85×10^{-7}, while the sensing area is about 130 μm long and 880 nm wide. The properties of the sensor can be tuned by changing the input/output branching angles or the twisted turns. The average tunability are 1.2 mW per

RIU per twisted turn and 1.8 mW per RIU per 5° branching angle change. The ultracompact, highly sensitive and tunable refractive index sensor would be useful in physics, biology, biochemistry, environmental, and toxicant sensing, while the easy, cheap, and fast twist technology would be promising in fabricating of multiterminal nanosensors.

7. Perspectives

PTT nanowires can be easily drawn by the one-step direct drawing technique and with very well surface smoothness, length uniformity, and mechanical strength. PTT nanowire-based devices including optical beam splitters, couplers, nanorings, tweezer/scissor-shaped structures, and a series of nanophotonic device arrays have been assembled. Their application in sensor has also been demonstrated. Apply elastic PTT nanowires to nanodiaply would be an interesting topic in the future. In addition, by adopting photosensitive materials such as rare-earth into the PTT melt, more versatile nanowires could be obtained and more promising nanophotonic devices could be realized.

8. References

[1] D. Appell, "Wired for success," Nature 419, 553-555 (2002).

[2] M. Law, D. J. Sirbuly, J. C. Johnson, J. Goldberger, R. J. Saykally, and P. Yang, "Nanoribbon waveguides for subwavelength photonics integration," Science 305, 1269-1273 (2004).

[3] L. Tong, R. R. Gattass, J. B. Ashcom, S. He, J. Lou, M. Shen, I. Maxwell, and E. Mazur, "Subwavelength-diameter silica wires for low-loss optical wave guiding," Nature 426, 816-819 (2003).

[4] L. Tong, L. Hu, J. Zhang, J. Qiu, Q. Yang, J. Lou, Y. Shen, J. He, and Z. Ye, "Photonic nanowires directly drawn from bulk glasses," Opt. Express 14, 82-87 (2006).

[5] J. C. Knight, G. Cheung, F. Jacques, and T. A. Birks, "Phase-matched excitation of whispering-gallery-mode resonances by a fiber taper," Opt. Lett. 22, 1129-1131 (1997).

[6] G. Brambilla, V. Finazzi, and D. J. Richardson, "Ultra-low-loss optical fiber nanotapers," Opt. Express 12, 2258-2263 (2004).

[7] G. Brambilla, F. Xu, and X. Feng, "Fabrication of optical fiber nanowires and their optical and mechanical characterisation," Electron. Lett. 42, 517-518 (2006).

[8] G. Brambilla, F. Koizumi, X. Feng, and D. J. Richardson, "Compound-glass optical nanowires," Electron. Lett. 41, 400-401 (2005).

[9] L. Tong, J. Lou, R. R. Gattass, S. He, X. Chen, L. Liu, and E. Mazur, "Assembly of silica nanowires on silica aerogels for microphotonic devices," Nano Lett. 5, 259-262 (2005).

[10] Y. Li and L. Tong, "Mach-Zehnder interferometers assembled with optical microfibers or nanofibers," Opt. Lett. 33, 303-305 (2008).

[11] S. A. Harfenist, S. D. Cambron, E. W. Nelson, S. M. Berry, A. W. Isham, M. M. Crain, K. M. Walsh, R. S. Keynton, and R. W. Cohn, "Direct drawing of suspended filamentary micro- and nanostructures from liquid polymers," Nano Lett. 4, 1931-1937 (2004).

[12] H. Liu, J. B. Edel, L. M. Bellan, and H. G. Craighead, "Electrospun polymer nanofibers as subwavelength optical waveguides incorporating quantum dots," Small 2, 495-499 (2006).
[13] J. R. Whinfield and J. T. Dickson, British Patent, 578, 079 (1946).
[14] H. H. Chuah, "Crystallization kinetics of poly(trimethylene terephthalate)," Polym. Eng. Sci. 41, 308-313 (2001).
[15] I. M. Ward, M. A. Wilding, and H. Brody, "The mechanical properties and structure of poly(m-methylene terephthalate) fibers," J. Polym. Sci., Polym. Phys. Ed. 14, 263-274 (1976).
[16] Process Economics Prcgrarn Report 227. 1.3-propanedwl and polytrimethylene terephthalate, SRI International (1999).
[17] J. S. Grebowicz and H. H. Chuah, "Thermal properties of poly(propylene terephthalate)," Research Awareness Bulletin, Shell Chemical Company, Houston, TX, 97-100 (1993).
[18] M. Pyda, A. Boller, J. Grebowicz, H. Chuah, B. V. Lebedev, and B. Wunderlich, "Heat capacity of poly(trimethylene terephthalate)," J. Polym. Sci., Polym. Phys. Ed. 36, 2499-2511 (1998).
[19] K. M. Lee, K. J. Kim, and Y. H. Kim, "Isothermal crystallization behavior and some physical parameters of poly(trimethylene terephthalate)," Polymer (Korea) 23, 56-65 (1999).
[20] J. M. Huang and F. C. Chang, "Crystallization kinetics of poly(trimethylene terephthalate)," J. Polym. Sci., Polym. Phys. Ed. 38, 934-941 (2000).
[21] J. M. Huang, M. Y. Ju, P. P. Chu, and F. C. Chang, "Crystallization and melting behaviors of poly(trimethylene terephthalate)," J. Polym. Res. 6, 259-266 (1999).
[22] K. Dangayach, H. Chuah, W. Gergen, P. Dalton, and F. Smith, "Poly(trimethylene terephthalate) - a new opportunity in engineering thermoplastic applications," Proceedings of the 55th SPE ANTEC'97 Conference, June 16–20, 1997, Chicago, IL, Society of Plastics Engineers, Brookfield, CT, 2097-2101 (1997).
[23] K. Dangayach, K. Ghosh, H. H. Chuah, B. Cristea, and W. P. Gergen, "PTT - a new polymer for engineering applications," ETP'99 Conference, June 7–9, 1999, Zurich, Switzerland, Maack Conference (1999).
[24] H. H. Chuah, "Intrinsic birefringence of poly(trimethylene terephthalate)," J. Polym. Sci.: Part B: Polym. Phys. 40, 1513-1520 (2002).
[25] X. B. Xing, H. Zhu, Y. Q. Wang, and B. J. Li, "Ultracompact photonic coupling splitters twisted by PTT nanowires," Nano Lett. 8, 2839-2843 (2008).
[26] J. Scheirs and T. E. Long, Modern polyesters: chemistry and technology of polyesters and copolyesters, John Wiley & Sons, 2003.
[27] R. Jakeways, I. M. Ward, M. A. Wilding, I. J. Desborough, and M. G. Pass, "Crystal deformation in aromatic polyesters," J. Polym. Sci., Polym. Phys. Ed. 13, 799-813 (1975).
[28] U. W. Gedde, Polymer Physics, Chapman & Hall, London, 1995.
[29] K. Nakamae, T. Nishio, K. Hata, F. Yokoyama, and T. Matsumoto, "Elastic moduli of the crystalline region of polytrimethylene terephthalate," J. Soc. Mater. Sci , Jpn., 35, 1066-1070 (1986).
[30] W. H. Jo, "Polymorphism and surface properties of poly(trimethylene terephthalate) (PTT)," Final Report, Shell Chemical Company, Houston, Tx, 1999.

[31] H. H. Chuah, "Orientation and structure development in poly(trimethylene terephthalate) tensile drawing," Macromolecules 34, 6985-6993 (2001).
[32] M. S. Khil, H. Y. Kim, M. S. Kim, S. Y. Park, and D.-R. Lee, "Nanofibrous mats of poly(trimethylene terephthalate) via electrospinning," Polymer 45, 295-301 (2004).
[33] W. He, B. J. Li, and E. Y.-B. Pun, "Wavelength, cross-angle, and core-diameter dependence of coupling efficiency in nanowire evanescent wave coupling," Opt. Lett. 34, 1597-1599 (2009).
[34] X. B. Xing, Y. Q. Wang, and B. J. Li, "Nanofiber drawing and nanodevice assembly in poly(trimethylene terephthalate)," Opt. Express 16, 10815-10822 (2008).
[35] X. B. Xing, H. Zhu, Y. Q. Wang, and B. J. Li, "Ultracompact photonic coupling splitters twisted by PTT nanowires," Nano Lett. 8, 2839-2843 (2008).
[36] Y. Q. Wang, H. Zhu, and B. J. Li, "Cascaded Mach-Zehnder interferometers assembled by submicrometer PTT wires," IEEE Photon. Technol. Lett. 21, 1115-1117 (2009).
[37] H. Zhu, Y. Q. Wang, and B. J. Li, "Tunable refractive index sensor with ultracompact structure twisted by poly(trimethylene terephthalate) nanowires," ACS Nano 3, 3110-3114 (2009).

10

Doping of Polymers with ZnO Nanostructures for Optoelectronic and Sensor Applications

Aga and Mu
Fisk University
USA

1. Introduction

The discovery of new materials with unique properties often leads to new technology. After discovery of conductive polymers at the end of the 1970s, it opened up a whole new research, which eventually led to a new technology of plastic electronics (Chiang et al., 1978). When interesting material properties are observed in the laboratory, efforts are made to understand their mechanisms, which leads to the fine control of the fabrication process of this new and potentially important material. If future applications are anticipated, optimization of material properties becomes a continuous and sometimes a lifelong process. Doping of materials is one of the ways of modifying their physical properties. For example, a pure silicon (Si), which has a very poor electrical conducting properties can be doped with boron or arsenic to make it a good hole or electron conductor respectively. This opened up a new class of doped-Si materials, which is now the basis of perhaps the largest global electronics industry.

Polymers, which consist of large molecules linked together in repeated fashion to form long chains, have naturally existed for a long time. Examples are tortoise shell, tar and horns. Today synthetic polymers are finding important applications in many areas. Polyolefins, epoxies and engineering resins are crucial materials for construction, commerce, transportation and entertainment. They are very appealing alternative materials because of the simple processing they offer such as drop casting, spray painting and printing. In addition, they almost provide low cost large-area scalability. In most applications however, polymeric materials are multicomponent systems. The integration of fillers such as minerals, ceramics, metals or even air, can generate an infinite variety of new materials with unique physical properties and possibly reduced production cost. Typically, when the filler in these multicomponent systems, which represents a minor constituent, has at least one dimension below 100 nm, the resulting material is termed, "polymer nanocomposite" (Winey & Vaia, 2007). Reseach in this area is very active and promising, mainly because of many different combinations of polymers and filler materials that can be explored. Few examples of fillers that were recently reported are $SrTiO_3$ (Umeda et al., 2009), fluorinated single walled carbon nanotubes (Bennett et al., 2009) and ZnS-coated CdSe nanocrystals (Kim et al., 2009).

The addition of a small amount of filler material into a host polymer can be considered a doping process if there is an intention of modifying the physical properties of the host. Some fillers, particularly with macroscopic dimensions, do not chemically react with the polymer so they cannot be regarded as dopants. Nanomaterials are attractive dopants to polymers because of their high surface reactivity, which is attributed to very large surface-to-volume ratio. Moreover, they possess intriguing properties associated with quantum confinement effects, making their interaction with different polymers a subject of great interest. They can also be dispersed into water or organic solvents, providing a simple process of doping polymeric solutions. It should be realized that nanomaterial-doped polymers should also be classified as polymer nanocomposite because they are multicomponents systems with a low dimensional non-polymer minor constituent embedded in a host polymer. When the nanomaterial dopant is inorganic, the term "hybrid nanocomposite" is specifically used for the reason that it becomes an inorganic-organic system. This manuscript focuses on the effect of doping polymers with ZnO nanostructures such as nanowires, nanorods, tetrapods and nanoparticles. Their combination may achieve three goals: 1) tailor the property of ZnO; 2) tailor the property of polymer or 3) create a hybrid nanocomposite with unique or enhanced properties. A discussion of how these goals are relevant for optoelectronic and sensor applications is presented.

2. ZnO as dopant to polymers

2.1 Technological importance of ZnO

ZnO has received much attention in recent years due to its diverse properties. It is a direct wide bandgap semiconductor (E_g = 3.4 eV) with large exciton binding energy (~60 meV), suggesting that it is a promising candidate for stable room temperature luminescent and lasing devices. It exhibits strong ultraviolet (Huang et al., 2001) and visible photoluminescence (Konenkamp et al., 2004). ZnO film can be used as transparent conducting electrode. It is also a piezoelectric material, with potential applications in surface acoustic wave filters (Wang et al., 2008). Ferromagnetism in ZnO has already been reported (Kim et al., 2009) so it is being considered a material for spintronics. With all these important properties, it becomes more attractive because of its abundance in nature. It can be fabricated practically into different forms like thin film and nanowire using a variety of methods such as solution-based approaches, vacuum deposition techniques chemical vapor deposition.

2.2 Fabrication of ZnO nanostructures for dispersion in solution

There are several ways of fabricating ZnO nanostructures on a substrate. Nanowires, nanorods, nanobelts, nanoneedles and nanosprings have been reported by many researchers. For doping polymers, it is advantageous for the nanostructures to be dispersed in a solution since most polymers are prepared via a solution process. ZnO nanostructures grown on a substrate can be scraped and dispersed into a solvent. This approach however does not produce enough amount of nanostructures. Hence, it is not a practical method for obtaining a usable concentration of ZnO in a solvent. Solution-based methods have been widely used to prepare ZnO nanostructures. Nanorods and nanowires with variable-aspect-ratio (length/diameter) have been prepared in alcohol/water solution by reacting a Zn^{2+} precursor with an organic weak base, tetramethylammonium hydroxide (Me_4NOH) at temperature window of 75 - 150 °C (Cheng et al., 2006). In another report, a room temperature wet-chemical approach was employed to synthesize highly regulated, monodispersed ZnO nanorods and derived hierarchical nanostructures such as hexagonally branched, reversed umbrella-type and cactus-like (Liu & Zeng, 2004). These hierarchical nanostructures are comprised of individual *c*-oriented nanorods. In the typical synthesis, the precursor solution of Zn was prepared by dissolving zinc nitrate and sodium hydroxide in deionized water. A sample of pure ethanol was added followed by ethylenediamine. The mixed solution was covered and kept at room temperature under constant stirring for 1-12 days to form the ZnO nanorods. ZnO nanoparticles with size ranging from 2 to 7 nm can also be prepared by addition of lithium hydroxide to an ethanolic zinc acetate solution (Meulenkamp, 1998).

One of the disadvantages of solution-based ZnO fabrication is that it involves successive processes that take a fair amount of time. After the synthesis, the product has to be thoroughly cleaned to remove all traces of chemicals used in the process. In this article, a simple, straightforward method to fabricate ZnO nanostructures with fast turn around is explored. It is based on the oxidation of zinc vapor while a high purity Zn metal is being thermally evaporated. Figure 1 is a diagram that illustrates this method. The Zn source is placed in a crucible, which is then inserted into a vertically-oriented quartz tube. The tube is flushed with argon to remove ambient air and its open end is sealed [Figure 1(a)]. It is inserted into a vertically-oriented furnace set at 600 °C to evaporate the Zn. When the source starts to evaporate, air is allowed to diffuse towards the crucible by opening the seal. With the correct diffusion rate of O_2 and evaporation rate of Zn, oxidation can be contained inside the crucible. Thus, the growth of ZnO nanostructures mainly occurs in the crucible. Figure 1(b) shows the actual crucible before and after the growth process. Typically, when the amount of Zn source is 0.03 g, it yields approximately 0.01 g of ZnO, which is the white material that fills the crucible in the picture. The product is like a cotton as shown by the SEM picture in Figure 1(c). It can be easily collected and dispersed in a solution. The entire fabrication process only takes less than 30 minutes. It is highly scalable and can be employed to mass-produce ZnO nanostructures.

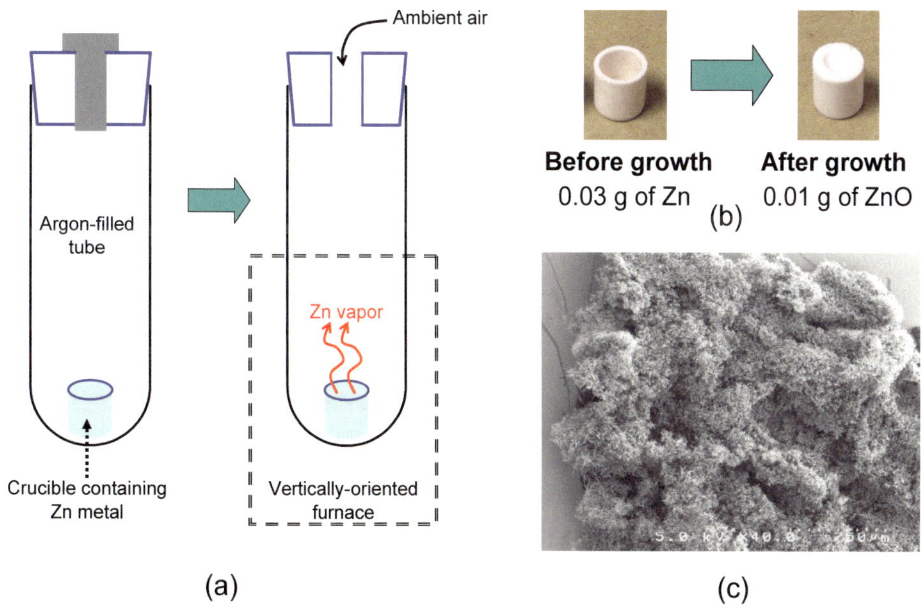

Fig. 1. (a) Diagram illustrating the simple fabrication process of ZnO nanostructures. (b) Photos of the actual crucible before and after the growth process. (c) SEM picture of the cotton-like ZnO product collected from the crucible.

The cotton-like ZnO product shown in Figure 1(c) consists of different nanostructures entangled together. They can be separated and size selected if desired. Two solvents were tried to disperse them: water and chlorobenzene. By ultrasonication, they were effectively segregated in both solvents. The solution became whitish after several minutes of agitation. When the procedure was stopped, ZnO structures slowly settled at the bottom and as expected, large entities sank first. This can be used as a simple size-selection method. For example, nanoparticles with lower diameter are the last ones to settle at the bottom. If the solution is allowed to be still for a long time after sonication, then get the top part of it, nanoparticles are obtained. They are shown in Figure 2(a). They are deposited on Si wafer by drop casting. If the large and unwanted entities are already removed, good nanowires can be extracted from the bottom of the solution. These nanowires are shown in Figure 2(b). Their length could exceed 10 μm.

(a) (b)

Fig. 2. SEM micrographs of (a) ZnO nanoparticles and (b) ZnO nanowires via a mass production method and deposited on Si wafer after a simple size-selection process.

3. Polymer-based light emitting devices

3.1 Basic requirements for efficient electroluminescence

Polymeric-based light emitting device is one type of OLED, which stands for organic light emitting diode. Photon is generated in the polymer emissive layer (EL), which should possess excellent photoluminescence (PL) properties. In an efficient device, the EL is sandwiched between an electron transport layer (ETL), which injects electrons into the LUMO (lowest unoccupied molecular orbital) of the EL, and a hole transport layer (HTL), which injects holes into the HOMO (highest occupied molecular orbital) of the EL. The injected electrons and holes undergo radiative recombination due to the applied electric field. The charge transporting layers serve as facilitator for efficient charge transfer from the electrodes to the EL. This is illustrated in Figure 3. When a negative potential is applied to the anode, which is a low work function metal, electrons are injected into the ETL. If the ETL has excellent electron mobility (*n*-type) and its LUMO align closely with the LUMO of EL and the work function of the anode, efficient electron charge transport towards the EL is achieved. On the other hand, the positive potential applied to the cathode, which is a high work function transparent conductor, results in the injection of holes into the HTL. Efficient hole transport towards the EL is achieved if the HTL has excellent hole mobility (*p*-type) and its HOMO is well aligned with the HOMO of the EL and the work function of the cathode. It should be realized that the realization of high efficient OLEDs depends not only on the electronic and optical properties of the EL material, but also on the control of charge transport in the device.

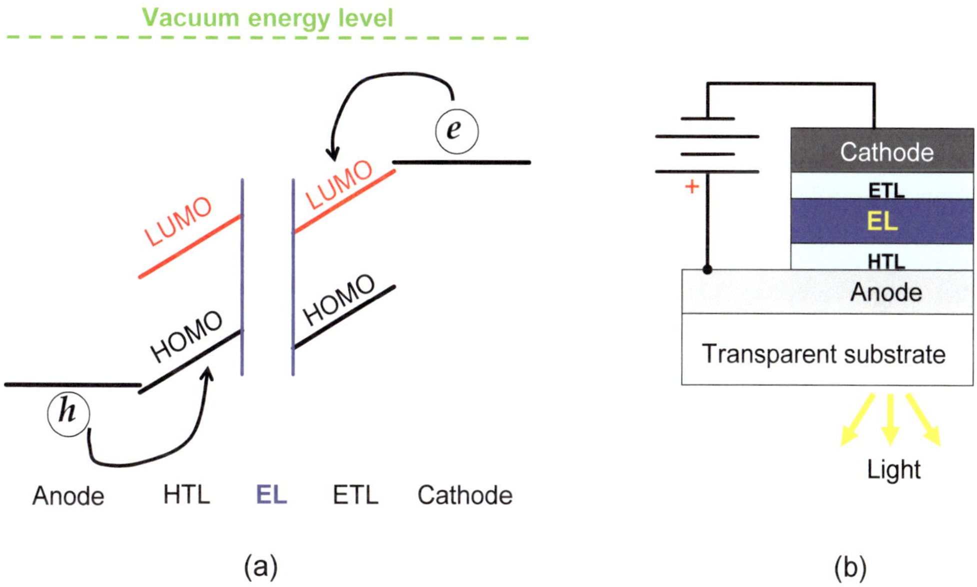

Fig. 3. (a) Operation of a polymeric OLED and its (b) device structure.

3.2 Control of charge transport

One of the benefits of doping polymers in OLED with ZnO nanostructures is the improvement in charge carrier transport. It has been reported that by incorporating ZnO nanorods with the EL material PVK:DCJTB, which represents poly(vinylcarbazole) doped with 4-(dicyanomethylene)-2-*t*-butyl-6-(1,1,7,7-tetramethyljulolidyl-9-enyl)-4H-pyran at 1%, low threshold voltage and higher electroluminescence efficiency is obtained (Zhang et al., 2006). The authors employed the device structure: ITO/EL/Alq_3/Al, with different concentrations of ZnO (1 wt%, 5 wt%, 10 wt%) in the EL material. They observed that the hole current density increases with increasing concentration of ZnO nanorods. This made the current injection more balanced and consequently enhanced the electroluminescence efficiency. They explained the increase in the hole current in terms of the effect of ZnO on the molecular chain of PVK. The nanorods can make the PVK chain more unwindable and possibly connect aligned molecular chains. Thus, current is not limited by hopping of carriers between chains. In addition, wurtzite ZnO has higher hole mobility (μ_h) than PVK. For low *p*-type conductivity at room temperature, it ranges from 5 to 50 cm^2/V-s (Norton et al., 2004). For PVK at low electric fields, it is in the order of 10^{-7} cm^2/V-s (Blom et al., 1997).
A similar observation has been reported on poly[2-methoxy,5-(2-ethylhexyloxy)-1,4-phenylenevinylene], which is commonly called MEH-PPV for short (Xu et al., 2007). By doping the MEH-PPV with ZnO nanocrystals or tetrapods and employing it to build field effect transistor, the authors showed higher *p* – type mobility in the polymer nanocomposite.

Pure MEH-PPV, which only exhibits *p* - channel behaviour, has μ_h value in the order of 10^{-4} cm^2/V-s. This increases with increasing concentration of ZnO nanostructures and saturates at 40 wt%. For nanocrystals and tetrapods, μ_h saturates at a value of about 0.08 cm^2/V-s and 0.15 cm^2/V-s respectively. It should be mentioned that doping with ZnO did not change the conductivity type, indicating that charge transport takes place in the polymer. This is not surprising because there is a large energy barrier for holes to be transferred from ZnO to MEH-PPV. The valence band edge for ZnO is 7.6 eV while the HOMO of MEH-PPV is 5.3 eV. When the density of traps was investigated, it was found that it decreases with increasing ZnO concentration. Interestingly, it also saturates at 40 wt%. This is a strong indication that the effect of ZnO is to reduce the density of traps in the polymer and consequently improve its μ_h. The authors further supported this interpretation by transient current measurement where a constant bias voltage was applied to the gate and the drain. The decrease in channel current with time was more dramatic for the pure polymer than the nanocomposite. This suggests that charge trapping is significantly reduced by ZnO doping. Other authors propose that the enhancement in μ_h is due to the superposition of several transport mechanisms in the nanocomposite including percolation through polymer-ZnO nanoparticle interface network (Aleshin et al., 2008).

3.3 Control of optical emission

The high surface reactivity of ZnO nanostructures with certain polymers may provide interactions that can modify polymer properties in a positive way. The interaction of ZnO quantum dots (QDs) with poly(vinyl alcohol) (PVA) has already been demonstrated (Sui et al., 2005). The authors prepared PVA/ZnO hybrid nanofibers with ZnO concentration of 10 wt% and employed differential scanning calorimetry (DSC) to study them with respect to pure PVA. It is known that the two exothermic peaks at about 305 °C and 500 °C are associated with the degradation of side chain (the scission of C-O) and main chain (the scission of C-C) with the delta enthalphy of 3760 and 1188 J g^{-1}, respectively (Koji et al., 1999). For PVA/ZnO nanofibers, the exothermic peak below 450 °C was not observed, while the peak around 500 °C was sharp and strong with delta enthalpy of 3305 J g^{-1}, suggesting that ZnO modifies the property of PVA. Their interaction, which is via the formation of an H bond and an O-Zn-O bond between PVA molecule and ZnO, is believed to be responsible for the novel luminescent properties of PVA/ZnO nanofibers.

Figure 4 shows a typical PL spectrum of ZnO nanowires (NWs) prepared by the physical vapor transport method described in section 2.2. It has two emission bands around 380 and 520 nm, which is consistent with another report using a polymer-assisted route (Li et al., 2003). The UV emission originates primarily from a mixture of free exciton and bound exciton related to impurity or defects (Shan et al., 2005). The visible emission comes from the transition between the electron near the conduction band and the deeply trapped hole at the V_O^{**} center which is an oxygen vacancy containing no electrons (Dijken et al., 2000). It is also attributed to the transition between the electron at [V_O^{*}, electron] or [V_O^{**}, two electrons] and the hole at vacancy associated with the surface defects. For pure PVA, the PL spectrum consists of two emission bands at about 364 and 440 nm, originating from its organic functional groups (Sui et al., 2005). In PVA/ZnO hybrid film, the PL spectrum is a superposition of the PL of each component. The 440 nm emission from PVA overlaps with the 550 nm from ZnO making the visible emission band broader. In hybrid nanofibers however, this visible emission is much intense possibly due to two factors. In the film, ZnO-

QDs can easily form aggregates resulting in bigger particle exhibiting bulk behavior. This is in fact one of the issues in polymer nanocomposites and ways to circumvent it are being explored (Sun et al., 2008). In the nanofiber, QDs align along the PVA matrix, resulting in better dispersion. Individual QD also interacts strongly with the PVA molecule. The other factor is the high surface to volume ratio of the nanofiber, which enhances the visible luminescence due to surface defects (Shalish et al., 2004).

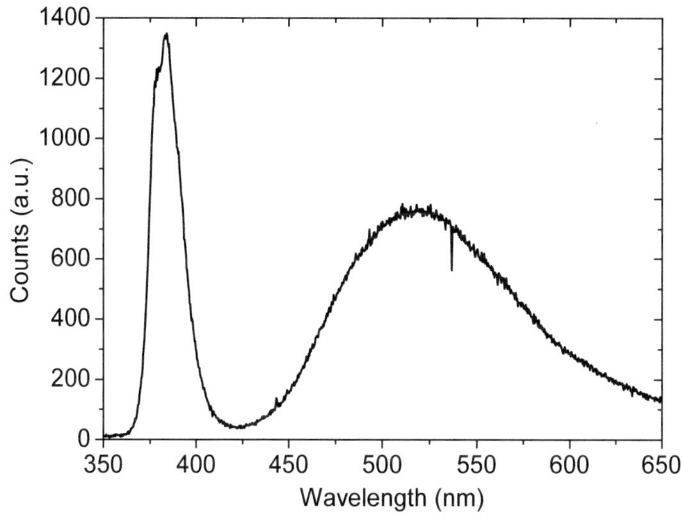

Fig. 4. Photoluminescence (PL) spectrum of ZnO nanowires for polymer doping, grown by physical vapour transport.

The interaction between ZnO and polymer can also provide several new radiative recombination mechanisms that may be utilized for broadband light emission or wavelength tunable emission. Since these mechanisms are influenced by several factors such as the individual properties of the organic and inorganic components as well as their relative concentration, control of emission properties may be done in a number of ways. The existence of new recombination mechanisms is strongly suggested by the generation of new PL emission lines (420, 460 & 480 nm) when MEH-PPV is doped with ZnO nanoparticles (Aleshin et al., 2009). In this work, the authors prepared the polymer nanocomposite film on a Si substrate with a 200 nm-thick SiO_2 layer via drop casting or spin coating. Gold and aluminum electrodes were then deposited on top with separation of 7 - 15 μm. With this device structure, they could investigate PL while applying an external electric field. The existence of new emission lines, which could not be generated when ZnO was replaced by Si nanoparticles, was attributed to the formation of interfacial states termed exciplex. These states are formed in a type II heterojunction, such as ZnO/MEH-PPV, where there is an accumulation of electrons and holes on the opposite side. Charge transport is blocked by the barriers that result from the offset of HOMO and conduction band edge of MEH-PPV and ZnO respectively. It was proposed that since the carriers cannot proceed across the heterojunction, they diffuse two-dimensionally along the interface until they encounter an oppositely charged carrier on the opposite side of the interface and these form an exciplex (Morteani et al., 2005). If bulk exciton in the ZnO or MEH-PPV is within the exciton diffusion length, it can excite exciplex emission. It should be mentioned that the influence of

ZnO nanoparticles is most dramatic at an optimum concentration of 10 wt%. This could be ascribed to the aggregation of nanoparticles which reduces interaction with polymer.
Another observed phenomena when MEH-PPV is doped with ZnO is the blue shift of its PL emission spectrum accompanied by quenching. At 2 wt% concentration of ZnO nanoparticles, 11 nm shift has been reported (Ton-That et al., 2008). This shift increases with increasing nanoparticle concentration. However, its dependence becomes weaker at concentration greater than 4%. Using Raman spectroscopy, the authors did not observe any changes in the conjugation length, which may have caused the blue shift in PL spectrum. Their results indicate that the chemical structure and the gap energy of MEH-PPV is not affected by the incorporation of ZnO nanoparticles. Thus, they explained their observation in terms of the effect of electric field produced by excess electrons on the nanoparticles surface (Musikhin et al., 2002). The surface of ZnO nanoparticles is a strong perturbation of the lattice where there exists a high concentration of both shallow and deep defect levels (Ton-That et al., 2008). X-ray photoelectron spectroscopy (XPS) analysis of their surface revealed that there is a large number of oxygen vacancies. These vacancies can accept electrons from MEH-PPV, creating an electric field outside the nanoparticles. As a result, the energy of the lowest exciton state is raised resulting in the blue shift of the luminescence.

4. Polymer-based photovoltaics

4.1 Challenges

The highest power conversion efficiency (η) demonstrated by a polymer-based solar cell is achieved by employing an active layer blend of poly(3-hexylthiophene) and (6,6)-phenyl C61 butyric acid methyl ester (Ma et al., 2005). For convenience, this blend will be referred as P3HT:PCBM. It has a strong optical absorbance in the limited wavelength range of 400 nm to 600 nm mainly due to the P3HT. Because of this narrow band absorption, more than 60% of the photons in the solar spectrum are not harnessed. In fact, even the ones absorbed are not 100% converted to photocurrent because not all electron-hole pairs become free charges. Exciton binding energy of P3HT is so large that it does not lead directly to free charge carriers. However, the high electron affinity of PCBM provides sufficient potential difference that breaks exciton and thus it becomes energetically favourable for the electrons to jump from the LUMO of P3HT to the LUMO of PCBM. This electron donor-acceptor mechanism implies that only excitons near the P3HT/PCBM interface dissociate. The maximum distance for dissociation to occur is determined by the exciton diffusion length (τ), which is around 10 nm for P3HT. In the blend, phase segregation results in the formation of isolated islands of P3HT and PCBM. The size of P3HT islands could be more than τ so not all excitons can dissociate. There is another factor in the exciton dissociation process that reduces η. Excited electrons in P3HT have excess energy that is dissipated quickly when they transfer to the PCBM. This energy loss, which depends on the energy alignment of the two materials, is reflected in the open circuit voltage (Koster et al., 2006).
In a photovoltaic device, separated holes and electrons need to reach their respective electrodes to have high η. This requires the blend to have high values of hole (μ_h) and electron (μ_e) mobilities. Good balance of mobilities is obtained using optimum ratio of P3HT and PCBM, which is estimated to be between 1:1 and 1:0.9 (Chirvase et al., 2004). This could yield values of $\mu_h \sim 6.5 \times 10^{-6}$ cm^2/V-s and $\mu_e \sim 5.0 \times 10^{-6}$ cm^2/V-s. However these numbers are still far inferior when compared to their inorganic counterpart, which raises the question of whether doping with inorganic nanostructures such as ZnO would be beneficial. In the

blend, PCBM does not provide direct conduction pathways for electrons towards the electrode because of the isolated island formation. Electrons undergo a hopping process between PCBM islands before reaching the electrode. Thus, they become prone to recombination with holes. This limits the thickness of the active layer. High efficient P3HT:PCBM devices are achieved with active layer thickness around 100 nm to 200 nm, which is not the optimum thickness for photon absorption. Increasing the thickness absorbs more photons but it also increases the electron-hole recombination and degrades charge transport. The final stage of charge transport from point of exciton dissociation to the electrode is the crossing of the organic-inorganic interface. Large contact resistance will reduce the fill factor of a device. Potential barriers at the interface originating from impurities or polymer damage during vacuum deposition of the metal electrodes may also limit the open circuit voltage. Controlling the organic-inorganic interfacial properties is an important and urgent challenge in organic electronics. Several ways have already been explored such as insertion of very thin interlayer (Brabec et al., 2002), tuning the work function of the inorganic electrode (Sharma et al., 2009) and surface doping of polymer (Mukherjee et al., 2007).

4.2 ZnO as electron acceptor

PCBM is the most widely used electron acceptor for P3HT in a bulk heterojunction (BHJ) solar cell. It has the right LUMO and HOMO levels for charge separation and exchange. Its electron mobility in the order of 10^{-2} $cm^2/V\text{-}s$, is sufficient for transporting photogenerated electrons towards the electrodes. However, higher mobilities are still demonstrated by inorganic materials. Wurtzite ZnO for example can have electron mobility of 200 $cm^2/V\text{-}s$. Its LUMO and HOMO levels are well positioned to accept electrons from P3HT. Thus, doping of conjugated polymers with ZnO nanostructures can be employed to fabricate BHJ solar cells. For example, P3HT doped with ZnO nanoparticles yielded a photovoltaic device with $\eta \sim 0.92\%$ at optimum concentration of 26% by volume (Beek et al., 2006). The values of the open circuit voltage (V_{OC}), short circuit current density (J_{SC}) and fill factor (FF) were 0.69 V, 2.19 mA/cm^2 and 55% respectively. The authors employed the device structure: ITO/PEDOT:PSS/P3HT:ZnO/Al, which is a common structure used by P3HT:PCBM active layer. They identified the major performance limiting factors as the poor contact of P3HT with the ZnO and the coarse morphology. They also observed that thermal annealing of the active layer was very important.

In another work, ZnO nanofibers were grown on the transparent electrode to achieve fixed morphology before the introduction of the P3HT (Olson et al., 2007). By spin-coating the P3HT on top of the ZnO nanostructure followed by annealing, intercalation into the voids between the nanofibers was induced. In this structure, a huge donor-acceptor interface area is formed between the P3HT and ZnO for exciton dissociation. Photogenerated electrons injected into the ZnO are transported directly towards the collecting electrode, providing pathways with higher electron mobility. The growth of ZnO nanofibers on ITO-coated glass was achieved via a low-temperature hydrothermal route from a solution of zinc nitrate precursor. In this device structure in which Ag is used as back electrode instead of Al, electrons are collected by the ITO while holes are collected by the Ag. The best device they obtained had V_{OC}, J_{SC} and FF of 0.44 V, 2.17 mA/cm^2 and 56% respectively corresponding to $\eta \sim 0.53\%$. Although this performance is relatively poorer than P3HT:PCBM devices ($\eta \sim 5\%$), it is expected to improve when the major mechanisms in the device are better understood. For example, the role of oxygen vacancies in the ZnO, which is the origin of visible emission, is not yet clear. Obtaining a balanced electron and hole mobilities in the active layer may also be a critical issue. Understanding the effect of ZnO on the effective Fermi level of the active layer, which determines the interfacial properties at the metal junction, is important. This may have a direct consequence on the V_{OC}. Lastly, the mechanism of aging in ambient air may give insights on how to optimize this device. The authors observed that right after fabrication, the device exhibited only little diode behavior. But after storing it in dark ambient air for 24 hours, better device performance was observed. In contrast, the device stored in dark argon environment decreased in performance. After 3 days in argon, the device performance continued to deteriorate whereas the performance of the device stored in air improved substantially. The improvement came from the increase in *FF* and V_{OC}. They explained this behavior in terms of the oxygen vacancies. These vacancies, which are intrinsic electron donors in ZnO, may be quenched when exposed to oxygen. As a result, semiconducting properties are improved as both the oxygen vacancy and carrier concentrations are reduced. This aging mechanism makes ZnO a very promising electron acceptor for BHJ solar cells, which require long term operation.

4.3 Effect of ZnO nanowire doping on the Fermi level of P3HT

In polymer-based photovoltaics, it is important to obtain the desired junction characteristic of the polymer-metal interface. Depending on the device structure, it may require an Ohmic or a Schottky junction. The junction characteristic, which has a direct influence on the charge transfer from the active layer to the electrode, is determined by the energy level alignment of the polymer and the metal. Fermi level (E_F) is the most fundamental parameter that is used to understand polymer-metal junction. It determines the work function (energy difference between E_F and vacuum level) of a material. It can be used to predict the built in potential (V_{bi}) across a junction in thermal equilibrium. In thin film photovoltaic devices, V_{bi} is an essential parameter. It influences charge dissociation, charge transport and charge collection. Most importantly, it sets the maximum V_{oc} of a single junction solar cell. Thus, measurement of V_{bi} is very important practically and it can be a tool to understand how the E_F of a material behaves when it is doped. In most situations, V_{bi} across a junction is not equal to the value predicted by the difference in the E_F of each component. During the formation of the junction, such as the deposition of the metal on the polymer or the drying of the polymer on a metal surface, chemical reaction may occur. This leads to the existence of new compounds in the middle of the junction, which modifies the expected V_{bi}. Thus, determination of actual V_{bi} in real devices is more important practically because it takes into account the effect of processing. In photovoltaics for example, measuring V_{bi} in P3HT-Al junction formed by thermal evaporation of Al on P3HT film, is more meaningful than using a similar junction formed by spin-coating P3HT on Al film. This is because the first system is the one being used in the development of organic solar cells.

In this manuscript, recent results are presented about the change in the Fermi level of P3HT when doped with ZnO nanowires. This system is already being explored as an active layer together with Al as back electrode (Beek et al., 2006). To measure V_{bi}, capacitance-voltage (C-V) measurement was employed using an LCR meter. Test diodes were prepared by drop-casting P3HT and doped P3HT (~ 1 μm thick) on Cr/Pt film then depositing Al on top via

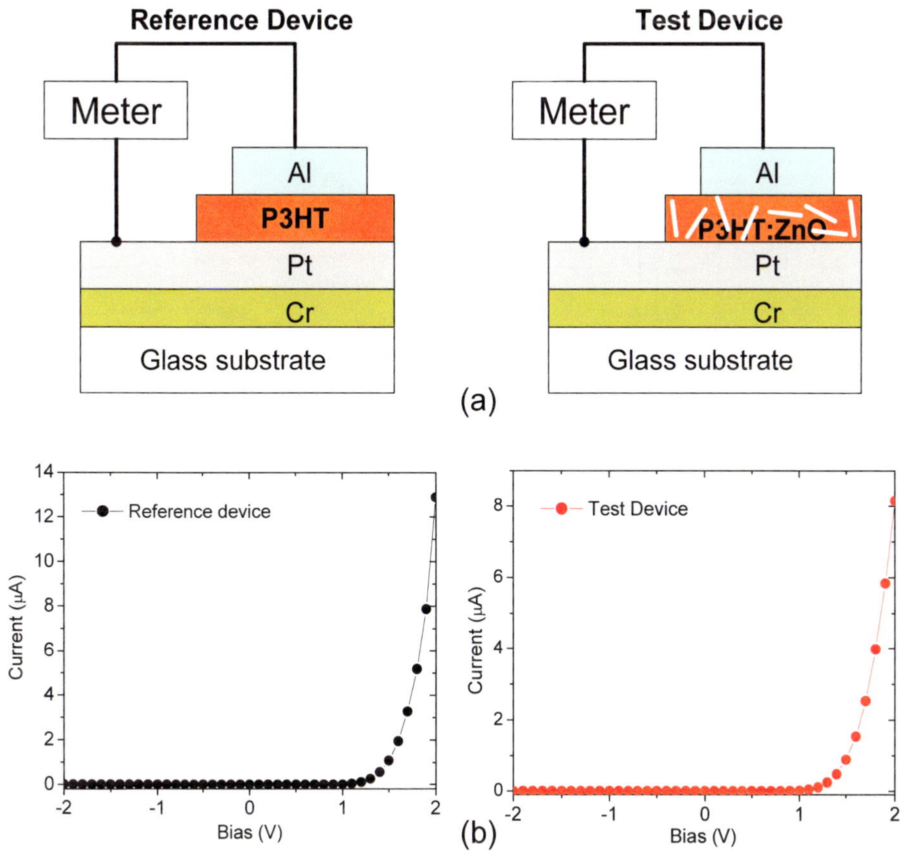

Fig. 5. (a) Device structure of the diodes used to measure the built in potential. (b) *I-V* characteristic of the reference (pure P3HT) device and the test (ZnO-doped P3HT) device.

resistive thermal evaporation. The device structure is shown on Figure 5(a). Both doped and undoped devices exhibited good rectifying behaviour as depicted by their *I-V* curves in Figure 5(b). The rectification is due to the Schottky junction formed by the polymer and Al. The high work function value of Pt (~6.35 eV) results in an Ohmic contact with the polymer. According to inorganic semiconductor theory, the capacitance C of the depletion layer is related to V_{bi} by the equation (Sze, 2007):

$$\frac{1}{C^2} = \frac{2(V_{bi} - V - kT/q)}{q\varepsilon_S N_D} \tag{1}$$

Where V is the applied bias voltage, kT/q is the thermal voltage whose value at room temperature is 0.0259 V, ε_s is the semiconductor permittivity and N_D is the donor impurity density. V_{bi} is determined by plotting this equation and finding the value of V where $1/C^2$ extrapolates to zero. This is illustrated in Figure 6. The behavior of $1/C^2$ is quite linear for both devices in the bias range of ±0.5 V, consistent with equation (1). The solid line is the linear fit for each data set. Its intercept plus the thermal voltage is equal to V_{bi}. The obtained value for the pure P3HT is 1.243 V. This is fairly close to the expected value based on the E_L

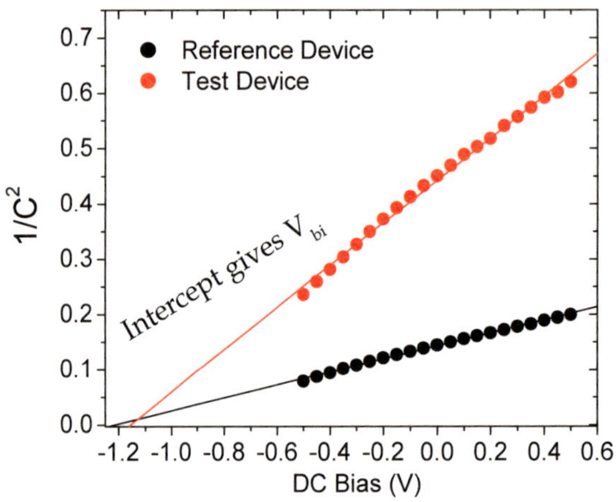

Fig. 6. Plot of $1/C^2$ where C is the depletion layer capacitance as a function of applied DC bias. The intercept of the linear fit gives the built in potential.

offset between P3HT and Al. The work function of P3HT is around 5.1 to 5.2 eV. For Al, it ranges from 4.0 to 4.3 eV. The ZnO-doped P3HT has V_{bi} of 1.181V, indicating that E_F of the P3HT moved farther from the valence band edge and closer to the E_F of Al. One possible explanation for this reduction is the increase in the free electron concentration in the polymer nanocomposite provided by the ZnO. In semiconductors, E_F lies closer to the conduction band edge for *n*-type conductivity and it lies closer to the valence band edge for *p* - type conductivity. The upward shift in E_F due to the ZnO may be the cause of the drop in V_{OC} for P3HT:ZnO nanofibers BHJ (Olson et al., 2007). When the authors compared the photovoltaic performance of ZnO bilayer and ZnO nanofiber devices, they observed higher V_{OC} in the bilayer device (500 mV versus 440 mV). Although the nanofiber device still had better η due to its higher J_{SC}. The results shown in Figure 6 may be helpful in optimizing BHJ solar cell based on P3HT and ZnO nanostructures. High values of J_{SC} and V_{OC} may be obtained simultaneously by tailoring the property of the active layer-metal interface or selecting alternative metals or alloys for back electrode.

5. Polymer-based conductivity sensors

5.1 Conductivity sensors for chemical detection

Conductivity sensors for chemical detection are very attractive because of their simple operation. The circuit design required to convert their response to electrical signals is relatively straightforward. These sensors are based on the change in the electrical conductivity of the sensing material due to its interaction with the stimuli. Device fabrication is also not difficult. For example, the sensing material may be deposited on interdigitated or two parallel electrodes, which form the connection for measuring the change in electrical resistance. Three common classes of sensing materials for conductivity sensors are conducting polymer composites, intrinsically conducting polymers and metal oxides (Arshak et al., 2004). The mechanism behind the change in conductivity is different for each class.

In conducting polymer composites, conducting particles such as polypyrrole and carbon black (Albert et al., 2000) are embedded in an insulating polymer matrix. The stimuli can cause expansion of the polymer and the polypyrrole particles resulting in the increase of electrical resistance. If the stimuli interact chemically with the polypyrrole, its intrinsic conductivity will also change, making the resistance change in the composite more difficult to interpret. The operation of intrinsically conducting polymers (ICP) such as polypyrrole, polythiophene and polyaniline, is based on the alteration of intrachain conductivity, intermolecular conductivity and ionic conductivity after interacting with the stimuli. Intrachain conductivity is detrmined by the backbone while intermolecular conductivity is due to electron hopping to different chains. Ionic conductivity is affected by proton tunneling induced by hydrogen bond interaction at the backbone and also by ion migration through the polymer. One reported application of ICP employed an array of conducting polymer sensors incorporating 3-Methylthiophene, Aniline and Pyrrole (Guadarrama et al., 2000). The authors evaluated the sensors to different wine varieties and considered the cross sensitivity of the polymeric films to moisture and ethanol. The operation of a metal oxide sensor is based on the dependence of the conductivity to O_2 molecules adsorbed on its surface. For *n*-type metal oxides such as ZnO or TiO_2, O_2 molecules capture electrons on the surface or at the grain boundaries resulting in the decrease of conductivity. Trapped electrons at the grain boundaries also produce potential barriers between grains that impede current flow. When exposed to reducing gas like H_2, CH_4, CO, C_2H_5 or H_2S, the adsorbed O_2 molecules react with the gas and release the captured electrons. This increases the conductivity of the sensing material. When exposed to oxidizing gas like O_2, NO_2 or Cl_2, the effect is opposite. Conductivity is decreased because the number of adsorbed O_2 molecules that can capture electrons is increased. For *p*-type metal oxides, their response to oxidizing gas is to increase in conductivity because the capture of electrons produces more holes. In contrast, their conductivity decreases when exposed to reducing gases.

There are still many challenges in conductivity gas sensors. For polymer composites, aging is a concern because this leads to sensor drift. Their applicability is only limited to certain gases because of lack of sensitivity. For ICP, challenges include understanding the mechanism of sensor response, high sensitivity to humidity and drift in conductivity with time. These sensors can also have short lifetimes due to oxidation of the polymer. For metal oxides, one challenge is the low sensitivity at room temperature. They require higher operating temperatures and thus heating element has to be incorporated with the sensor (Roy & Basu, 2002) This increases the power consumption, which is not very appealing for handheld applications. Good selectivity is another challenge for metal oxide conductivity sensors. Their behavior to reducing gases is practically the same. For example, the conductivity of a sensor changed by Δ when exposed to ethanol vapor with concentration C_1. The same response can be realized when it is exposed to methanol with a different concentration C_2. Thus, the identification of a particular reducing vapor, with a background of different others is almost impossible. Due to the mentioned challenges, there is a continuous effort to develop new materials for conductivity sensors. Reliable sensors with appropriate levels of selectivity and sensitivity will always be in demand. In the standpoint of defense and homeland security, new hazardous compounds or biomaterials that need to be detected is expected to arise. New applications such as self-powered sensors will require excellent energy efficient sensing materials without sacrificing sensitivity and selectivity.

5.2 Sensing with P3HT doped with ZnO nanowires

There is a lack of work in the exploration of ZnO doped polymers as conductivity sensors for chemical detection. Interestingly, only P3HT, which is mostly used in photovoltaics, has been reported to have promising sensing behavior when blended with ZnO-NWs (Saxena et al., 2007). As separate materials, P3HT and ZnO-NWs can both be used as a conductivity gas sensor. But due to the difference in their conductivity type, they have contrasting response to a specific stimulus. For example, they both exhibit high sensitivity to NO_2 and H_2S (Saxena et al., 2007). P3HT decreases in resistance upon exposure to NO_2, which is oxidizing gas, while it behaves oppositely for H_2S, a reducing gas. This characteristic is due to its p - type conductivity. The adsorption of an oxidizing compound on P3HT surface captures electrons leading to increased hole carrier concentration. Conversely, the adsorption of a reducing compound leads to the decrease in hole carrier concentration via the release of captured electrons. For n - type ZnO, the effect of oxidizing and reducing compounds is to decrease and increase conductivity respectively.

Doping of P3HT with oxygen defficient ZnO nanowires results in the reduction of P3HT. This is well supported by GI-XRD, XPS and FTIR studies (Saxena et al., 2007). The authors also observed decrease in the P3HT conductivity after ZnO doping suggesting that its hole carrier concentation drops when ZnO donates electrons. This interaction may be used to tailor the response of polymers to different stimuli, opening up the possilibity of tunable polymer-based conductivity sensors. For example, the authors have demonstrated that highly reduced P3HT resulting from ZnO doping, has enhanced sensitivity to NO_2 because this chemical can easily pick up electrons from P3HT to get adsorbed as NO^-_2 ions. They have obtained room temparature sensors that can detect NO_2 in the 0 - 10 ppm range with very high selectivity. In contrast, the sensitivity of ZnO-doped P3HT to a reducing H_2S is weakened because it is difficult to further reduce the P3HT.

6. Conclusion

As separate materials, polymers and ZnO are already considered to be important technologically. They have found several commercial applications in optoelectronics and sensors. Despite that, research activities involving these materials are still in full swing. For example, there is a lot of effort in developing high efficient polymeric solar cells and light emitting diodes. One of the big challenges in this area is long term stability. For ZnO, most of its expected applications, such as UV light emitters, spin functional devices, chemical sensors, surface acoustic wave devices and transparent conductors, are still in the laboratory level. This is mainly because ZnO is prone to material defects that it becomes very difficult to obtain reproducible device performance and reliability. Clearly, there is still a considerable work to be done with ZnO alone, but it is worthwhile to broaden its potential as a technological material. In this manuscript, a more unique role of ZnO, which is as a dopant to polymers, has been presented. This is an area of research, which is just beginning to be explored. Relevant publications are still limited but they report some intriguing observations that may have novel optoelectronic and sensor applications. The interaction of ZnO with polymers may provide ways of obtaining unique or enhanced optical and electronic properties in polymer nanocomposites. Thus, by doping polymers with ZnO, new applications may be realized without losing the benefits offered by polymers in terms of processing, scalability and mechanical flexibility.

7. References

Albert, K.; Lewis, N.; Schauer, C.; Sotzing, G.; Stitzel, S.; Vaid, T. & Walt, D. (2000). Cross-reactive chemical sensor arrays. *Chemical Reviews,* Vol. 100, No. 7, (June 2000) 2595 - 2626, ISSN 1520-6890

Aleshin, A.; Shcherbakov, I; Alexandrova, E. & Lebedev, E. (2008). Effect of electric field on the photoluminescence of polymer-inorganic nanoparticle composites. *Solid State Communications,* Vol. 146, No. 3-4 , (April 2008) 161 - 165, ISSN 0038-1098

Aleshin, A.; Alexandrova, E. & Shcherbakov, I. (2009). Hybrid active layers from a conjugated polymer and inorganic nanoparticles for organic light emitting devices with emission colour tuned by electric field. *Journal of Physics D: Applied Physics,* Vol. 42, No. 10, (May 2009) 105108, ISSN 1361-6463

Arshak, K.; Moore, E.; Lyons, G.; Harris, F. & Clifford, S. (2004). A review of gas sensors employed in electronic nose applications. *Sensor Review,* Vol. 24, No. 2, (April 2004) 181 - 198, ISSN 0260-2288

Beek, W.; Wienk, M. & Janssen, R. (2006). Hybrid solar cells from regioregular polythiophene and ZnO nanoparticles. *Advanced Functional Materials,* Vol. 16, No. 8, (May 2006) 1112-1116, ISSN 1616-3028

Bennett, G.; Greenbaum, S. & Owens, F. (2009). NMR and Raman spectroscopic characterization of single walled carbon nanotube composites of polybutadiene. *Journal of Material Research,* Vol. 24, No. 7, (July 2009) 2215 - 2220, ISSN 0884-2914

Blom, P.; de Jong, M. & van Munster, M. (1997). Electric-field and temperature dependence of the hole mobility in poly(*p* - phenylene vinylene). *Physical Review B,* Vol. 55, No. 2, (January 1997) 656 - 659, ISSN 1538-4489

Brabec, C.; Shaheen, S.; Winder, C. & Sariciftci, N. (2002). Effect of LiF/metal electrodes on the performance of plastic solar cells. *Applied Physics Letters,* Vol. 80, No. 7, (February 2002) 1288-1290, ISSN 1077-3118

Cheng, B.; Shi, W.; Russell-Tanner, J.; Zhang, L. & Samulski, E. (2006). Synthesis of variable-aspect-ratio, single-crystalline ZnO nanostructures. *Inorganic Chemistry,* Vol. 45, No. 3, (January 2006) 1208 - 1214, ISSN 0020-1669

Chiang, C.; Druy, M.; Gau, S.; Heeger, A.; Louis, E. ; MacDiarmid, A. & Shirakawa, H. (1978). Synthesis of highly conducting films of derivatives of polyacetylene, $(CH)_x$. *Journal of the American Chemical Society,* Vol. 100, No. 3, (February 1978) 1013 - 1015, ISSN 1520-5126

Chirvase, D.; Parisi, J.; Hummelen, J. & Dyakonov, V. (2004). Influence of nanomorphology on the photovoltaic action of polymer-fullerene composites. *Nanotechnology,* Vol. 15, No. 9, (August 2004) 1317-1323, ISSN 1361-6528

Dijken, A.; Meulenkamp, E.; Vanmaekelbergh, D. & Meijerink, A. (2000). The kinetics of the radiative and nonradiative processes in nanocrystalline ZnO particles upon photoexcitation. *Journal of Physical Chemistry B,* Vol. 104, No. 8, (February 2000) 1715 - 1723, ISSN 1520-6106

Guadarrama, A.; Fernandez, J.; Iniguez, M.; Souto, J. & de Saja, J. (2000). Array of conducting polymer sensors for the characterisation of wines. *Analytica Chemica Acta,* Vol. 411, No. 1-2, (May 2000) 193 - 200, ISSN 0003-2670

Huang, M.; Mao, S.; Feick, H.; Yan, H.; Wu, Y.; Kind, H.; Weber, E.; Russo, R. & Yang, P. (2001). Room-temperature ultraviolet nanowire nanolasers. *Science,* Vol. 292, No. 5523, (June 2001) 1897 - 1899, ISSN 1095-9203

Kim, J.; Ingrosso, C.; Fakhfouri, V.; Striccoli, M.; Agostiano, A.; Curri, M. & Brugger, J. (2009). Inkjet-printed multicolor arrays of highly luminescent nanocrystal-based nanocomposites. *Small,* Vol. 5, No. 9, (February 2009) 1051 - 1057, ISSN 1613-6829

Kim, D.; Yang, J. & Hong, J. (2009). Ferromagnetism induced by Zn vacancy defect and lattice distortion in ZnO. *Journal of Applied Physics,* Vol. 106, No. 1, (July 2009) 013908, ISSN 1089-7550

Koji, N.; Tomonori, Y.; Kenji, I. & Fumio, S. (1999). Properties and structure of poly(vinyl alcohol)/silica composites. *Journal of Applied Polymer Science,* Vol. 74, No. 1, (October 1999) 133 - 138, ISSN 1097-4628

Konenkamp, R.; Word, R. & Schlegel, C. (2004). Vertical nanowire light emitting-diode. *Applied Physics Letters,* Vol. 85, No. 24, (December 2004) 6004 - 6006, ISSN 1077-3118

Koster, L.; Mihailletchi, V. & Blom, P. (2006). Ultimate efficiency of polymer/fullerene bulk heterojunction solar cells. *Applied Physics Letters,* Vol. 88, No. 9, (March 2006) 093511, ISSN 1077-3118

Li, Z.; Xiong, Y. & Xie, Y. (2003). Selected-control synthesis of ZnO nanowires and nanorods via a PEG-assisted route. *Inorganic Chemistry,* Vol. 42, No. 24, (October 2003) 8105 - 8109, ISSN 0020-1669

Liu, B. & Zeng, C. (2004). Room temperature solution synthesis of monodispersed single-crystalline ZnO nanorods and derived hierarchical nanostructures. *Langmuir,* Vol. 20, No. 10 , (April 2004) 4196 – 4204, ISSN 0743-7463

Ma, W.; Yang, C.; Gong, X.; Lee, K. & Heeger, A. (2005). Thermally stable, efficient polymer solar cells with nanoscale control of the interpenetrating network. *Advanced Functional Materials,* Vol. 15, No. 10, (October 2005) 1617-1622, ISSN 1616-3028

Meulenkamp, E. (1998). Synthesis and Growth of ZnO nanoparticles. *Journal of Physical Chemistry B,* Vol. 102, No. 29 , (June 1998) 5566 – 5572, ISSN 1520-6106

Morteani, A.; Ho, P.; Friend, R.; & Silva, C. (2005). Electric field-induced transition from heterojunction to bulk charge recombination in bilayer polymer light-emitting diodes. *Applied Physics Letters,* Vol. 86, No. 16, (May 2005) 163501, ISSN 1077-3118

Mukherjee, A.; Thakur, A.; Takashima, W. & Kaneto, K. (2007). Minimization of contact resistance between metal and polymer by surface doping. *Journal of Physics D: Applied Physics,* Vol. 40, No. 6, (March 2007) 1789-1793, ISSN 1361-6463

Musikhin, S.; Bakueva, L.; Sargent, E. & Shik, A. (2002). Luminescent properties and electronic structure of conjugated polymer-dielectric nanocrystal composites. *Journal of Applied Physics,* Vol. 91, No. 10, (May 2002) 6679-6683, ISSN 1089-7550

Norton, D.; Heo, Y.; Ivill, M.; Ip, K.; Pearton, S.; Chisholm, M. & Steiner, T. (2004). ZnO: growth, doping & processing. *Materials Today,* Vol. 7, No. 6, (June 2004) 34 – 40, ISSN 1369-7021

Olson, D.; Shaheen, S.; Collins, R. & Ginley, D. (2007). The effect of atmosphere and ZnO morphology on the performance of hybrid poly(3-hexylthiophene)/ZnO nanofiber photovoltaic devices. *Journal of Physical Chemistry C,* Vol. 111, No. 44, (October 2007) 16670 - 16678, ISSN 1932-7455

Roy, S. & Basu, S. (2002). Improved zinc oxide film for gas sensor applications. *Bulletin of Material Science,* Vol. 25, No. 6, (November 2002) 513 - 516, ISSN 0250-4707

Saxena, V.; Aswal, D.; Kaur, M.; Koiry, S.; Gupta, S.; Yakhmi, J.; Kshirsagar, R. & Deshpande, S. (2007). Enhanced NO_2 selectivity of hybrid poly(3-hexylthiophene): ZnO-nanowire thin films. *Applied Physics Letters,* Vol. 90, No. 4, (February 2007) 043516, ISSN 1077-3118

Shalish, I.; Temkin, H. & Narayanamurti, V. (2004). Size-dependent surface luminescence in ZnO nanowires. *Physical Review B,* Vol. 69, No. 24, (June 2004) 245401, ISSN 1538-4489

Shan, W.; Walukiewicz, W.; Ager III, J.; Yu, K.; Yuan, H.; Xin, H.; Cantwell, G. & Song, J. (2005). Nature of room-temperature photoluminescence in ZnO. *Applied Physics Letters,* Vol. 86, No. 19, (May 2005) 191911, ISSN 1077-3118

Sharma, A.; Hotchkiss, P.; Marder, S. & Kippelen, B. (2009). Tailoring the work function of indium tin oxide electrodes in electrophosphorescent organic light-emitting diodes. *Journal of Applied Physics,* Vol. 105, No. 8, (April 2009) 084507, ISSN 1089-7550

Sui, X.; Shao, C. & Liu, Y. (2005). White-light emission of polyvinyl alcohol/ZnO hybrid nanofibers prepared by electrospinning. *Applied Physics Letters,* Vol. 87, No. 11, (September 2005) 113115, ISSN 1077-3118

Sun, D.; Sue, H-J. & Miyatake, N. (2008). Optical properties of ZnO quantum dots in epoxy with controlled dispersion. *Journal of Physical Chemistry C,* Vol. 112, No. 41, (September 2008) 16002 - 16010, ISSN 1932-7455

Sze, S. (2007). *Physics of Semiconductor Devices,* John Wiley & Sons, Inc. ISBN-13: 978-0-471-14323-9, Hoboken NJ USA

Ton-That, C.; Phillips, M. & Nguyen, T-P. (2008). Blue shift in the luminescence spectra of MEH-PPV films containing ZnO nanoparticles. *Journal of Luminescence,* Vol. 128, No. 12, (July 2008) 2031 - 2034, ISSN

Ton-That, C.; Phillips, M.; Foley, M.; Moody, S. & Stampfl, A. (2008). Surface electronic properties of ZnO nanoparticles. *Applied Physics Letters,* Vol. 92, No. 26, (July 2008) 261916, ISSN 1077-3118

Umeda, J.; Sakamoto, W. & Yogo, T. (2009). Synthesis and field-responsive properties of $SrTiO_3$ nanoparticle/polymer hybrid. *Journal of Material Research,* Vol. 24, No. 7, (July 2009) 2221 - 2228, ISSN 0884-2914

Wang, Q.; Pflugl, C.; Andress, W.; Ham, D. & Capasso, F. (2008). Gigahertz surface acoustic wave generation on ZnO thin films deposited by radio frequency magnetron sputtering on III-V semiconductor substrates. *Journal of Vacuum Science Technology B,* Vol. 26, No. 6, (November 2008) 1848 - 1851, ISSN 1520-8567

Winey, K. & Vaia, R. (2007). Polymer nanocomposites. *MRS Bulletin,* Vol. 32, No. 4, (April 2007) 314 - 322, ISSN 0883-7694

Xu, Z-X.; Roy, V.; Stallinga, P.; Muccini, M.; Toffanin, S.; Xiang, H-F. & Che, C-M. (2007). Nanocomposite field effect transistors based on zinc oxide/polymer blends. *Applied Physics Letters,* Vol. 90, No. 22, (June 2007) 223509, ISSN 1077-3118

Zhang, T.; Xu, Z.; Qian, L.; Tao, D.; Teng, F. & Xu, X. (2006). Influence of ZnO nanorod on the luminescent and electrical properties of fluorescent dye-doped polymer nanocomposite. *Optical Materials,* Vol. 29, No. 2, (November 2006) 216 - 219, ISSN 0925-3467

11

A Review on Electronic Transport Properties of Individual Conducting Polymer Nanotubes and Nanowires

Yun-Ze Long[1], Zhaojia Chen[2], Changzhi Gu[2], Meixiang Wan[3], Jean-Luc Duvail[4], Zongwen Liu[5] and Simon P. Ringer[5]

[1]*College of Physics Science, Qingdao University*
[2]*Institute of Physics, Chinese Academy of Sciences*
[3]*Institute of Chemistry, Chinese Academy of Sciences*
[4]*Institut des Matériaux Jean Rouxel, CNRS, Université de Nantes*
[5]*Australian Key Centre for Microscopy and Microanalysis, The University of Sydney*
[1,2,3]*People's Republic of China*
[4]*France*
[5]*Australia*

1. Introduction

Quasi-one-dimensional nanostructures, such as carbon nanotubes, inorganic semi-conducting nanotubes/wires, and conjugated polymer nanotubes/wires, have drawn considerable attention in the past 20 years due to their importance for both fundamental research and potential applications in nanoscale devices (Kuchibhatla *et al.*, 2007; Xia *et al.*, 2003; MacDiarmid, 2002). Since the electrical conductivity of conjugated polymers can be increased by many orders of magnitude from 10^{-10}-10^{-5} to 10^{3}-10^{5} S/cm upon doping (MacDiarmid, 2002), conducting polymer nanotubes and nanowires (e.g., polyacetylene, polyaniline (PANI), polypyrrole (PPY), and poly(3,4-ethylenedioxythiophene) (PEDOT), poly(*p*-phenylenevinylene) (PPV)), are promising materials for fabricating polymeric nanodevices such as field-effect transistors (Aleshin, 2006), actuators (Jager et al., 2000), bio- and chemical sensors (Huang *et al.*, 2003; Ramanathan *et al.*, 2004), nano light emitting diodes, electrochromic displays (Cho *et al.*, 2005), artificial muscles, and solar cells, etc. (Zhang & Wang, 2006)

By now, conducting polymer nanotubes and nanowires can be prepared by various methods such as the template-guided synthesis (Martin, 1994), template-free method (Wan, 1999), interfacial polymerization (Huang *et al.*, 2003), electrospinning (MacDiarmid, 2001), dilute polymerization (Chio & Epstein, 2005), reverse emulsion polymerization method (Zhang *et al.*, 2006), etc. The template method of polymerization proposed by Martin *et al.* is an effective technique to synthesize polymer micro-/nanotubes and wires with controllable length and diameter (Cai & Martin, 1989; Cai *et al.*, 1991; Parthasarathy & Martin, 1994; Martin, 1994 & 1995). The disadvantage of this method is that a post-synthesis process is needed in order to remove the template. The template-free method developed by Wan *et al.*

is a simple self-assembly method without an external template (Wan *et al.*, 1999 & 2001; Zhang *et al.*, 2004; Huang *et al.*, 2005; Wan, 2008 & 2009; Long *et al.*, 2003). By controlling synthesis conditions such as temperature and molar ratio of monomer to dopant, polyaniline and polypyrrole nanostructures can be prepared by in-situ doping polymerization in the presence of protonic acids as dopants. The self-assembled formation mechanism in this approach is that the micelles formed by dopant and/or monomer-dopant act as soft templates in the process of forming tubes. The interfacial polymerization method proposed by Kaner et al. (Huang *et al.*, 2003; Huang & Kaner, 2004) involves step polymerization of two monomers or agents, which are dissolved respectively in two immiscible phases so that the reaction takes place at the interface between the two liquids. Interfacial polymerization has been used to prepare various polymers, such as polyaniline nanofibers and nanotubes. Electrospinning is an effective approach to fabricate long polymer fibers using strong electrostatic forces (MacDiarmid, 2001; Tan *et al.*, 2008). For instance, submicron fibers of doped polyaniline blended with polyethylene oxide or pure polyaniline have been prepared by this technique (MacDiarmid *et al.*, 2001; Cárdenas *et al.*, 2007). It should be noted that various spectroscopic results have indicated that the polymer micro- and nanostructures produced by these methods are usually partially crystalline; in other words, the small metallic regions of aligned polymer chains are interspersed with amorphous regions where the chains are disordered. The crystalline fraction depends on synthesis methods and synthesis conditions. At present, fabrication of highly crystalline and metallic polymer nanotubes and nanowires is still a challenge.

In order to fulfill the potential applications of conducting polymer nanotubes and wires, it is necessary to understand the electronic transport properties of individual polymer tubes/wires. The electrical characterization of individual conducting polymer nanotubes/ wires has made significant progress during the last decade. There are several strategies for measuring the conductivity of the template-synthesized fibres. The easiest and usual way is to leave the fibres in the pores of the template membrane and measure the resistance across the membrane (Cai & Martin, 1989; Cai *et al.*, 1991; Parthasarathy & Martin, 1994; Martin, 1995; Granström and Inganäs, 1995; Duchet *et al.*, 1998; Mativetsky and Datars, 2002; Duvail *et al.*, 2002 & 2004). Provided the number and diameter of the fibres are known, the measured trans-membrane resistance can be used to calculate the conductivity of a single fiber. However, this method may result in huge uncertainties on values due to the unknown number of connected fibres. Another way is to measure the resistances of compressed pellet or films (membrane removed) of the polymer nanofibres (Parthasarathy & Martin, 1994; Spatz *et al.*, 1994; Orgzall *et al.*, 1996). In fact, all these approaches did not realized the conductivity measurement of an individual fiber directly. Recently, the conductivity measurement of single polyaniline or polypyrrole tube/wire was achieved based on a conductive tip of an atom-force microscope (Park *et al.*, 2002; Park *et al.*, 2003; Saha *et al.*, 2004; Liu *et al.*, 2006). In this two-probe geometry, the contact resistance can be minimized by applying a significant pressure of the tip onto the nanotube/wire. A common approach was generally realized by dispersing nanotubes/wires on patterned micro- or nano-electrodes prepared by photo-lithography, electron-beam lithography and focused-ion beam deposition, followed by the subsequent searching of nanofibers just lying on the two or four electrodes only (Kim *et al.*, 1999; MacDiarmid *et al.*, 2001; Park *et al.*, 2001; Park *et al.*, 2003; Lee *et al.*, 2004; Aleshin *et al.*, 2004; Samitsu *et al.*, 2005; Kim *et al.*, 2005; Joo *et al.*, 2005; Aleshin, 2006; Gence *et al.*, 2007 & 2008; Callegari *et al.*, 2009). Particularly, focused-ion beam assisted deposition technique has been employed to attach metal microleads on isolated nanotubes/wires directly (Long *et al.*, 2003a, 2004b & 2005b; Zhang *et al.*, 2006; Huang *et al.*, 2006; Long *et al.*, 2006c; Duvail *et al.*, 2007; Lu *et al.*, 2007; Long *et al.*, 2008a & 2009b). There are also reports demonstrating a directed electrochemical nanowire assembly technique for the fabrication and measurement of polymer nanowire arrays between pre-patterned electrodes (Ramanathan *et al.*, 2004). All these recent investigations contribute significantly to identify and understand the specific electrical behaviour of conjugated polymer nanwires and nanotubes in comparison to the bulk materials. Though a lot of efforts have been done, there are still some key questions needed to be clarified, for example, the effects of the nanocontacts on the electrical measurements, the differences in electrical properties between that of polymer nanotubes/wires and that of their bulk counterparts, the possibility of tuning and controlling the electrical properties of individual nanotubes/wires and so on. These questions are very important to fabrication and characterization of nanodevices based on individual nanofibers.

In this chapter we provide a brief review of recent advances in the study of electronic transport properties (e.g., size effect in electrical conductivity, nonlinear current-voltage characteristics, small magnetoresistance effect, and nanocontact resistance effect) of individual conducting polyaniline, polypyrrole and PEDOT nanotubes/wires.

2. Experimental

2.1 Preparation and characterization

The results reported in this review have been measured for conducting polymer nanotubes and nanowires prepared by template-free self-assembly method and template-guided method. The protonic acids doped polyaniline and polypyrrole nanotubes/wires were prepared by the template-free self-assembly method. The polyaniline nanotubes are chosen as an example to introduce the synthesis procedure. Aniline monomer was distilled under reduced pressure. Ammonium persulphate as an oxidant and camphor sulfonic acid (CSA) as a dopant were used without any further treatment. In a typical synthesis procedure, aniline monomer (0.002 mol) and CSA (0.001 mol) were mixed in distilled water (10 ml) with stirring. The mixture reacted and formed a transparent solution of CSA-aniline salt. Before oxidative polymerization, the solution was cooled in an ice bath. Then an aqueous solution of ammonium persulphate (0.002 mol in 5 ml of distilled water) cooled in advance was added slowly into the above cooled CSA-aniline salt solution. After all the oxidant was added, the mixture was allowed to react for 15 h in the ice bath. The precipitates were then filtered and washed with distilled water and ethanol for several times, and finally dried at room temperature in a dynamic vacuum for 24 h. It was noted that if the synthesis temperature was changed to -10 °C, polyaniline microspheres would be obtained (Long *et al.*, 2004a). *p*-toluene sulfonic acid and 8-hydroxyquinoline-5-sulfonic acid doped polypyrrole tubes/wires were carried out along similar lines (Huang *et al.*, 2005). The PEDOT nanowires were prepared in templates of polycarbonate track-etched membranes (Duvail *et al.*, 2002, 2004 & 2008b). After the polymerization, polycarbonate (membrane template) was removed by dissolution with a flow of dichloromethane and the nanowires were dispersed onto a SiO_2 wafer.

The resulting polyaniline, polypyrrole, and PEDOT nanotubes/wires were characterized by elemental analysis, field-emission scanning electron microscopy (SEM), transmission electron microscopy, infrared and Raman spectroscopy, x-ray diffraction, x-ray photoelectron spectra and electron spin resonance (Zhang *et al.*, 2004; Huang *et al.*, 2005; Duvail *et al.*, 2002 & 2004).

Fig. 1 shows the SEM and TEM images of the obtained polyaniline and polypyrrole nanotubes. The outer diameter is about 80-400 nm for the self-assembled polyaniline and polypyrrole tubes/wires and 20-190 nm for the template-synthesized PEDOT wires. It was found that the polymer tubes/wires prepared by the template-free self-assembly method show a partially crystalline character according to the x-ray diffraction patterns. This partially crystalline feature was also proved by specific heat (Long *et al.*, 2004b) and magnetic susceptibility (Long *et al.*, 2006b) studies on polymer nanotubes/wires.

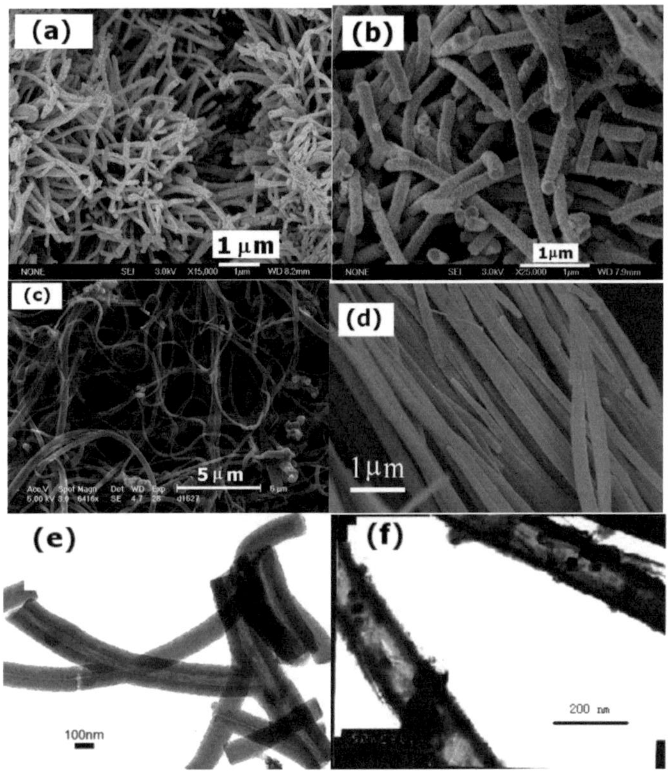

Fig. 1. Typical SEM images of self-assembled polyaniline nanotubes (a and b) and polypyrrole nanotubes (c and d). Typical TEM images of polyaniline nanotubes (e) and polypyrrole nanotubes (f).

2.2 Fabrication of Pt microleads and electrical measurements

The method used to attach Pt microleads on an isolated polymer nanotube/wire was described in previous publications (Long *et al.*, 2003a & 2005b). First, polymer nanotubes/wires were ultrasonically dispersed in ethanol for template-free prepared nanotubes/wires and in dichloromethane for template-prepared PEDOT nanowires. Then, a drop of solution was placed on an insulating SiO_2/Si wafer. After the evaporation of the solvent, an electron microscope was used to find an appropriately isolated nanotube/wire on the wafer. At last, two pairs of Pt microleads typically 0.5 μm in width and 0.4 μm in thickness were fabricated by FIB deposition (Dual-Beam 235 FIB System from FEI Company, working voltage of the system is 5 kV for the electron beam and 30 kV for the focused-ion beam, respectively, current of the focused-ion beam is very small, 1-10 pA, to minimize the modification of the conjugated polymer). Fig. 2 shows the individual polymer nanotubes/wires and the attached Pt microleads. It was noticed that the resistance of Pt microleads (less than 1 kΩ) is negligibly small compared with that of a single polymer nanotube/wire (several tens or hundreds of kΩ typically).

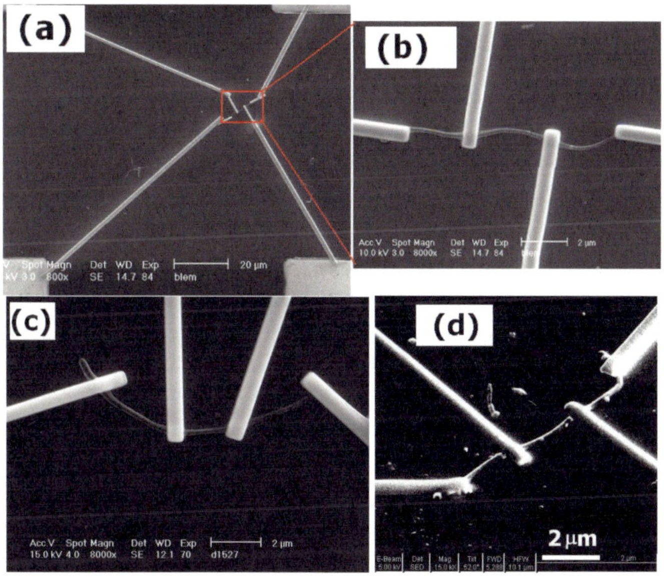

Fig. 2. Typical SEM images of an isolated polyaniline nanotube/wire (a, b and c) and polypyrrole nanotube/wire (d), and the attached platinum microleads. (Long *et al.*, 2003a & 2005b)

The electrical connection between the Pt microleads and the sample holder was made by highly conductive silver paste and gold wires and the electrical measurements of individual polymer nanotube/wire were carried out using a Physical Property Measurement System from Quantum Design and a Keithley 6487 picoammeter/voltage source, or a Keithley 236 source-measure unit in an Oxford helium gas flow cryostat covering a wide temperature range from 300 down to 2 K. The four-probe resistance was measured by applying a very small current in the linear part of the *I*-*V* characteristics. The two-probe resistance was determined under V_{bias} = 0.02 V, while no rectifying behaviour has been measured in our samples. The *I*-*V* curves were obtained by scanning the voltage from -4 to 4 V with a step of 0.02 V. The d*I*/d*V* curves were numerically derived from the corresponding *I*-*V* curves. The same polymer nanotube/wire was used for the four-probe measurement first and then for the two-probe measurement. The resistance of the polymer nanotube/wire with a given diameter was measured at least two times, for example, under cooling and during heating with a good reproducibility. In addition, for nanotubes/wires with a given diameter, two or more individual nanowires were measured to check the reproducibility.

3. Electronic transport properties

3.1 Diameter and temperature dependent electrical conductivity

The dependence of electrical conductivity on the diameter of the polymer nanotubes/wires (prepared by the template method) at room temperature has been widely reported (Cai *et al.*, 1991; Parthasarathy & Martin, 1994; Martin *et al.*, 1995; Granström & Inganäs, 1995; Duchet *et al.*, 1998). It was found that the room-temperature conductivities of nanotubes/wires of conducting polypyrrole, polyaniline et al. can increase from 10^{-1}-10^{0} to about 10^{3} S/cm with the decrease of their outer diameters from 1500 to 35 nm. The possible reason can be ascribed to the enhancement of molecular and supermolecular ordering (alignment of the polymer chains). For PEDOT nanowires prepared by template method, the room-temperature conductivities of the nanowires with diameters of 190, 95-100, 35-40, and 20-25 nm are about 11.2, 30-50, 490-530, and 390-450 S/cm, respectively (Duvail *et al.*, 2007 & 2008a). For polypyrrole nanotubes prepared by template-free method, as shown in Fig. 3, it was found that the polypyrrole tube with a 560-400 nm outer diameter is poorly conductive and the room-temperature conductivity is only 0.13-0.29 S/cm. When the outer diameter decreased to 130 nm, the conductivity of the single nanotube increased to 73 S/cm (Long *et al.*, 2005b). Such conductivity dependence on diameter was observed not only for template-synthesized polymer tubes/wires but also for self-assembled polypyrrole tubes, which indicates that the polymer tubes/wires prepared by these two different methods have similar structural characteristic: the smaller the diameter, the larger the proportion of ordered polymer chains.

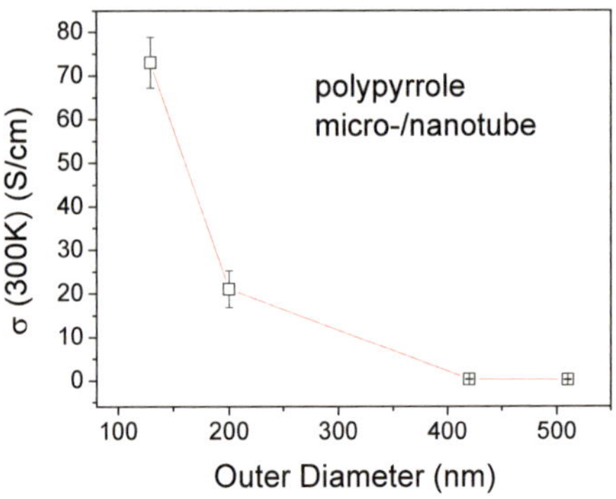

Fig. 3. Diameter dependence of room-temperature conductivity of individual polypyrrole micro-/nanotubes prepared by template-free self-assembly method. (Long *et al.*, 2005b)

Since the electrical properties of conducting polymers are strongly influenced by the effect of disorder and temperature, three different regimes (namely, insulating, critical, and metallic regimes close to the metal-insulator transition) have sorted out based on the extent of disorder and conductivity dependence on temperature (Yoon *et al.*, 1994; Menon *et al.*, 1998; Heeger, 2002). In the insulating regime, for a three-dimensional system, the temperature dependent resistivity usually follows Mott variable-range hopping (VRH) model: $\rho(T) = \rho_0\exp(T_M/T)^{1/4}$. At lower temperatures, when the Coulomb interaction between charge carriers is significant, ρ(T) usually follows Efros-Shklovskii (ES) VRH: $\rho(T) = \rho_0\exp(T_{ES}/T)^{1/2}$. In the critical regime, for a three-dimensional system close to the metal-insulator transition, the resistivity follows the power-law dependence: $\rho(T)\propto T^{-\beta}$, where β lies within the range of $0.3<\beta<1$. In the metallic regime, the sample shows a positive temperature coefficient of the resistivity at low temperatures (for example, below 10-20 K for metallic polypyrrole films, Menon *et al.*, 1998).

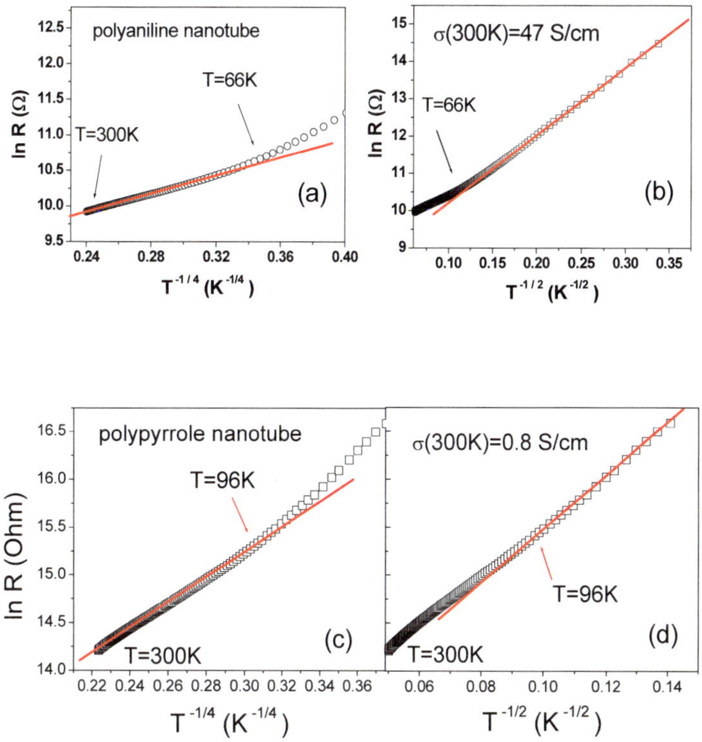

Fig. 4. The dependence of resistance on temperature of a single polyaniline nanotube (σ_{RT} = 47 S/cm) and a single polypyrrole nanotube (σ_{RT} = 0.8 S/cm) in the insulating regime of the metal-insulator transition: (a) and (c) plotted as ln $R(T)$ versus $T^{-1/4}$; (b) and (d) plotted as ln $R(T)$ versus $T^{-1/2}$; the temperature dependence of resistance follows 3D-VRH at higher temperatures and ES-VRH at lower temperatures. (Long *et al.*, 2005b)

Long *et al.* reported temperature dependent resistivity of a single polyaniline nanotube with average outer and inner diameters of 120 nm and 80 nm, which falls in the insulating regime of metal-insulator transition (Long *et al.*, 2005b). The tube's room-temperature conductivity is 47 S/cm. It was found that the resistivity follows three-dimensional (3D) Mott-VRH above 66 K, and follows ES-VRH model below 66 K, as shown in Figs. 4a and 4b. Here it is noted that from the view point of electrons, the polymer tube/wire with an outer diameter of 120 nm is still three dimensional because the localization length of carriers (L_C < 20 nm) is much smaller than the wall thickness or the diameter of the submicrotube. Similar smooth crossover from Mott-VRH to ES-VRH has also been observed in a single polypyrrole microtube (room-temperature conductivity, 0.8 S/cm) at around 96 K (as shown in Figs. 4c and 4d, Long et al., 2005b). However, the crossover temperature (T_{cros} ~ 66-96 K) and the characteristic ES temperature (T_{ES} ~ 316-780 K) of a single polymer tube/wire are much higher than those of a polyaniline pellet or a polypyrrole film (T_{cros} < 15 K and T_{ES} ~ 29-56 K), which could be possibly due to enhanced strong Coulomb interaction in polymer nanotubes/wires at low temperatures.

In addition, with the decrease of disorder or diameter of polymer nanotubes/wires, Long *et al.* found that a 130-nm polypyrrole nanotube with room-temperature conductivity of 73 S/cm is lying close to the critical regime of metal-insulator transition (Long *et al.*, 2005b). Its resistivity follows the power-law dependence: $\rho(T) \propto T^{-\beta}$, as shown in Fig. 5. The fit yields a β value of 0.488. Duvail *et al.* reported that a 100-nm PEDOT nanowire (σ_{RT} ~ 50 S/cm) fell in the critical regime with a β value of ~ 0.78 (Duvail *et al.*, 2007). Furthermore, the 35-40 nm template-prepared PEDOT nanowire (σ_{RT} ~ 490 S/cm) displays a metal-insulator transition at about 32 K, indicating that the nanowire is lying in the metallic regime (Duvail *et al.*, 2007). However, for a PEDOT nanowire with a diameter of 20-25 nm, though its conductivity is relatively high at room temperature (~ 390-450 S/cm), the nanowire shows very strong temperature dependence (R(10K)/R(300K) ~ 10^5) or insulating behavior at low temperature. This is possibly due to a confining effect since the value of the diameter (20-25 nm) becomes equal or close to the localization length of electrons (Lc ~ 20 nm). In such a case, localization of electrons induced by Coulomb interaction or small disorder must be taken into account for explaining this insulating behavior at low temperatures.

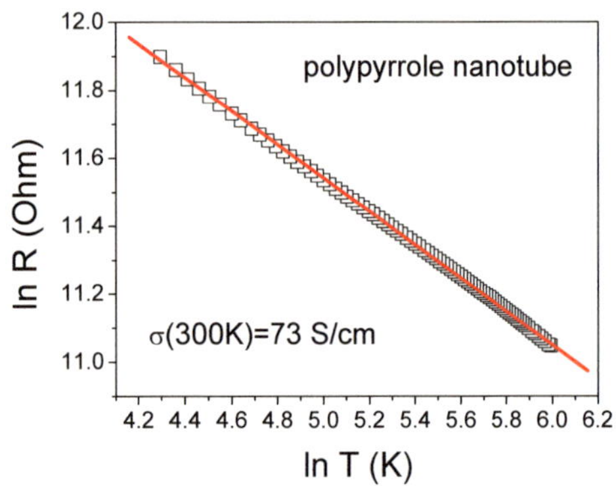

Fig. 5. The dependence of resistance on temperature (ln $R(T)$ versus ln T plot) for a single 130-nm polypyrrole nanotube in a conductive state close to the critical regime of the metal-insulator transition. (Long *et al.*, 2005b)

3.2 Nonlinear I-V characteristics

The current-voltage (*I-V*) characteristics of individual polymer nanowires/tubes such as the polyacetylene, polyaniline, polypyrrole, and PEDOT have been explored extensively in the past ten years (Park *et al.*, 2001; Park *et al.*, 2003; Kaiser et al., 2002, 2003 & 2004; Aleshin *et al.*, 2004; Long *et al.*, 2005b). With lowering temperature, a transition from linear to nonlinear *I-V* characteristics is usually observed (Fig. 6a), and a clear zero bias anomaly (i.e., Coulomb gap-like structure) gradually appears on the differential conductance (dI/dV) curves (Fig. 6b). Similar transition has also been reported in carbon nanotubes (Kang *et al.*, 2002) and inorganic compound nanowires such as CdS nanoropes (Long *et al.*, 2005a & 2008b), $K_{0.27}MnO_2 \cdot 0.5H_2O$ nanowires (Long *et al.*, 2008c), ZnO (Ma *et al.*, 2005) and SnO_2 (Ma *et al.*, 2004) nanowires.

Up to now, several theoretical models such as the space-charge limited current, fluctuation-induced tunneling (Kaiser *et al.*, 2003 & 2004; Kaiser and Park, 2005), Coulomb gap (Kang *et

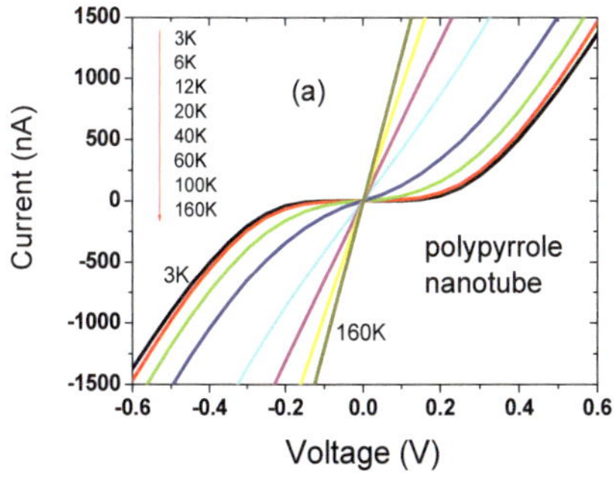

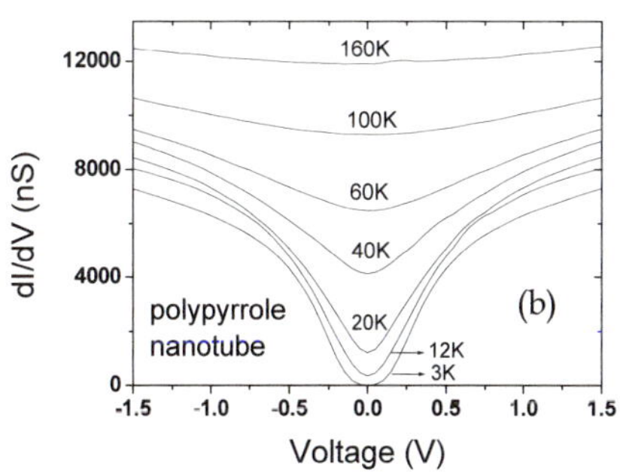

Fig. 6. *I-V* characterisitics (a) and the corresponding differential conductance (d*I*/d*V*) curves (b) of a single polypyrrole nanotube at low temperatues.

al., 2002; Ma *et al.*, 2004; Long *et al.*, 2005b & 2008c), Coulomb blockade (Saha, 2002; Aleshin *et al.*, 2005; Long *et al.*, 2008b), Lüttinger liquid (Aleshin *et al.*, 2004), Wigner crystal (Rahman and Sanyal, 2007) models, etc. have been considered to explain the conduction mechanism of quasi-one dimensional nanofibers. Saha and Aleshin *et al.* reported single-electron tunneling or Coulomb-blockade transport in conducting polypyrrole and helical polyacetylene nanofibers (Saha, 2002; Aleshin *et al.*, 2005) separately. In addition, power-law behaviors for both *I-V* characteristics and electrical conductance G(T) have been reported recently in polyacetylene fibers (Aleshin *et al.*, 2004) and polypyrrole wires/tubes (Rahman and Sanyal, 2007), which are characteristics of one-dimensional systems composed of several Lüttinger liquids or Wigner crystals connected in series, owing to electron-electron interactions (repulsive short-range electron-electron interactions or long-rang Coulomb interactions). Particularly, Kaiser et al. (Kaiser *et al.*, 2004; Kaiser and Park, 2005) recently proposed a generic expression (extended fluctuation-induced tunneling and thermal excitation model) for the nonlinear *I-V* curves based on numerical calculations for metallic conduction interrupted by small barriers:

$$G=I/V=G_0\cdot\exp(V/V_0)/\{1+h[\exp(V/V_0)-1]\} \quad (1)$$

where G_0, V_0 and h are parameters: G_0 is the temperature-dependent zero-bias conductance; V_0 is the voltage scale factor, which strongly depends on the barrier energy. Kaiser *et al.* showed that this expression could give a very good description of the observed nonlinearities in polyacetylene nanofibers, vanadium pentoxide nanofibers, etc. Here, one question arises for the Kaiser expression: Is it still appropriate to fit the nonlinear I-V characteristics of individual polymer nanowires/tubes if the Coulomb interactions are strong and should be taken into account?

The Kaiser expression has been used by Long *et al.* (Long *et al.*, 2009b; Yin *et al.*, 2009) to numerically calculate the *I-V* characteristics of individual polyaniline nanotube, polypyrrole nanotubes, PEDOT nanowires, CdS nanorope, and $K_{0.27}MnO_2$ nanowire, as shown in Fig. 7a. The fitting results indicate that except at low temperatures and low bias voltages, the Kaiser generic expression can give a good description of the *I-V* characteristics of individual nanotubes/wires, because the Kaiser expression (extending the Sheng model or fluctuation-induced tunneling and thermal excitation model) has well included the microstructure feature and the conduction feature of conjugated polymer nanofibers (quasi-one-dimensional metallic conduction interrupted by small barriers). Apparent deviation from the Kaiser expression has been evidenced in the low-temperature *I-V* curves as shown in Fig. 7b).

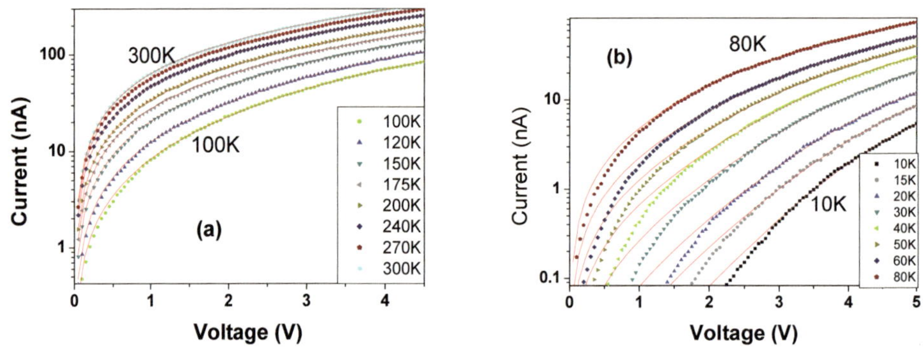

Fig. 7. *I-V* characterisitics of single polypyrrole nanotube with fits to expression (2), at temperatures (a) ranging rom 300 K to 100 K and (b) from 80 K to 10 K. (Long *et al.*, 2009b)

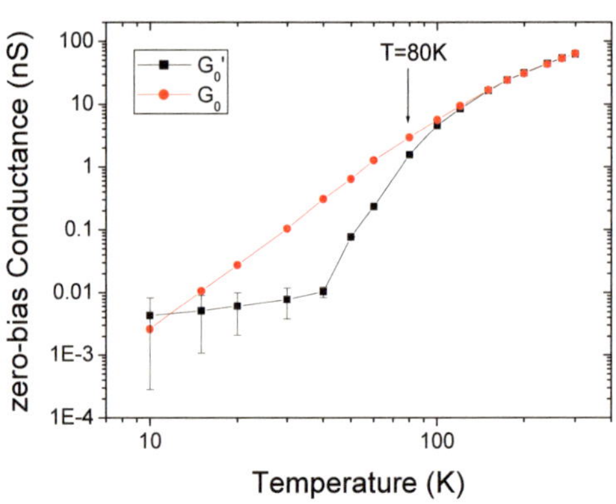

Fig. 8. Zero-field conductance versus temperature, where G_0 is determined from the fitting data and G_0' is determined from the experimental data. (Long *et al.*, 2009b)

Particularly, we compare the values of zero-bias conductance determined from the fitting parameter (G_0) with that determined from experimental measurements (G_0', obtained from the *I-V* curve or the differential conductance). As shown in Fig. 8, the fitting parameter G_0 decreases smoothly with temperature lowering, but the experimental value G_0' sharply decreases below 80-100 K and deviates from G_0, although it becomes superposable to G_0 for temperature equal and larger to 100 K. We note that the deviation temperature (about 80 K) is close to and consistent with the crossover temperature (66-96 K) for the crossover from Mott-VRH to ES-VRH, as shown in Fig. 4.

We propose that one possible reason for the deviation is that the Kaiser expression does not include the contributions from the Coulomb-gap occurring in density of states near Femi level and/or enhanced Coulomb interactions due to nanosize effects, which become important at low temperatures and voltages. (Long *et al.*, 2009b & 2005b; Yin *et al.*, 2009)

3.3 Magnetoresistance

The magnetoresistance (MR, defined as MR=ΔR(*H*)/R(0)=[R(*H*)-R(0)]/R(0)) of bulk films of conducting polymers have been extensively studied in the past 20 years (Menon *et al.*, 1998). For example, polyaniline, polypyrrole, PEDOT films, and polyaniline composites usually exhibit a positive magnetoresistance at low temperatures (T<10 K) and MR$\propto H^2$ (*H* is not very large). The mechanism generally involved is the shrinkage of localised wavefunctions of electrons in the presence of a magnetic field or electron-electron interactions (Menon *et al.*, 1998). Whereas highly conductive polyacetylene films usually show a negative magnetoresistance at low temperatures, which is mainly attributed to the weak localization effects (Menon *et al.*, 1998; Kozub *et al.*, 2002). Up to date, only a few papers have reported the magnetoresistance of polymer nanotubes/wires (Kim *et al.*, 1999; Park *et al.*, 2001; Kozub *et al.*, 2002; Long *et al.*, 2006a, 2006c & 2009c).

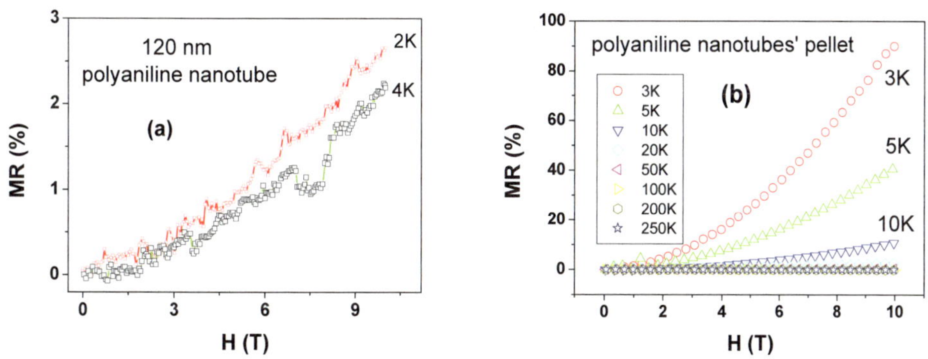

Fig. 9. The magnetoresistance curves for different temperatures of (a) a single polyaniline nanotube and (b) a pellet of polyaniline nanotubes. (Long *et al.*, 2006a)

Long *et al.* reported that the magnetoresistance of single polyaniline nanotube and single PEDOT nanowire is positive below 10 K and increases as H^2 up to 9 T. Typically, a positive magnetoresistance is expected for hopping conduction, because applying a magnetic field results in a contraction of the overlap of the localized state wavefunctions and thus an increase in the average hopping length. This corresponds to a positive magnetoresistance at sufficiently low temperatures. The theory of positive magnetoresistance has been developed for two cases, without and with electron-electron interactions. In both cases the weak-field MR with strong temperature dependence can be expressed as $\ln(R(H)/R(0))\propto H^2 \cdot T^{-3/4}$ (Shklovskii & Efros 1984). However, the magnetoresistance of a single nanotube/wire is much smaller than that of the nanotube/wire pellet at 9 T: MR<5% (2K) for the single nanotube/wire (Fig. 9a), and MR ~ 90% (3K) for the polyaniline nanotubes' pellet (Fig. 9b, Long *et al.*, 2006a). In addition, when the temperature increases, the magnetoresistance of the single nanotube/wire becomes smaller and close to zero. No evident transition from positive magnetoresistance to negative one was observed. In contrast to that of single nanotube/wire, pellets of polyaniline and polypyrrole nanotubes/wires show a relatively larger positive magnetoresistance at low temperatures. With temperature increasing, there is a transition from a positive magneto-resistance to a small negative magnetoresistance at about 60 K. The results indicate that the magnetoresistance in the bulk pellet samples made of polymer nanotubes/wires is dominated by a random network of inter-fibril contacts.

The small magnetoresistance effect in individual polymer nanotube/wire has been confirmed in other samples. For example, the low-temperature magnetoresistance (MR~0.1%) in a polyacetylene nanofiber network is rather smaller than that in a bulk polyacetylene film (Kim *et al.*, 1999; Park *et al.*, 2001). A single gold/polyaniline microfiber shows a small positive magnetoresistance (MR<4.1%) below 6 K (Long *et al.*, 2006c). The reason for this weak magnetoresistance effect in individual polymer nanotube/wire is possibly due to the elimination of inter-nanotube/wire contacts, small size and, relatively high conducitivity of individual polymer nanotube/wire. (Park *et al.*, 2001; Long et al., 2006a)

3.4 Nanocontact resistance

The contact resistance is often encountered when we study electronic transport in an individual polymer nanowire/tube or polymer nanofiber-based nano-devices. As we know, there are two major factors responsible for the contact resistance magnitude: geometry and insulating layers (potential barriers) between the contacting surfaces. The resistance of a contact is inversely proportional to its area, and is dependent on the surface stiffness and the force holding the two surfaces together.The insulating layers (potential barriers) between the polymer nanowire and the metal electrode are usually formed due to their different energy levels or work functions. A bad (insulating or semiconducting) electronic contact may possess a strongly temperature dependent contact resistance, and thus can seriously complicate or even dominate the measured resistance. In this section, we discuss two kinds of nanocontact resistance: between two crossed polymer nanowires/tubes and between polymer nanowire/tube and metal microlead.

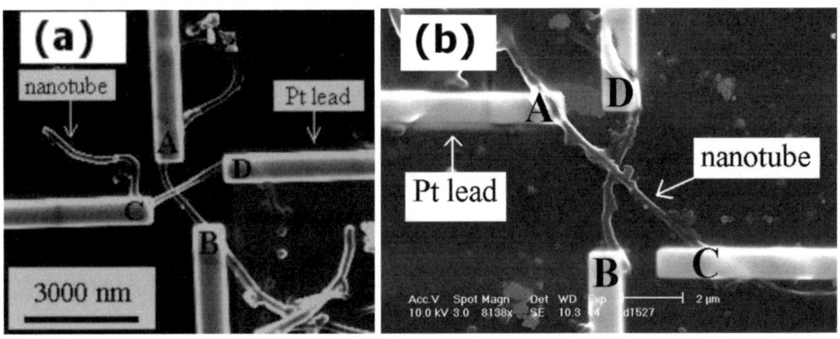

Fig. 10. SEM images showing two crossed polyaniline nanotubes and their attached Pt microleads. (Long *et al.*, 2003c)

The nanocontact resistance between two crossed polymer nanotubes/wires has been studied by Long et al. (Long *et al.*, 2003c & 2009a). It was found that the inter-tubular junction resistance of two crossed polyaniline nanotubes (Fig. 10) is very large, about 500 kΩ at room temperature, which is nearly 16 times larger than the intra-tube (intrinsic) resistance of an individual PANI nanotube (about 30 kΩ, Long *et al.*, 2003c). This result explains straightforwardly why an individual polyaniline nanotube has a much higher room-temperature conductivity (30.5 S/cm) than that of a pellet of polyaniline nanotubes where the measured resistance is dominated by the inter-fibril resistance (0.03 S/cm). For crossed PEDOT nanowires, the junction resistance (between the two nanowires) at room temperature can vary from 885-1383 kΩ for one sample and to 370-460 MΩ for another sample, which is respectively comparable or much larger than the intrinsic resistance of the PEDOT nanowires. In addition, the contact resistance shows a stronger temperature dependence (R(72K)/R(300K) is about 120~141) and could be fitted by a thermal fluctuation-induced tunneling (FIT) model (Long *et al.*, 2009a). It should be noted that the nanojunction resistance is comparable to the intrinsic resistance of polymer nanotube/wire and shows large sample-sample variations. The possible reasons could be attributed to the contamination of the nanotube/wire surfaces (polycarbonate for template-prepared PEDOT nanowires, solvent impurities or water adsorption), the variation of the junction area between the two nanotubes/ wires, and the self-formation conditions of the junction. It has to be mentioned that no special effort was made to control the formation of the junction between the two crossed nanotubes/wires during fabrication.

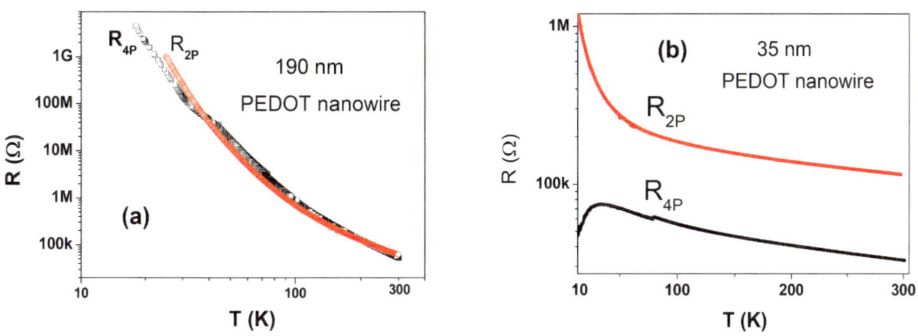

Fig. 11. The temperature dependence of the four-probe resistance (R_{4P}) and the two-probe resistance (R_{2P}) of (a) an individual PEDOT nanowire with a diameter of 35 nm, which falls in the metallic regime of the metal-insulator transition, and (b) an individual PEDOT nanotube with a diameter of 190 nm, which falls in the insulating regime of the metal-insulator transion (Long *et al.*, 2010)

The nanocontact resistance between a polymer PEDOT nanowire and a platinum microlead prepared by FIB deposition has also been studied by Long *et al.* (Long *et al.*, 2010). It was found that the nanocontact resistance (determined from four-probe resistance and two-probe resistance of the same nanowire) is in the magnitude of 10 kΩ at room temperature and can reach 10 MΩ at low temperatures, which, in some cases, is comparable to the intrinsic resistance of the PEDOT nanowires. For a semiconducting polymer nanowire in the insulating regime of the metal-insulator transition, the four-probe resistance is quite close to the two-probe resistance because the contact resistance is much smaller than the intrinsic resistance of the polymer nanowire as shown in Fig. 11a (Long *et al.*, 2008a & 2010). However, for a nanowire that falls in the metallic regime of the metal-insulator transition (for example, the 35 nm PEDOT nanowire as shown in Fig. 11b, Long *et al.*, 2010), the metallic nature of the measured polymer fibers could be over shadowed by the two-probe measurement although the nanowire shows a relatively high electrical conductivity at room temperature (390-450 S/cm). It can be attributed to the nanocontact resistance is much larger than the intrinsic resistance of the nanowire especially at low temperatures. We note that, for individual RuO_2 nanowires (Lin *et al.*, 2008), the temperature dependence of the two-probe resistance indicates that the nanowire is semiconducting, whereas the four-probe resistance dependence of the same nanowire shows the measured nanowire is metallic. So, in order to explore the intrinsic electronic transport properties of individual nanowires, especially in the case of metallic nanowires, the four-probe electrical measurement is necessary because nanocontact resistance cannot be excluded in a two-probe measurement.

4. Conclusion

During the past 15 years, significant progress has been made in synthesis, structural and electrical characterizations, and applications of conducting polymer nanotubes/wires. In this chapter, a brief review of the recent advances in electronic transport properties of individual conducting polymer nanotubes/wires prepared by both the template-free self-assembly method and the template method is presented. Results with broad interest have been discussed. For example, it was found that the electrical conductivity of the individual polymer tubes/wires increases by several times of magnitudes with decreasing outer diameter (size effect in electrical conductivity). The crossover from Mott to Efros-Shklovskii variable-range hopping conduction was observed at a relatively high transition temperature in single nanotubes/wires (enhanced Coulomb interaction effect). The low-temperature magnetoresistance of a single polymer tube/wire is positive and quite smaller than those of the nanotube/wire pellets (small magnetoresistance effect). The intrinsic resistance of an individual nanotube is much smaller than the contact resistance of two crossed nanotubes (nanocontact resistance effect). In addition, individual polymer tubes/wires show obvious transition from linear to nonlinear *I-V* curves at low temperature, and a clear zero-bias anomaly with Coulomb gap-like structure appeared on the differential conductance curves at low temperatures. These results indicate that the electrical properties of isolated conducting polymer tubes/wires are different from those of bulk polymer pellets or films in some cases due to their nanoscale diameters. However, in order to eliminate nanocontact resistance and reveal the intrinsic electronic transport properties of an individual nanotube/wire, it is still quite important to develop new or improved conductivity measurement approaches on a single nanofibre. Furthermore, due to the complicated microstructures of conducting polymers, there are still problems and challenges to fulfill their applications in nanoscale devices, such as whether fully metallic conducting polymer nanotubes/wires, which show metallic behavior from room-temperature down to low temperatures, can be prepared through improving their molecular or supramolecular ordering. In addition, reproducibility and/or controllability of individual polymer nanotubes/wires are also a problem, since their electrical properties are sensitive to many factors such as doping level, extent of disorder, diameter, temperature, aging effect, etc.

5. Acknowledgements

This work was financially supported by the National Natural Science Foundation of China (Grant Nos.: 10604038, 10374107, 50825206 and 10910101081) and the Program for New Century Excellent Talents in University of China (Grant No.: NCET-07-0472). Part of this work has been supported by the Communauté urbaine de Nantes, France. The authors also wish to acknowledge the support from the National Center for Nano Science and Technology of China.

6. References

Aleshin, A.N. (2006). Polymer nanofibers and nanotubes: Charge transport and device applications. *Adv. Mater.*, Vol. 18, No. 1, pp. 17-27, ISSN: 0935-9648

Aleshin, A.N.; Lee, H.J.; Jhang, S.H.; Kim, H.S.; Akagi, K. & Park, Y.W. (2005). Coulomb-blockade transport in quasi-one-dimensional polymer nanofibers. *Phys. Rev. B*, Vol. 72, No. 15, 153202 (4pp), ISSN: 1098-0121

Aleshin, A.N.; Lee, H.J.; Park, Y.W. & Akagi, K. (2004). One-dimensional transport in polymer nanofibers. *Phys. Rev. Lett.*, Vol. 93, No. 19, 196601 (4pp), ISSN: 0031-9007

Cai, Z.H.; Lei, J.T.; Liang, W.B.; Menon, V. & Martin, C.R. (1991). Molecular and super-molecular origins of enhanced electronic conductivity in template-synthesized polyheterocyclic fibrils. 1. Supermolecular effects. *Chem. Mater.*, Vol. 3, No. 5, pp. 960-967, ISSN: 0897-4756

Cai, Z.H. & Martin, C.R. (1989). Electronically conductive polymer fibers with mesoscopic diameters show enhanced electronic conductivities. *J. Am. Chem. Soc.*, Vol. 111, pp. 4138-4139, ISSN: 0002-7863

Callegari, V.; Gence, L.; Melinte, S. & Demoustier-Champagne, S. (2009). Electrochemically template-grown multi-segmented gold-conducting polymer nanowires with tunable electronic behavior. *Chem. Mater.*, Vol. 21, No. 18, pp. 4241-4247, ISSN: 0897-4756

Cárdenas, J.R.; de França, M.G.O.; de Vasconcelos, E.A.; de Azevedo, W.M. & da Silva Jr, E.F. (2007). Growth of sub-micron fibres of pure polyaniline using the electrospinning technique. *J. Phys. D.: Appl. Phys.*, Vol. 40, pp. 1068-1071, ISSN: 0022-3727

Chiou, N.R. & Epstein, A.J. (2005). Polyaniline nanofibers prepared by dilute polymerization. *Adv. Mater.*, Vol. 17, pp. 1679-1683, ISSN: 0935-9648

Cho, S.I.; Kwon, W.J.; Choi, S.J.; Kim, P.; Park, S.A.; Kim, J.; Son, S.J.; Xiao, R.; Kim, S.H. & Lee, S.B. (2005). Nanotube-based ultrafast electrochromic display. *Adv. Mater.*, Vol. 17, No. 2, pp. 171-175, ISSN: 0935-9648

Duchet, J.; Legras, R. & Demoustier-Champagne, S. (1998). Chemical synthesis of polypyrrole: structure-properties relationship. *Synth. Metals*, Vol. 98, pp. 113-122, ISSN: 0379-6779

Duvail, J.L.; Dubois, S.; Demoustier-Champagne, S.; Long, Y. & Piraux, L. (2008a). Physical properties of magnetic metallic nanowires and conjugated polymer nanowires and nanotubes. *Int. J. Nanotechnol.*, Vol. 5, Nos. 6/7/8, pp. 838-850, ISSN: 1475-7435

Duvail, J.L.; Long, Y.Z.; Cuenot, S.; Chen, Z.J. & Gu, C.Z. (2007). Tuning electrical propeties of conjugated polymer nanowires with the diameter. *Appl. Phys. Lett.*, Vol. 90, No. 10, 102114 (3pp), ISSN: 0003-6951

Duvail, J.L.; Long, Y.; Rétho, P.; Louarn, G.; Dauginet De Pra, L. & Demoustier-Champagne, S. (2008b). Enhanced electroactivity and electrochromism in PEDOT nanowires. *Mol. Cryst. Liq. Cryst.*, Vol. 485, pp. 835-842, ISSN: 1542-1406

Duvail, J.L.; Rétho, P.; Fernandez, V.; Louarn, G.; Molinié, P. & Chauvet, O. (2004). Effects of the confined synthesis on conjugated polymer transport properties. *J. Phys. Chem. B*, Vol. 108, No. 48, pp. 18552-18556, ISSN: 1520-6106

Duvail, J.L.; Rétho, P.; Garreau, S.; Louarn, G.; Godon, C. & Demoustier-Champagne, S. (2002). Transport and vibrational properties of poly(3,4-ethylenedioxythiophene) nanofibers. *Synth. Metals*, Vol. 131, pp. 123-128, ISSN: 0379-6779

Gence, L.; Callegari, V.; Faniel, S.; Vlad, A.; Dutu, C.; Melinte, S.; Demoustier-Champagne, S. & Bayot, V. (2008). Size related transport mechanisms in hybrid metal-polymer nanowires. *Phys. Stat. Sol. (a)*, Vol. 205, No. 6, pp. 1447-1450, ISSN: 1862-6300

Gence, L.; Faniel, S.; Gustin, C.; Melinte, S.; Bayot, V.; Callegari, V.; Reynes, O. & Demoustier -Champagne, S. (2007). Structural and electrical characterization of hybrid metal-polypyrrole nanowires. *Phys. Rev. B*, Vol. 76, No. 11, 115415 (8pp), ISSN: 1098-0121

Granström, M. & Inganäs, O. (1995). Electrically conductive polymer fibres with mesoscopic diameters: 1. Studies of structure and electrical properties. *Polymer*, Vol. 36, No. 15, pp. 2867-2872, ISSN: 0032-3861

Heeger, A.J. (2002). Semiconducting and metallic polymers: the fourth generation of polymeric materials. *Synth. Metals*, Vol. 125, pp. 23-42, ISSN: 0379-6779

Huang, J.X. & Kaner, R.B. (2004). A general chemical route to polyaniline nanofibers. *J. Am. Chem. Soc.*, Vol. 126, No. 3, pp. 851-855, ISSN: 0002-7863

Huang, J.X.; Virji, S.; Weiller, B.H. & Kaner, R.B. (2003). Polyaniline nanofibers: facile synthesis and chemical sensors. *J. Am. Chem. Soc.*, Vol. 125, No. 2, pp. 314-315, ISSN: 0002-7863

Huang, K.; Wan, M.X.; Long, Y.Z.; Chen, Z.J. & Wei, Y. (2005). Multi-functional polypyrrole nanofibers via a functional dopant-introduced process. *Synth. Metals*, Vol. 155, pp. 495-500, ISSN: 0379-6779

Huang, K.; Zhang, Y.J.; Long, Y.Z.; Yuan, J.H.; Han, D.X.; Wang, Z.J.; Niu, L. & Chen, Z.J. (2006). Preparation of highly conductive, self-assembled gold/polyaniline nano-cables and polyaniline nanotubes. *Chem. Eur. J.*, Vol. 12, pp. 5314-5319, ISSN: 0947-6539

Jager, E.W.H.; Smela, E. & Inganäs, O. (2000). Microfabricating conjugated polymer actuators. *Science*, Vol. 290, pp. 1540-1545, ISSN: 0036-8075

Kaiser, A.B.; Iorns, T.M.; Park, J.G.; Kim, B.; Lee, S.H. & Park, Y.W. (2002). Variation with temperature of the *I-V* characteristics of polyacetylene nanofibres. *Curr. Appl. Phys.*, Vol. 2, No. 4, pp. 285-288, ISSN: 1567-1739

Kaiser, A.B.; Rogers, S.A. & Park, Y.W. (2004a). Charge transport in conducting polymers: Polyacetylene nanofiberes. *Mol. Cryst. Liq. Cryst.*, Vol. 415, pp. 115-124, ISSN: 1542-1406

Kaiser, A.B.; Park, J.G.; Kim, B.; Lee, S.H. & Park, Y.W. (2004b). Polypyrrole micro-line: current-voltage characteristics and comparison with other conducting polymers. *Curr. Appl. Phys.*, Vol. 4, No. 5, pp. 497-500, ISSN: 1567-1739

Kaiser, A.B. & Park, Y.W. (2003). Comparison of tunnelling conduction in polyacetylene nanofibres, CDW and SDW systems. *Synth. Metals*, Vol. 135, Nos. 1-3, pp. 245-247, ISSN: 0379-6779

Kaiser, A.B. & Park, Y.W. (2005). Current-voltage characteristics of conducting polymers and carbon nanotubes. *Synth. Metals*, Vol. 152, pp. 181-184, ISSN: 0379-6779

Kang, N.; Hu, J.S.; Kong, W.J.; Lu, L.; Zhang, D.L.; Pan, Z.W. & Xie, S.S. (2002). Consistent picture of strong electron correlation from magnetoresistance and tunneling conductance measurements in multiwall carbon nanotubes. *Phys. Rev. B*, Vol. 66, No. 24, 241403(R) (4pp), ISSN: 1098-0121

Kim, B.H.; Park, D.H.; Joo, J.; Yu, S.G. & Lee, S.H. (2005). Synthesis, characteristics, and field emission of doped and de-doped polypyrrole, polyaniline, poly(3,4-ethylenedioxythiophene) nanotubes and nanowires. *Synth. Metals*, Vol. 150, pp. 279-284, ISSN: 0379-6779

Kim, B.K.; Kim, Y.H.; Won, K.; Chang, H.; Choi, Y.; Kong, K.J.; Rhyu, B.W.; Kim, J.J. & Lee, J.O. (2005). Electrical properties of polyaniline nanofibre synthesized with biocatalyst. *Nanotechnology*, Vol. 16, No. 8, pp. 1177-1181, ISSN: 0957-4484

Kim, G.T.; Burghard, M.; Suh, D.S.; Liu, K.; Park, J.G.; Roth, S. & Park, Y.W. (1999). Conductivity and magnetoresistance of polyacetylene fiber network. *Synth. Metals*, Vol. 105, pp. 207-210, ISSN: 0379-6779

Kozub, V.I.; Aleshin, A.N.; Suh, D.S. & Park, Y.W. (2002). Evidence of magnetoresistance for nanojunction-controlled transport in heavily doped polyacetylene. *Phys. Rev. B*, Vol. 65, No. 22, 224204 (5pp), ISSN: 1098-0121

Kuchibhatle, S.V.N.T.; Karakoti, A.S.; Bera, D. & Seal, S. (2007). One dimensional nanostructured materials. *Prog. Mater. Sci.*, Vol. 52, No. 5, pp. 699-913, ISSN: 0079-6425

Lee, H.J.; Jin, Z.X.; Aleshin, A.N.; Lee, J.Y.; Goh, M.J.; Akagi, K.; Kim, Y.S.; Kim, D.W. & Park, Y.W. (2004). Dispersion and current-voltage characteristics of helical polyacetylene single fibers. *J. Am. Chem. Soc.*, Vol. 126, No. 51, pp. 16722-16723, ISSN: 0001-4842

Lin, Y.H.; Chiu, S.P. & Lin, J.J. (2008). Thermal fluctuation-induced tunneling conduction through metal nanowire contacts. *Nanotechnology*, Vol. 19, No. 36, 365201 (7pp), ISSN: 0957-4484

Liu, L.; Zhao, Y.M.; Jia, N.Q.; Zhou, Q.; Zhao, C.J.; Yan, M.M. & Jiang, Z.Y. (2006). Electrochemical fabrication and electronic behavior of polypyrrole nano-fiber array devices. *Thin Solid Films*, Vol. 503, pp. 241-245, ISSN: 0040-6090

Long, Y.Z.; Chen, Z.J.; Wang, N.L.; Ma, Y.J.; Zhang, Z.; Zhang, L.J. & Wan, M.X. (2003a). Electrical conductivity of a single conducting polyaniline nanotube. *Appl. Phys. Lett.*, Vol. 83, No. 9, pp. 1863-1865, ISSN: 0003-6951

Long, Y.Z.; Chen, Z.J.; Wang, N.L.; Zhang, Z.M. & Wan, M.X. (2003b). Resistivity study of polyaniline doped with protonic acids. *Physica B*, Vol. 325, pp. 208-213, ISSN: 0921-4526

Long, Y.Z.; Zhang, L.J.; Ma, Y.J.; Chen, Z.J.; Wang, N.L.; Zhang, Z. & Wan, M.X. (2003c). Electrical conductivity of an individual polyaniline nanotube synthesized by a self-assembly method. *Macromol. Rapid Commun.*, Vol. 24, No. 16, pp. 938-942, ISSN: 1022-1336

Long, Y.Z.; Chen, Z.J.; Ma, Y.J.; Zhang, Z.; Jin, A.Z.; Gu, C.Z.; Zhang, L.J.; Wei, Z.X. & Wan, M.X. (2004a). Electrical conductivity of hollow polyaniline microspheres synthesized by a self-assembly method. *Appl. Phys. Lett.*, Vol. 84, No. 12, pp. 2205-2207, ISSN: 0003-6951

Long, Y.Z.; Luo, J.L.; Xu, J.; Chen, Z.J.; Zhang, L.J.; Li, J.C. & Wan, M.X. (2004b). Specific heat and magnetic susceptibility of polyaniline nanotubes: inhomogeneous disorder. *J. Phys.: Condens. Matter*, Vol. 16, pp. 1123-1130, ISSN: 0953-8984

Long, Y.Z.; Xiao, H.M.; Chen, Z.J.; Wan, M.X.; Jin, A.Z. & Gu, C.Z. (2004c). Electrical conductivity of individual polypyrrole microtube. *Chin. Phys.*, Vol. 13, No. 11, pp. 1918-1921, ISSN: 1009-1963

Long, Y.Z.; Chen, Z.J.; Wang, W.L.; Bai, F.L.; Jin, A.Z. & Gu, C.Z. (2005a). Electrical conductivity of single CdS nanowire synthesized by aqueous chemical growth. *Appl. Phys. Lett.*, Vol. 86, No. 15, 153102 (3pp), ISSN: 0003-6951

Long, Y.Z.; Zhang, L.J.; Chen, Z.J.; Huang, K.; Yang, Y.S.; Xiao, H.M.; Wan, M.X.; Jin, A.Z. & Gu, C.Z. (2005b). Electronic transport in single polyaniline and polypyrrole microtubes. *Phys. Rev. B*, Vol. 71, No. 16, 165412 (7pp), ISSN: 1098-0121

Long, Y.Z.; Chen, Z.J.; Shen, J.Y.; Zhang, Z.M.; Zhang, L.J.; Huang, K.; Wan, M.X.; Jin, A.Z.; Gu, C.Z. & Duvail, J.L. (2006a). Magnetoresistance studies of polymer nanotube/wire pellets and single polymer nanotubes/wires. *Nanotechnology*, Vol. 17, No. 24, pp. 5903-5911, ISSN: 0957-4484

Long, Y.Z.; Chen, Z.J.; Shen, J.Y.; Zhang, Z.M.; Zhang, L.J.; Xiao, H.M.; Wan, M.X. & Duvail, J.L. (2006b). Magnetic properties of conducting polymer nanostructures. *J. Phys. Chem. B*, Vol. 110, No. 46, pp. 23228-23233, ISSN: 1520-6106

Long, Y.Z.; Huang, K.; Yuan, J.H.; Han, D.X.; Niu, L.; Chen, Z.J.; Gu, C.Z.; Jin, A.Z. & Duvail, J.L. (2006c). Electrical conductivity of a single Au/polyaniline microfiber. *Appl. Phys. Lett.*, Vol. 88, No. 16, 162113 (3pp), ISSN: 0003-6951

Long, Y.Z.; Duvail, J.L.; Chen, Z.J.; Jin, A.Z. & Gu, C.Z. (2008a). Electrical conductivity and current-voltage characterisitics of individual conducting polymer PEDOT nanowires. *Chin. Phys. Lett.*, Vol. 25, No. 9, 3474-3477, ISSN: 0256-307X

Long, Y.Z.; Wang, W.L.; Bai, F.L.; Chen, Z.J.; Jin, A.Z. & Gu, C.Z. (2008b). Current-voltage characteristics of an individual helical CdS nanowire rope. *Chin. Phys. B*, Vol. 17, No. 4, 1389-1393, ISSN: 1674-1056

Long, Y.Z.; Yin, Z.H.; Chen, Z.J.; Jin, A.Z.; Gu, C.Z.; Zhang, H.T. & Chen, X.H. (2008c). Low-temperature electronic transport in single $K_{0.27}MnO_2{\cdot}0.5H_2O$ nanowires: enhanced electron-electron interaction. *Nanotechnology*, Vol. 19, No. 21, 215708 (5pp), ISSN: 0957-4484

Long, Y.Z.; Duvail, J.L.; Wang, Q.T.; Li, M.M. & Gu, C.Z. (2009a). Electronic transport through crossed conducting polymer nanowires. *J. Mater. Res.*, Vol. 24, No. 10, pp. 3018-3022, ISSN: 0884-2914

Long, Y.Z.; Yin, Z.H.; Li, M.M.; Gu, C.Z.; Duvail, J.L.; Jin, A.Z. & Wan, M.X. (2009b). Current-voltage characteristics of individual conducting polymer nanotubes and nanowires. *Chin. Phys. B*, Vol. 18, No. 6, pp. 2514-2522, ISSN: 1674-1056

Long, Y.Z.; Duvail, J.L.; Chen, Z.J.; Jin, A.Z. & Gu, C.Z. (2009c). Electrical properties of isolated poly(3,4-ethylenedioxythiophene) nanowires prepared by template synthesis. *Polym. Adv. Technol.*, Vol. 20, pp. 541-544, ISSN: 1042-7147

Long, Y.Z.; Duvail, J.L.; Li, M.M.; Gu, C.Z. & Liu, Z. (2010). Electrical conductivity studies on individual conjugated polymer nanowires: two-probe and four-probe results. *Nanoscale Res. Lett.*, DOI: 10.1007/s11671-009-9471-y, ISSN: 1931-7573

Lu, G.W.; Li, C.; Shen, J.Y.; Chen, Z.J. & Shi, G.Q. (2007). Preparation of highly conductive gold-poly(3,4-thylenedioxythiophene) nanocables and their conversion to poly(3,4-ethylenedioxythiophene) nanotubes. *J. Phys. Chem. C*, Vol. 111, No. 16, pp. 5926-5931, ISSN: 1932-7447

Ma, Y.J.; Zhang, Z.; Zhou, F.; Lu, L.; Jin, A.Z. & Gu, C.Z. (2005). Hopping conduction in single ZnO nanowires. *Nanotechnology*, Vol. 16, No. 6, pp. 746-749, ISSN: 0957-4484

Ma, Y.J.; Zhou, F.; Lu, L. & Zhang, Z. (2004). Low-temperature transport properties of individual SnO_2 nanowires. *Solid State Commun.*, Vol. 130, pp. 313-316, ISSN: 0038-1098

MacDiarmid, A.G. (2002). Synthetic metals: a novel role for organic polymers. *Synth. Metals*, Vol. 125, pp. 11-22, ISSN: 0379-6779

MacDiarmid, A.G.; Jones, Jr., W.E.; Norris, I.D.; Gao, J.; Johnson, Jr., A.T.; Pinto, N.J.; Hone, J.; Han, B.; Ko, F.K.; Okuzaki, H. & Llaguno, M. (2001). Electrostatically-generated nanofibers of electronic polymers. *Synth. Metals*, Vol. 119, pp. 27-30, ISSN: 0379-6779

Martin, C.R. (1994). Nanomaterials: A membrane-based synthetic approach. *Science*, Vol. 266, No. 5193, pp. 1961-1966, ISSN: 0036-8075

Martin, C.R. (1995). Template synthesis of electronically conductive polymer nanostructures. *Accounts Chem. Res.*, Vol. 28, No. 2, pp. 61-68, ISSN: 0001-4842

Mativetsky, J.M. & Datars, W.R. (2002). Morphology and electrical properties of template-synthesized polypyrrole nanocylinders. *Physica B*, Vol. 324, pp. 191-204, ISSN: 0921-4526

Menon, R.; Yoon, C.O.; Moses, D. & Heeger, A.J. (1998). In: *Handbook of Conducting Polymers*, Skotheim, T.A.; Elsenbaumer, R.L. & Reynolds, J.R. (Eds.), 2nd Edition, 85-136, Marcel Dekker, ISBN: 0824700503, New York

Orgzall, I.; Lorenz, B.; Ting, S.T.; Hor, P.H.; Menon, V.; Martin, C.R. & Hochheimer, H.D. (1996). Thermopower and high-pressure electrical conductivity measurements of template synthesized polypyrrole. *Phys. Rev. B*, Vol. 54, No. 23, pp. 16654-16658, ISSN: 1098-0121

Rahman, A. & Sanyal, M.K. (2007). Observation of charge density wave characteristics in conducting polymer nanowires: Possibility of Wigner crystallization. *Phys. Rev. B*, Vol. 76, No. 4, 045110 (6pp), ISSN: 1098-0121

Park, J.G.; Kim, B.; Lee, S.H. & Park, Y.W. (2003). Current-voltage characteristics of polypyrrole nanotube in both vertical and lateral electrodes configuration. *Thin Solid Films*, Vol. 438-439, pp. 118-122, ISSN: 0040-6090

Park, J.G.; Kim, G.T.; Krstic, V.; Kim, B.; Lee, S.H.; Roth, S.; Burghard, M. & Park, Y.W. (2001). Nanotransport in polyacetylene single fiber: Toward the intrinsic properties. *Synth. Metals*, Vol. 119, pp. 53-56, ISSN: 0379-6779

Park, J.G.; Lee, S.H.; Kim, B. & Park, Y.W. (2002). Electrical resistivity of polypyrrole nanotube measured by conductive scanning probe microscope: The role of contat force. *Appl. Phys. Lett.*, Vol. 81, No. 24, pp. 4625-4627, ISSN: 0003-6951

Park, J.H.; Yu, H.Y.; Park, J.G.; Kim, B.; Lee, S.H.; Olofsson, L.; Persson, S.H.M. & Park, Y.W. (2001). Non-linear I-V characteristics of MEH-PPV patterned on sub-micrometer electrodes. *Thin Solid Films*, Vol. 393, pp. 129-131, ISSN: 0040-6090

Parthasarathy, R.V. & Martin, C.R. (1994). Template-synthesized polyaniline microtubules. *Chem. Mater.*, Vol. 6, No. 10, pp. 1627-1632, ISSN: 0897-4756

Ramanathan, K.; Bangar, M.A.; Yun, M.; Chen, W.; Mulchandani, A. & Myung, N.V. (2004). Individually addressable conducting polymer nanowires array. *Nano Lett.*, Vol. 4, No. 7, pp. 1237-1239, ISSN: 1530-6984

Saha, S.K. (2002). Room-temperature single-electron tunneling in conducting polypyrrole nanotube. *Appl. Phys. Lett.*, Vol. 81, pp. 3645-3647, ISSN: 0003-6951

Saha, S.K.; Su, Y.K.; Lin, C.L. & Jaw, D.W. (2004). Current-voltage characteristics of conducting polypyrrole nanotubes using atomic force microscopy. *Nanotechnology*, Vol. 15, pp. 66-69, ISSN: 0957-4484

Samitsu, S.; Shimomura, T.; Ito, K.; Fujimori, M.; Heike, S. & Hashizume, T. (2005). Conductivity measurements of individual poly(3,4-ethylenedioxythiophene)/poly(styrenesulfonate) nanowires on nanoelectrodes using manipulation with an atomic force microscope. *Appl. Phys. Lett.*, Vol. 86, No. 23, 233103 (3pp), ISSN: 0003-6951

Spatz, J.P.; Lorenz, B.; Weishaupt, K.; Hochheimer, H.D.; Menon, V.; Parthasarathy, R.; Martin, C.R.; Bechtold, J. & Hor, P.H. (1994). Observation of crossover from three- to two-dimensional variable-range hopping in template-synthesized polypyrrole and polyaniline. *Phys. Rev. B*, Vol. 50, No. 20, pp. 14888-14892, ISSN: 1098-0121

Tan, J.S.; Long, Y.Z. & Li, M.M. (2008). Preparation of aligned polymer micro/nanofibers by electrospinning. *Chin. Phys. Lett.*, Vol. 25, No. 8, pp. 3067-3070, ISSN: 0256-307X

Wan, M.X. (2008). A template-free method towards conducting polymer nanostructures. *Adv. Mater.*, Vol. 20, pp. 2926-2932, ISSN: 0935-9648

Wan, M.X. (2009). Some issues related to polyaniline micro-/nanostructures. *Macromol. Rapid. Commun.*, Vol. 30, pp. 963-975, ISSN: 1022-1336

Wan, M.X.; Huang, J. & Shen, Y.Q. (1999). Microtubes of conducting polymers. *Synth. Metals*, Vol. 101, pp. 708-711, ISSN: 0379-6779

Wan, M.X.; Liu, J.; Qiu, H.J.; Li, J.C. & Li, S.Z. (2001). Template-free synthesized microtubules of conducting polymers. *Synth. Metals*, Vol. 119, pp. 71-72, ISSN: 0379-6779

Xia, Y.N.; Yang, P.D.; Sun, Y.G.; Wu, Y.Y.; Mayers, B.; Gates, B.; Yin, Y.D.; Kim, F. & Yan, H.Q. (2003). One-dimensional nanostructures: Synthesis, characterization, and applications. *Adv. Mater.*, Vol. 15, No. 5, pp. 353-389, ISSN: 0935-9648

Yin, Z.H.; Long, Y.Z.; Gu, C.Z.; Wan, M.X. & Duvail, J.L. (2009). Current-voltage characterisitics in individual polypyrrole nanotube, poly(3,4-ethylene-dioxythiophene) nanowire, polyaniline nanotube, and CdS nanorope. *Nanoscale Res. Lett.*, Vol. 4, No. 1, pp. 63-69, ISSN: 1931-7573

Yoon, C.O.; Reghu, M.; Moses, D. & Heeger, A.J. (1994). Transport near the metal-insulator transition: Polypyrrole doped with PF_6. *Phys. Rev. B*, Vol. 49, No. 16, pp. 10851-10863, ISSN: 1098-0121

Zhang, D.H. & Wang, Y.Y. (2006). Synthesis and applications of one-dimensional nano-structured polyaniline: An overview. *Mat. Sci. Eng. B - Solid*, Vol. 134, No. 1, pp. 9-19, ISSN: 0921-5107

Zhang, L.J.; Long, Y.Z.; Chen, Z.J. & Wan, M.X. (2004). The effect of hydrogen bonding on self-assembled polyaniline nanostructures. *Adv. Funct. Mater.*, Vol. 14, No. 7, pp. 693-698, ISSN: 1616-301X

Zhang, X.Y.; Lee, J.S.; Lee, G.S.; Cha, D.K.; Kim, M.J.; Yang, D.J. & Manohar, S.K. (2006). Chemical synthesis of PEDOT nanotubes. *Macromolecules*, Vol. 39, pp. 470-472, ISSN: 0024-9297

12

Conjugated Polymer and Hybrid Polymer-Metal Single Nanowires: Correlated Characterization and Device Integration

L. Gence[1], V. Callegari[1], S. Melinte[1], S. Demoustier-Champagne[1], Y. Long[2], A. Dinescu[3] and J.L. Duvail[4]

[1]*CeRMiN, Université catholique de Louvain, Louvain-la-Neuve*
[2]*College of Physics, Qingdao University, Qingdao*
[3]*National Institute for R&D in Microtechnologies, Bucharest*
[4]*Institut des Matériaux Jean Rouxel, CNRS, Université de Nantes, Nantes*
[1]*Belgium*
[2]*People's Republic of China*
[3]*Romania*
[4]*France*

1. Introduction

More and more in recent years, investigators in industry and academia are taming the structural and electronic properties of conjugated polymer-based nanowires (NWs) and nanotubes (NTs) that display an impressive applicative potential from bio-environmental sensing and energy harvesting to ultra-low power electronic devices and high density memories. Today, the methods used to synthesize conjugated polymer (CP) NWs span from scanning probe lithography [Lim & Mirkin (2002)] and mechanical stretching [He et al. (2001)] to electrospinning [Reneker (2000)] and templated electrochemical [Jerôme & Jerôme (1998); Martin (1994; 1996)] and wetting procedures [Massuyeau (2009)].

Intriguingly, while CPs present vast technological opportunities and can be easily combined to inorganic materials, hybrid metal-CP NWs have been scarcely studied [Gence (2007); Hernandez (2004); Park (2004)]. They can be advantageously produced by high-throughput template strategies that offer a very good control over the composition and spatial distribution of the different NW segments [Duvail (2008); Liang (2002)]. Nevertheless, the use of single hybrid metal-CP NWs in real applications asks for properly identifying their intrinsic behavior. First, this exigency stems from the obvious requirement of tuning their specific properties in post-synthesis and post-assembly technological steps [Jang (2006); Tran et al. (2009); Yoon et al. (2007)]. Second, the unique opto-electronic properties of single hybrid metal-CP NWs are expected to be closely linked to their intricate architecture and to emerge more exquisitely than those of pure CP NWs [Aleshin (2004); Cao (2008); Duvail (2007)].

The aim of this chapter is not to give a complete overview of the literature addressing CP NWs; the focus is on state-of-the-art techniques for synthesizing, characterizing and integrating hybrid metal-CP NWs. We first present the fabrication of well-shaped and mechanically robust hybrid metal-CP NWs and in the next section we describe the morphology and the microstructure of the resulting NWs. Several advanced techniques such as Raman spectroscopy and electron spin resonance, that enhance our ability to harness promising properties of CP-based NWs, are also presented and discussed. The following section includes research efforts to decrypt electronic properties of CP-based NWs and hybrid metal-CP NWs. Methods for contacting single NWs are reviewed and critically compared. The subsequent section introduces microdevices specially designed for the correlated characterization of single CP-based NWs. The last section refers to the integration and the potential applications of CP-based NWs.

2. Fabrication by the template method

We first address the major issues for preparing well-shaped and mechanically robust hybrid metal-CP NWs and elaborate on a highly reproducible route based upon an all-electrochemical template-based strategy [Reynes & Demoustier-Champagne (2005)].

2.1 Template synthesis

The synthesis of NWs can be achieved by several techniques that can be regrouped into two main families, depending if there are based or not on the use of a template. The template-free method is relatively simple and inexpensive but its control over the morphology and diameter of the NWs is poor compared to the template method [Nalwa (2006); Wan (2008)]. The template method is very efficient for achieving a controlled growth of NWs: the shape, size and orientation of the produced structures are precisely defined by the template [Duvail (2008)]. It allows the synthesis of complex multi-segmented NWs, core-multishell, and coaxial NW structures [Hurst (2006); Kovtyukhova (2004); Lorcy (2009)]. The most used templates are membranes made of ion track-etched polycarbonate (PC) and porous anodic aluminium oxide (AAO); yet, other materials such as Si or polymethylmethacrylate (PMMA) can be used as templates. Typical PC templates [available, for example, at www.it4ip.be] have a mean pore diameter ranging from 30 to 200 nm, pore densities ranging from 10^8 to 10^9 pores.cm^{-2} and thicknesses around 20 μm. Fig. 1(a) gives schematic views of a typical PC template used for synthesizing CP and hybrid metal-CP NWs.

Among the large variety of CPs, three representative species have been included in this chapter: poly[3,4-ethylene-dioxythiophene] (PEDOT), polypyrrole (PPy), and polyaniline (PANI). These CPs are by far the most extensively studied as they exhibit remarkable properties such as environmental stability, interesting redox behavior and the ability to exhibit high electrical conductivities [Sadki (2000)]. While the polymerization of these CPs can be performed by both chemical and electrochemical reactions, the electrochemical polymerization is well adapted for the synthesis within templates. Furthermore, multi-segmented NWs [e.g. combining CP segments with noble (Au, Pt) or transition (Ni, Co metal blocks] can be easily produced using an all-electrochemical process, just by changing the plating solution and adapting the parameters of the deposition of each segment [Callegari (2009); Chung (2005)].

A crucial point of the electrochemical synthesis of the CPs within a template is that, depending on the synthesis parameters, NWs or NTs can be obtained [Cho & Lee (2008);

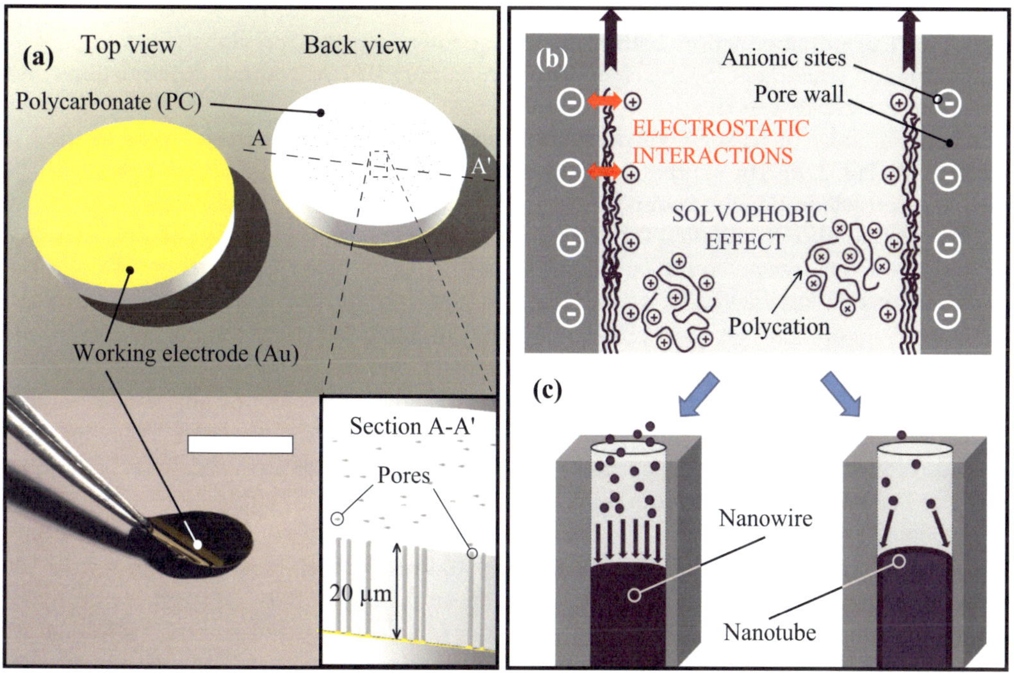

Fig. 1. (a) Top and back views of a flexible PC template used for synthesizing nanowires. In the left inset, the scale bar represents 5 mm. (b) Scheme of the growth mechanism for polymer NTs and NWs within a single pore. (c) Usual relations between synthesis conditions and morphology: Fast reaction rate and insufficient monomer supply lead to NTs. Slow reaction rate and sufficient monomer supply direct NWs.

Xiao (2007)][see Fig. 1.(b)]. While CP NTs could be interesting for different applications such as energy storage or drug release, NWs are needed for obtaining hybrid multi-segmented NWs with good metal-CP electrical interfaces. As revealed in the sequel, the determination of the exact geometry of metal-CP interfaces requires correlated scanning electron microscopy (SEM) and transmission electron microscopy (TEM) investigations.

The mechanism behind the two possible geometrical configurations is still matter of debate. For explaining the growth of nanotubes inside a template, it has been proposed by Martin and coworkers [Martin (1994; 1996)] that solvophobic and electrostatic interactions between CPs and the template occur within the pores of the template, favouring thus the tubular geometry [see Fig. 1.(b)]. A recent study [Xiao (2007)] has shown that the morphological transition between the two configurations can be explained by a mechanism based on monomer diffusion and reaction kinetics. Roughly, as exemplified in Fig. 1.(c), a slow polymerization reaction performed in presence of a sufficient monomer supply leads to the formation of NWs, while a faster reaction associated to an insufficient supply of monomer produces preferentially produces NTs.

2.2 Pure PEDOT and hybrid tri-segmented Au-PPy-Au NWs

Here, we briefly present the synthesis of pure PEDOT NWs and detail the fabrication of hybrid metal-CP NWs, taking the case of a tri-segmented Au-PPy-Au NW. Specifically, the

3,4-ethylenedioxythiophene (EDOT) monomer was provided by Bayer AG and used as received. The polymerization bath was an aqueous solution, previously developed by Sakmeche and coworkers [Sakmeche (1996)], containing 0.07 *M* sodium dodecylsulfate, 0.1 *M* $LiClO_4$ and 0.05 *M* EDOT.

A schematic view of tri-segmented NW synthesized by an all-electrochemical method is given in the Fig. 2.(a). The synthesis is performed by three sequential electrodepositions. The first step consists in the electrodeposition of a gold segment by cyclic voltammetry (CV). A cyanide free solution containing 0.1 *M* KCl, 0.1 *M* K_2HPO_4 and 0.03 *M* $HAuCl_4.3H_2O$ (Acros) in deionized (DI) water could be used [Reynes & Demoustier-Champagne (2005)]. The electrodeposition is achieved by cycling the potential from 0.7 to 0 V at a typical scan rate of 200 $mV.s^{-1}$. Figure 2.(b) gives a cyclic voltammogram recorded during the deposition of the first gold segment within a 100 nm pore diameter PC template. The typical growth rate is about 5 $nm.s^{-1}$. It is important to notice that the gold electrodeposition rate is highly dependent on the pore diameter and decreases with increasing the pore diameter. The second, PPy segment can be deposited into the pores by the electropolymerization of the pyrrole. The synthesis of PPy onto noble metal segments (Au, Pt) could be realized via CV by sweeping the potential from 0 to 0.85 V. A common electrolyte is an aqueous solution containing 0.1 *M* $LiClO_4$ and the monomer (5 to 100 *mM* range). In some cases, 7×10^{-4} *M* sodium dodecyl sulfate is added and used as surfactant. Figure 2.(c) gives the cyclic voltammograms corresponding to the deposition of the PPy segment at the first (full curve) and 200th cycle (dashed curve). Then, the third Au segment is electrodeposited under similar conditions to those used for the first segment. As shown in the Fig. 2.(d), a drastic change of the CV curve shape, accompanied by a strong increase of the current, indicates that deposition overflows the nanopores. This allows to detect easily the end of the synthesis. After the synthesis of each NW segment, the sample is rinsed in DI water for at least 30 minutes in order to remove the remaining plating solution left inside the pores of the membrane.

Remarkably, the resulting metal-PPy-metal NWs are characterized by two morphologically different metal-PPy interfaces as revealed by TEM observations and schematized in the insets of the Fig. 2.(a). The metal-onto-PPy interface is mechanically more robust than the PPy-onto-metal interface thanks to a higher adhesion surface. Several strategies have been envisaged to improve the mechanical strength of these interfaces. First, we optimized electrochemical parameters for the growth of the Au-onto-PPy interface by systematically investigating the effects of the potential sweep rate and the monomer concentration on its mechanical strength [Gence (2007)]. Concerning the bottom PPy-onto-Au interface, as its morphology cannot be modified just by playing on electrodeposition parameters, its mechanical strength could be enhanced by increasing the adhesion between the two different materials. Among the different possible strategies, we first investigated the use of self-assembled alkylthiol monolayers (SAMs) onto gold [Lahav (2006)]. We used 11-mercaptoundecanoic and 3-mercaptopropionic acids. The SAM is inserted between the gold and the PPy segment to create an electrostatic bond between the negatively charged carboxylate groups of the chemisorbed monolayer and the positively charged growing polycation as shown in the Fig. 3.(a). On top of the 11-mercaptoundecanoic acid SAM no CP growth has been observed. This is probably due to the presence of long aliphatic chains that passivated the Au segments. Unfortunately, even if the CP growth occurred on the thinner chemisorbed SAM (3-mercaptopropionic acid), no qualitative improvement of the mechanical strength has been detected upon this chemical modification of the interface. Nevertheless, the modification of interfaces by the use of SAMs is essential for the synthesis of hybrid transition metal-CP NWs.

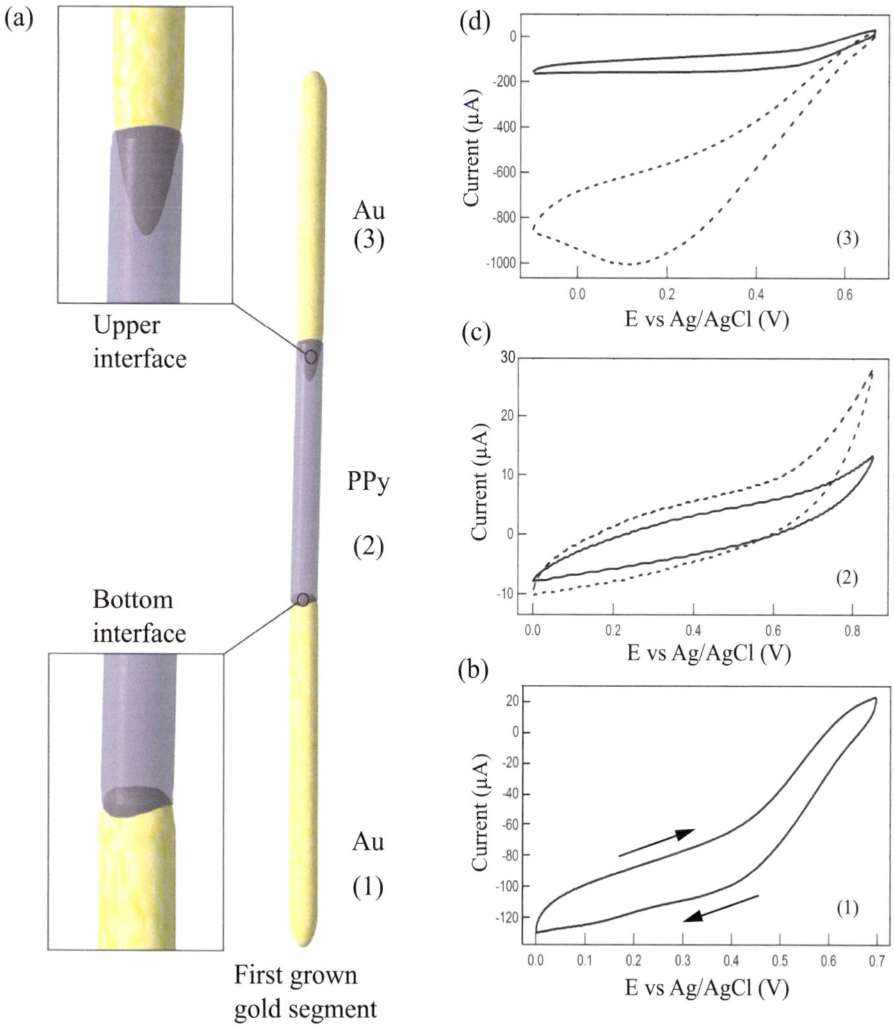

Fig. 2. (a) Schematic view of a tri-segmented Au-PPy-Au nanowire. The insets are magnified views of both polymer-metal interfaces. (b-d) Typical cyclic voltammograms obtained for the synthesis of each nanowire segment. (b) Cyclic voltammogram collected during the deposition of the first gold segment. (c) Cyclic voltammograms recorded during the deposition of the PPy segment. First cycle (full curve) and 200th cycle (dashed curve) are given. (d) Cyclic voltammograms acquired during the deposition of the third grown gold segment before (full curve) and after overflowing (dashed curve).

2.3 Hybrid NWs with transition metals

The synthesis of bi- and tri-segmented NWs including transition metals is now described. Bi-segmented, transition metal-onto-CP NWs can be easily elaborated by changing the electrochemical bath. While reducing the metal ions on top of the CP segment, the CP cannot be dissolved. Yet, it can be reduced, thus exhibiting a lower doped level state. The reverse sequence, i.e. the electropolymerization of usual CPs on transition metals, including Ni and Co, is less straightforward to achieve. Indeed, the oxidation potentials of these metals are much lower than the one of conjugated monomers and thus, dissolution of the metal occurs before electropolymerization begins. Figure 3.(c) gives typical chronoamperograms for the electrodeposition of Co and Ni segments. The length of the

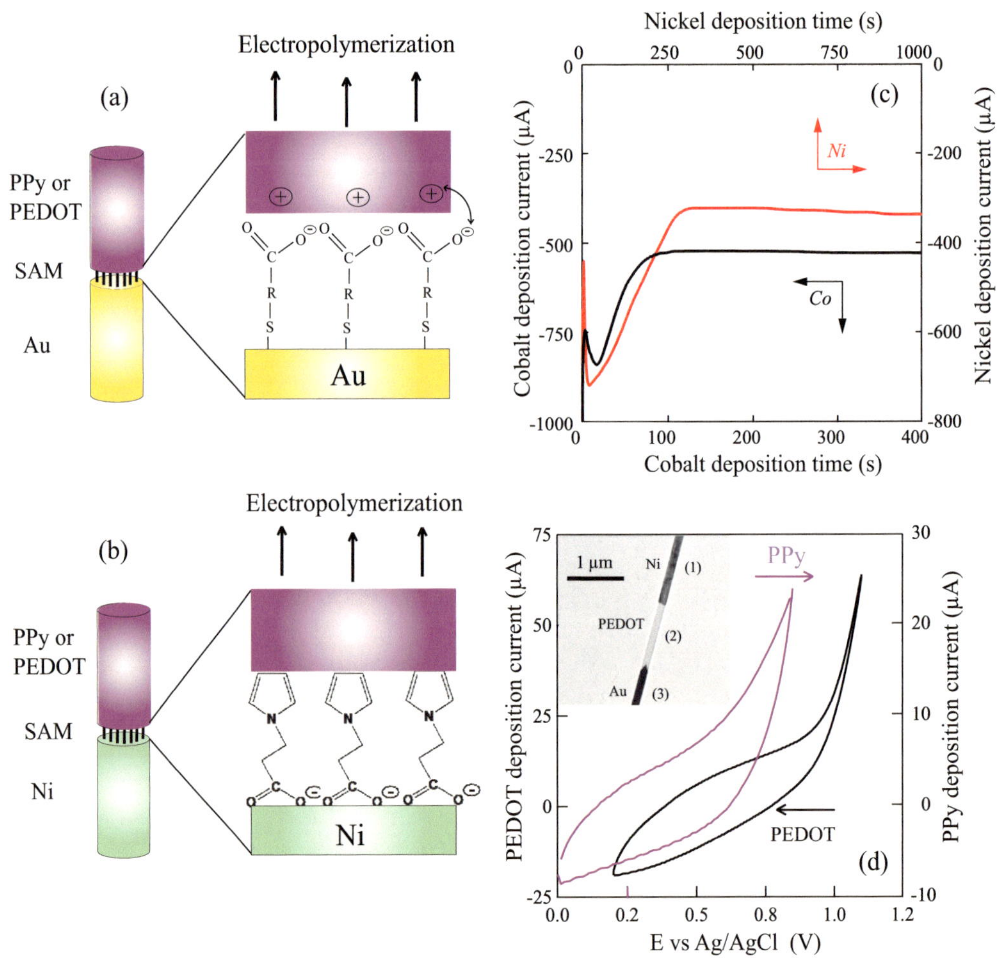

Fig. 3. (a, b) Schematic views of the modified metallic (Au, Ni) interfaces using SAMs. In the case of Ni (b), the presence of the SAM improves the electropolymerization process. (c) Typical chronoamperograms for the electrodeposition of Co (Ni) NW segments. (d) Cyclic voltammograms recorded during the deposition of PPy (PEDOT) segment on top of the first grown Ni segment modified by a SAM. The inset displays a typical 110 nm diameter Ni-PEDOT-Au NW as seen by TEM.

transition metal segments can be easily controlled by the electrodeposition time: the growth rates for Ni and Co, observed for 110 nm diameter NWs, are 0.5 and 0.7 $\mu m.min^{-1}$. Specifically, the Ni segment could be electrodeposited by chronoamperommetry at −1.05 V from an electrochemical bath containing Ni sulfate hexahydrate ($NiSO_4.6H_2O$) and boric acid (H_3BO_3), while Co is electrodeposited from a solution containing cobalt sulfate heptahydrate ($CoSO_4.7H_2O$) by applying a constant potential of -0.95 V.
Towards preparing hybrid transition metal-CP NWs, a major problem to overcome is to suppress or limit the metal re-dissolution at the oxidative potential required for the electrodeposition of the CP. It is therefore necessary to find synthesis conditions that will strongly passivate the metal without impeding electropolymerization. Our approach consists in the use of a SAM. The 3-(Pyrrol-1-yl) propanoic acid has been chosen due to its facile synthesis and because it contains two end-functional groups, a carboxylate group at one extremity and a pyrrole group at the other end, that can specifically interact with the Ni substrate and the growing CP, respectively [see Fig. 3.(b)]. Prior to the electrodeposition of the CP, the SAM is deposited by immersion of the PC template containing the Ni segment in a 3-(Pyrrol-1-yl) propanoic acid solution for more than 18 hours. The CP segment (PPy or PEDOT) could be then synthesized by CV using standard electrodeposition parameters. Typical cyclic voltammograms recorded during the deposition of the second, (PPy or PEDOT) segments on top of the first grown Ni segments modified by SAMs are given in Fig. 3.(d). In order to fabricate tri-segmented nanowires, a third metallic block can be synthesized on top of the CP segment, by CV (e.g. Au) or by chronoamperommetry (e.g. Co). Following this procedure, 100-nmdiameter Ni-(PPy or PEDOT)-Au and Ni-(PPy or PEDOT)-Co NWs have been successfully synthesized. The inset to the Fig. 3.(d) gives an example of a 100-nm-diameter Ni-PEDOT-Au NW as observed by TEM: the first grown Ni segment is clearly distinguishable. Despite the use of the 3-(Pyrrol-1-yl) propanoic acid SAM, the percentage of unbroken NWs containing transition metal segments remains low compared to the tri-segmented NWs containing only noble metals.

3. Structural characterization

In this section, we present the morphology of several classes of CP NWs, including pure PEDOT NWs and hybrid metal-CP NWs, as studied by SEM, TEM, atomic force microscopy (AFM), Fourier transform Raman spectroscopy and electron spin resonance (ESR).
The morphology of PEDOT NWs is captured in SEM and AFM images reported in Fig. 4.(a, c, d) and Fig. 4.(b), respectively. Once the PC membrane has been removed, PEDOT - as well as other CP NWs - tend to cluster into bundles because of their long length and flexibility. This behavior is reinforced by the NW anchoring on the bottom metallic electrodes (see below). This is an important point to take into account for the integration of CP NWs into devices as well-separated NWs or NTs can be required for high-sensitivity applications. The unambiguous discrimination between NW bundles and large diameter NWs can be easily obtained via AFM studies; the AFM image Fig. 4.(b) reveals five closely-packed PEDOT NWs. In this case, the diameter is determined from the height shift to avoid the convolution by the tip which has to be considered when measuring the lateral size. Finally, a SEM study is not sufficient to determine whether NWs or NTs have been fabricated for given synthesis conditions. As an example, a TEM investigation has shown that the template electropolymerization of PEDOT under the conditions described above results in the

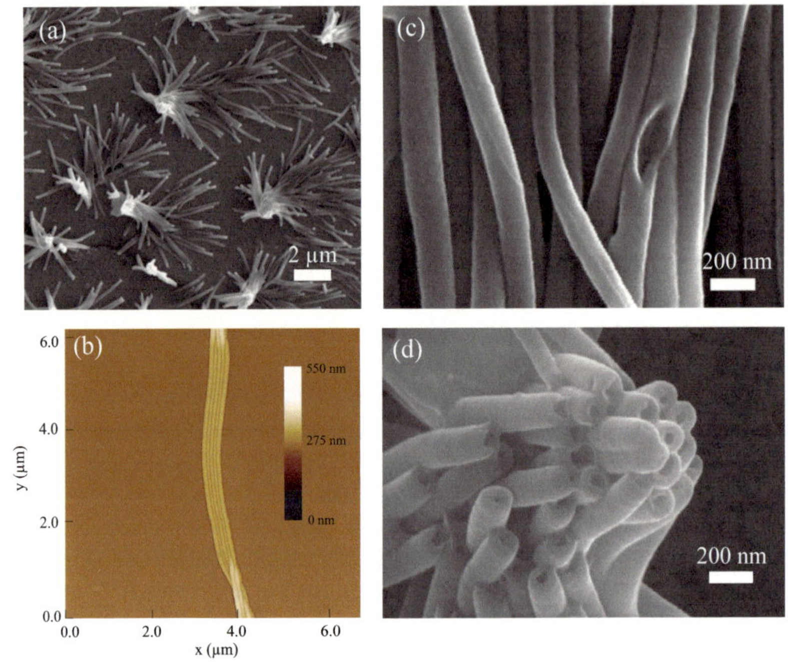

Fig. 4. Morphological characterization of PEDOT NWs after the removal of the PC template. (a) SEM picture of 3–4 μm long NWs. (b) AFM topography image of a rope of NWs. (c) SEM image revealing 100-120 nm diameter NWs. (d) SEM view of a standing up bundle of NWs.

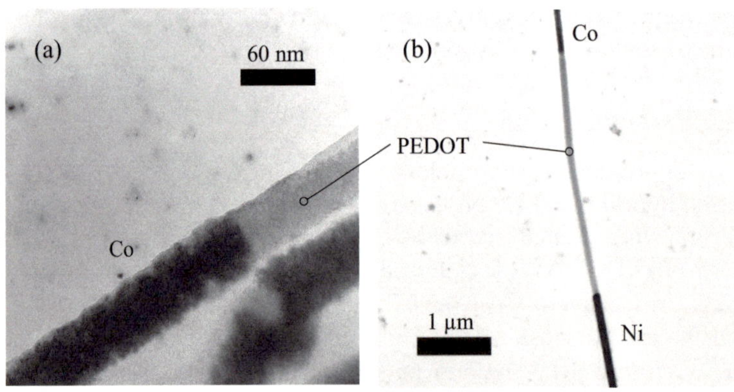

Fig. 5. (a) TEM picture of a 60 nm diameter Co-onto-PEDOT NW. (b) TEM picture of a 100 nm diameter tri-segmented Ni-PEDOT-Co NW.

fabrication of PEDOT NWs, while the opened top-ends of the synthesized nano-objects, observed by SEM [Fig. 4.(b)], suggest NTs. In the case of hybrid metal-CP NWs, the metal and CP segments are unambiguously distinguishable by SEM and TEM without any particular contrasting treatment. For example, in a classic SEM picture, tri-segmented metal-(PPy or PEDOT)-metal NWs appear composed of a dark polymer segment inserted between two bright metal segments. In TEM micrographs (Figs. 5 and 6), the polymer block appears more transparent than the metallic segments, due to its lower electronic density.

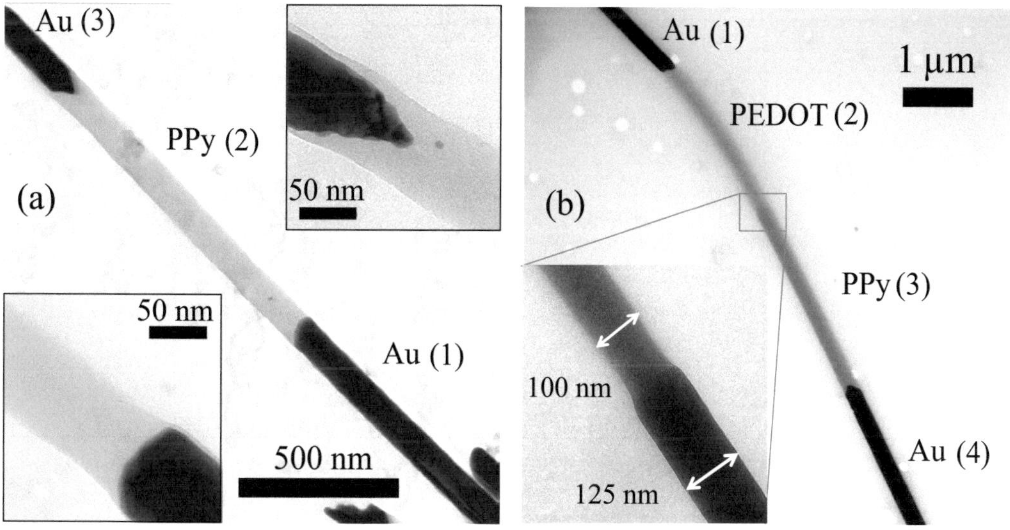

Fig. 6. (a) TEM image of the two Au-PPy interfaces in Au-PPy-Au NWs. The lower and upper insets give closer views of the PPy-onto-Au (first grown) and Au-onto-PPy (second grown) interfaces. (b) TEM picture of a 110 nm diameter tetra-segmented Au-PEDOT-PPy-Au NW. The inset shows a zoom of the polymer heterojunction.

Figure 6 presents TEM images of tri- and tetra-segmented metal-CP NWs. As displayed in the two insets of Fig. 6.(a), the fabricated Au-PPy-Au NWs have two morphologically different Au-PPy interfaces. In terms of mechanical robustness, the PPy-onto-Au [referred as (1)] interface is smoother and more fragile than the Au [referred as (3)]-onto-PPy interface. Figure 6.(b) illustrates typical Au-PEDOT-PPy-Au NWs. During the synthesis, for enhancing the overall mechanical robustness of the NWs, the deposition sequence was determined according to careful observations on the shape and strength of the different Au-CP interfaces. The PEDOT block was first deposited on the Au segment, as the strong chemical bond between Au-S compensates for the weakness of the flat interface. The PPy block was then electrodeposited onto the PEDOT segment, the strength of the metal-onto-PPy interface being ensured by its meniscus shape. The inset to Fig. 6.(b) gives a magnified view of the PPy-onto-PEDOT heterojunction. It appears that the PEDOT segment shrank more than the PPy segment during the solvent evaporation in a sampling process.
Another unique characterization tool for CP-based NWs is Fourier transform Raman spectroscopy. In the past, the influence of the diameter (35, 75, 100 and 150 nm) on the PEDOT structure was studied on electrodeposited NWs at +0.80 V. All spectra were recorded with an excitation line at 676 nm after dissolution of the membrane [Fig. 7.(a)]. The bands located at 1424 cm^{-1} and at 1495 cm^{-1} are attributed, respectively, to the symmetric and antisymmetric C=C vibrations of the thiophene rings. Interestingly, the intensity ratio $I(1495\ cm^{-1})/I(1424\ cm^{-1})$ increases significantly when the diameter decreases. This variation has been attributed to an increase of the conjugation length when the NW diameter decreases [Duvail (2004)]. Additional information on the nature of the charge (polaronic, bipolaronic) carriers could be obtained from ESR studies. Figure. 7.(b) compares room temperature ESR spectra of PEDOT NWs (accomodated into PC membranes) with diameters approximately equal to 200, 100, and 50 nm (as measured by SEM) and normalized

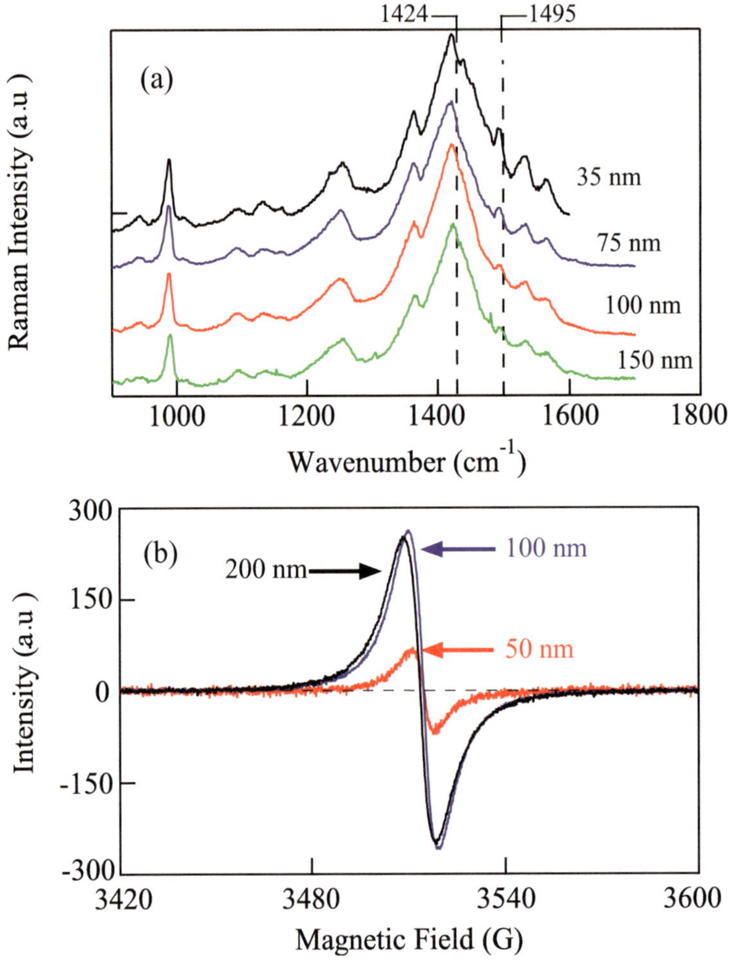

Fig. 7. (a) Effect of the PEDOT NW diameter on the Raman spectra (λ_{exc} = 676 nm). Curves are shifted for clarity. (b) ESR spectra (f_{exc} = 9.4 GHz) of PEDOT NWs.

to the same EDOT mass. A quantitative analysis of the Lorentzian-like peaks (directly proportional to the spin density) has been carried out. Subsequently, these results have been correlated to the doping level of the NWs estimated from XPS data [Duvail (2004)], revealing that the bipolarons are the dominant charge carriers in these NWs and their contribution increases when the diameter decreases. This conclusion corroborates Raman studies and suggests a strong molecular and supermolecular improvement when the NW diameter decreases.

4. Electrical properties

In this section, we review different techniques for contacting NWs grown by the template method and compare their limitations and advantages. The electrical characterization can be performed in a vertical configuration with multiple NWs contained within the template or in a planar configuration, with NWs freed from the template and dispersed on a substrate.

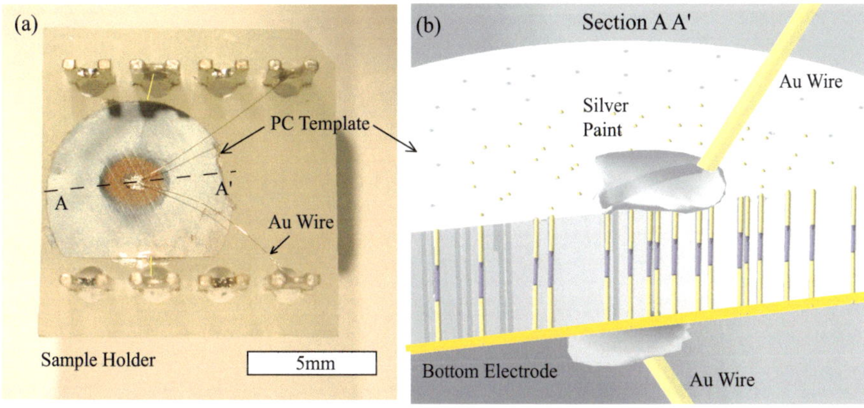

Fig. 8. (a) Optical micrograph of multiple NWs contacted in a vertical configuration within the PC template. (b) Schematic view (for tri-segmented NWs) of the section AA′ defined in the panel (a).

4.1 Vertical contacting techniques

The vertical configuration is quite convenient and has been frequently employed those last years for the electrical and thermal characterization of NWs [Duvail (2002); Heremans (2002; 2003); Gence (2008); Lorcy (2009)] as it is very fast and requires no complex processing of the sample to be measured. Figure 8.(a) gives an optical micrograph of a PC template mounted on a sample holder and electrically contacted immediately after the electrochemical synthesis. The NWs enclosed within the template are mechanically supported by the PC membrane and are protected from the effect of both oxygen and ambient humidity [Chtanko (2004)]. However, four main drawbacks arise from this method. First, the exact number of connected NWs is not known. An estimation of the number of connected NWs can be made from the porosity of the template and the surface of the electrical contact. The typical values of porosity and hand-made contact surface are 10^9 pores/cm^2 and 5×10^{-4} cm^2. Assuming that all pores are filled, this results in a typical number of connected NWs of about 5×10^5. Among this huge number of NWs, by taking into account the calibration of the synthesis parameters, one expects a wide distribution of the physico-chemical characteristics. This distribution results in an averaging effect on the electrical measurements and contributes in masking the intrinsic properties of single NWs. Second, the length of each segment can differ quite significantly between different NWs prepared in the same PC membrane. As the diameter can differ slightly (10 - 20%) from one NW to the other, the intrinsic conductivity (calculated by using the geometrical parameters) can vary by a factor of 2. Third, the resulting two-contact geometry can not avoid the contact resistance to be included in the measurements, in contrast to the four-contact geometry. Fourth, the measurement of multiple NWs contained within a template does not allow any interaction of the NWs with the environment or with any chemical, electrical or physical agent. For all the above reasons, the use of NWs freed from the template is the best choice for the characterization of their intrinsic properties. For application-oriented purposes, the integration of NWs with microelectromechanical systems or electronic devices is required and the NWs have to be precisely aligned and positioned onto microfabricated contacts. Fortunately, various techniques have been specifically developed for the assembly of single NWs onto electrodes [Cao (2008)] and many others are under close scrutiny.

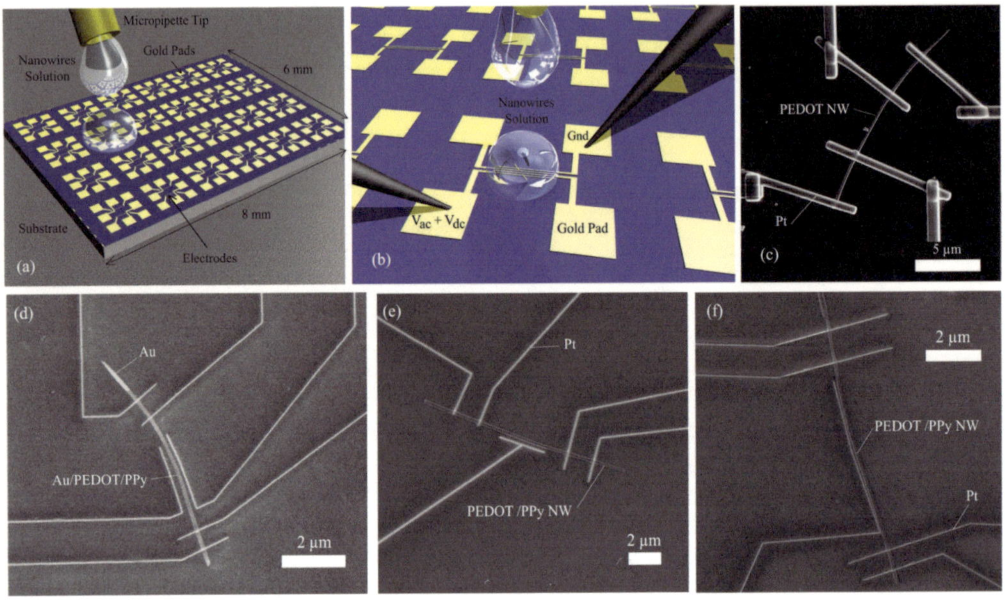

Fig. 9. (a) Schematic 3D view of an electrical device designed for integrating the NWs by a bottom contact approach. (b) Zoom on a single NW deposition assisted by an external electric field. (c-f) SEM micrographs of CP NWs connected by the top contact approach: (c) FIB and (d-f) EBID techniques.

4.2 Planar contacting techniques

For planar contacting techniques, the NWs have to be gently freed from their matrix and the first step consists in the removal of the working electrode from the back of the templates. Generally, the working electrode is made of a few hundred nm Au layer that can be etched with iodine solutions. The PC templates are easily dissolved in dichloromethane, while AAO templates require typically NaOH solutions. The techniques used for contacting single NWs in planar configurations [see Fig. 9.(a)] can be divided in: (i) bottom-contact methods (where NWs are deposited on top of predefined electrodes) and (ii) top-contact methods (where the electrodes are deposited on top of the NWs). In both cases, the substrate includes a high quality insulating layer (Si_3N_4 or SiO_2). For the bottom-contact method, metallic (Au, Pt) electrodes are defined on this top insulating layer by photolithography or electron-beam lithography (EBL). Figure 9.(a) gives a schematic 3D view of a typical device designed for integrating NWs by the bottom-contact method via dropcast deposition. Dielectrophoretic trapping of NWs [see Fig. 9.(b)] has been known for a long time [Smith (2000); Dong (2005)] and used very recently for the precise alignment of hybrid metal-CP NWs in solution [Park (2004); Cao (2008)]. For the top-contact approach, typically, EBL and related methods are employed for the definition of the electrodes on top of the NWs. For example, Pt electrodes could be directly deposited by electron beam induced deposition (EBID) or by focused ion beam (FIB) on top of the NWs [Duvail (2007); Shen (2006)]. Micrographs of CP and hybrid metal-CP NWs contacted by Pt electrodes fabricated via FIB and EBID depositions are given in Fig. 9.(c) and Fig. 9.(d-f), respectively.

The main advantage of the top-contact approach is to allow chemical and physical cleaning of NW-electrode interface which is not possible for bottom-contacted NWs. Indeed, in the bottom-contact approach, once the NW is deposited onto the electrodes and after the solvent has dried, some PC residues, previously dissolved in the solution are left on the substrate. These residues, when present at the NW-electrodes interfaces, can alter dramatically the electrical contact between the deposited NW and the electrodes in a non reproducible way and cannot be controlled by any treatment. On the other hand, in the framework of the top-contact approach, the NW-electrodes interfaces can be rinsed in pure solvent repetitively for many hours. Another possibility is to use a short cleaning oxygen plasma before the deposition of the electrodes (Au or Pt) for lift-off processing. Nevertheless, the main drawbacks of the direct writing techniques are the likely modification of the intrinsic doping of the NWs (e.g. dedoping associated to EBID manipulation and additional doping during FIB integration). Specifically, for the bottom-contact approach, the expected physisorption of the CP NWs on the metal electrodes can induce a significant contact resistance to inject charge current with a possibly non-ohmic behaviour. The opposite case is met for FIB top-contacts. The ion implantation, although it is reduced by careful choice of experimental parameters, results quite systematically in ohmic contacts. As a consequence, the Pt lines deposited by FIB have to be separated by typically 2 μm or more to avoid overlapping of the surrounding residual carbon-based deposit. Concerning contacts made by EBID on top of CP NWs, our attempts generally failed to get reliable ohmic contacts.

4.3 Electrical properties of pure PEDOT NWs

The motivation for measuring isolated NWs instead of arrays of NWs comes from the opportunity to determine unambiguously the conductivity σ through a four-probe geometry. Figure 10.(a) gives the diameter dependence of σ at room-temperature for pure PEDOT NWs. An increase by a factor 50 in σ is evidenced when the diameter decreases from 190 to 25 nm. This result confirms the tendency generally reported for arrays of template-synthesized CP NWs. The T-dependence of the resistivity ρ, normalized by ρ(300 K), for PEDOT NWs with different diameters is shown Fig. 10.(b). The diameter appears to induce dramatic differences in the electrical behaviour. The resistivity ratio ρ(20 K)/ρ(300 K), a useful empirical parameter for quantifying the extent of disorder, is reported as a function of the diameter in the inset to Fig. 10.(b). A decrease by a factor about 100 takes place when the diameter decreases from 190 to 35 nm. A precise analysis of the T-dependence of the resistivity has shown that the 190 nm diameter NWs are in an insulating regime (three-dimensional Mott variable range hopping), the 100 nm diameter NWs are in the critical regime of the metal-insulator transition and the 35 nm ones are on the metallic side of the transition. According to correlated investigations of the molecular and supermolecular ordering and the doping level (by Raman spectroscopy, electron spin resonance and X-ray photoelectron spectroscopy), the insulating-to-metal transition has been attributed to a strong structural improvement when the NW diameter decreases [Duvail (2004; 2007)]. The mechanisms responsible for this remarkable structural improvement are most likely the presence of anionic sites at the nanopore surface and the enhanced role of the solvophobic effect when the (electro)polymerization is confined at nanoscale, as initially proposed by C. R. Martin [Martin (1994; 1996)]. However, for NW diameters smaller than 35 nm, the resistivity ratio increases by four orders of magnitude. Interestingly, the electron conduction in the 25 and 30 nm NWs follows the Efros-Shklovskii law. This result suggests that electron-electron interactions play an important role for the charge transport at low T in CP NWs [Aleshin (2004)].

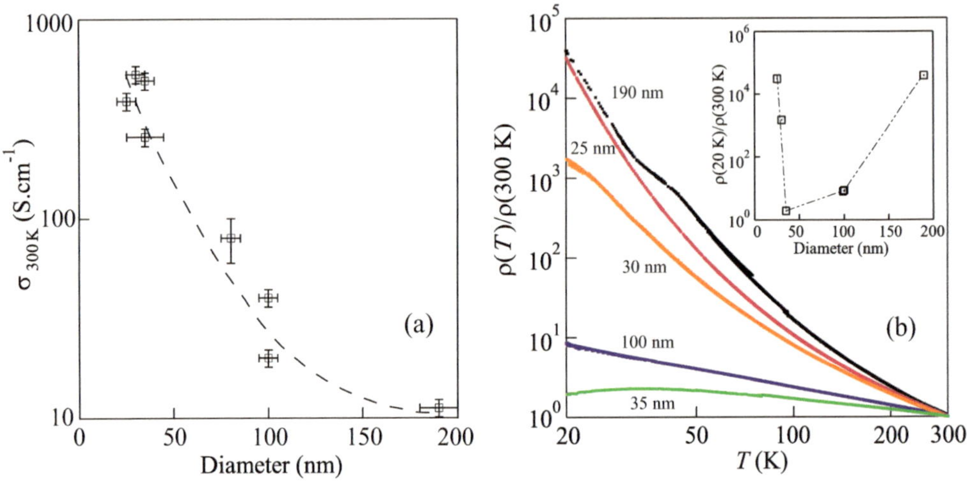

Fig. 10. (a) Diameter dependence of the room-temperature conductivity of PEDOT NWs. (b) Log-Log plot of the temperature dependence of the resistivity (normalized by the room T resistivity) of PEDOT NWs. The inset gives the diameter dependence of the resistivity ratio $\rho(20\text{ K})/\rho(300\text{ K})$.

4.4 Electrical properties of hybrid tri-segmented NWs

We next present electrical transport data for hybrid Au-PPy-Au NWs. The current-voltage $I-V$ spectroscopy is a powerful technique to gain insight into various transport phenomena such as tunnelling and rectification [Aleshin (2004); Gence (2007; 2008)]. Typical current-voltage $I-V$ characteristics are given in Fig. 11.(a-c) for different tri-segmented Au-PPy-Au NWs with diameter ranging from 40 to 160 nm at room and low temperature T. In all samples, the $I-V$ plots are symmetrical and show no rectification effect. For samples with diameter $\phi \geq 50$ nm, the $I-V$ characteristics are ohmic between room temperature and approximatively $T \approx 120$ K. As shown in Fig. 11, the non-linearity of the $I-V$ curves increases with decreasing T, signaling that a peculiar conduction mechanism occurs at very low temperatures [Long (2005)]. Noteworthy, for the 40 nm samples the ohmic region extends down to much lower temperatures $T \approx 25$ K. The resistance values $R(T)$ are obtained from the $(dI/dV)^{-1}$ values at zero bias. For all investigated samples, the resistance monotonically increases with decreasing T, indicating that PPy has a non-metallic behavior. This is coherent with the room temperature conductivity of ≈ 0.04 S.cm^{-1} and it is comparable to the bulk insulating PPy prepared under the same conditions.

Several experimental studies have shown that the Mott variable-range-hopping (VRH) regime is an appropriate model of charge transport in CP NWs as it is well-known for bulk CPs [Aguilar-Hernández & Potje-Kamloth (1999); Park (2004); Bufon (2005)]. In this model, the charge transport is due to thermally activated tunneling among states that are localized in a constant density of states [Mott & Davis (1979)]. In the VRH model, the resistance follows the relation $ln[R(T)] \propto (T_0/T)^{1/(d+1)}$, where T_0 is the Mott temperature and d is the dimensionality of the system. The best fits to our data have been obtained with $d = 3$ [see Fig.11.(d and e)]. This suggests that the three-dimensional (3D) VRH is the appropriate model of transport for tri-segmented Au-PPy-Au NWs.

The peculiarity of the ϕ = 40-nm-diameter tri-segmented NWs can be highlighted using the empirical parameter $R_R = R(77\text{ K})/R(290\text{ K})$. This parameter has been frequently used to characterize the extent of disorder in CPs such as PPy or PEDOT [Yoon (1994); Duvail (2007)]. In Fig.11.(f), we have plotted the parameter R_R as a function of the diameter for different NWs. The parameter R_R is found to be four orders of magnitude smaller for the 40 *nm* diameter samples than for the 70 nm diameter specimens. This suggests that the ϕ = 40 *nm* Au-PPy-Au NWs exhibit a dramatically lower extent of disorder compared to higher diameter tri-segmented NWs. Moreover, it indicates that the resistance of the PPy-Au junctions can be neglected, at least when measured at temperatures down to 77 K.

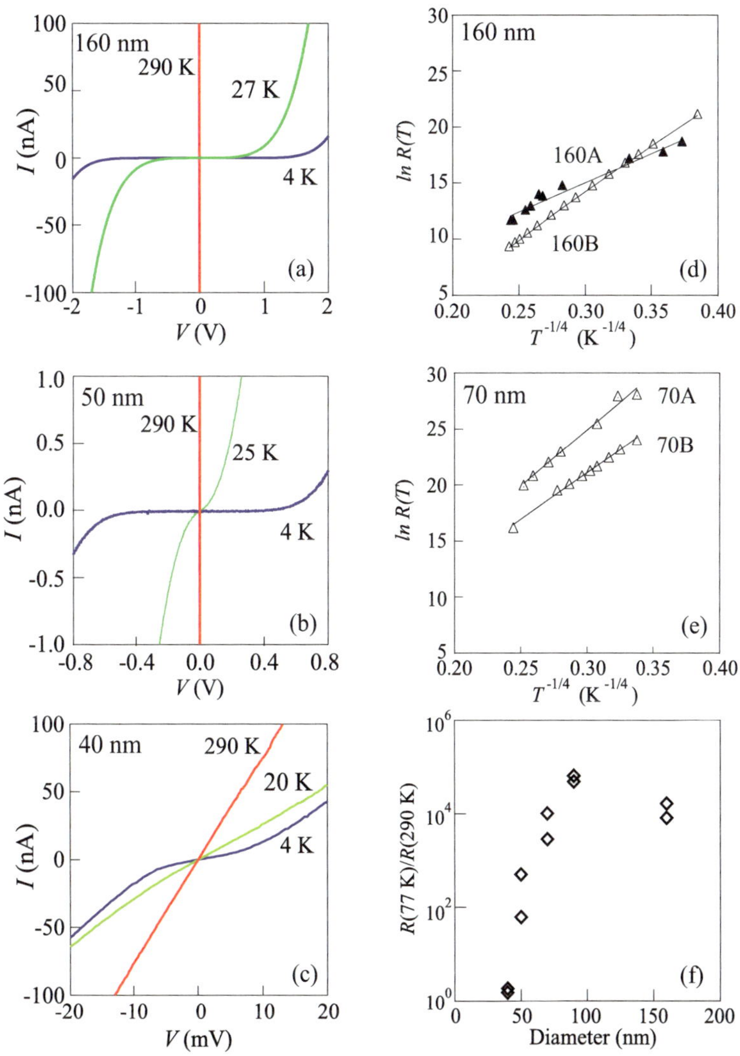

Fig. 11. (a-c) $I - V$ characteristics of tri-segmented Au-PPy-Au NWs with diameters of 160, 50 and 40 nm. (d-e) 3D-VRH plots for 160 and 70 nm diameter samples. The solid lines are the 3D-VRH fits to the data. (f) Diameter dependence of the resistance ratio R_R observed for tri-segmented Au-PPy-Au NWs.

4.5 Tunable electronic behavior of hybrid metal-CP NWs

Hybrid metal-CP NWs are attractive candidates for nanodiodes and transistors [Pinto (2003); Park (2004); Merlo & Frisbie (2004); Pinto (2009)] as well as photonic [Guo (2008); Camposeo (2009)] and electrochromic devices [Cho & Lee (2008)]. This high applicative potential comes from the fact that they make profit from intrinsic properties of CPs (low-cost, flexibility, environmental stability, high biocompatibility) concomitantly with other properties stemming from their low-dimensionality (large surface-to-volume ratio and enhanced properties compared to bulk materials). In particular, the literature data show clearly the current interest towards the engineering of novel CP-based NW devices with tunable electronic properties. We present here a new class of tetra-segmented Au-PEDOT-PPy-Au NWs able to switch their electrical characteristics in function of the redox state of the two CP blocks. To illustrate the principle, we use a chemical treatment performed by immersing the samples in a 0.5 M aqueous solution of $FeCl_3$ for 1 h. We contrast the behavior observed for the tetra-segmented NWs to that observed for tri-segmented NWs.

Figures 12.(a) and 12.(c) give schematic view of tri-segmented (Au-PEDOT-Au) and tetra-segmented (Au-PEDOT-PPy-Au) NWs, respectively. The $I - V$ characteristics of a tri-segmented NW have been measured in a four-probe configuration, before and after $FeCl_3$ treatment, and are given in Fig. 12.(b). The $I - V$ curves are symmetrical and linear. While

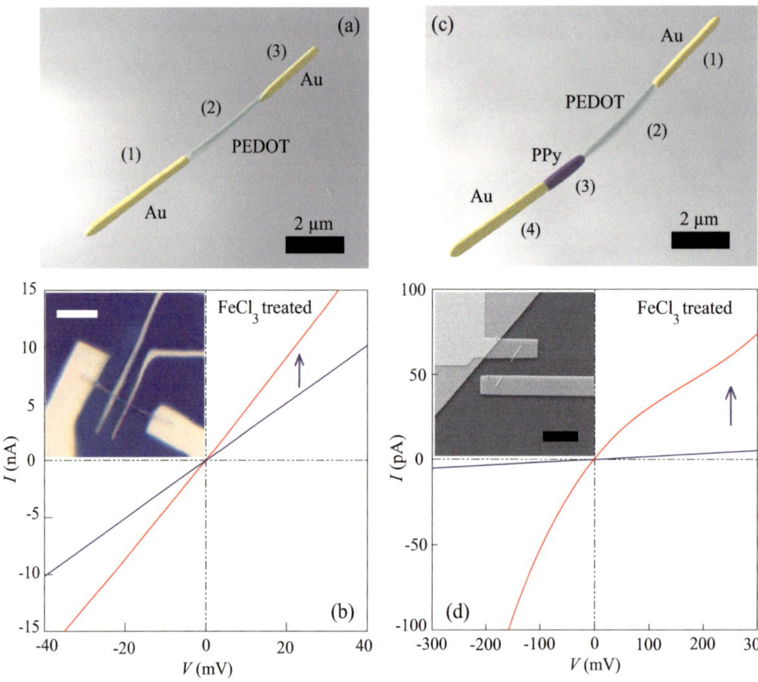

Fig. 12. (a) Schematic view of a tri-segmented Au-PEDOT-Au NW. (b) Four-point $I - V$ characteristics of a single tri-segmented Au-PEDOT-Au NW before (blue curve) and after (red curve) $FeCl_3$ treatment. The inset shows the four probe configuration used for contacting the tri-segmented NW. The scale bar is 5 μm. (c) Schematic view of a tetra-segmented Au-PEDOT-PPy-Au NW. (d) Corresponding $I - V$ characteristics before (blue curve) and after (red curve) $FeCl_3$ treatment. The inset shows a typical tetra-segmented device and the scale bar is 5 μm.

the blue curve corresponds to the as-synthesized NW, the red curve has been obtained after $FeCl_3$ treatment. The observed increase of electrical conductivity of the PEDOT segment could be understood by a shift in the dopant anion equilibrium within the PEDOT [Callegari (2009)]. The chloride ions inserted during the chemical treatment effectively adjust the low dimensional carrier hopping network on the backbone of the polymer inducing an enhanced charge transport.
Similarly to Au-PEDOT-Au NWs, we performed the same experiment with Au-PEDOT-PPy-Au NWs. Remarkably, after the chemical treatment, tetra-segmented NWs exhibit a highly nonlinear $I - V$ characteristic (red curve), accompanied by a current increase. Furthermore, the $I - V$ curve is asymmetric, i.e. the gain of current depends on the sign of the bias voltage: roughly, the gain is enhanced by a factor of 2 at negative bias compared to positive bias. These observations recall the behavior of other types of hybrid NWs that have been proposed in the literature as innovative structures for chemical sensing.

5. Correlated characterization

The correlation of multiple analyses performed on the same, well-defined multi-segmented NWs could lead to a deeper understanding of their structure-function relationships. The aim of this section is to present microdevices specially designed for the correlated characterization of single hybrid metal-CP NWs but it could obviously be applied to other class of hybrid NWs such as hybrid inorganic NWs [Wang (2004)]. We address here the NWs assemblies with membrane-based micro-electromechanical systems. While structural analysis are performed by combining SEM, TEM and AFM microscopies, the microstructure and the elemental composition are investigated by selected area electron diffraction (SAED) and energy dispersive X-ray spectroscopy (EDS) measurements.

5.1 Platforms for correlated characterization

We have used Si micromachining techniques for batch-producing devices based on Si_3N_4 membranes on top of which are deposited the NWs to be analyzed. They are compatible with both bottom and top-contact planar approaches for contacting single NWs (see section 4.2). Figure 13.(a) gives the schematic view of these membrane-based platforms. The relatively small dimensions of the devices are designed for compatibility with various sample holders for correlated characterization. Essentially, 200 μm-thick Si wafers are oxidized and a 80 nm-thick low pressure chemical vapor deposition Si_3N_4 layer is then deposited. Afterwards, 100 nm thick Au pads are defined by optical lithography. Openings in the nitride layer on the back side of the wafer are created by a double side alignement photolithography followed by dry etching of the exposed areas. Membranes are then released by an anisotropic etching of the Si all through the wafer. The top-contact planar approach is often preferred for device integration, as it allows custom design of the electrodes for each addressed NW. As an example, Fig. 13.(b) gives a topographic AFM image of a single PEDOT NW connected in a transistor configuration. Naturally, in the case of NW transistors, the use of membrane-based devices could allow a comprehensive study of the accord between their structural characteristics - as observed by correlated analyses (TEM, micro Raman, AFM, SAED) - and the measured transistor performances such as the on/off current ratio, the threshold voltage or the gate leakage current [Lieber & Wang (2007)].

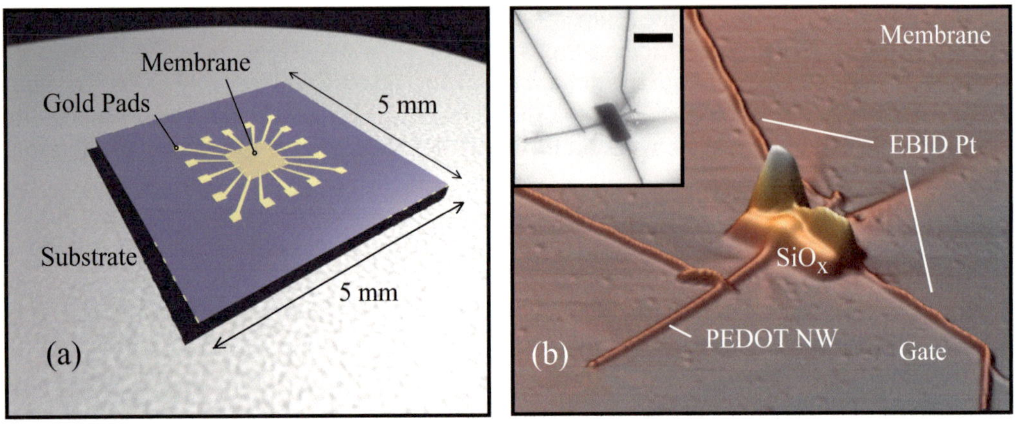

Fig. 13. (a) 3D schematic view of a membrane-based device. (b) AFM image of a single PEDOT NW connected in a transistor configuration on top of a membrane-based device. The *Pt* and SiO_x elements have been produced by electron-beam induced deposition. The inset is a TEM image of the same NW. The scale bar is 2 μm.

5.2 Correlated analysis of hybrid tetra-segmented NWs

A better demonstration of the correlated characterization is offered by hybrid metal-CP NWs. A TEM picture of a Au-PEDOT-PPy-Au NW deposited on top of a Si_3N_4 membrane and connected with multiple electrodes is given in the Fig. 14.(a). The geometries of both metal-CP interfaces have been determined by high magnification TEM (not shown here). By the shape of the menisci and the diameter of the segments, it is often possible to guess the nature of the CP blocks. The precise knowledge of the respective positions of the PPy and PEDOT segments in tetra-segmented NWs is important for understanding their subtle electronic behavior, as exemplified in the section 4.5. This identification can be established by EDX spectroscopy. Figure 14.(c) shows EDS spectra recorded on the four spots marked in Fig. 14.(a). Because these spectra have been recorded on top of a Si_3N_4 membrane, they all contain peaks at *K* and *L* X-ray emission energies corresponding to Si and N [Bearden & Burr (1967)]. Interestingly, the spectra reveal the presence of S atoms, characteristic for PEDOT, only in the spot III, while the presence of chlorine is identified in spots I and II. The positions of the PEDOT and PPy segments can thus be distinctly determined.

The panel (b) to Fig. 14 gives SAED patterns taken at four different positions on the device. While the SAED obtained through the Si_3N_4 membrane and the Au electrodes exhibit the expected amorphous and polycristalline patterns, the PEDOT and PPy both display amorphous patterns. This component of the correlated characterization tool box gives important information about the polymer chain organization of the produced CP segments. Polycristalline structures have been observed recently for PPy and PEDOT NWs synthesized electrochemically [Wang (2007); Lee (2008)]. Hence, the data acquired from the SAED operation of suspended membrane platforms accommodating hybrid metal-CP NWs can be crucial for further optimization of the CP segments' morphology. Figure 15 presents the four probe $I-V$ characteristics of different segments of a Au-PEDOT-PPy-Au NW contacted with several electrodes (A to F) and shown in the Fig. 15.(b). The respective values for the electrical conductivity are estimated from the dI/dV values at zero bias. The low value of the conductivity σ_{CD} = 0.2 mS/cm, deduced for the segment CD, is in agreement with the rather

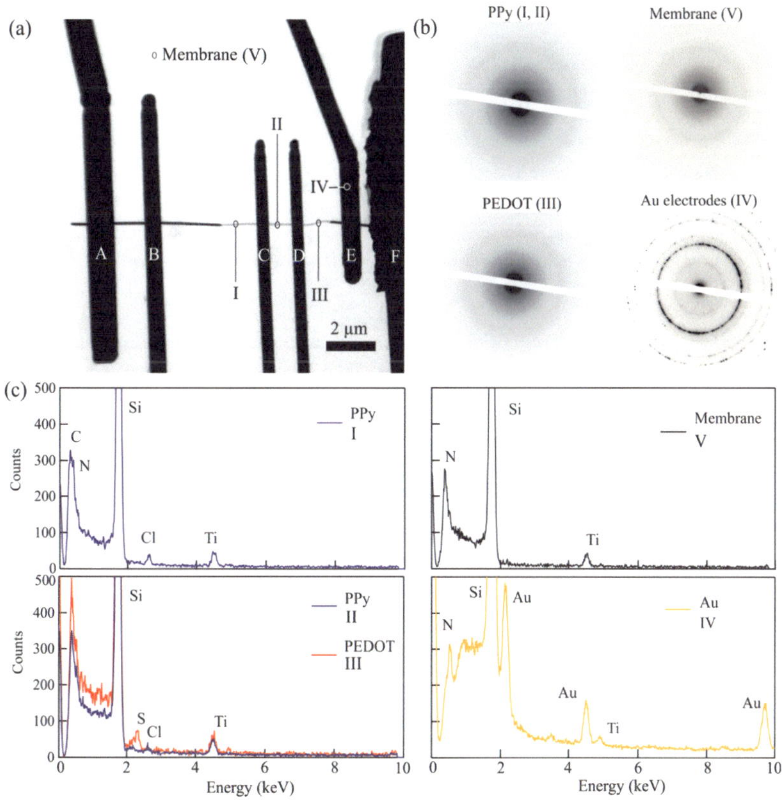

Fig. 14. (a) TEM picture of a Au-PEDOT-PPy-Au NW deposited on top of a silicon nitride membrane and connected with several electrodes. (b) Selected area electron diffraction patterns taken at three positions along the NW, membrane, and on top of the electrodes. (c) Energy dispersive X-ray spectra measured at four different positions on the device.

low PPy doping ratio generally observed in Au-PPy-Au NW [Gence (2007)]. More important, the data suggest that the Au-PEDOT and PPy-Au interfaces are not equivalent in terms of electrical conductivities. The Au-PPy interface presents a conductance roughly 30 times smaller than the Au-PEDOT interface. Therefore, electrical data analysis of hybrid metal-CP NWs requires utmost care: normally, the conductivities of the metallic segments and the associated metal-CP interfaces are much larger than the conductivities of the CP segments and the interfaces between two CPs. This, obviously, is not always the case. By illustrating this electrical dissymmetry of the metal-CP interfaces, the $I - V$ spectroscopy is heralded as the key component of the correlated characterization tool box that could enable the rational design of hybrid metal-CP NWs.

6. Integration and applications

The successful assembly of CP-based NWs into hierarchical structures is a key step necessary towards commercially attractive materials and devices. An illustrative example, sorted out from emerging niche applications, is molecular thermoelectricity. Innovative, low-cost thermoelectric converters, combining CPs and metals, could open new vistas in the

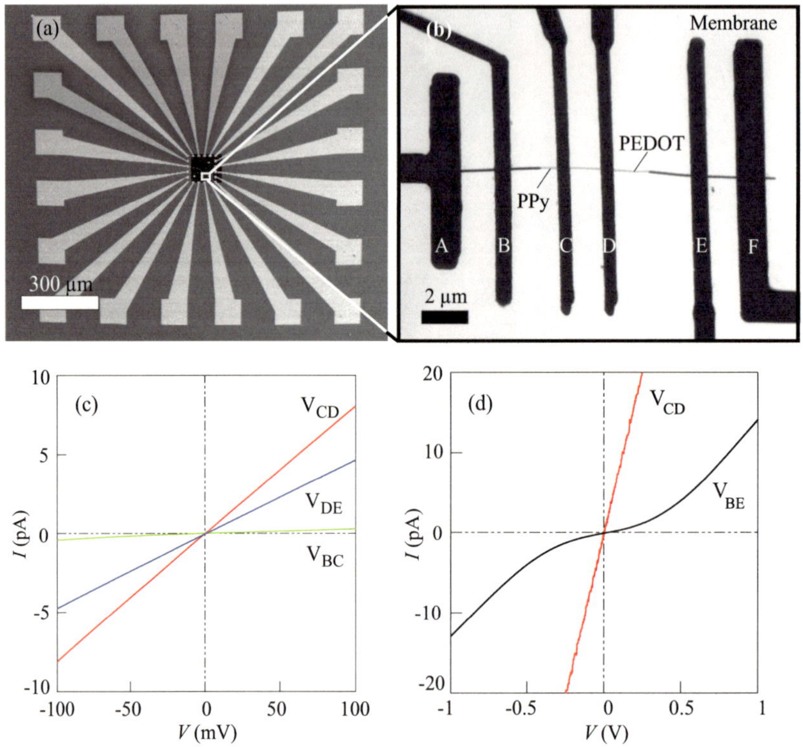

Fig. 15. (a) SEM picture of a membrane-based device used for the correlated characterization of a 110 nm diameter Au-PEDOT-PPy-Au NW. (b) TEM picture of the nanowire contacted with several electrodes. (c) $I - V$ characteristics of three different segments of the NW measured by the four-point technique. (d) Comparison of high bias $I - V$ characteristics of the NW segments BE (V_{BE}) and CD (V_{CD}). The resistance of the interfaces contributes to the higher resistance measured for the segment BE.

search for inexpensive sources of energy. Clearly, the level of integration depends on the envisaged applications; CP-based NWs, processed in mass quantities, are often incorporated in functional layers and they not require complex integration schemes. On one hand, CP designed as functional low-dimensional coatings of NWs or NTs [Lorcy (2009)] are following an easy way towards applications in life sciences (e.g. bio or chemical sensors). On the other hand, CP-based NWs used in electronic devices (e.g. flexible memory devices or thermoelectric converters), have to be integrated into a specific area of the device where physical, chemical or electrical interactions with the environment are permitted. This integration means that NWs have to be manipulated individually and placed on specific sites. Many techniques based on the use of the electric or magnetic field, fluidic flow or chemical assembly have been proposed for addressing the integration of template synthesized NWs. Interestingly, this integration can be achieved by combining classical top-down and template methods. Precisely, it consists in synthesizing the NWs directly on the final device, where they are requested. In this respect, recent work has demonstrated that PANI NWs can be polymerized within PMMA templates [Yunus et al. (2009); Vlad (2009)] in both planar and vertical architectures.

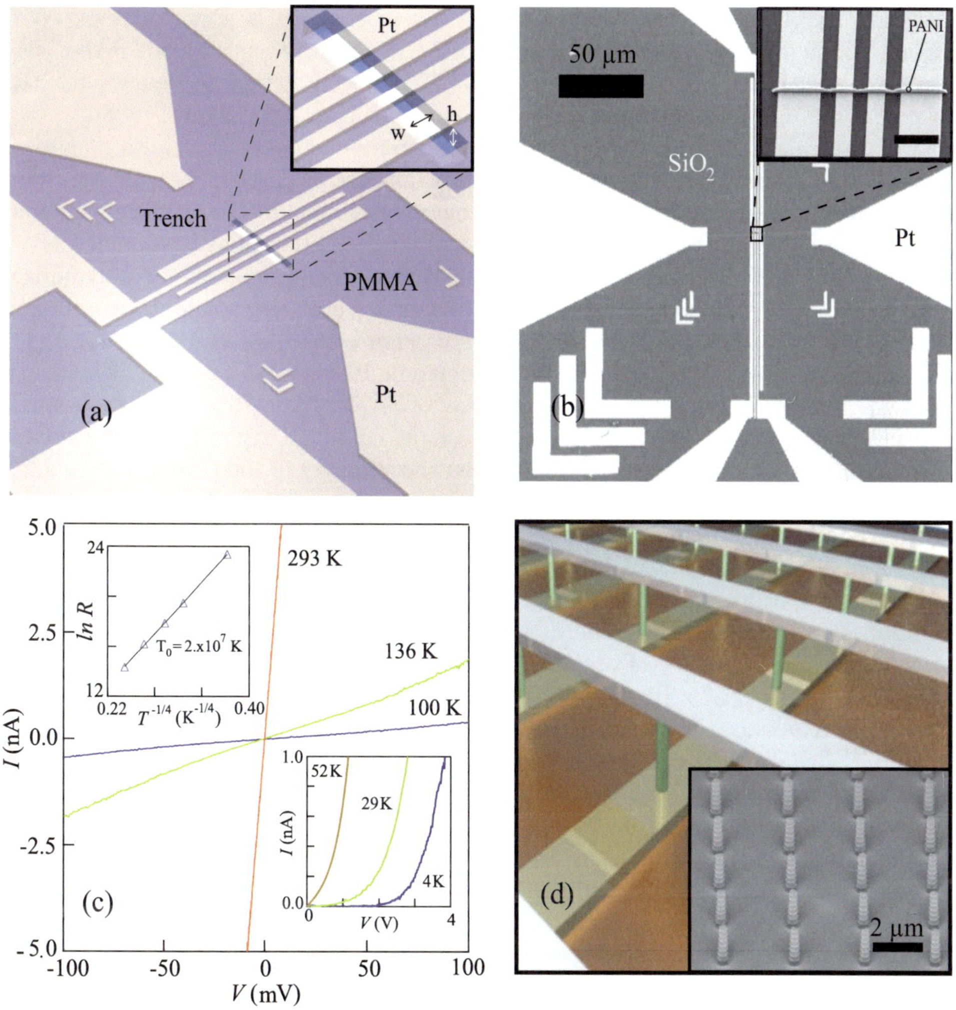

Fig. 16. (a) 3D schematic view of a PMMA template defined on top of four Pt electrodes. (b) SEM image of Pt electrodes used for synthesizing a single PANI NW defined within a PMMA template. The scale bar is 50 μm. The inset gives a closer zoom of the NW obtained after template removal. The scale bar is 2 μm. (c) Temperature dependent $I - V$ characteristics of the single PANI NW shown in (a). The lower inset shows the low-T data. The upper inset gives a 3D-VRH plot of the resistance. (d) 3D view of a cross bar architecture using single PANI NWs as active elements. The inset is a SEM picture of a highly ordered array of PANI NWs with diameter modulations.

Figure 16.(a) is a schematic view of a PMMA template defined on top of four Pt electrodes for planar NW integration. Here the template is created by EBL from a 500 nm thick PMMA layer; trenches are typically 200 nm wide and about 10 μm long. Figure 16.(b) presents a SEM micrograph of a device used for synthesizing a single PANI NW within a nano-engineered PMMA trench as a template and the inset offers a zoom on the NW after template removal. The temperature dependent $I - V$ characteristics of this PANI NW are given in Fig. 16.(c), revealing that the 3D VRH conduction regime dictates the charge transport in these planar integrated NWs, obtained via electroless polymerization.

In many cases, the bottom-up approach based on pure chemistry principles fails to reach extended, technologically relevant lengths for functional hybrid assemblies. On the contrary, hierarchical strategies employing standard Si manufacturing allow the achievement of a full superstructural control over the 3D spatial positioning of different self-assembled NWs for developing new artificial nanostructures. In this respect, femtomol-resolved polymerization of aniline on nano-engineered Pt reactors allows nano- and microscale device integration with single NW patterning and growth resolution. Figure 16.(d) offers a schematic view of a cross bar architecture for vertical integration of single PANI NWs as active elements [Vlad (2009)]. The inset displays a highly ordered array of PANI NWs with diameter modulations. The multiplexing feature included in these systems could be exploited for the controlled switching of individual NWs, while the peculiar morphology of the NWs could be adapted to plasmonics and photonics purposes.

A unique opportunity for macroscale self-assembly is offered by the structural engineering of CP-based NWs [Pokroy (2009)]. In a seminal paper, Park and co-workers, by controlling the dissolution of the template, created robust self-assembled mesoscopic metal-CP amphiphiles in which the segregated inorganic hydrophilic segments (Au), connected to the hydrophobic soft segments of the NWs (PPy), display unusual architectures such as bundles, tubes, and sheets [Park (2004)]. In fact, new and versatile functionalities of such hybrid architectures can be achieved easily by chemical treatment such as reversible acid-base doping-dedoping or redox reactions [He (2003); Chiou (2007); Vlad (2009)]. Another way to reach useful functionalities for CP-based NWs is to modify the CP segments by metallic nanoparticles. The figure 17.(a) gives a TEM image of a PEDOT NW functionalized by Au nanoparticles: the Au clusters are chemisorbed onto the PEDOT NW surface due to the presence of S atoms. Such modified NWs could be useful for biosensors applications [Tseng (2005)]. Furthermore, functionalities of CP architectures can be targeted by taking advantage of the peculiar morphology of the constituents. For instance, the axial modulation of aligned PANI NWs can be exploited to fabricate in a single step, self-aligned 3D arrays of vertically stacked Au nanorings (see Fig. 17.(b)).

7. Perspectives

Hybrid metal-CP NWs are promising high-tech materials - as they exhibit enhanced performances compared to their bulk counterparts [Hernandez (2004)] and are of direct interest for developing novel multifunctional systems for convergent micro-nano-bio-technology applications. Among many remaining challenges, we can mention 1) novel methodologies to fabricate highly-ordered architectures - for example, combining electroless and electrochemical depositions with self-assembly; 2) characterization strategies at nanoscale of hybrid NW based superstructures as demanded by molecular electronics, spintronics, plasmonics and photonics; 3) the use of CPs adaptive features for controlled

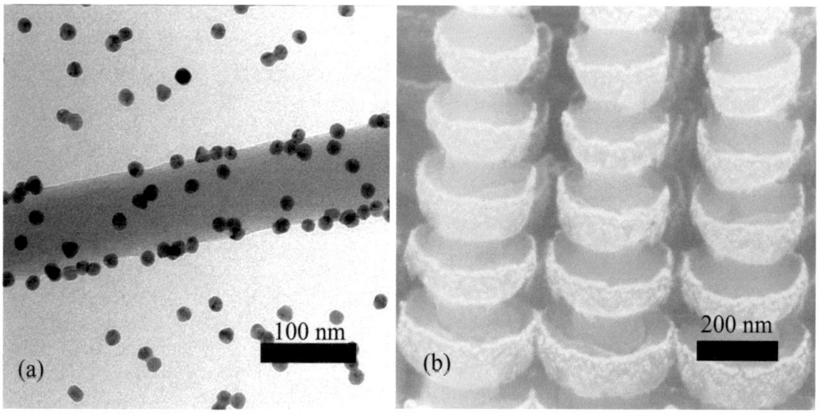

Fig. 17. (a) TEM picture of a single PEDOT NW functionalized by Au nanoparticles. The scale bar is 100 nm. (b) Au nanorings on PANI NWs obtained by physical evaporation through self-aligned stencil masks, provided here by the diameter modulations.

switching of device properties - at nanoscale - in the presence of external stimuli; 4) protocols for the manipulation of recognition properties of the constitutive segments towards engineering hybrid materials containing biomolecules like DNA or DNA/protein conjugates.
We are convinced that the greatest potential of electroactive CP-based NWs, by virtue of their responsiveness to magnetic, thermal and optical stimuli, is in interfacing the macroscopic world to the human physiological environment. Clearly, integrated hybrid metal- CP NWs into electrical impedance measurement platforms could prove highly beneficial in eliciting desired cellular responses with respect to biomolecules, including chemotaxis processes [Bagorda & Parent (2008)]. Relying on polypyrrole for exemplification, nanostructured 2D and 3D architectures with tunable electronic conductivity could be at the heart of interactive scaffolds that can be employed in tissue regeneration. As PPy has been extensively evaluated for various life science applications, encompassing amperometric biosensors, enzyme-based multicomponent electrodes in analyte detection systems [Li & Lin (2007)], or integrated DNA and peptide arrays on Si chips [Mailley et al. (2005)], there are now great prospects for the use of PPy-based NWs in high-tech applications, like enantiomeric sensoring and chiral molecules separation [Huang et al. (2008)] or biocompatible nanoactuators [Lee (2008); Lim (2008)].
In conclusion, we believe that the challenges proposed here will bring new developments in the engineering of hybrid metal-CP NWs, together with the understanding of the mechanisms underlying a rational use of their molecular functions. In all likelihood, the combined efforts of engineers, chemists, and physicists will lead to breakthroughs in the development of artificial molecular recognition biosensor systems and opto-electronic applications.

8. Acknowledgements

We acknowledge the help of Drs. S. Faniel, A. Vlad and S. Cuénot. This work was partially supported by the Belgian FRS - FNRS, NANOMOL and TINTIN projects - Communauté Française de Belgique and the NANOTIC - Feeling and NANOTIC - Cite projects of the Belgian Walloon Region and the CNano Project.

9. References

Aguilar-Hernández, J. & Potje-Kamloth, K. (1999). Optical and electrical characterization of a conducting polypyrrole composite prepared by in situ electropolymerization, *Phys. Chem. Chem. Phys* 1: 1735–1742.

Aleshin, A. N. *et al.* (2004). One-dimensional transport in polymer nanofibers, *Physical Review Letters* 93(19): 196601.

Bagorda, A. & Parent, C. A. (2008). Eukaryotic chemotaxis at a glance, *J. Cell. Sci.* 121(16): 2621–2624.

Bearden, J. A. & Burr, A. F. (1967). Reevaluation of x-ray atomic energy levels, *Rev. Mod. Phys.* 39(1): 125–142.

Bufon, B. *et al.* (2005). Relationship between chain length, disorder, and resistivity in polypyrrole films, *The Journal of Physical Chemistry B* 109(41): 19191–19199.

Callegari, V. *et al.* (2009). Electrochemically template-grown multi-segmented gold-conducting polymer nanowires with tunable electronic behavior, *Chemistry of Materials* 21(18): 4241–4247.

Camposeo, A. *et al.* (2009). Laser emission from electrospun polymer nanofibers, *Small* 5(5): 562–566.

Cao, Y. *et al.* (2008). Electrical transport and chemical sensing properties of individual conducting polymer nanowires, *Nano Letters* 8(12): 4653–4658.

Chiou, N.-R. *et al.* (2007). Growth and alignment of polyaniline nanofibres with superhydrophobic, superhydrophilic and other properties, *Nature Nanotechnology* 2(6): 354–357.

Cho, S. I. & Lee, S. B. (2008). Fast electrochemistry of conductive polymer nanotubes: Synthesis, mechanism, and application, *Acc. Chem. Res.* 41(6):699-707.

Chtanko, N. *et al.* (2004). Etched single-ion-track templates for single nanowire synthesis, *The Journal of Physical Chemistry B* 108(28): 9950.

Chung, H.-J. *et al.* (2005). Cobalt-polypyrrole-cobalt nanowire field-effect transistors, *Applied Physics Letters* 86(21): 213113.

Dong, L. *et al.* (2005). Dielectrophoretic manipulation and electrical characterization of gold nanowires, *Nanotechnology* 16: 1500–1505.

Duvail, J. L. *et al.* (2002). Charge carrier transport in organic semiconductors, *Synthetic Metals* 131: 123.

Duvail, J. L. *et al.* (2004). Effects of the confined synthesis on conjugated polymer transport properties, *The Journal of Physical Chemistry B* 108(48): 18552–18556.

Duvail, J. L. *et al.* (2007). Tuning electrical properties of conjugated polymer nanowires with the diameter, *Applied Physics Letters* 90(10): 102114.

Duvail, J. L. *et al.* (2008). Physical properties of magnetic metallic nanowires and conjugated polymer nanowires and nanotubes, *International Journal of Nanotechnology* 5(6): 838–850.

Gence, L. *et al.* (2007). Structural and electrical characterization of hybrid metal-polypyrrole nanowires, *Physical Review B (Condensed Matter and Materials Physics)* 76(11): 115415.

Gence, L. *et al.* (2008). Size related transport mechanisms in hybrid metal-polymer nanowires, *Physica Status Solidi (a)* 205(6): 1447–1450.

Guo, Y. *et al.* (2008). Light-controlled organic/inorganic p-n junction nanowires, *Journal of the American Chemical Society* 130: 9198.

He, H. X. (2003). Discrete conductance switching in conducting polymer wires, *Phys. Rev. B* 68(4): 045302.

He, H. X., Li, C. Z. & Tao, N. J. (2001). Conductance of polymer nanowires fabricated by a combined electrodeposition and mechanical break junction method, *Applied Physics Letters* 78(6): 811–813.
Heremans, J. P. *et al.* (2002). Thermoelectric power of bismuth nanocomposites, *Phys. Rev. Lett.* 88(21): 216801.
Heremans, J. P. *et al.* (2003). Resistance, magnetoresistance, and thermopower of zinc nanowire composites, *Phys. Rev. Lett.* 91(7): 076804.
Hernandez, R. M. *et al.* (2004). Template fabrication of protein-functionalized gold-polypyrrole-gold segmented nanowires, *Chemistry of Materials* 16(18): 3431.
Huang, J., Wei, Z. & Chen, J. (2008). Molecular imprinted polypyrrole nanowires for chiral amino acid recognition, *Sensors and Actuators B: Chemical* 134(2): 573 - 578.
Hurst, S. J. *et al.* (2006). Multisegmented one-dimensional nanorods prepared by hard template synthetic methods, *Angew. Chem. Intl. Ed.* 45(17): 2672–2692.
Jang, J. (2006). Conducting polymer nanomaterials and their applications, *Adv Polym Sci.* 199: 189–259.
Jerôme, C. & Jerôme, R. (1998). Electrochemical synthesis of polypyrrole nanowires, *Angew. Chem. Intl. Ed.* 37(18): 2488.
Kovtyukhova, N. *et al.* (2004). Coaxially gated in-wire thin-film transistors made by template assembly, *Journal of the American Chemical Society* 126(40): 12738.
Lahav, A. *et al.* (2006). Core-shell and segmented polymer-metal composite nanostructures, *Nano Letters* 6(9): 2166.
Lee, J. I. *et al.* (2008). Highly aligned ultrahigh density arrays of conducting polymer nanorods using block copolymer templates, *Nano Letters* 8(8): 1530.
Li, J. & Lin, X. (2007). Glucose biosensor based on immobilization of glucose oxidase in poly(o-aminophenol) film on polypyrrole-Pt nanocomposite modified glassy carbon electrode, *Biosensors and Bioelectronics* 22(12): 2898–2905.
Liang, L. *et al.* (2002). Highly diastereo- and enantioselective intramolecular amidation of saturated C-H bonds catalyzed by ruthenium porphyrins, *Angewandte Chemie International Edition* 41(18): 3465.
Lieber, C. M. & Wang, Z. L. (2007). Functional nanowires, *MRS Bulletin* 32(3): 99.
Lim, J.-H. & Mirkin, C. A. (2002). Electrostatically driven dip-pen nanolithography of conducting polymers, *Advanced Materials* 14(20): 1474–1477.
Lim, J. K. *et al.* (2008). Actuation of self-assembled two-component rodlike nanostructures, *Nano Letters* 8(12): 4441.
Long, Y. *et al.* (2005). Electronic transport in single polyaniline and polypyrrole microtubes, *Phys. Rev. B* 71(16): 165412.
Lorcy, J.M. *et al.* (2009). Coaxial nickel/poly(p-phenylene vinylene) nanowires as luminescent building blocks manipulated magnetically, *Nanotechnology* 20(40): 405601.
Mailley, P., Roget, A. & Livache, T. (2005). Conducting polymers for DNA sensors and DNA chips: from fabrication to molecular detection, Persp. In Bioanalysis 1: 297–330.
Martin, C. R. (1994). Nanomaterials: A membrane-based synthetic approach, *Science* 266(5193): 1961–1966.
Martin, C. R. (1996). Membrane-based synthesis of nanomaterials, *Chemistry of Materials* 8(8): 1739.
Massuyeau, F. *et al.* (2009). Elaboration of conjugated polymer nanowires and nanotubes for tunable photoluminescence properties, *Nanotechnology* 20(15): 155701.
Merlo, J. A. & Frisbie, C. D. (2004). Field effect transport and trapping in regioregular polythiophene nanofibers, *J. Phys. Chem. B* 108:19169.

Mott, N. F. & Davis, E. A. (1979). *Electronic Processes in Non-Crystalline Materials*, Clarendon Press, Oxford.

Nalwa, H. S. (ed.) (2006). *Polymeric Nanostructures and Their Applications*, American Scientific Publishers.

Park, S. *et al.* (2004). Hybrid organic-inorganic, rod-shaped nanoresistors and diodes, *Journal of the American Chemical Society* 126: 11772.

Pinto, N. J. *et al.* (2003). Electrospun polyaniline/polyethylene oxide nanofiber field-effect transistor, *Applied Physics Letters* 83(20): 4244–4246.

Pinto, N. J. *et al.* (2009). Rectifying junctions of tin oxide and poly(3-hexylthiophene) nanofibers fabricated via electrospinning, *Applied Physics Letters* 94(8): 083504.

Pokroy, B. *et al.* (2009). Self-organization of a mesoscale bristle into ordered, hierarchical helical assemblies, *Science* 323(5911): 237–240.

Reneker, D. H. *et al.* (2000). Bending instability of electrically charged liquid jets of polymer solutions in electrospinning, *Journal of Applied Physics* 87(9): 4531–4547.

Reynes, O. & Demoustier-Champagne, S. (2005). Template electrochemical growth of polypyrrole and gold-polypyrrole-gold nanowire arrays, *Journal of The Electrochemical Society* 152(9): D130–D135.

Sadki, S. *et al.* (2000). The mechanisms of pyrrole electropolymerization, *Chem. Soc. Rev.* 29: 283.

Sakmeche, N. *et al.* (1996). Anionic micelles: a new aqueous medium for electropolymerization of poly(3,4-ethylenedioxythiophene) films on Pt electrodes, *Chem. Commun.* pp. 2723–2724.

Shen, J. *et al.* (2006). Electrical properties of a single electrochemically template-synthesized polypyrrole nanowire, *Applied Physics Letters* 88(25): 253106.

Smith, P. A. *et al.* (2000). Electric-field assisted assembly and alignment of metallic nanowires, *Applied Physics Letters* 77(9): 1399–1401.

Tran, H. D., Li, D.& Kaner, R. B. (2009). One-dimensional conducting polymer nanostructures: Bulk synthesis and applications, *Advanced Materials* 21(1): 1–13.

Tseng, R. J. *et al.* (2005). Polyaniline nanofiber/gold nanoparticle nonvolatile memory, *Nano Letters* 5(6): 1077.

Vlad, A. *et al.* (2009). Highly ordered conjugated polymer nanoarchitectures with three-dimensional structural control, *Nano Letters* 9(8): 2838 - 2843.

Wan, M. (2008). A template-free method towards conducting polymer nanostructures, *Advanced Materials* 20(15): 2926–2932.

Wang, J. G. *et al.* (2004). Microstructure and interdiffusion of template-synthesized Au/Sn/Au junction nanowires, *Nano Letters* 4(7): 1313.

Wang, Y. *et al.* (2007). Individually addressable crystalline conducting polymer nanowires in a microelectrode sensor array, *Nanotechnology* 18(42): 424021.

Xiao, R. *et al.* (2007). Controlled electrochemical synthesis of conductive polymer nanotube structures, *Journal of the American Chemical Society* 9(3): 40–50.

Yoon, C. O. *et al.* (1994). Transport near the metal-insulator transition: Ppy doped with PF_6, *Phys. Rev. B* 49(16): 10851–10863.

Yoon, H., Hong, J.-Y. & Jang, J. (2007). Charge-transport behavior in shape-controlled poly(3,4-ethylenedioxythiophene) nanomaterials: Intrinsic and extrinsic factors, *Small* 3(10): 1774–1783.

Yunus, S., Attout, A. & Bertrand, P. (2009). Controlled aniline polymerization strategies for polyaniline micro- and nano self-assembling into practical electronic devices, *Langmuir* 25(3): 1851.

13

Photoferroelectric Nanowires

Marian Nowak
Silesian University of Technology
Poland

1. Introduction

The presented in (Gruverman & Kholkin, 2006; Scott, 2006) advances in the quickly developing field of nanoscale ferroelectrics are made because they are important for many applications as well as for the fundamental physics questions. This review summarizes results of investigations of nanowires of antimony sulfoiodide (SbSI) -type materials (i.e. SbSI, antimony selenoiodide (SbSeI), and antimony sulfoselenoiodide ($SbS_xSe_{1-x}I$)). This class of materials represents the semiconducting ferroelectrics (Fridkin, 1980; Gerzanich et al., 1982; Dittrich et al., 2000) known also as photoferroelectrics (Fridkin, 1979). Since photons in a semiconductor generate excess free carriers, and induce a change of its electronic state, one may observe a lot of interesting phenomena in these materials. Obviously, the presented new materials as the other one-dimensional semiconductor nanostructures (Xia et al., 2003) should receive considerable attention from the scientific and engineering communities due to their potentially useful novel electronic and optical properties.

The first description of the synthesis of SbSI was given almost two centuries ago (Henry & Garot, 1824) but the crystal structure of this and the other ternary chalcohalides formed from the group 15-16-17 elements was established much later (Dönges, 1950). However, the intensive investigation of SbSI started after discoveries of its photoconductivity (Nitsche & Merz, 1960) and its ferroelectric properties (Fatuzzo et al., 1962). An unusually large number of interesting properties of SbSI has been found. Among them there are the pyroeletric, pyro-optic, piezoeletric, electromechanical, electrooptic and nonlinear optical effects. Due to these properties it is an attractive and suitable material for thermal imaging, light modulator, ferroelectric field effect transistor, gas sensors, piezoelectric elements used in certain types of electromechanical sensors and actuators, temperature auto stabilized nonlinear dielectric elements (TANDEL), time-controlling devices and other applications (see e.g. Refs. in (Nowak et al., 2008; Nowak et al., 2009d). The SbSI is also taken into consideration as a valuable material for photonic crystals (see Refs. in (Nowak et al., 2008)).

It should be noted that quaternary compounds formed as solid solutions from the group 15-16-17 elements possess additional very interesting feature: their energy band gaps and physical properties are tailored with stoichiometric composition. For example, in $SbS_{1-x}Se_xI$ mixed crystals the strong monotonous decrease of the Curie temperature with the increase of Se content had been observed (Nitsche et al., 1964).

Being a promising material with potential applications, SbSI-type materials were synthesized in a variety of ways and prepared in different forms: bulk crystals, polycrystalline samples, ceramics, and thin-films (see e.g. Refs. in (Nowak et al., 2008;

Nowak et al., 2010b). The attempts have been made to produce SbSI nanocrystals. The SbSI quantum dots were synthesized in the Na_2O-B_2O_3-SiO_2 organic modified silicates matrix by the sol-gel technique (Yuhuan et al., 1999). Nanocrystals of SbSI were produced also in organically modified TiO_2 glass (Hui et al., 2000; Yuhuan et al., 2001; Hui et al., 2002). The SbSI nanorods were prepared by hydrothermal method (Wang et al., 2001). It should be mentioned that the SbSI nanocrystals were obtained by ball milling (Gomonnai et al., 2003; Voynarovych et al. 2003; Gomonnai et al. 2004), too. Recently, the ultrasound irradiation was applied to induce the 1D growth of nanowires of ternary and quaternary chalcohalides formed from the group 15-16-17 elements. Up to now this sonochemical technique was established for direct preparation of SbSI (Nowak et al. 2008), SbSeI (Nowak et al. 2009b), and $SbS_{1-x}Se_xI$ (Nowak et al. 2010b) nanowires. It was also applied for growing SbSI in multi-walled carbon nanotubes (CNTs) (Nowak et al. 2009c).

2. Sonochemical preparation of photoferroelectric nanowires

In sonochemistry powerful ultrasound is used to stimulate chemical reactions and physical changes in liquids. It is successfully applied to produce nano-structured metals, alloys, oxides, carbides and sulfides, or nanometer colloids (Gedanken, 2004). Ultrasound irradiation can be used at room temperature and ambient pressure to promote heterogeneous reactions that normally occur only under extreme conditions of hundreds of atmospheres and degrees (Li et al., 1999). An acoustic pressure wave consists of alternate compressions and rarefactions in the transmitting medium along the wave propagation direction. When a large negative pressure is applied to a liquid, intermolecular van der Waals forces are not strong enough to maintain cohesion and small cavities or gas-filled microbubbles are formed. The rapid nucleation, growth and implosive collapse of these micrometer-scale bubbles constitutes the phenomenon of cavitation. According to the thermal "hot spot" theory, extreme local temperatures and pressures are produced inside the cavitating bubbles and at their interfaces when they collapse. The effective temperature of the resulting transient, local "hot spots" was estimated to be in the range of 5200±650 K (Suslick et al., 1986). Assuming such value, the pressure during collapse, as inferred from the van der Waals equation, would be approximately 1700 atm. These exceptional local conditions can be used to generate nanostructured SbSI-type materials.

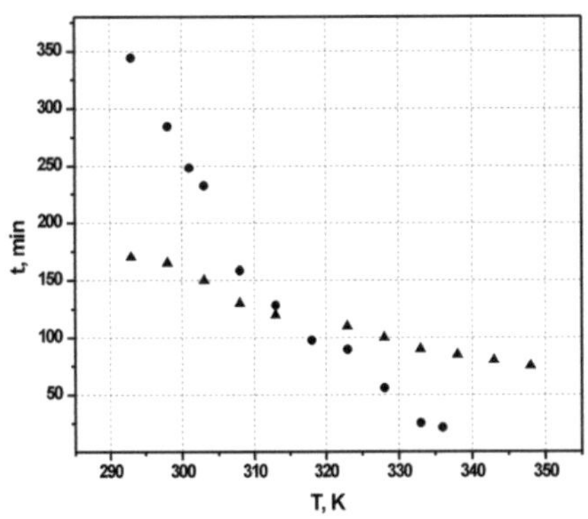

Fig. 1. Comparison of duration of the sonochemical synthesis of SbSI in ethanol (▲) and methanol (●) as a function of temperature of the bath (after (Starczewska et al., 2009))

The SbSI, SbSeI, and $SbS_{1-x}Se_xI$ were prepared sonically from the constituents (the elements antimony, sulfur, selenium and iodine). Ethanol or methanol served as the solvent for this reaction. In a typical procedure, the elemental mixture with appropriate stoichiometric ratio and with total mass of a few grams was immersed at room temperature and ambient pressure in 8-40 ml of alcohol, which was contained in a 54 ml Pyrex glass cylinder of 20 mm inside diameter. The vessel was closed during the experiment to prevent volatilization of the precipitant in long time tests. The cylinder was partly submerged in water in an ultrasonic reactor (frequency 35 kHz, with 80 W electrical power and 2 W/cm^2 power density guaranteed by the manufacturer). The sonolysis was carried out at various temperatures (Fig. 1) but usually at 323 K. The level of the alcohol slurry inside the tube was the same as that of the water in the sonication bath, in order to obtain reproducible sonochemical yields.

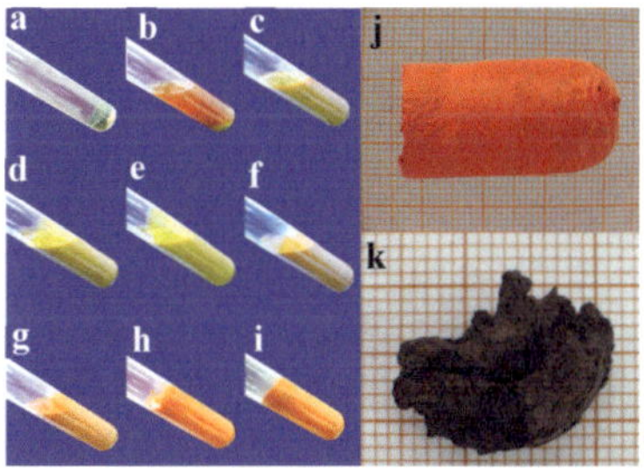

Fig. 2. Change of color and consistence during the sonication of Sb, S and I in ethanol (a-i) and the obtained SbSI (j) as well as SbSeI (k) xerogels (a- dry elements before the process; b- test-tube with the elements in ethanol at the beginning of the process; c- after 20 sec. of sonication, T=323 K; d- 3 min.; e- 6 min.; f- 26 min.; g- 48 min.; h- 75 min.; i- SbSI ethanogel solidified after 110 min.) (after (Starczewska et al., 2008; Nowak et al., 2010c))

During the sonications sols were formed (Fig. 2). The oxidation-reduction potential Eh=0.15 V and negative logarithm of the activity of the hydrogen ion pH=1.2 of the Sb-S-I-ethanol sol after 20 minutes of sonication was measured (Nowak et al., 2010b). It was observed that the color of the slurry changed gradually indicating the growth process of the SbSI, SbSeI and $SbS_{1-x}Se_xI$. For example, in the case of SbSI the color changed from red (before sonication) into olive, green, yellow and then into red-orange after 45 min of sonication, indicating the growth process of SbSI nanorods (Fig. 2). To control this process, measurements of optical diffusive reflectance $R_d(\lambda)$ of the sample were performed. It was assumed that the sonochemical process is finished when the spectral characteristics of $R_d(\lambda)$ do not change with time. It is noteworthy that the SbSI, SbSeI and $SbS_{1-x}Se_xI$ sols converted into gels after relatively short sonications (see e.g. Fig. 1). Their colors depended on the molar composition and changed from red-orange for SbSI to brown for SbSeI (Fig. 2). The time necessary for complete gelation depends on the molar composition, the used solvent, and the temperature of the water in the ultrasonic bath (see e.g. Fig. 1). The temperature dependences of acoustic impedances do not allow to explain the differences in the observed times of ultrasonic synthesis of SbSI gels in ethanol and methanol. Probably, these differences are due to different temperature dependences of solubilities of the components in the growth of SbSI nanowires in ethanol and methanol.

When the alcohol was evaporated from the SbSI or SbSeI gels during the drying, the relatively big (of a few cm^3 volume), rigid but brittle samples of a so-called xerogels were obtained (see e.g. Fig. 2). During this process the observed SbSeI syneresis was a few times larger (Nowak et al., 2009b) than the about 40 % decrease of volume of the dried SbSI ethanogel (Nowak et al., 2008). It means that the SbSeI ethanogel is much less rigid than the SbSI ethanogel. The SbSeI xerogel is also more brittle than the SbSI xerogel. The masses of the samples after drying were practically equal to the sum of masses of the elements used for the sonochemical synthesis.

Recently (Nowak et al., 2010b), to extract the eventual remaining substrates the $SbS_{1-x}Se_xI$ gels were eight times rinsed with ethanol and centrifuged. Each time the liquid above the sediment was replaced with pure ethanol to wash the precipitates. At the end the centrifuged product was covered by colorless ethanol. As in the cases of sonochemical preparation of other nanocrystalline products, e.g. SbSI and SbSeI, this alcohol was evaporated in air at 313 K, so the $SbS_{1-x}Se_xI$ xerogels were obtained.

In (Nowak et al., 2009c) the SbSI was prepared ultrasonically in CNTs from the constituents (the elements Sb, S and I). Methanol served as the solvent for this reaction. In a typical procedure, the elemental mixture with stoichiometric ratio of e.g. 0.380 g Sb, 0.099 g S and 0.394 g I, was immersed with 0.282 g of CNTs in 40 ml absolute methanol. The used experimental set up and the applied procedure were the same as the described for sonochemical preparation of alone SbSI-type nanowires. The sonication was continued for 3 h at 323 K. When it was finished a dark sol was obtained. It was centrifuged to extract the products. Then the liquid above the sediment was replaced with pure methanol to wash the precipitates. The centrifugation and washing were performed 5 times. At the end methanol was evaporated from the sample and a brown-purple CNTs filled with SbSI were obtained.

3. Mechanism of sonochemical preparation of photoferroelectric nanowires

3.1 Process route of SbSI-type nanowires growing

The transient high-temperature and high-pressure field produced during ultrasound irradiation provide a favorable environment for the 1D growth of the SbSI-type nanocrystals, though the bulk solution surrounding the collapsing bubbles is at ambient temperature and atmospheric pressure. It seems that $SbS_{1-x}Se_xI$ as well as SbSI and SbSeI belong to the many solid materials (Xia et al., 2003) that naturally grow into 1D nanostructures, and this habit is determined by the highly anisotropic bonding in the crystallographic structure. The probable reaction route of $SbS_{1-x}Se_xI$ synthesis and the mechanism of formation of its nanowires using elemental Sb, S, Se and I in the presence of ethanol under ultrasonic irradiation can be summarized as follows (Nowak et al., 2010b):

1. iodine, I_2, dissolved in ethanol reacts with antimony and forms the antimony triiodide, SbI_3, also dissolved in ethanol

$$2Sb+3I_2 \rightarrow 2SbI_3 \tag{1}$$

2. dehydrogenation, dehydration as well as decomposition of ethanol in or close to the cavitation bubbles leads to the formation of hydrogen and water as the main products (Gutiĕrrez & Henglein, 1988; Mizukoshi et al., 1999)

$$C_2H_5OH \leftrightarrow C_2H_4O+H_2,\ C_2H_5OH \leftrightarrow C_2H_4+H_2O,\ C_2H_5OH \leftrightarrow C_2H_5O+H \tag{2}$$

3. the sonolysis of water yields the H• and OH• radicals

$$H_2O\ )))\ H^{\bullet}+OH^{\bullet} \tag{3}$$

4. the ultrasonic irradiation facilitates the reduction of chalcogens (S and Se) to the active forms of S^{-2} and Se^{-2} (see e.g. Refs. in (Li et al., 1999; Gedanken, 2004)) that react with the in-situ generated H• radicals forming H_2S and H_2Se (Li et al., 2003)

$$S+2H^{\bullet}\rightarrow H_2S,\ Se+2H^{\bullet}\rightarrow H_2Se \quad (4)$$

5. the released H_2S and H_2Se react with SbI_3 to yield SbSI and SbSeI molecules

$$SbI_3+H_2S\rightarrow SbSI+H_2+I_2 \quad (5a)$$

$$SbI_3+H_2Se\rightarrow SbSeI+H_2+I_2 \quad (5b)$$

6. the created SbSI and SbSeI molecules, under the microjets and shockwaves formed at the collapse of the bubbles are pushed towards each other and are held by chemical forces. Therefore, the nuclei of SbSI, SbSeI and $SbS_{1-x}Se_xI$ are formed as a result of the interparticle collisions (see e.g. (Gedanken, 2004));
7. the freshly formed nuclei in the solution are unstable and have the tendency to connect with each other and self-assemble to form double chain-type structures. These $[(SbSI)_\infty]_2$, $[(SbSe)I)_\infty]_2$ or $[(SbS(Se)I)_\infty]_2$ structures consist of two chains related by a two-fold screw axis and linked together by a short and strong Sb–S (Gerzanich et al. 1982) or Sb-Se bonds (Voutsas & Rentzeperis, 1986). High temperature, local turbulent flow associated with cavitation, and acoustic streaming greatly accelerate mass transport in the liquid phase and are favorable for the self-assembly of the SbSI, SbSeI and $SbS_{1-x}Se_xI$ nuclei;
8. the SbSI, SbSeI or $SbS_{1-x}Se_xI$ chains can be readily crystallized into 3D lattice of nanowiskers through van der Waals interactions. Induced by this structure, crystallization tends to occur along the c-axis, favoring the stronger covalent bonds over the relatively weak, inter-chain van der Waals forces (Molnar et al., 1965). Thus, this solid material has a tendency to form highly anisotropic, 1D structures;
9. the aggregated SbSI, SbSeI or $SbS_{1-x}Se_xI$ nanowires produce larger species. Ultrasound can also promote chemical reaction and crystal growth by mixing heterogeneous phases involving the dispersion of an insoluble solid reactant, e.g. SbSI, in a liquid medium. During the sonication time, the surface state of the nanowires might change: the dangling bonds, defects, or traps decrease gradually, and the species grow until the surface state becomes stable; surface corrosion and fragmentation by ultrasound irradiation affect the formation of regular nanowires.

Compounds of low volatility, which are unlikely to enter cavitation bubbles, experience a high-energy environment resulting from the pressure changes associated with the propagation of the acoustic wave or with shock waves; or they can react with radical species generated by sonolysis of the solvent. A nanocrystalline product is expected if the reaction takes place at the interface (Suslick et al., 1986). In the presented case, the reagents Sb, S and Se are much less volatile than the ethanol and the iodine, so they stay in the interfacial region of the cavitation bubbles to yield SbSI, SbSeI or $SbS_{1-x}Se_xI$ nanowires. The fine crystallinity of the products, which was confirmed by the HRTEM and SAED results (chapter 4), strongly supports this hypothesis.

The reactions (5a) and (5b), describing the synthesis of SbSI, SbSeI and $SbS_{1-x}Se_xI$, were used in different methods of preparation of bulk SbSI-type crystals, i.e.: by reaction of dry H_2S with dry heated SbI_3, by passing gaseous H_2S through the solution of SbI_3, by mixing H_2S

aqueous solutions with SbI_3 solution in HI (modification included the use of thioacetamide to generate in-situ the H_2S), by hydrothermal synthesis with $(NaH_2)_2CS$ or Na_2S used to generate in-situ the H_2S and with Na_2Se used to generate in-situ the H_2Se (see Refs. in (Nowak et al., 2010b)). The reaction (5a) was also used for the preparation of SbSI quantum dots in borosilicate gel (Yuhuan et al., 1999). According to (Popolitov & Litvin, 1970) formation of SbSI and SbSeI chains in solutions with pH<7 may be associated with ceaseless serial linking of hypothetical $[Sb_2I_2(H_2S)_4{}^{4-}]$ or $[Sb_2I_2(H_2Se)_4{}^{4-}]$ type complexes.

Solid solubility is a feature of many metallic and semiconducting systems, being favored when the components have similarities in crystal structure and atomic (ionic) diameter as the SbSI and SbSeI (see e.g. the Refs. in (Voutsas & Rentzeperis, 1986)). Therefore, the homogeneous $SbS_{1-x}Se_xI$ solid solutions exist. It has been established in powder X-ray diffraction of bulk crystals (Spitsyna et al., 1975; Voutsas & Rentzeperis, 1986) that $SbS_{1-x}Se_xI$ is obtained by statistical substitution of S atoms by Se. The fact that $SbS_{1-x}Se_xI$ forms a continuous solid solution has been proved by investigations of its optical transmission (Pouga et al., 1973). It is obvious, that the ultrasound irradiation provides appropriate conditions for statistical substitution of S and/or Se atoms in the $SbS_{1-x}Se_xI$ crystal structure. Of considerable importance is the increase in entropy associated with the formation of a disordered solid solution from the S and Se components random arranged in the matrix.

The sonochemical formation of SbSI nanowires is accompanied with the formation of other species, e.g. iodine can be sonochemically oxidized to triiodide (I_3^-) by OH• radicals produced during cavitation (Entezari & Kruus, 1996). However, the oxidative power of the I_3^- ion is lower than that of iodine molecules, I_2, itself (Lindsjö, 2005). So, the I_3^- seems to be not essential for the presented sonochemical preparation of $SbS_{1-x}Se_xI$.

It is well known that the liquid used during sonochemical preparation of nanomaterials strongly affects the yield of the sonochemical process and the properties of the produced material. The sonication depends on such properties of the liquid as e.g. the viscosity, surface tension, vapor pressure and sound speed. Methanol has greater vapour pressure and evaporation rate than ethanol but smaller decomposition rate during sonolysis (Mizukoshi et al., 1999). However, one should take into account not only the ultrasonical properties of the liquid. It was confirmed experimentally (Lebedev et al., (2003) that the mechanisms of interaction with crystal surface of the ions (e.g. HS^-) solvated e.g. by alcohol molecules and water are fundamentally different, because the chemical properties of the ions in solution can be strongly modified by the surrounding solvation shells. The solute reactivity depends on the following solvent's properties: dielectric constant, dipole moment, molecule polarizability, etc. The variation in the dielectric constant leads to a change in the nucleophilic properties of the reacting ions during bond formation. Such a variation can affect the charge transfer between bonding atoms, the bond length, and the covalent character of bonding (Lebedev et al., (2003).

3.2 Filling of CNTs with SbSI nanowires

As in the case of alone SbSI-type nanowires, the transient high-temperature and high-pressure field produced during ultrasound irradiation provide a favorable environment for the growth of SbSI nanocrystals from elements inside multi-walled CNTs in methanol, though the bulk solution surrounding the collapsing bubbles is at ambient temperature and atmospheric pressure. It was suggested in (Nowak et al., 2009c) that the filling is induced via capillarity. Hence, the reaction route presented above remains valid for this case. Discussion of the crystal habit, the anisotropy of growth kinetics, and the critical role of unusual chain-type structure of SbSI played in the formation of the nanowires was presented above. The application of sonochemistry to prepare CNTs filled with SbSI is also justified by the fact that ultrasonication is often used in an attempt to cut the outer caps of CNTs (Eletskii, 2004).

4. Morphology, composition and crystal structure of SbSI-type nanowires

The scanning electron microscope (SEM) micrographs of SbSI xerogels obtained after drying the SbSI gels sonochemically prepared in ethanol and methanol as well as xerogels of SbSeI and $SbS_{0.5}Se_{0.5}I$ sonochemically prepared in ethanol (Fig. 3) show that morphologies of the SbSI-type xerogels are very similar. In all cases the samples existed as a porous gels composed of nanowires. It was found that the SbSI nanocrystals amounts only 4.7 % of volume of the sonicated SbSI xerogel (Nowak et al., 2008). The surface areas of the sonochemically prepared SbSI and SbSeI were estimated as about 75 m^2/g (Nowak et al., 2008) and 14÷34 m^2/g (Nowak et al., 2009b), respectively.

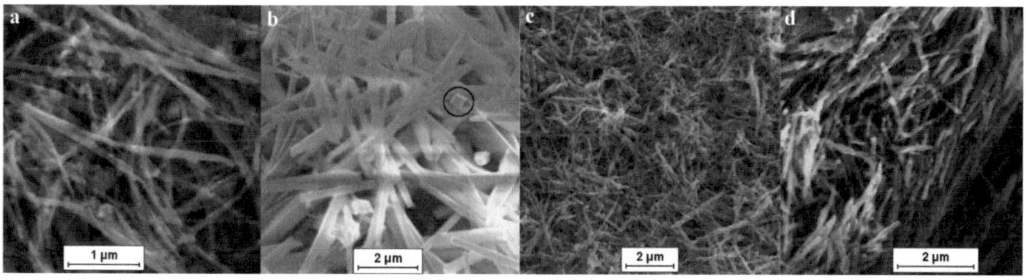

Fig. 3. Typical SEM micrographs of xerogels of SbSI sonochemically prepared in ethanol (a after (Starczewska et al., 2008)) and methanol (b after (Starczewska et al., 2009)) as well as xerogels of SbSeI (c after (Nowak et al., 2009b)) and $SbS_{0.5}Se_{0.5}I$ (d after (Nowak et al., 2010b)) sonochemically prepared in ethanol

The lateral dimensions of the SbSI nanowires produced sonochemically in methanol were found in wider range (10-300 nm (Starczewska et al., 2009)) than the ones of SbSI nanowires in ethanogel (10-50 nm (Nowak et al., 2008)). The SbSeI ethanogel (Nowak et al., 2009b) was composed of nanowires with average lateral dimensions of 20-50 nm. The $SbS_{1-x}Se_xI$ nanowires described in (Nowak et al., 2010b) had lateral dimensions of 10-50 nm. The average lengths of the SbSI, SbSeI, and $SbS_{1-x}Se_xI$ nanowires produced in ethanol and methanol were very similar (the lengths were up to several micrometers). Taking into account the mechanical properties mentioned in the last chapter, it was concluded (Nowak et al., 2009b) that the SbSeI nanowires are connected each other with much weaker forces than the SbSI ones. However, the crosslinking of the nanowires in SbSI-type xerogels still needs evaluation.

Circle in Fig. 3b marks rhomboidal cross section of one of the wider SbSI nanowires. The lateral dimensions of this nanowire are equal 280(4) nm, 263(7) nm, 294(4) nm and 252(3) nm. The angles between the appropriate surfaces are equal 92.7(6)°, 86.3(5)°, 91.3(5)° and 89.7(7)°. If one assumes that the observed nanowire is oriented perpendicularly to the figure, these values correspond to the angles between the (100) and (010) surfaces of bulk SbSI crystals (Starczewska et al., 2009).

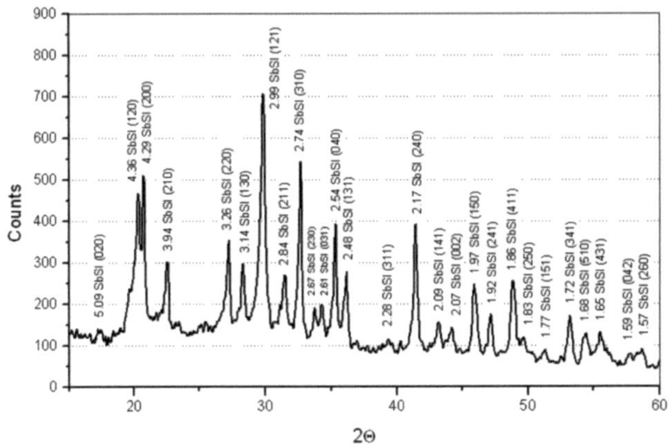

Fig. 4. The powder XRD pattern of orthorhombic phase of dried SbSI ethanogel at T=298 K (after (Nowak et al., 2008)

The reported powder X-ray diffraction (XRD) patterns of the SbSI-type gels prepared by the sonochemical method (see e.g. Fig. 4) indicated that the products were obtained with high purity and well crystallized. The crystal structure of SbSI ethanogel (Nowak et al. 2008) was well described by the *Pnam* crystal symmetry that is appropriate for the paraelectric bulk SbSI. The XRD investigations of SbSI sonochemically produced in methanol showed the coexistence of phases with $Pma2_1$ and *Pnam* crystal symmetry, hence the coexistence of ferroelectric and paraelectric domains (Starczewska et al., 2009). The values of the determined cell parameters are given in Table 1. It should be underlined that these lattice parameters are very close to the data reported for bulk SbSI (Dönges, 1950; Kikuchi et al., 1967). Probably, the XRD data obtained for SbSI methanogel at 298 K represent a structure of just below or very near the transition temperature. The coexistence of the ferroelectric and paraelectric phases of bulk SbSI near the Curie point was discussed in (Starczewska et al., 2009). From the differences in XRD data, it was concluded (Starczewska et al., 2009) that SbSI gels produced sonochemically in methanol and ethanol can have different Curie temperatures. The reason for this can be the differences in solvation processes in ethanol and methanol.

Liquids used during sonication	Phases and cell parameters determined in the case of SbSI nanowires sonicated in different liquids		
	Ethanol [a)]	Methanol [b)]	
Phases	Pnma	*Pnma*	*$Pma2_1$*
a , nm	0.858	0.8587(3)	0.8473(6)
b , nm	1.017	1.0190(6)	1.0478(5)
c , nm	0.414	0.4146(7)	0.4247(8)

Table 1. Phases and cell parameters determined at temperature 298 K for sonochemically prepared SbSI ([a)]- after (Nowak et al., 2008); [b)]- after (Starczewska et al., 2009)

All diffraction peaks in XRD pattern of sonochemical prepared SbSeI (Nowak et al., 2009b) were indexed to be a pure orthorhombic phase for SbSeI with the cell constants a=0.86862(9) nm, b=1.0.3927(3) nm, and c=0.41452(9) nm very close to the data reported for bulk SbSeI crystals (Voutsas & Rentzeperis, 1982). However, the XRD patterns (Fig. 5) of $SbS_{1-x}Se_xI$ xerogels prepared by the sonochemical method (Nowak et al., 2010b) suggested the presence of additional phase. The red vertical line in Fig. 5 indicates the position of the main peak observed in the XRD pattern of antimony subiodide (Sb_3I) described in (Nowak et al., 2010d). Therefore, one can conclude that the Sb_3I can be present in $SbS_{1-x}Se_xI$ samples prepared sonochemically from the elemental Sb, S, Se and I in ethanol.

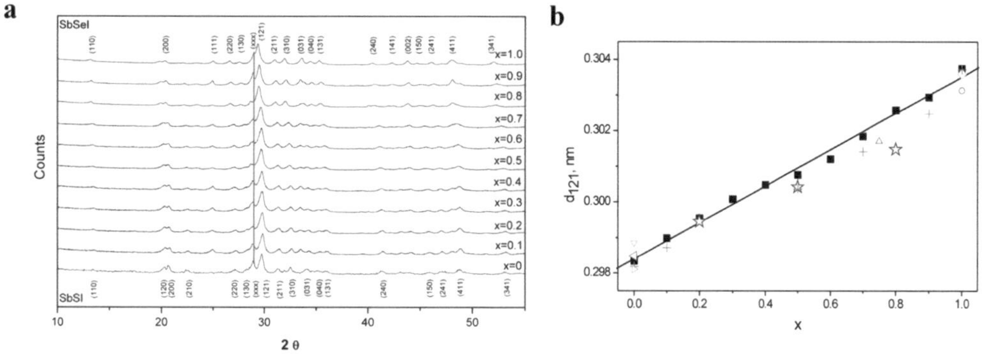

Fig. 5. (a) Influence of molar composition on the powder XRD patterns of orthorhombic phase of $SbS_{1-x}Se_xI$ nanowires (after (Nowak et al., 2010b)); (b) The distance between (121) planes determined from the diffraction peak vs. the Se concentration in $SbS_{1-x}Se_xI$ nanowires (■-(Nowak et al., 2010b); ○- (JCPDS Card 76-1354)); ◇- (JCPDS PDF 01-075-1723); △- (JCPDS PDF 01-072-2366); ◁ - (JCPDS Card 74-2245; JCPDS Card 74-2246; JCPDS Card 74-1195; JCPDS Card 74-1196, 2000; JCPDS Card 74-2244; JCPDS Card 75-0781; JCPDS Card 88-0985); (▽- JCPDS Card 74-2210); ✭- (Spitsyna et al., 1975); + - (Belayev et al., 1970); ▷- (JCPDS Card 21-0050); black line represents the least square fitted linear dependence (6)) (after (Nowak et al., 2010b))

Positions of XRD peaks depend on the molar composition of investigated $SbS_{1-x}Se_xI$ xerogel (Figs. 5a). The distances between (121) planes (determined from the diffraction peaks) were well fitted (Fig. 5b) by a linear function of the molar composition of $SbS_{1-x}Se_xI$ nanowires

$$d_{121}(x)=A_{121}+B_{121}\cdot x \qquad (6)$$

where A_{121}=0.2984(1) nm, B_{121}=0.00511(16) nm. The obtained results are well compared to the bulk values of d_{121} reported for different compositions of $SbS_{1-x}Se_xI$ (Fig. 5b).

Typical transmission electron microscopy (TEM) image of an individual needle-shaped nanowire from sonochemically prepared SbSI-type xerogel is presented in Fig. 6a. The selected area electron diffraction (SAED) pattern (inserted at the upper left corner of Fig. 6a) of this nanowire indicates its good single-crystalline structure appropriate for the orthorhombic structure of bulk SbSI-type crystals (Voutsas & Rentzeperis, 1986). Figure 6b presents TEM image of an individual multiwalled CNT filled sonochemically with SbSI in methanol. The corresponding SAED (Fig. 6b) recorded on the end of this CNT indicates the interplanar spacings appropriate for CNTs as well as SbSI crystals (see Table 2). The TEM of a relatively thick CNT sonochemically filled with SbSI reveal that the product consists of coaxial nanocables (Fig. 6c). The lateral dimensions of the nanocables were in the range from 30 to 200 nm, and their lengths reached up to several micrometers (Nowak et al., 2009c).

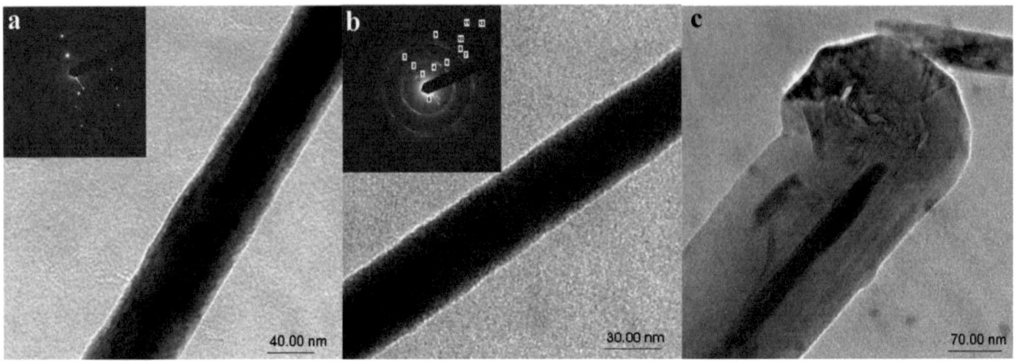

Fig. 6. Typical TEM images of (a) individual nanowire from sonochemically prepared $SbS_{0.75}Se_{0.25}I$ (after (Nowak et al., 2010b)), and relatively thin (b) and thick (c) multiwalled CNTs filled ultrasonically with SbSI in methanol (after (Nowak et al., 2009c)). The corresponding SAED patterns are inserted at the upper left corners of the images. The description of diffraction pattern presented in Fig. 6b is given in Table 2.

Sign	Results of the SAED	Literature data			
		for C [a)]		for SbSI [b)]	
	d_{hkl} (nm)	d_{hkl} (nm)	(hkl)	d_{hkl} (nm)	(hkl)
1 reflex	0.4360	-- --	-- --	0.43402 0.42450	(120) (200)
4 reflex	0.3732	--	--	0.38465	(011)
5 reflex	0.3470	0.33950	(002)	0.35036	(111)
2 reflex	0.2189	0.21390	(100)	0.21663	(330)
6 reflex	0.2089	0.20402	(101)	0.20800	(002)
3 reflex	0.1446	-- --	-- --	0.14244 0.14244	(422) (531)
7 reflex	0.1190	-- --	-- --	0.12031 0.11960	(143) (181)
8 circle	0.1210	0.12350	(110)	--	--
9 circle	0.1155	0.11606 0.11464 0.11316	(112) (105) (006)	-- -- --	-- -- --
10 circle	0.1036	0.10425	(202)	--	--
11 circle	0.07874	0.07954	(122)	--	--
12 circle	0.07038	0.07720	(206)	--	--

Table 2. Comparison of interplanar spacings determined by SAED (Fig. 6B) of multiwalled CNT filled with SbSI ultrasonically in methanol with literature data for CNTs and SbSI bulk crystals ([a)]- after (JCPDS Card 75-1621); [b)]- after (JCPDS Card 74-0149)).

The good fitting of SAED patterns recorded on individual SbSI and SbSeI nanowires with the simulated diffraction patterns (Fig. 7) indicates that the sonochemically produced nanowires exhibit a single-crystalline structure with a preferred growth oriented along the [001] direction. The observed rod type morphology of the SbSI-type nanowires is possibly due to mentioned in chapter 3.1 the inherent chain-type structure and growth habit of them.

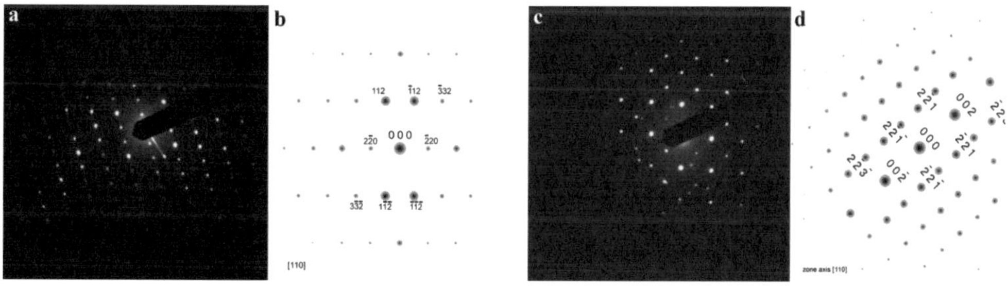

Fig. 7. SAED patterns of individual nanowires from sonochemically made SbSI methanogel (a- after (Starczewska et al. 2009)) and SbSeI ethanogel (c- after (Nowak et al., 2009b)) in the orientation close to the [110] zone axis with their simulated diagrams (b and d).

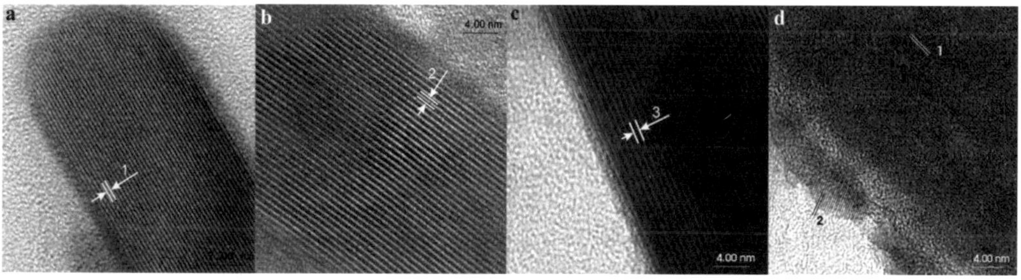

Fig. 8. Typical HRTEM images of individual nanowires from sonochemically prepared SbSI (a), $SbS_{0.75}Se_{0.25}I$ (b), SbSeI (c), and CNT filled with SbSI (d) (description in the text) (after (Nowak et al. 2010b) and (Nowak et al., 2009c))

Material	HRTEM investigations		Literature data for bulk crystals			
	Sign	d_{hkl}, nm	hkl	Material	d_{hkl}, nm	JCPDS Card
SbSI ethanogel	--	0.2962(22) [d)]	(201)	SbSI	0.29540	74-1195
	(1) in Fig. 8a	0.654(8) [a)]	(110)		0.65203	
	--	0.651(4) [b)]				
	--	0.642(2) [c)]				
	--	0.6556(22) [d)]				
	--	0.6569(35) [e)]				
SbSI methanogel	--	0.652(1) [f)]				
	--	0.421(1) [f)]	(001)		0.4100	
	--	0.377(1) [f)]	(011)		0.38005	
$SbS_{0.75}Se_{0.25}I$	(2) in Fig. 8b	0.641(5) [a)]				
SbSeI	(3) in Fig. 8c	0.661(7) [a)]	(110)	SbSeI	0.66648	01-075-1723
	--	0.4467(9) [g)]	(120)		0.4459	01-075-1723
CNT filled with SbSI	(1) in Fig. 8d	0.319(2) [h)]	(220)	SbSI	0.32494	74-1195
	(2) in Fig. 8d	0.209(2) [h)]	(100)	C	0.21390	75-1621

Table 3. Comparison of interplanar spacings evaluated from HRTEM images of SbSI-type nanowires and SbSI in an individual CNT with the data for bulk crystals ([a)] (Nowak et al. 2010b); [b)] (Nowak et al. 2008); [c)] (Starczewska et al. 2008); [d)] (Szperlich et al., 2009); [e)] (Nowak et al., 2009e); [f)] (Starczewska et al. 2009); [g)] (Nowak et al., 2009b); [h)] (Nowak et al., 2009c))

Figure 8 presents typical HRTEM images of sonochemically prepared SbSI-type nanowires and an individual CNT sonochemically filled with SbSI. They exhibit good crystalline and clear (110) lattice fringes parallel to the rod axis appropriate for the orthorhombic structure of bulk crystals. It indicates that the growth velocity in [001] direction is much larger than in the [110] one. Figure 8d shows also the lattice fringes of the CNT walls. Table 3 presents good conformity of the interplanar spacings evaluated from HRTEM images of SbSI-type nanowires with the XRD data of bulk crystals. The difference between d_{110} spacings reported in Table 3 for SbSI can be due to the little different positioning of the nanowires.

Figure 8 reveals that the SbSI-type nanowires are completely surrounded by fuzzy shells of practically constant thickness. The 5.4(2) nm and 6 nm thick fuzzy shells on the surfaces of SbSI nanowires sonochemically produced in ethanol has been reported in (Nowak et al., 2009e) and (Nowak et al., 2008), respectively. Similar but thinner fuzzy shell of about 2,77(8) nm was observed on the surface of SbSI nanowires sonochemically produced in methanol (Starczewska et al., 2009). The SbSeI nanowires sonochemically produced in ethanol had 7.1(2) nm thick fuzzy shell (Nowak et al., 2009b). This is probably due to the amorphous species adsorbed on the surface of the crystalline nanowires during their growing.

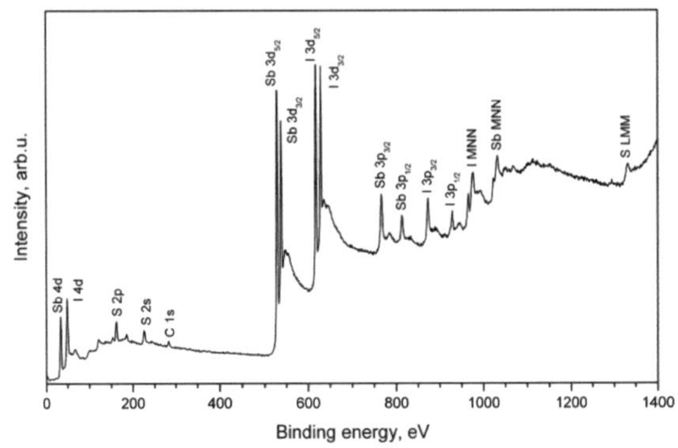

Fig. 9. XPS spectrum of SbSI ethanogel (after (Nowak et al., 2009e))

The presented in (Nowak et al., 2009e) results of the X-ray photoelectron spectroscopy (XPS) described just the surface layer of the SbSI nanowires sonochemically prepared in ethanol. They do not show any traces of O, OH groups, H_2O or other impurities, only a small amount of carbon (C 1s peak at 284.5 eV) is visible (Fig. 9). The carbon element may result from the rest gases in the vacuum chamber and/or can be bonded on the surface during preparation of the sample for XPS (Nowak et al., 2009e).

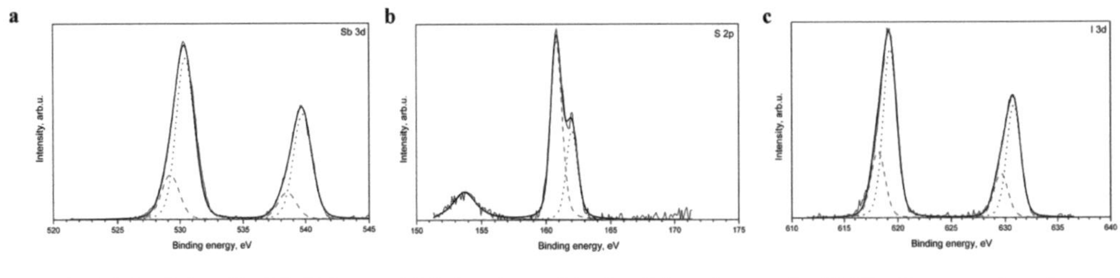

Fig. 10. XPS of the Sb 3d, S 2p and I 3d spin-orbit doublets from the SbSI ethanogel and Gaussian XPS contributions of them (after (Nowak et al., 2009e))

Table 4 shows binding energies of Sb, S and I in SbSI ethanogel prepared sonochemically. The overlapping peaks of S $2p_{3/2}$ and S $2p_{1/2}$ have been fitted with Gaussian functions (Fig. 10). Every line of spin-orbit doublets of the Sb 3d and I 3d states in SbSI nanowires, has been split into two Gaussian components (Fig. 10). Comparing parameters of these components (Table 5) with literature data, one can see that, the chemical shifts of binding energies of peaks in the XPS spectrum of SbSI ethanogel prepared sonochemically are different from those of bulk SbSI.

Material	Binding energy (eV)					
	Sb $3d_{5/2}$	Sb $3d_{3/2}$	S $2p_{3/2}$	S $2p_{1/2}$	I $3d_{5/2}$	I $3d_{3/2}$
SbSI nanowires[a]	530.3	539.7	160.8	162.0	619.1	630.7
powdered SbSI crystals[b]	529.5	538.9	161.2		618.8	630.3
(110) plane of single crystal at 293 K[c]	529	538	161	162	618.2	629.2
Sb_3I nanoparticles[d]	530.7	540.1	--	--	619.5	630.9
Pure elements[e]	528.3	537.6	164.0	165.2	619.3	630.8

Table 4. Comparison of energies of peaks in the XPS spectrum of SbSI ethanogel prepared sonochemically with the data reported for single crystals of SbSI and Sb_3I nanoparticles ([a]- (Nowak et al., 2009e); [b]- (Ikemoto, 1981); [c]- (Grigas et al., 2004); [d]- (Nowak et al., 2010d); [e]- (Moulder et al., 1995); calibrated to C 1s = 284.5 eV).

Material	Line	Chemical shift in comparison with binding energy in pure element (eV)					
		Sb		S		I	
		$3d_{5/2}$	$3d_{3/2}$	$2p_{3/2}$	$2p_{1/2}$	$3d_{5/2}$	$3d_{3/2}$
SbSI ethanogel[a]	Total	+2.0	+2.1	-3.2	-3.2	-0.2	-0.1
	Main line	+2.2	+2.2	-3.1	-3.2	0.0	0.0
	Additional line	+0.9 (0.28)	+0.9 (0.26)			-1.2 (0.39)	-1.2 (0.39)
powdered SbSI crystals[b]		+1.2	+1.26	~(-3.3)		-0.5	-0.5
(110) plane of SbSI crystal[c]		+1		-3		-0.8	
SbSI crystals[d]		+1		-3		-0.8	
SbSI crystals[e]		+0.8		-3		-0.8	
Sb_3I nanoparticles[f]	Total	+2.4	+2.5	--	--	+0.1	+0.1
	Main line	+2.6	+2.6	--	--	+0.2	+0.2
	Additional line HE	+0.8 (0.35)	+0.5 (0.35)	--	--	-1.7 (0.40)	-1.8 (0.40)
	Additional line LE	-1.4 (0.11)	-2.1 (0.11)	--	--	-4.0 (0.10)	-4.1 (0.10)

Table 5. Comparison of the chemical shifts of binding energies of peaks in the XPS spectrum of SbSI ethanogel prepared sonochemically with the data reported for SbSI and Sb_3I (LE – low energy line; HE – high energy line; [a]- (Nowak et al., 2009e); [b]- (Ikemoto, 1981); [c]- (Grigas et al., 2004); [d]- (Grigas et al., 2007); [e]- (Grigas & Talik, 2003); [f]- (Nowak et al., 2010d); In parenthesis there are relative intensities (in comparison with the intensity of the main line)).

The XPS measurements revealed the chemical shift in the SbSI ethanogel of Sb states of +(2.0÷2.1) eV to a higher binding energy, and S states of -3.2 eV and I states of -(0.1÷0.2) eV to a lower binding energy (Table 4 and 5). These shifts suggest a charge transfer from Sb to S and I - more to S than to I. One can see that probably in surface layer of SbSI nanowire the Sb atom transfers more electrons than the Sb in bulk SbSI crystals (Table 5). As a result, the surface layer of SbSI nanowire crystal is more ionic one while in SbSI crystals are less ionic. It means that the bonding environments of the atoms at surfaces of the sonochemically prepared SbSI nanowires and at SbSI single crystals grown from a vapour phase are different.

Taking into account that the splitting of spin-orbit doublets into different components may be induced by the contributions of atoms with different ion charges, the main Gaussian components of each peak in XPS spectra of surface layer of SbSI nanowires (see Fig. 10 and Table 5) were assigned (Nowak et al., 2009e) to the antimony subiodide. Due to relatively good coincidence between chemical shifts observed for the additional lines in XPS of SbSI ethanogel and the data reported for SbSI single crystals (see Table 5), these contributions were recognized as appropriate ones of SbSI (Nowak et al., 2009e).

Element		Sb	S	I	C
Concentration determined in the XPS investigations (at. %)	all detected elements	38	25	28	9
	components without C	42	27	31	--
	components without SbSI and C	15	--	4	--
	components in hypothetical Sb_xI_y	79	--	21	--
Concentration determined in the EDX investigations (at. %)	all elements	36	30	34	--
	components without SbSI	6	--	4	--
	components in hypothetical Sb_xI_y	60	--	40	--

Table 6. Atomic concentration of components of the sonochemically prepared SbSI ethanogel determined by XPS and EDX (after (Nowak et al., 2008e)

The XPS measured (Nowak et al. 2009e) atomic concentrations of Sb, S, I and C in SbSI ethanogel are compared in Table 6 with the results of energy dispersive X-ray (EDX) analysis averaged over the investigated sample (Nowak et al., 2008). Due to the large excess of Sb and the smaller excess of I detected by XPS chemical analysis (Table 6), probably a separate phase of Sb and I is present on the surface of the SbSI. It is improbable that the excess Sb and/or I are present within the ordered nanowires, since the mentioned above interplanar spacing corresponds to stoichiometric bulk SbSI. In fact, the sonochemically produced crystalline SbSI-type nanowires are completely surrounded by fuzzy shells (see e.g. Fig. 8). The growth of these nanowires from elements (Sb, I, and S or Se) in ethanol needs transport of the components to the surface of the growing crystal structure. It is very probable that the transported components are composed of small clusters. Before they are built into the crystal structure of the nanowire, they should be adsorbed on its surface. The composition of such surface layer may be different than the bulk of the SbSI-type nanowires, due to different solubilities, transport parameters and chemical activity of Sb, S and I (as well as their ions and hypothetical compounds) in ethanol. It gives the possible explanation of the difference between the XPS measured concentration of the essential components of the SbSI nanowires (Table 6) and their concentrations in an ideal SbSI crystal (i.e. the 33.33 at % of Sb, S and I).

From the XPS data it was concluded (Nowak et al., 2009e) that the surface layer of SbSI nanowires contains SbSI (the sulfur is mainly inside it) and some additional material - the hypothetical antimony subiodide (Sb_xI_y) with the atomic ratio of Sb:I of about 79:21. However, it should be noticed that this composition is comparable with composition of the sonochemically prepared Sb_3I (Nowak et al., 2010d). Comparison of the XPS data presented in Tables 4 and 5 proves that the last compound can exist in surface layer of sonochemically prepared SbSI nanowires. Ratios of the hypothetical Sb_xI_y to SbSI components determined by XPS and EDX investigations are about 53 % and 20 %, respectively (Table 6). However, one should have in mind that the total XPS and EDX signals contain different contributions from the surface layer and bulk of the investigated material. Essentially, the XPS gives better information on the surface properties than the EDX technique.

The EDX (Fig. 11b) measured atomic concentrations of 0.46:0.21:0.33 of Sb, S and I in SbSI grown in CNT (Nowak et al., 2008c) are very similar to the concentrations determined by XPS in sonochemically prepared SbSI ethanogel (Table 6). It seems that the additional antimony subiodide phase exists between the CNT walls and the grown inside SbSI. Due to the large excess of Sb and the deficit of Se detected by EDX (the atomic ratio of 0.41:0.26:0.33 for Sb, Se and I (Fig. 11a)), probably the same separate phase of Sb and I is present and presumably it is adsorbed on the surface of the SbSeI nanowires (Nowak et al., 2009b). It should be underlined that the EDX analysis performed on SbSI methanogel with larger lateral dimensions of nanowires and about twice thinner fuzzy shells confirmed an elemental atomic ratio of 0.35:0.34:0.31 for Sb, S and I averaged over the sample (Starczewska et al. 2009).

The EDX analysis (Fig. 12) performed on individual nanowire of $SbS_{0.75}Se_{0.25}I$ with very thin fuzzy shell confirmed an elemental atomic ratio of 0.33:0.28:0.07:0.32 for Sb, S, Se and I (Nowak et al., 2010b). The characteristic peaks of Cu in Fig. 12 have their origin in sample support.

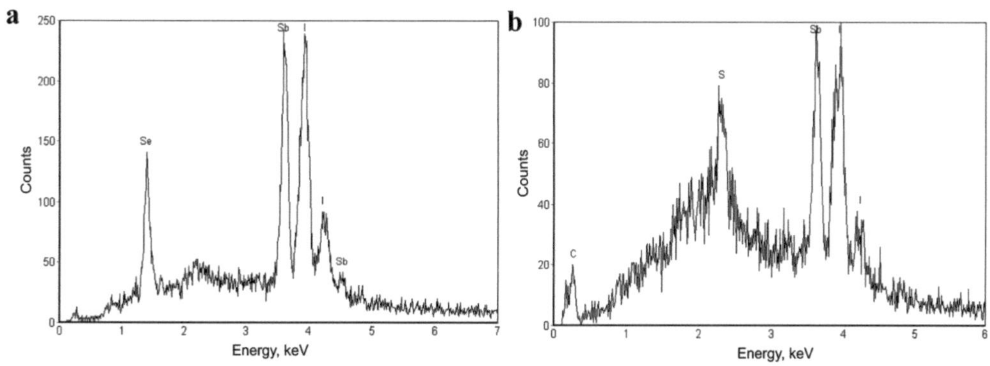

Fig. 11. The EDX spectra of SbSeI xerogel sonochemically prepared in ethanol (a after (Nowak et al., 2009b)) and of dried multiwalled CNTs filled sonically in methanol with SbSI (b after (Nowak et al., 2009c))

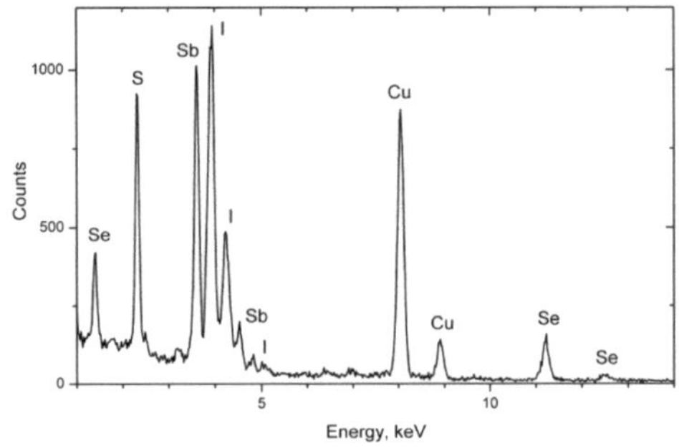

Fig. 12. The EDX spectrum of an individual nanowire from sonochemically prepared $SbS_{0.75}Se_{0.25}I$ (after (Nowak et al., 2010b)

5. Optical properties of photoferroelectric nanowires

5.1 Diffuse reflectance and optical energy gap of SbSI-type nanowires

The spectrum of diffuse reflectance signal R_d (the ratio of intensities of radiation reflected in diffuse manner from the sample and from the known standard) of SbSI xerogel is presented in Fig. 13. Usually the R_d values are converted to the Kubelka-Munk function (Philips-Invernizzi et al., 2001) proportional to the absorption coefficient of light (α)

$$F_{K-M}(R_d) = \frac{(1-R_d)^2}{2R_d} \sim \alpha \tag{7}$$

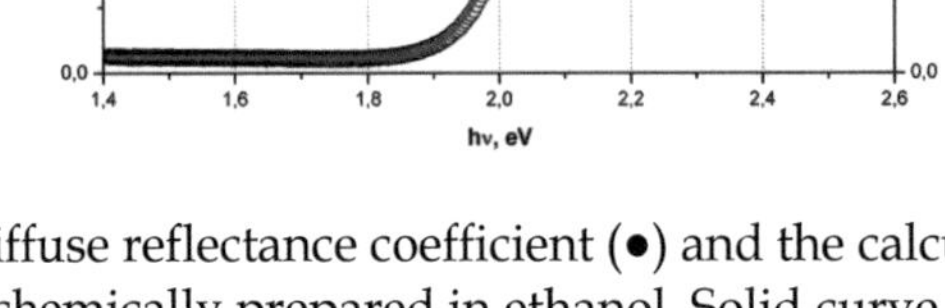

Fig. 13. Spectra of the diffuse reflectance coefficient (•) and the calculated F_{K-M} function (O) of SbSI nanowires sonochemically prepared in ethanol. Solid curve represents the fitted theoretical dependence (values of the fitted parameters are given in Table 7) (after (Nowak et al., 2009a))

It has been proved (Nowak et al., 2009a) that the best method of determining energy band gap (E_g) is based on simultaneous fitting of many mechanisms of absorption to the spectral dependence of F_{K-M} function evaluated from the diffuse reflectance spectroscopy (DRS). This method allows determining not only E_g and the main mechanism of absorption but also gives information on the coexisting phenomena. Therefore, the spectral dependence of the F_{K-M} function should be fitted by minimization of the following least square function

$$\chi^2 = \sum_{i=1}^{n} [(F_{KM}(h\nu_i) - B\sum_j \alpha_j(h\nu_i)]^2 \tag{8}$$

where i represents photons of different energy, α_j describes various mechanisms of light absorption, B is the proportionality factor.

Figure 13 presents the spectrum of F_{KM} of the investigated SbSI nanowires and the least square fitted theoretical dependence appropriate for the sum of indirect forbidden absorption without excitons and phonon statistics (α_1), Urbach ruled absorption (α_2), and constant absorption term (α_3) (see Refs. in (Nowak et al., 2009a)):

$$\alpha_1 = A_{60}(h\nu - E_{gIf})^3 \qquad \text{for } h\nu > E_{gIf} \tag{9}$$

$$\alpha_2 = A_U \exp\left[\frac{h\nu}{E_U}\right] \tag{10}$$

$$\alpha_3 = A_0 \tag{11}$$

where E_{gIf} represents the indirect forbidden energy gap, E_U is the Urbach energy, A_{60} is constant parameter proportional to the probability of photon absorption due to fundamental indirect forbidden transition in the investigated SbSI nanowires, and A_U is constant parameter proportional to the probability of absorption of photons due to Urbach transitions. The constant absorption term A_0 is an attenuation coefficient that is considered as the sum of the scattering and/or absorption independent of hν near the absorption edge.

	Values obtained for				
Fitted parameters	SbSI nanowires[a)]	SbSI nanowires[b)]	SbSI nanowires[c)]	CNTs filled with SbSI[d)]	SbSI crystals[b)]
E_{gIf}, eV	1.81(1)	1.829(27)	1.854(3)	1.871(1)	1.790(31)
$B\cdot A_{60}$, $1/(eV^3m)$	45.28(24)	54.83(14)	157(1)	95.6(7)	35.20(27)
A_{125} , 10^{12} m^{-2}		--	--	8.24(1)	
A_0, m^{-1}	0.0488(1)	0.0321(13)	0.0213(1)	2.702(3)	0.0141(17)
E_U, eV	0.571(7)	0.1031(16)	0.1470(20)	--	0.1027(43)
$B\cdot A_U$, 10^{-9} m^{-1}	$279(6)\cdot10^3$	0.3570(19)	84.4(3)	--	0.2523(67)

Table 7. Comparison of the values of SbSI ethanogel parameters (a)- (Nowak et al., 2009a); b)- (Nowak et al., 2008)), SbSI methanogel (c)- (Starczewska et al., 2009)), multiwalled CNTs filled with SbSI in methanol (d)- (Nowak et al., 2009c)), and powdered SbSI single crystals (b)) that were determined from the fitting of the spectrum of Kubelka-Munk function evaluated from the measured diffuse reflectance

The fitting presented in Fig. 13 is rather good. Values of the fitted parameters (Table 7) are well compared with other literature data. The value E_{gIf}=1.81(1) eV of the indirect forbidden energy band gap of SbSI nanowires is close to the value evaluated for powdered SbSI single crystals (E_{gIf}=1.79(31) eV (Nowak et al., 2008)). It is also comparable to the bulk values of band gap of SbSI: 1.82 eV at 301 K (Fridkin, 1980) for the light polarized in [001] direction, i.e. in the direction of growth of SbSI nanowires. The E_g for light polarized in perpendicular direction is slightly bigger (Fridkin, 1980). The greater Urbach absorption in the case of SbSI produced in methanol (Table 7) is probable due to the larger amount of crystal defects and additional electron states in its energy gap or due to the coexistence of different phases.

Figure 14a shows the large (about 100 nm) shift to the longer wavelengths of the absorption edge with the increase of Se content in $SbS_{1-x}Se_xI$ nanowires sonochemically prepared in ethanol. Figure 14b presents the rather good fitting of the F_{K-M} spectra with theoretical dependences appropriate for the sum of indirect forbidden absorption without excitons and phonon statistics (9), Urbach ruled absorption (10) and constant absorption term (11). It should be underlined that the same mechanisms of light absorption exist in $SbS_{1-x}Se_xI$ nanowires for any molar composition ($0 \le x \le 1$). The values of indirect forbidden energy gaps determined for the sonochemically produced $SbS_{1-x}Se_xI$ xerogel can be well fitted (Fig. 15) with a linear function of the molar composition

$$E_{gIf}(x) = A + B \cdot x \tag{12}$$

where A=1,852(5) eV, B=-0,210(9) eV (Nowak et al., 2010b). Figure 15 presents also comparison of the obtained results with the literature data for bulk $SbS_{1-x}Se_xI$.

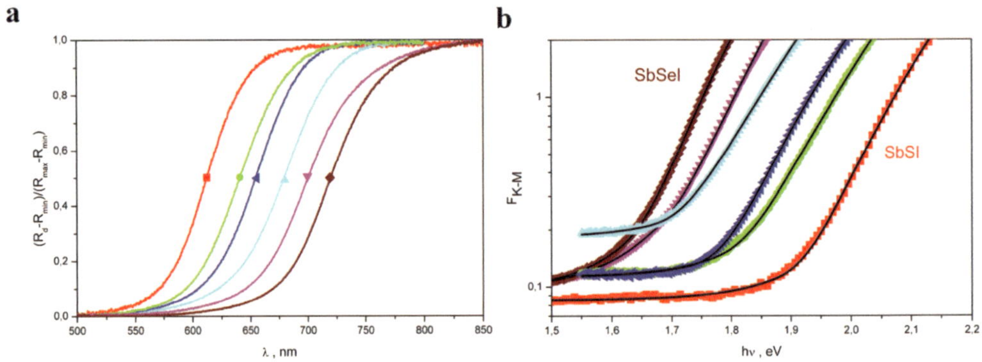

Fig. 14. Comparison of the diffuse reflectance spectra (A) and the calculated spectra of Kubelka-Munk functions (B) of $SbS_{1-x}Se_xI$ nanowires with different molar compositions (■- SbSI, ●- $SbS_{0.8}Se_{0.2}I$, ◄- $SbS_{0.6}Se_{0.4}I$, ▲-$SbS_{0.4}Se_{0.6}I$, ▼- $SbS_{0.2}Se_{0.8}I$, ♦- SbSeI; R_{min}, R_{max} – minimum and maximum value of the coefficient of diffuse reflectance in the investigated spectra range). Solid curves in Fig. 14B represent the fitted theoretical dependences for the sum of indirect forbidden absorption without excitons and phonon statistics, Urbach ruled absorption, and constant absorption term (values of the fitted energy gaps are given in Fig. 15) (after (Nowak et al., 2010b))

In Fig. 16a the diffuse reflectance spectrum of multiwalled carbon nanotubes (CNTs) filled sonochemically in methanol with SbSI is compared with the spectrum registered for hollow CNTs in methanol. In the first case one can see the characteristic, for semiconducting materials, edge of fundamental absorption around 615 nm. However, the diffuse reflectance decreases also with increasing wavelengths (Fig. 16a), probable due to the large amount of free carriers absorbing light. The best fitting of the spectrum of calculated F_{K-M} function was obtained for the sum of indirect forbidden absorption without excitons and phonon statistics (9), constant absorption term (11) and free carrier absorption (α_4) (Nowak et al., 2009c)

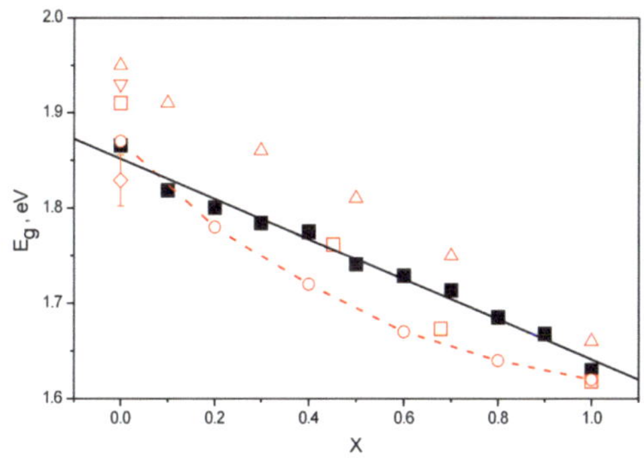

Fig. 15. Energy gap of $SbS_{1-x}Se_xI$ vs. molar composition (■- for nanowires (Nowak et al., 2010b); □- (Belayev et al., 1968); ○-(Park et al., 1990); △,▽- for plane polarized light with electric field perpendicular and parallel to the c-axis, respectively (Turjanica et al., 1969), ◇- for nanowires (Nowak et al., 2008); line represents the least square fitted linear dependence (12)) (after (Nowak et al., 2010b))

$$\alpha_4 = A_{125}\lambda^2 \tag{13}$$

where A_{125} is a constant parameter, and λ is the wavelength of radiation. The fitting presented in Fig. 16b is rather good. The determined values of parameters are given in Table 7. The nanocrystalline SbSI filling the CNTs is a semiconductor with little larger energy gap than the alone SbSI nanowires (Table 7). To test if this shift is due to the quantum size effect (Hui, 2000) the synthesis of SbSI in CNTs of smaller diameters is needed.

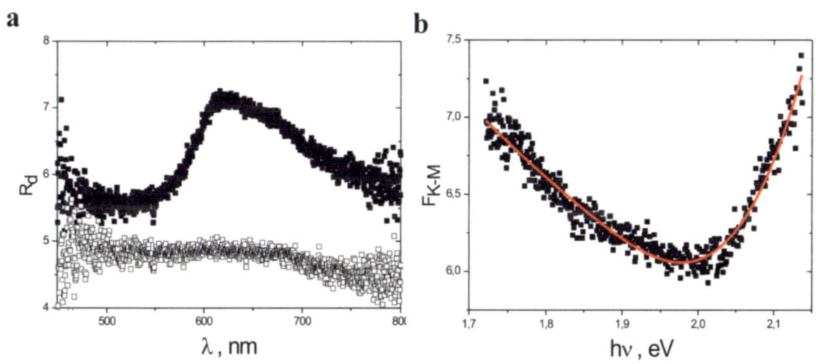

Fig. 16. (a) Comparison of the diffuse reflectance spectra of the multiwalled CNTs filled with SbSI (■) and of the empty multiwalled CNTs (□) in methanol. (b) Fitting of the spectrum of calculated $F_{K\text{-}M}$ function of the multiwalled CNTs filled with SbSI. Solid curve represents the least square fitted theoretical dependence (values of the fitted parameters are given in Table 7) (after (Nowak et al., 2009c)

It is known (see e.g. (Eletskii, 2004)) that a metal atom intercalated inside the internal cavity of a CNT displays a tendency towards the transfer of some part of the valence electrons to the outer surface of the nanotube, where unoccupied electronic states exist. As a result of such a transfer there arises an additional mechanism of electrical conduction, related to the travel of an electron about those states in CNTs. The observed free carrier absorption of light in the case of CNTs filled with SbSI is evoked by the CNTs material because it is absent in the case of alone SbSI nanowires (Table 7).

5.2 Temperature dependence of optical energy gap and spontaneous polarization

In Fig. 17 one can see the shift of fundamental absorption edge to longer wavelengths with increasing temperature of about 30 μm thick film of SbSI ethanogel.

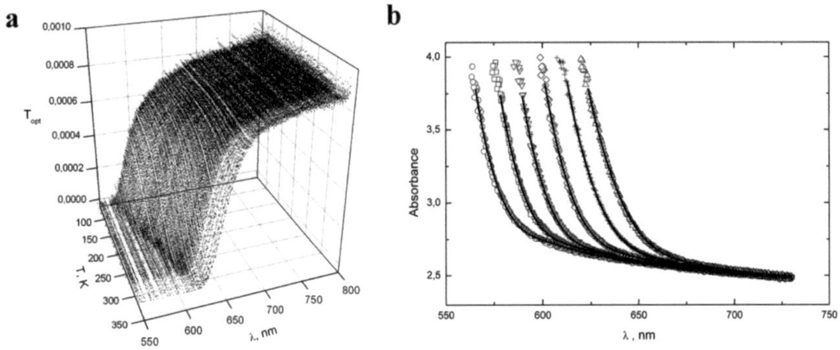

Fig. 17. (a) Influence of temperature on spectral characteristics of optical transmittance of a film of SbSI xerogel. (b) Typical spectra of absorbance of SbSI xerogel at different temperatures (○ - 80 K; □ - 150 K; ▽- 200.1 K; ◇- 250.1 K; +-296.8 K; △- 344.1 K; p=1.33 Pa). Solid curves represent the fitted theoretical dependences (values of the fitted parameters are given in Table 8) (after (Nowak & Szperlich, 2010))

T, K	Fitted parameters				
	E_{gIf}, eV	A_{60}, $1/(eV^3m)$	E_U, eV	A_U, 1/m	A_0, 1/m
80	2.066(5)	405(4)	0.2564(1)	$1.088(1)\cdot10^{-4}$	2.4011(6)
150	2.019(4)	423(3)	0.2329(1)	$5.051(5)\cdot10^{-5}$	2.4160(5)
200.1	1.976(3)	417(2)	0.2063(1)	$1.722(2)\cdot10^{-5}$	2.4258(4)
250.1	1.926(3)	369(2)	0.1709(1)	$2.627(3)\cdot10^{-6}$	2.4391(4)
296.8	1.879(3)	302(1)	0.1424(5)	$3.203(3)\cdot10^{-7}$	0.2448(4)
344.1	1.833(3)	220(1)	0.1195(4)	$3.076(3)\cdot10^{-8}$	0.2449(4)

Table 8. Influence of temperature on absorption parameters determined by fitting the absorbances spectra of SbSI xerogel (after (Nowak & Szperlich, 2010))

Figure 17b presents typical spectral dependences of SbSI xerogel absorbance at different temperatures fitted using numerical minimization of function (8). As in the DRS investigations of SbSI nanowires, the best results have been obtained for the sum of indirect forbidden absorption without excitons and phonon statistics (9), Urbach ruled absorption (10), and the constant absorption term (11). It should be underlined that these mechanisms of absorption of light do not change in the temperature range from 80 K to 344 K, i.e. in the ferroelectric and paraelectric phase of the SbSI. The fitting presented in Fig. 17b is rather good. Values of the fitted parameters are given in Table 8 and Figs. 18 and 19.

The optical energy gap of SbSI xerogel was determined for unpolarized light because the nanowires are variously directed in the investigated samples (see Fig. 8). For light with electric field normal to the SbSI c-axis of bulk-size crystals the optical absorption edge shifts towards greater energies than for perpendicularly polarized light (Fig. 18a). Since, the determined E_{gIf} values of SbSI nanowires should be appropriate for the case of plane polarized illumination with electric field parallel to the SbSI c-axis.

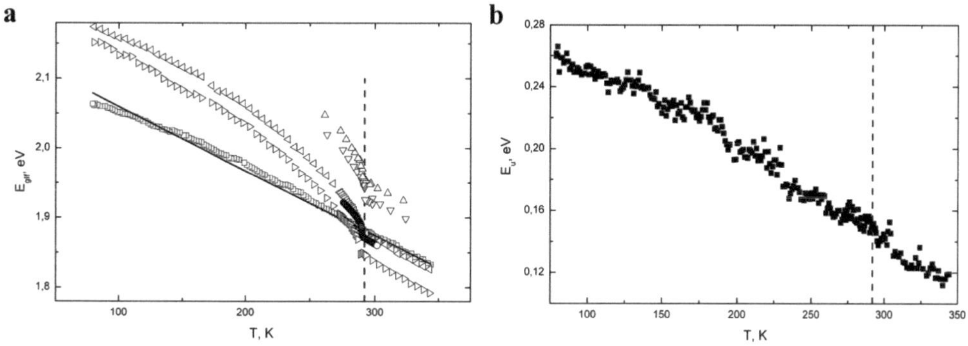

Fig. 18. Temperature dependences of (a) indirect forbidden energy gap (□) and (b) Urbach energy of SbSI nanowires and E_g of bulk SbSI crystals (△,▽- E_g defined by the izoabsorption energy corresponding α=150 cm^{-1} for plane polarized light with electric field normal ($\vec{E} \perp \vec{c}$) and parallel ($\vec{E} \parallel \vec{c}$) to ferroelectric c-axis, respectively (Ishikawa, 1980); ○ - izoabsorption energy corresponding α=150 cm^{-1} for $\vec{E} \perp \vec{c}$ (Audzijonis et al., 2008); ◁,▷- E_{gIf} for $\vec{E} \perp \vec{c}$ and $\vec{E} \parallel \vec{c}$, respectively (Nowak et al., 2003)). Solid line represents dependence (14) least square fitted to the experimental data in case of paraelectric phase (T_c=292 K < T < 344 K). Values of the fitted parameters are given in Table 9. The vertical dash lines show the Curie temperature (Szperlich et al., 2009) for SbSI nanowires (after (Nowak & Szperlich, 2010))

The mechanisms of optical absorption in SbSI xerogel evaluated using the presented absorbances and DRS data are the same (compare Tables 7 and 8). It should be underlined that the indirect forbidden absorption was reported for bulk-size SbSI by many investigators (see e.g. (Park et al., 1995)). Although, many other investigators described the optical properties of bulk-size SbSI using the Urbach absorption (see e.g. (Audzijonis et al., 2008)). Comparison of the temperature dependence of the E_g of SbSI xerogel wit the data reported for bulk-size SbSI is difficult. Values of E_g reported for the same temperature are different (see e.g. Fig. 18a). Even the value of E_{gIf} of SbSI xerogel at 296.8 K (Table 8) is slightly bigger than the one derived from DRS at 297 K (Table 7). First of all, it can be a consequence of various methods used to evaluate the E_g (see e.g. (Nowak et al., 2009a)). Secondly, the E_g values depend not only on the actual phase of SbSI but also on the quality and history of the sample. For example, it was found that E_g is a function of spontaneous as well as induced electric polarizations of SbSI which in turn depend on the structure of ferroelectric domains and defects existing in crystal structure (see e.g. (Gerzanich et al., 1982)).

It is well known that the ferroelectric phase $Pma2_1$ disappears in SbSI crystals near room temperature (see chapter 4). The $E_{gIf}(T)$ dependences of SbSI xerogel (Szperlich et al., 2009) were least square fitted below and above T_c=292 K using the linear formula

$$E_{gIf}(T) = E_{g0} + \beta T . \tag{14}$$

Values of the determined parameters E_{g0} and $\beta=(dE_g/dT)_p$ are given in Table 9. The slope of $E_{gIf}(T)$ in the paraelectric phase of SbSI xerogel is in good conformity with data reported for bulk-size SbSI (compare Table 9 with the data cited in (Nowak & Szperlich, 2010)). The slope of $E_{gIf}(T)$ in the ferroelectric phase of SbSI xerogel is smaller than most of the slopes reported for bulk-size SbSI (see Fig. 18a and compare Table 9 with the data cited in (Nowak & Szperlich, 2010)). Only the $(dE_g/dT)_p$ reported in (Zeinally et al., 1974) for unpolarized illumination is quite comparable with the value determined for ferroelectric SbSI xerogel.

Fitted parameters	Ferroelectric phase (241 K < T < T_c=292 K)		Paraelectric phase (T_c=292 K < T < 344 K)	
	cooling	heating	cooling	heating
E_{g0}, eV	2.195(3)	2.189(4)	2.152(4)	2.152(5)
$(dE_g/dT)_p$, 10^{-4} eV/K	-10.8(1)	-10.5(2)	-9.3(1)	-9.2(2)

Table 9. Values of the determined parameters E_{g0} and $(dE_g/dT)_p$ of the linear temperature dependence (14) of E_{gIf} of SbSI xerogel (after (Nowak & Szperlich, 2010))

There is no discontinuity of the forbidden energy gap of SbSI xerogel larger than ΔE_{gIf}=0.0015 eV at T_c (Figs. 18a). Only the jump of $(dE_g/dT)_p$ in Curie point (Table 9) was observed. According to (Fridkin, 1966), it suggests that the ferroelectric phase transition in SbSI nanowires has the properties of the second-order phase transition. Unfortunately, there is very confused situation concerning the type of ferroelectric phase transition in bulk SbSI. Some authors have suggested that it is a first-order transition having features relevant to a second-order one. However, many other investigators have concluded that this transition is a first-order one (see Refs. in (Nowak & Szperlich, 2010)).

The Urbach energy (Fig. 18b) as well as the probability of Urbach absorption of photons (Fig. 19b) in SbSI xerogel increase with decreasing the temperature. According to (Cody et al., 1981), it should be due to the increase of disorder in the investigated material. Additional investigations are necessary to obtain the information about the possible disorder and the domain structure of the nanowires forming SbSI xerogel.

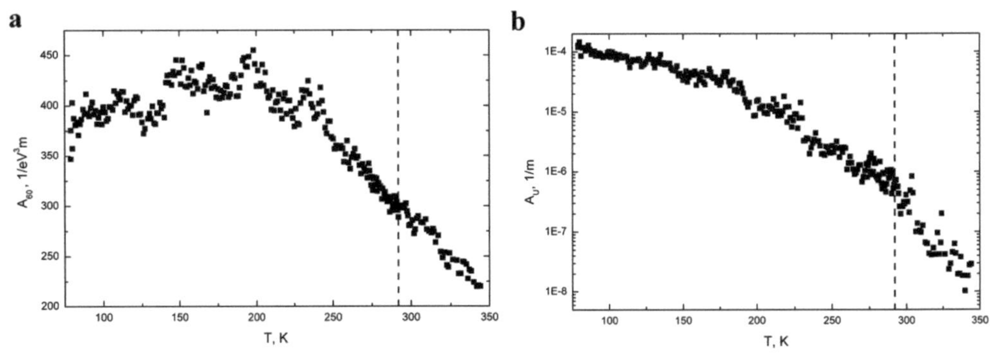

Fig. 19. Temperature dependences of the probability of absorption of photons due to fundamental, indirect, forbidden transitions (a) and of the probability of Urbach absorption of photons (b) in SbSI nanowires. The vertical dash lines show the temperature of the ferroelectric transition in SbSI nanowires (after (Nowak & Szperlich, 2010))

It is interesting that the temperature dependence of the probability of photon absorption due to fundamental, indirect forbidden transition in SbSI nanowires (Fig. 19a) shows two different slops below and above 240 K. A low temperature phase transition at ~233 K in bulk-size SbSI was suggested from the analysis of some types of measurements (see e.g. (Fridkin, 1980) and Refs. in (Nowak & Szperlich, 2010)). These evidences, however, are not conclusive. There is no anomaly in the static dielectric constant, and thus if this transition is indeed real, it does not involve a soft IR optic mode (Samara, 1975). Hence, two explanations exist for the mechanism of low-temperature anomalies in SbSI (Gerzanich et al., 1982):

- the anomalies are the result of the second-order phase transition due to the mutual influence of the phonons and defects existing in the SbSI crystal chains,
- the low temperature anomalies in SbSI near 233 K are regarded as due to a process associated with the interaction between defects and domain walls.

The anomalies of spontaneous polarization at low temperature are absent in monodomain bulk SbSI crystals (Nakonechnyi et al., 1979). Since, in (Nowak & Szperlich, 2010) it was concluded that the SbSI nanowires are probably polydomain structures.

It should be noted that the change of the forbidden energy gap of SbSI nanowires near the Curie point (at 292 K) can be explained by the influence of spontaneous polarization (P_s) of this material (Nowak & Szperlich, 2010)

$$\Delta E_{gs} = E_f(T) - E_p(T) = \beta_3^X P_S^2(T) \tag{15}$$

where E_f is the energy gap determined in the ferroelectric phase, E_p is the value of energy gap evaluated by extrapolation of the straight line that fits $E_g(T)$ in the paraelectric region to the ferroelectric region, and β_3^X is the c-axis component of polarization potential independent of T for a free crystal. The values of β_3^X were determined using (15) and the experimentally determined temperature dependences of E_g and P_s of bulk SbSI in e.g. (Ishikawa, 1980; Žičkus et al., 1984).

Taking the value β_3^X =1.64 eVm^{4}C^2 as the appropriate for the case of plane polarized radiation with the electric field parallel to the ferroelectric c-axis of SbSI (Nowak & Szperlich, 2010), the pertinent parameters from Table 9, and using (15), the P_s values of the SbSI nanowires have been estimated for each temperature (Fig. 20). It should be underlined that the trend of temperature dependence of the spontaneous polarization near Curie temperature is the same in both cases of bulk-size SbSI and SbSI nanowires (see Fig. 20 and the figures published e.g. in (Ishikawa, 1980; Toyoda, 1986). Such calculated P_s of SbSI xerogel is about three times smaller than the one of SbSI single crystals (Fig. 20). Usually, the values of P_s reported for the bulk-size SbSI are in the range from 0.11 C/m^2 at 283 K (Popolitov et al., 1969) to 0.30 C/m^2 at 200 K (Žičkus et al., 1984). The spontaneous polarization 0.012 C/m^2 was reported for thin films of SbSI (Agasiev et al., 1973). The smaller P_s value in the SbSI xerogel is obviously caused by disordering in its structure. It can be explained by the fact that for minimizing electrostatic energy of the sample a polydomain configuration is more favorable than a monodomain particle (Kretschmer & Binder, 1979). Discussing these effects one should also remember that without the poling dc field (so, in the conditions of transmittance measurements of SbSI ethanogel (Nowak & Szperlich, 2010)) the polarization of bulk SbSI was smaller than 0.025 C/m^2 (Imai et al., 1966).

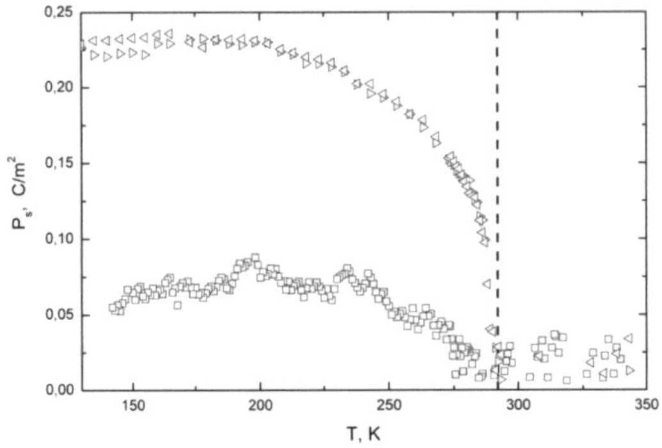

Fig. 20. Comparison of the temperature dependences of spontaneous polarization in SbSI nanowires (□) with data evaluated for bulk-size SbSI crystals (◁,▷- values calculated using data reported in (Nowak et al., 2003) for plane polarized light with electric field normal and parallel to ferroelectric c-axis, respectively). The vertical dash line shows the Curie temperature (Szperlich et al., 2009) of SbSI nanowires (after (Nowak & Szperlich, 2010))

5.3 Infrared absorbance of SbSI nanowires

The infrared (IR) absorbance of SbSI nanowires fabricated by sonochemical method in ethanol (Fig. 21a) is quite different from the one (Fig. 21c) of the SbSI methanogel. Both absorbance spectra have diffusive character due to the great porosity of the investigated samples. In the spectra range from 750 cm^{-1} to 4000 cm^{-1} the bulk crystalline SbSI practically does not show any absorbance lines (Fig. 21a). The IR absorbances of SbSI ethanogel and methanogel show some shifts of the observed bands in comparison with the IR spectra of ethanol (Fig. 21b) and methanol (Fig. 21c). The comparison of the IR absorbance spectra of as prepared SbSI ethanogel as well as SbSI methanogel and the samples heated to different temperatures is given in Fig. 22. The wavenumbers of main IR absorbance bands shown in Fig. 22 are presented in Table 10.

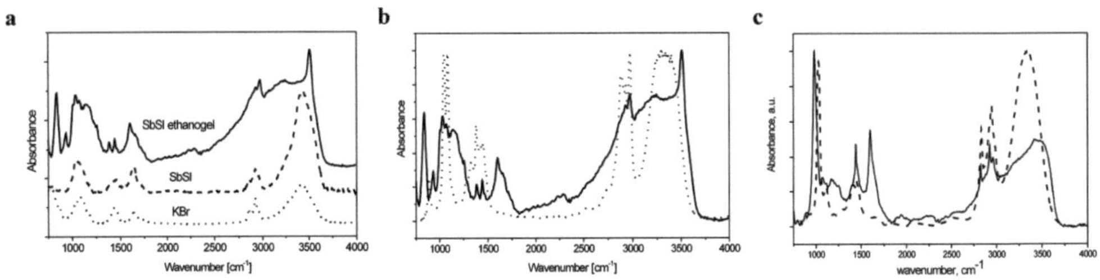

Fig. 21. Comparison of (a) IR absorbance spectra (without the background): — SbSI ethanogel as prepared, -- powdered SbSI single crystals embedded in KBr, ··· KBr alone (after (Starczewska et al., 2008)); (b) IR absorbance spectra of the SbSI ethanogel as prepared (—) and the spectra of ethanol (···) (after (Starczewska et al., 2008)); (c) Comparison of the IR absorbance spectra of the SbSI methanogel as prepared (—) with the spectra of methanol (--) (after (Starczewska et al., 2009)); All spectra were recorded at 300 K.

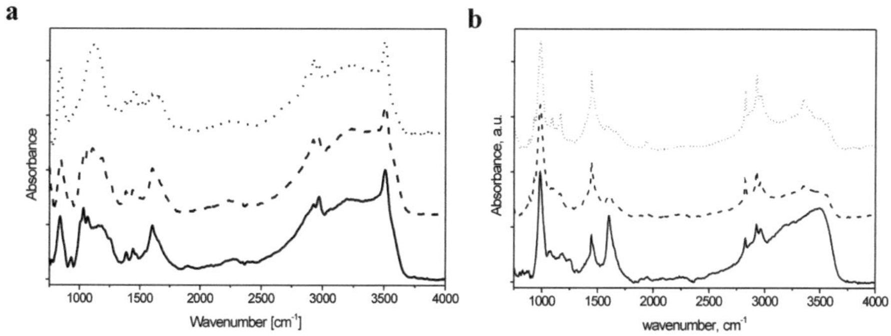

Fig. 22. Comparison of the IR absorbance spectra (without the background) of the SbSI ethanogel (a after (Starczewska et al., 2008)) and SbSI methanogel (a after (Starczewska et al., 2008)) heated to different temperatures: — as prepared, -- dried in vacuum (10^{-4} mbar) at T=353 K, ··· annealed in vacuum (10^{-4} mbar) at T=393 K. Drying and annealing were performed during 8 hours. All spectra were recorded at 300 K

Generally, in the case of SbSI nanowires sonochemically prepared in ethanol, the IR bands can be attributed to the vibrations inside the adsorbed C_2H_5OH molecules, C_2H_5O species, C_2H_4 molecules, OH groups, and H_2O molecules (Table 10). Probably these species are produced during the sonolysis. The postannealing eliminates the adsorbed water. However, Fig. 22a and Table 10 clearly illustrate the presence of OH, CH_2 and CH_3 groups on the surface of SbSI nanowires underwent high-temperature treatment in ultrahigh vacuum. In (Starczewska et al., 2008) it was shown that ethanol adsorbs dissociatively over SbSI nanowire surfaces via scission of the O–H bond to form two types of adsorbed ethoxide species (monodentate and bidentate). The reason for this behaviour is due to the phenomenon called the "proximity effect". Bonds tend to break when they are in relatively close proximity to the surface. The atomic arrangement has been considered almost consistent with the ionic model $Sb^{3+}S^{2-}I^{-1}$ (Kikuchi et al., 1967). The rather short distances between negative ions, on the other hand, suggest a partial covalency of the bond (Kikuchi et al., 1967). The adsorption of alcohols on ionic surfaces, like SbSI, is believed to involve dipole-induced dipole bonds. The dipole is strongest near the oxygen, and consequently the interaction with the surface is maximized if the oxygen of the alcohol is directed toward the surface. The formation of Sb-O bond is much easier from this configuration and thus O-H scission dominates. The adsorbed C_2H_5O may be further dehydrated to ethylene. The C_2H_4 is absorbed on the SbSI nanowire surface, too. Summarizing, one observes the ethanol dehydration over the surface of SbSI nanowires. In samples of gas desorbed from sonochemically prepared SbSI ethanogel and heated at temperature range 368 K to 373 K the ethanol and ethylene were detected using gas chromatographs (Starczewska et al., 2008).
Dehydration of alcohols is a characteristic reaction on acidic sites, i.e. Brönsted and Lewis sites, of transition metal oxides. Thus, C_2H_4 formation indicates the presence of higher acidic sites on the SbSI nanowires. Accordingly, S^{-2} or I^{-1} can act as donor site, whereas surface coordinatively unsaturated Sb^{3+} can act as electron acceptor (Starczewska et al., 2008).
It was mentioned in (Nowak et al., 2009d) that the ultrahigh vacuum conditions (10^{-8} Pa) and/or the used Al K_α monochromatic radiation (with 1486.6 eV energy) could affect desorption or photodesorption of ethanol, ethoxide species, ethylene and water from the surface of sonochemically fabricated SbSI nanowires. The XPS data (Nowak et al., 2009d) revealed no such species, at least within the detection limit of the analysis.

Wavenumbers, cm^{-1}			Mode description after	Wavenumbers, cm^{-1}			Mode description after
Methanogel of SbSI				Ethanogel of SbSI			
as prepared	dried	annealed		as prepared	dried	annealed	
				837	842	837	$\rho(CH_2)$ in C_2H_4
885	888	885	$\rho(CH_3)$ in CH_3I				
	930	929	$\gamma(CO)$				
				936	933	926	$\omega(CH_2)$ in C_2H_4
983	984	984	$\nu(CO)$	1037 1074 1105	1029 1064 1108	 1072 1112	$\nu(CO)$
1075	1082	1091	$\rho_{ip}(CH_3)$				
1170	1162	1165	$\rho_{ip}(CH_3)$	1143	1148	1146	$\rho_{ip}(CH_3)$
				1255	1244	1269	$\delta(OH)$ or $\tau(CH_2)$
				1384	1384	1384	$\delta_s(CH_3)$
1444	1444	1443	$\delta_s(CH_3)$ or $\delta_{as}(CH_3)$	1441	1442	1447	$\delta(CH_2)$ in C_2H_4 or $\delta_{as}(CH_3)$
1601	1603		$\delta(H_2O)$	1602	1602		$\delta(H_2O)$
2825	2825	2825	$\nu_s(CH_3)$				
		2857	$\nu_{as}(CH_3)$		2853	2855	$\nu_s(CH_2)$ and $\nu_s(CH_3)$
2925	2925	2925	$\nu_{as}(CH_3)$ or $2\delta_{as}(CH_3)$	2928	2925	2925	$\nu_{as}(CH_3)$ or $\nu_{as}(CH_2)$
2962	2962	2961	$\nu_{as}(CH_3)$	2972	2967	2963	$\nu_{as}(CH_3)$
3419	3349	3347	$\nu(OH)$ in associated -OH; broad absorption band	3244	3234	3197	$\nu(OH)$ in associated -OH;broad absorption band
		3563	$\nu(OH)$ non-bounded	3510	3510	3508	$\nu(OH)$ non-bounded

Table 10. IR peak positions of the SbSI methanogel (Starczewska et al., 2009) and ethanogel (Starczewska et al., 2008) prepared in different conditions: as prepared, dried at 353 K in vacuum (10^{-4} mbar), annealed at 393 K in vacuum (10^{-4} mbar)

In the case of SbSI nanowires sonochemically prepared in methanol the IR bands (Fig. 22 and Table 10) can be attributed to the vibrations inside the adsorbed CH_3OH molecules, CH_3O species, dimethyl ether molecules, OH groups, and H_2O molecules. Probably these species are produced during the sonolysis. In (Starczewska et al., 2009) it was shown that methanol adsorbs dissociatively over SbSI nanowire surfaces via scission of the O–H bond as in the case of ethanol. Such scission can form different types of adsorbed methoxide species: the monodentate and bidentate. The reason for this behaviour is the mentioned above "proximity effect". Hence, the surfaces of SbSI nanowires prepared sonically in methanol are covered with methoxide species and dimethyl ether (Starczewska et al., 2009).

It should be underlined that in contrary to the OH groups strongly absorbed on surfaces of SbSI ethanogel (see Fig. 22a), these groups are absent in the case of SbSI methanogel nanowires ensemble that underwent high-temperature treatment in ultrahigh vacuum (Fig. 22b). Because molecular adsorbates can compensate the surface polarization charges, providing a mechanism for reducing the depolarization fields, the electric properties of SbSI methanogel should be different from the properties of SbSI ethanogel. Additionally, in the case of SbSI ethanogel the adsorbed C_2H_5OH molecules were reported (Starczewska et al., 2008). Surprisingly, the whole molecules of CH_3OH were not adsorbed on the surfaces of SbSI nanowires or the amount of them was very small (Starczewska et al., 2009). Hence, the methanol decomposes more effectively than ethanol during the sonification or due to the adsorption process on SbSI nanowires.

6. Electrical and photoelectrical properties of photoferroelectric nanowires

6.1 Type of majority electric carriers in SbSI nanowires

In (Nowak et al., 2009d) the XPS was used to investigate the valence band (VB) in the sonochemically prepared SbSI nanowires at room temperature under ultrahigh vacuum conditions ($p=10^{-8}$ Pa). In the XPS spectrum of the VB region of the SbSI nanowires there are two main bands (Fig. 23). The lower binding energy band is over four times larger than the higher binding energy band. The valley between these bands is very deep. The overall shape of this structure is well compared to the VB in bulk SbSI single crystals and corresponds well to the calculated density of states in VB (Ikemoto, 1981; Grigas et al., 2007). According to the theoretical calculations the lower binding energy is the p-band and the higher binding energy band is s-band in VB. The experimental XPS of the valence band is the integral picture of all electronic states in the VB.

Since the optical energy gap of the SbSI ethanogel equals E_{gIf}=1.81(1) eV (Table 7) and the VB determined in XPS measurements is located about 0.5 eV from the Fermi level (Fig. 23), the surface layer of sonochemically prepared SbSI nanowires is of p-type. It is dissimilar to

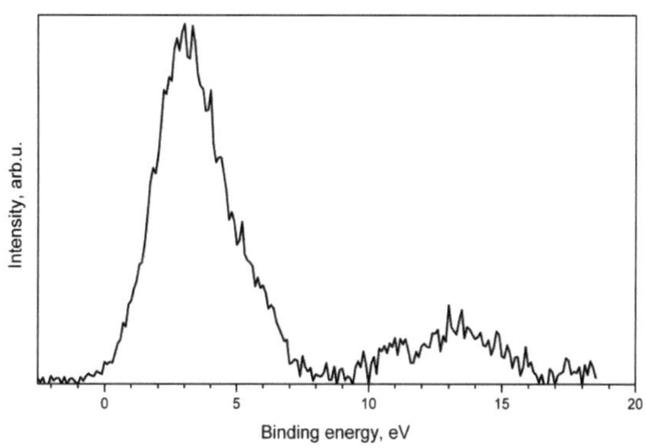

Fig. 23. XPS of the valence band of SbSI ethanogel (after (Nowak et al., 2009d)).

the n-type bulk SbSI single crystals that had VB separated by gaps of 1.8 eV and 2.3 eV from the Fermi level at temperatures 330 K and 215 K, respectively (Grigas et al., 2007). However, in XPS measurements of powdered SbSI crystals (Ikemoto, 1981), the VB was pinned to the Fermi level as should be in the case of p-type material. As to the nature of the conductive carriers in p-type SbSI (so, also in the sonochemically prepared SbSI ethanogel), it was proposed that the iodine vacancies in the SbSI lattice play the role of acceptors (Irie, 1973), or that some of the S^{2-} ions that replace the I^- ions play the part of acceptors (Toyoda & Ishikava, 1970).

6.2 Electrical conductivity of SbSI nanowires

Figure 24 shows the electric current response on the switch on of the dc bias voltage applied to the sonochemically produced SbSI ethanogel. The characteristic decrease of the current with time was noted. The saturated current showed an ohmic behavior with applied field E from -16 kV/m to 16 kV/m (Szperlich et al., 2009). One should notice that when the bias voltage was switch off the current changed its direction and then the drop in the negative current was observed (Fig. 24).

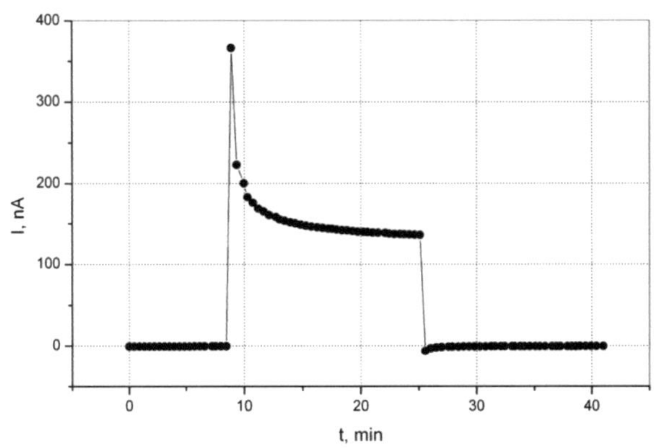

Fig. 24. Current response on the switch on and switch off of voltage applied to the sample of SbSI ethanogel (E=3300 V/m, T=295 K, p=1 atm, humidity 50 %) (after (Szperlich et al., 2009))

Unfortunately, even in the case of bulk SbSI single crystals the mechanism of electrical conduction is not fully understood. Investigations have shown that at contact of such ferroelectric with a metal a Schottky barrier is formed (Wemple et al., 1967). The surface states play a significant role in this case (Zhdan & Artobolevskaya, 1971). Their density may exceed 10^{14} $cm^{-2}eV^{-1}$ (Pipinys et al., 2004). In (Audzijonis et al., 2001) the electrical conductivity of bulk SbSI was explained employing Frenkel emission and a phonon-assisted tunneling. It was also supposed that the field induced phonon-assisted tunneling mechanism may dominate the charge transport process in the bulk SbSI (Pipinys et al., 2004). It was found that the current in bulk SbSI could be limited by the residual polarization charge or by the emission current flowing from the injecting electrode (Kosman & Sleptsov, 1971). The complicated structure of the sonochemically prepared SbSI ethanogel needs a quite new theoretical description of the electric conductivity of this material.

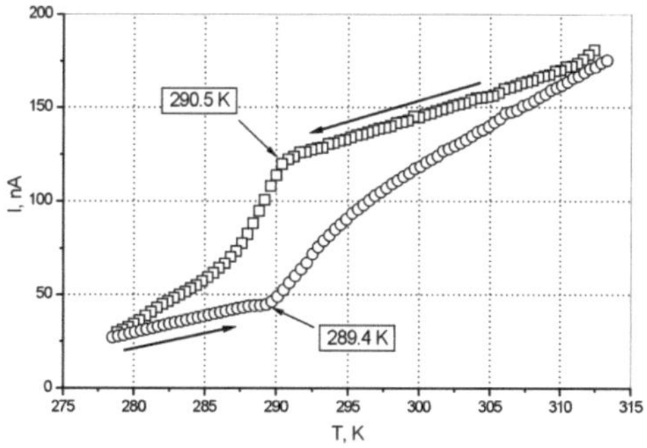

Fig. 25. Temperature dependence of electric current intensity in a sample of SbSI ethanogel upon heating (○) and cooling (□) (E=3300 V/m, p=1 atm, humidity 70 %) (after (Szperlich et al., 2009))

Temperature dependence of electric current in SbSI ethanogel proves its semiconductor properties (Fig. 25). The observed hysteresis of the electric current intensity in the heating-cooling cycle proves that this material is a ferroelectric with Curie temperature near 292 K. The difference between the electric current intensities upon cooling and heating attains maximum near the temperature 291.4 K that corresponds with the Curie temperature in bulk SbSI crystals (see chapter 6.3). The slope of the temperature dependence of electric current intensity changes radically near temperatures 289.4 K (under heating) and 290.5 K (under cooling), so near the Curie temperature (Fig. 25). This dependence was least squares fitted in the paraelectric and ferroelectric regions using the following relation

$$I = I_A \cdot \exp(-\frac{\Delta E_A}{k_B T}) \tag{16}$$

where ΔE_A represents activation energy, I_A is the proportionality factor, k_B and T have the obvious meanings. Table 11 summarizes the fitted values of ΔE_A for ferroelectric and paraelectric phases during heating and cooling of the sample.

Phase	Fitted activation energy, eV	
	cooling	heating
Ferroelectric	0.785(15)	0.346(6)
Paraelectric	0.105(9)	0.243(3)

Table 11. Comparison of the activation energies in temperature dependences of electric conductivity of SbSI ethanogel evaluated for ferroelectric and paraelectric phases during heating and cooling of the sample (after (Szperlich et al., 2009)

The thermal activation energy of the electric conductivity of bulk SbSI crystals was evaluated e.g. in (Popik & Betsa, 1988). The values of ΔE_A were different in the ferroelectric and paraelectric phases as it is in the case of sonochemically prepared SbSI ethanogel (Table 11). Unfortunately, the values of the reported ΔE_A are in very wide range (e.g. from 0.01 eV to 1.6 eV (Popik & Betsa, 1988)). It can be affected by very complicated structure of the electron levels in the energy gap of SbSI. Hence, it is difficult to present an interpretation of ΔE_A determined in the case of the sonochemically prepared SbSI ethanogel.

6.3 Dielectric properties of SbSI nanowires near the Curie point

The temperature dependences of dielectric constant and loss tangent of the sonochemically prepared SbSI ethanogel (Fig. 26) are typical for the temperature region near Curie point of ferroelectrics. The maximum of dielectric constant $\varepsilon=1.6\cdot10^4$ was observed for $T_c=292(1)$ K (Szperlich et al., 2009). The loss tangent of this material is relatively high and also shows its maximum (about 4.3) near T=292(1) K. Such a Currie temperature is characteristic for the single crystals of SbSI (Ishikawa, 1980; Toyoda, 1986).

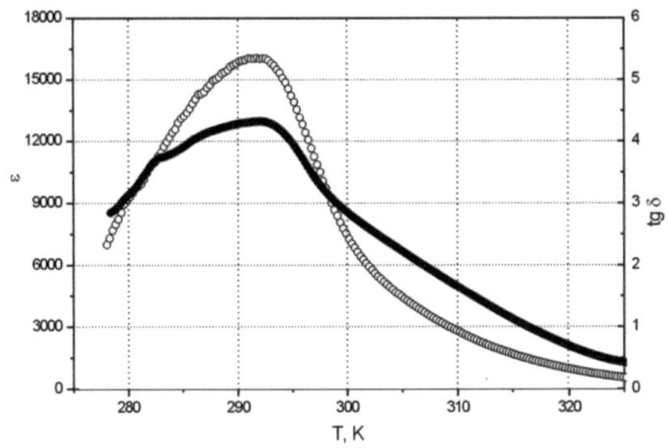

Fig. 26. Temperature dependence of dielectric constant (○) and loss tangent (●) of SbSI ethanogel at E=660 V/m and f=1 kHz (p=1 atm, humidity 70 %) (after (Szperlich et al., 2009))

The determined value of dielectric constant of the sonochemically prepared SbSI ethanogel is lower than the peak dielectric constant $\varepsilon=6.2\cdot10^4$ measured along the polar axis of the best SbSI crystals at $T_c=291$ K (Ishikawa et al., 1984). However, it should be noted that the peak dielectric constant measured in the case of SbSI ethanogel is higher than values reported for the best thin films of SbSI ($\varepsilon=5.2\cdot10^3$ at $T_c=292.3$ K for the pulsed laser deposited and annealed SbSI 4 μm films (Kotru et al., 2000)). One should remember that SbSI is a highly anisotropic material and the permittivity of SbSI single crystal of the polar direction is about 2000 larger than that of the perpendicular direction (Merz & Nitsche, 1964). The investigated SbSI ethanogel has disordered nanowires and very large porosity (see chapter 4), so the determined averaged value of its dielectric constant does not represent the exact value for an individual SbSI nanowire.

The loss tangent of the sonochemically prepared SbSI ethanogel is much higher than the values reported for bulk crystals and thin films of SbSI. Probably, it is due to the complicated structure of the investigated samples of SbSI ethanogel. However, it should be underlined that the great value of the loss tangent sometimes can be a positive feature of ferroelectrics, for example in TANDEL (Rittenmyer et al., 1988).

6.4 Photoconductivity of SbSI nanowires

In (Nowak et al., 2010a) about 5 μm thick samples of as synthesized sonochemically SbSI ethanogel were deposited upon interdigitated finger electrodes on alumina substrates with heaters and temperature detectors. The ethanol was evaporated from them in air at 313 K and a so-called xerogel films were obtained (made up of single crystalline SbSI nanowires with diameters of about 10-50 nm and lengths reaching up to several micrometers). To normalize the photoconductive current of the SbSI ethanogel for constant illumination, the experimental results (Fig. 27) were least square fitted with semiempirical power equation

$$I_{PC}(h\nu,T) = A(h\nu,T)\, I_0^{\gamma(h\nu,T)} \tag{17}$$

where I_0 - incident light intensity, $\gamma(h\nu,T)$, $A(h\nu,T)$ - coefficients which depend on photon energy and temperature. The values of power coefficients $\gamma(h\nu,T)$ (Fig. 28), evaluated using this fitting, are smaller than one and indicate the nonlinear recombination of carriers in SbSI nanowires with the increase of excess carrier concentration. Using the determined values of $A(h\nu,T)$ and $\gamma(h\nu,T)$, the spectral characteristics of $I_{PC}(h\nu,T)$ were normalized for various constant light intensities (Fig. 27b). These characteristics are typical for the case of surface recombination of the excess carriers (Gärtner, 1957).

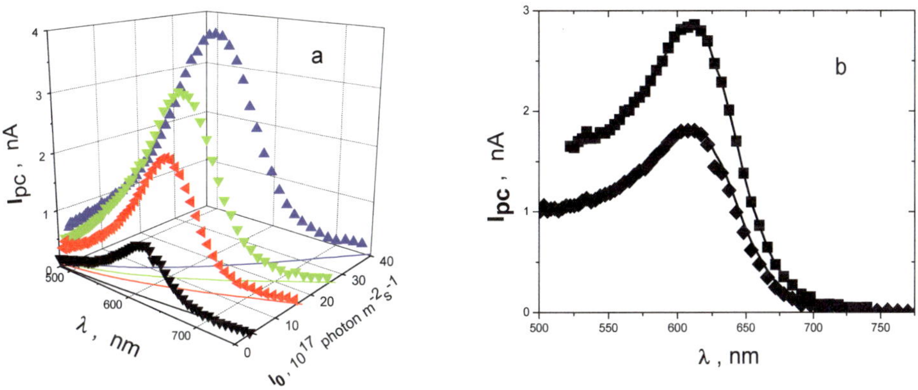

Fig. 27. (a) Experimental data of spectral dependences of photoconductive currents in SbSI ethanogel at 323.7 K (p=1.33 Pa). Solid curves show the spectral dependences of illumination intensities; (b) Normalized for different light intensities spectral characteristics of photoconductivity (◆- $I_0=5\cdot10^{17}$ photons $m^{-2}s^{-1}$; ■- $I_0=10^{18}$ photons $m^{-2}s^{-1}$; T=323.7 K). Solid curves represent theoretical dependence (18) calculated for the values of the fitted β (Fig. 28b) as well as W, S_1 and S_2 (Fig. 29) (after (Nowak et al., 2010a))

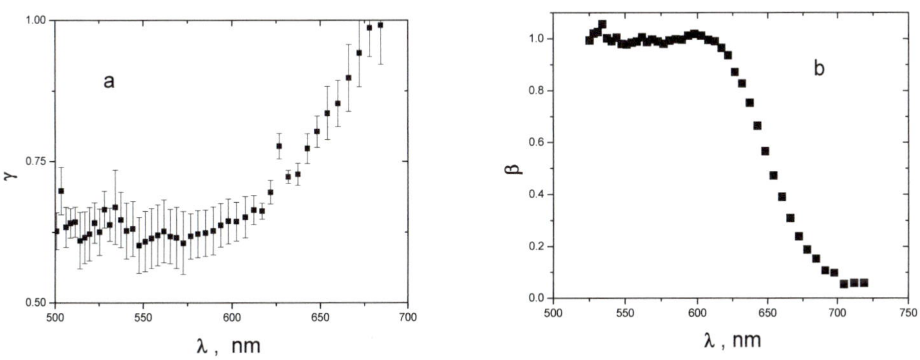

Fig. 28. Spectral dependences of (a) power coefficient from equation (17) least-squares fitted to the experimental data of photoconductivity of SbSI ethanogel and (b) the quantum efficiency coefficient for photogeneration of carriers (T=323.7 K) (after (Nowak et al., 2010a))

The change of recombination velocity with the increase of excess carrier concentration evokes the cumbersome theoretical description of nonlinear dependence of I_{PC} on the illumination intensity. However, in the first approximation of this problem, one can take into consideration different values of recombination parameters (i.e. ambipolar carrier diffusion length L, surface recombination velocities s_1 and s_2 at the front and back surfaces of the sample) for each intensity of illumination (Szałajko & Nowak, 2007). In such a case, the I_{PC} can be described by the theory presented in for example (Gärtner, 1957)

$$I_{PC} = A_{pc}(1-R_d)I_0\beta \frac{K}{W^2-K^2}\frac{1}{W}\left[\frac{(K-S)(W-S)\left(1-e^{-W}\right)e^{-K}+(K+S)(W+S)\left(1-e^{W}\right)}{(W+S)(W+S)e^{W}-(W-S)(W-S)e^{-W}}\right.$$

$$\left. -\frac{(K-S)(W+S)\left(1-e^{-W}\right)e^{-K}+(K+S)(W-S)\left(1-e^{-W}\right)}{(W+S)(W+S)e^{W}-(W-S)(W-S)e^{-W}}+\frac{W}{K}\left(1-e^{-K}\right)\right] \tag{18}$$

where $S=s_1w/D=s_2w/D$ is dimensionless surface recombination velocity on the front and back surfaces, $K=\alpha w$ is the dimensionless absorption coefficient, w the sample thickness, $W=w/L$ the dimensionless sample thickness, β the quantum efficiency coefficient for photogeneration of carriers, A_{pc} is a proportionality constant dependent on geometrical dimensions of the sample, the electron and hole mobilities, the ambipolar diffusion coefficient of carriers, the electric field strength, and elementary charge.

In (Nowak et al., 2010a) values of $\alpha(h\nu,T)$ and $R_d(h\nu,T)$ have been determined experimentally. The results are very similar to the presented in chapters 5.1 and 5.2. Normalized spectral dependences of photoconductivity (e.g. the presented in Fig. 27b) have been least-squares fitted using equation (18) for different light intensities. The fitting was performed for photon energies characteristic for intrinsic photogeneration, i.e. for quantum efficiency coefficient equal $\beta=1$. Then the real value of β has been evaluated (Fig. 28b) applying the method described in (Szałajko & Nowak, 2007). It allowed determining also the influence of illumination intensity on recombination parameters of carriers in SbSI nanowires (Fig. 29).

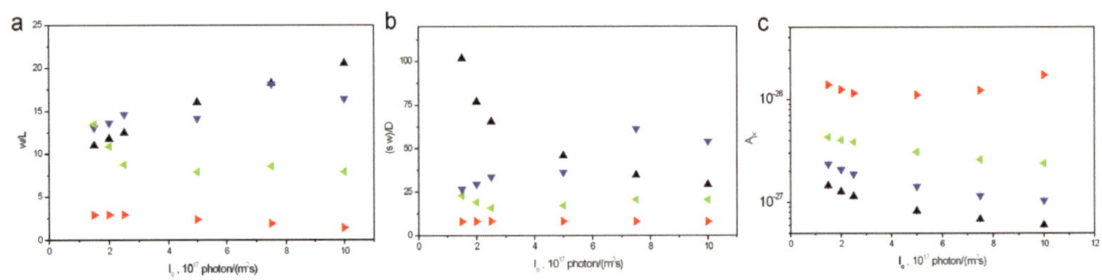

Fig. 29. Influence of illumination intensity on ambipolar diffusion length (a), surface recombination velocity (b), and A_{pc} factor (c) in aligned SbSI nanowires at different temperatures (▲- 263 K; ▼- 283 K; ◄- 303 K; ►- 323 K; p=1.33 Pa) (after (Nowak et al., 2010a))

The nonlinear recombination of carriers in SbSI nanowires needs future investigations that should result in the determination of mechanism of the recombination as well as energetic positions, concentrations and cross-sections of electron states in the energy gap of this material. Probably, a quite new model of photoconductivity in xerogels composed of nanowires is necessary.

7. Conclusions

The SbSI-type well crystallized nanowires, belonging to the class of photoferroelectrics, can be prepared using very simple sonochemical synthesis from elements at low temperature and ambient pressure. Such bottom-up technique is convenient, fast, mild, efficient (with a yield approaching 100 %), and an environmentally friendly route of production of nanowires with uniform shape and high crystallinity in a single step. It may be predicted that upscaling of this method will lead to large quantities of SbSI-type nanowires with uniform morphology and high purity. It seems that this approach can be extended to the preparation of some other ternary and quaternary nanomaterials formed from the group 15-16-17 elements. It should be remember that the physical properties of such quaternary compounds, formed as solid solutions, can be tailored with stoichiometric composition. It should be an advantage of using such material in different devices.

Up to now the polar solvents (ethanol and methanol) were used in the sonochemical synthesis of SbSI, SbSeI and $SbS_{1-x}Se_xI$ nanowires. Temperature of the synthesis was low but still higher than the ones of ferroelectric phase transitions of these materials. It should be interesting to investigate if the unpolar liquids (e.g. toluene) as well as temperature of sonolysis lower than the Curie temperature of the product (i.e. lower than 291 K for SbSI) can influence the rate of preparation of SbSI-type nanowires and the properties of the products. The growing of materials in ferroelectric state should be interesting at least for fundamental investigations on nanoscale ferroelectricity and on mechanisms of crystal growth. In contradistinction to all techniques used for growth of the bulk SbSI-type crystals, the sonochemistry makes such an approach rather easy and may be effective.

One of the features of the presented SbSI-type nanomaterials is that they are formed sonochemically as xerogels. The high surface-to-volume ratios associated with these nanostructures, possessing semiconducting as well as ferroelectric properties, make them very interesting for the sensing of important molecules or for using as electrodes. The catalytic properties of SbSI-type nanomaterials, e.g. suitable for alcohols decomposition, should be investigated. However, further studies on the cross-linking of SbSI-type nanowires in sonochemically prepared xerogels and on the role of the surface layers of these nanowires are needed to establish the xerogels more fracture-tough. Further studies on the electrophysical properties of the presented as well as the other ternary and quaternary SbSI-type nanomaterials are needed. Especially, the impedance spectroscopy should give information on the structure of the sonochemically prepared xerogels.

The presented sonochemical synthesis of nanophase SbSI in carbon nanotubes should be extended to the preparation of some other nanomaterials formed from the group 15-16-17 elements within CNTs. Such hybrid objects should be distinguished in their properties from both hollow nanotubes and the encapsulated substances, which permits one to purpose-tailor "nanowires" and "nanocables" with unique physical and chemical properties (Friedrichs et al., 2005). Special interest is in investigations on SbSI-type nanowires in single wall CNTs of smaller diameters. It may give an insight into quantum size effects in them.

This review summarizes all so far published results of investigations on SbSI-type photoferroelectric nanowires. One can see that the properties of the 1D SbSI-type materials still remain little known. Meantime, nanowires of SbSI-type should provide promising materials for fundamental investigations on nanoscale ferroelectricity and piezoelectricity as well as materials for some applications. Probably, they may be useful in nanoscale nonvolatile memory applications. It is important because the Authors of (Yun et al., 2002) have suggested that ferroelectric $BaTiO_3$ nanowires may be used to fabricate nonvolatile memory devices with an integration density approaching 1 terabit/cm^2. Especially, investigations are necessary to obtain the information about the domain structure of the ferroelectric SbSI-type nanowires.

8. Acknowledgements

The author is indebted to Dr. P. Szperlich of the Silesian University of Technology, Katowice (Poland) for helpful discussions and his aid in preparing this paper. This review was partially supported by the MNiSzW (Poland) under contract No. N N507 1577 33.

9. References

Agasiev, A.A.; Zeinally, A.Kh. & Silvestrov, V.G. (1973). SbSI thin films with a ferroelectric phase transition, *Kristallografiya* 18 (1973) 1293-1294, ISSN: 0023-4761

Audzijonis, A.; Pipinys, P.; Lapeika, V. & Žaltauskas, A. (2001). Optical and electrical properties of SbSI single crystals, *Lith. J. Phys.* 41 (2001) 294-298, ISSN: 1648-8504

Audzijonis, A.; Žigas, L.; Žaltauskas, R.; Sereika, R. & Pauliukas, A. (2008). Origin of the Optical Anomalies Near the Ferroelectric Phase Transition in SbSI and SbSBr Crystals, *Ferroelectrics Lett. Sect.* 35 (2008) 51-61, ISSN: 0731-5171

Belayev, A. D.; Krivshich, V.V.; Micelyuk, E. G.; Slivka, V. Yu.; Turyanitza, I. D. & Chepur, D. V. (1968). Some electrical, optical and photoelectrical properties of $Sb_xBi_{1-x}SJ$ and $SbS_xSe_{1-x}J$ crystals, *Ukrainskii Fiz. Zhurnal* 13 (1968) 854-856, ISSN: 0202-3628

Belayev, A. D.; Miselyuk, E. G.; Slivka, V. Yu.; Turyanitza, I. D. & Chepur, D. V. (1970). On determination of the lattice parameters and Curie points of solid solutions of compounds $A^{V}B^{VI}C^{VII}$, *Ukrainskii Fiz. Zhurnal* 15 (1970) 497-502, ISSN: 0202-3628

Bhalla, A.S.; Newnham, R.E.; Cross, L.E.; Dougherty, J.P. & Smith, W.A. (1981). Pyroelectricity in SbSI. *Ferroelectrics* 33, no.1-4 (1981) 3-7, ISSN: 0015-0193

Cody, G.D.; Tiedje, T.; Abeles, B.; Brooks, B. & Goldstein, Y. (1981). Disorder and the Optical-Absorption Edge of Hydrogenated Amorphous Silicon, *Phys. Rev. Lett.* 47 (1981) 1480-1483, ISSN: 0031-9007

Dittrich, H.; Karl, N.; Kűck, S. & Schock W. (2000), Ternary compounds, organic semiconductors;, In: *Semiconductors, Landolt-Börnstein Condensed Matter III/41E*, Madelung, O. (Ed.), Springer-Verlag, ISBN: 3540667814, Berlin

Dönges, E. (1950). Über Chalkogenohalogenide des dreiwertigen Antimons und Wismuts. I.. Über Thiohalogenide des dreiwertigen Antimons und Wismuts, *Z. anorg. Allg Chem.* vol. 263 (1950) pp. 112-132, ISSN: 0044-2313

Eletskii, A. V. (2004), Sorption properties of carbon nanostructures, *Physics- Uspekhi* 47 (2004) 1119-1154, ISSN: 0038-5670

Entezari, M.H. & Kruus, P. (1996). Effect of frequency on sonochemical reactions II. Temperature and intensity effects, *Ultrason. Sonochem.* 3 (1996) 19-24, ISSN: 1350-4177

Fatuzzo, E.; Harbeke, G.; Merz, W. J.; Nitsche, R.; Roetschi, H. & Ruppel, W. (1962). Ferroelectricity in SbSI, *Phys. Rev.* 127, no 6 (1962) 2036-2037, ISSN: 0031-899X

Fridkin, V.M. (1966). Effects evoked by electron-phonon interaction in phase transition of ferroelectric-semiconductor, *Pisma Zhur. Eksp. Teor. Fiz.* 3 (1966) 252-255, ISSN: 0320-0116

Fridkin, V.M. (1979). *Photoferroelectrics*, Springer-Verlag, ISBN-13: 9780387094182 , New York

Fridkin, V.M. (1980). *Ferroelectric semiconductors*, Consultants Bureau, ISBN: 0306109573, New York

Friedrichs, S.; Falke, U. & Green, M. L. H. (2005), Phase Separation of LaI_3 inside Single-Walled Carbon Nanotubes, (2005). *ChemPhysChem* 6 (2005) 300-305, ISSN: 1439-4235

Gedanken, A. (2004), Using sonochemistry for the fabrication of nanomaterials, *Ultrason Sonochem.* 11 (2004) 47–55, ISSN: 1350-4177

Gärtner, W. (1957). Spectral distribution of the photomagnetoelectric effect in semiconductors: Theory, *Phys. Rev.* 105 (1957) 823-829, ISSN: 0031-899X

Gerzanich, E.I.; Lyakhovitskaya, V.A.; Fridkin, V.M. & Popovkin, B.A. (1982). SbSI and other ferroelectric $A^VB^{VI}C^{VII}$ materials, In: *Current topics in materials science*, vol. 10, Kaldis, E. (Ed.), 55-190, North-Holland, ISBN: 0444863214, Amsterdam

Gomonnai, A.V.; Azhniuk, Yu.M.; Vysochanskii, Yu.M.; Prits, I.P.; Voynarovych, I.M.; Maior, M.M. & Lopushansky, V.V. (2004). Raman scattering in chalcogenide-based ferroelectrics: from bulk to nanoscale, *Phys. Stat. Sol. C* 1, no. 11 (2004) 3166-3169, ISSN: 1862-6351

Gomonnai, A.V.; Voynarovych, I.M.; Solomon, A.M.; Azhniuk, Yu.M.; Kikineshi, A.A.; Pinzenik, V.P.; Kis-Varga, M.; Daroczy, L. & Lopushansky, V.V. (2003). X-ray diffraction and Raman scattering in SbSI nanocrystals, *Mater. Res. Bull.* 38 (2003) 1767-1772, ISSN: 0025-5408

Grekov, A.A.; Korchagina, N.A. & Rogach, E.D. (1979). Clock unit based on ferroelectric device, *Prib. Tekh. Eksp.* 22, no.4 (1979) 262-263, ISSN: 0032-8162

Grigas, J. & Talik, (2003). X-ray photoelectron spectroscopy of ferroelectrics, *Abstracts of 10th European Meeting on Ferroelectrics*, Cambridge U.K. 2003, 143

Grigas, J.; Talik, E. & Lazauskas, V. (2004). X-ray photoelectron spectroscopy of ferroelectric semiconductor SbSI crystals, *Lith. J. Phys.* 44 (2004) 427-438, ISSN: 1648-8504

Grigas, J.; Talik, E. & Lazauskas, V. (2007). X-ray photoelectron spectroscopy of ferroelectrics, *Ferroelectrics* 347 (2007) 86-100, ISSN: 0015-0193

Gruverman, A. & Kholkin, A. (2006). Nanoscale ferroelectrics: processing, characterization and future trends, *Rep. Prog. Phys.* 69 (2006) 2443-2474, ISSN: 0034-4885

Gutiěrrez, M. & Henglein, A. (1988). Sonolytic decomposition of poly(vinylpyrrolidone), ethanol, and tetranitromethane in aqueous solution, *J. Phys. Chem.* 92 (1988) 2978-2981, ISSN: 0021-9606

Henry son & Garot, D'un produit résultant de l'action réciproque du sulfure d'antimoine et de l'iode, Journal de Pharmacie 10 (1824) 511-24

Hui, Ye.; Yuhuan, Xu. & Mackenzie, J.D. (2000). Semiconducting ferroelectric SbSI quantum dots in organically modified TiO_2 matrix, *Proc. SPIE* 3943 (2000) 95-101, ISSN: 0277-786X

Hui, Ye.; Ligong, Y. & Peifu Gu. (2002). Semiconducting ferroelectric SbSI quantum dots in amorphous matrix: preparation and nonlinear optical properties. *Proc. SPIE* 4918 (2002) 99-104, ISSN: 0277-786X

Ibanez, A.; Jumas, J.C.; Olivier-Fourcade, J.; Philippot, E. & Maurin, M. (1983). On chalcogenide-iodates of antimony SbXI (X=S, Se, Te): structure and Mossbauer spectra of ^{121}Sb, *J. Sol. State Chem.* 48, no.2 (1983) 272-283, ISSN: 0022-4596

Ikemoto, I. (1981). X-ray photoelectroc spectroscopic studies of SbSI, *Bull. Chem. Soc. Jap.* 54 (1981) 2519-2520, ISSN: 1348-0634

Imai, K.; Kawada, S. & Ida, M. (1966). Anomalous pyroelectric properties of SbSI single crystals, *J. Phys. Soc. Japan* 21 (1966) 1855-1860, ISSN: 0031-9015

Irie, K. (1973). Excitation of trapped electrons in SbSI, J. Phys. Soc. Japan 34 (1973) 1530-1535, ISSN: 0031-9015

Irie, K. (1978). Dielectric properties of SbSI, *Ferroelectrics* 21 (1978) 395-397, ISSN: 0015-0193

Ishikawa, K. (1980). Electric field effect on the absorption edge in SbSI, *Japan. J. Appl. Phys.* 19, no.7 (1980) 1301-1309, ISSN: 0021-4922

Ishikawa, K.; Tomoda, W. & Toyota, K. (1984). Crystal growth of SbSI from the vapour phase, *J. Crystal Growth* 69, no.2-3 (1984) 399-403, ISSN: 0022-0248

JCPDS PDF No. 01-072-2366. *Antimony Selenide Sulfide Iodide*

JCPDS PDF No. 01-075-1723. *Antimony Selenide Iodide*

JCPDS Card file 21-0050. *Antimony Iodide Sulfide*

JCPDS Card File 74-0149. *Antimony Sulfide Iodide*

JCPDS Card file 74-1195. *Antimony Sulfide Iodide*

JCPDS Card file 74-1196. *Antimony Sulfide Iodide*

JCPDS Card file 74-2210. *Antimony Sulfide Iodide*

JCPDS Card file 74-2244. *Antimony Sulfide Iodide*

JCPDS Card file 74-2245. *Antimony Sulfide Iodide*

JCPDS Card file 74-2246. *Antimony Sulfide Iodide*

JCPDS Card file 75-0781. *Antimony Sulfide Iodide*

JCPDS Card file 75-1621. *Carbon*

JCPDS Card File 76-1354. *Antimony Sulfide Iodide*

JCPDS Card file 88-0985. *Antimony Sulfide Iodide*

Kikuchi, A.; Oka, Y. & Sawaguchi, E. (1967). Crystal structure determination of SbSI, *J. Phys. Soc. Japan* 23 (1967) 337-354, ISSN: 0031-9015

Kosman, M.S. & Sleptsov, A.I. (1971). Injection currents and accumulation of polarization charge in SbSI, *Fizika Tverdogo Tela* 13, no.10 (1971) 3084-3086, ISSN: 0367-3294

Kotru, S.; Liu, W. & Pandey, R.K. (2000). PLD growth of high vapor pressure antimony sulpho-iodide ferroelectric films for IR applications, ISAF 2000. *Proceedings of the 2000 12th IEEE International Symposium on Applications of Ferroelectrics, IEEE.* Part vol. 1, 2001, pp.231-4 vol. 1. Piscataway, NJ, ISSN: 1099-4734

Kretschmer, R. & Binder, K. (1979). Surface effects on phase transitions in ferroelectrics and dipolar magnets, *Phys. Rev.* B 20 No 3 (1979) 1065-1076, ISSN: 1098-0121

Lebedev, M. V.; Mayer, T. & Jaegermann, W. (2003). Sulfur adsorption at GaAs(1 0 0) from solution: role of the solvent in surface chemistry, *Surf. Sci.* 547 (2003) 171–183, ISSN: 0039-6028

Li, B.; Xie, Y.; Huang, J. & Qian, Y. (1999). Sonochemical synthesis of silver, copper and lead selenides, *Ultrason. Sonochem.* 6 (1999) 217-220, ISSN: 1350-4177

Li, H.-L.; Zhu, Y.-C.; Chen, S.-G.; Palchik, O.; Xiong, J.-P.; Koltypin, Yu.; Gofer, Y. & Gedanken, A. (2003). A novel ultrasound-assisted approach to the synthesis of CdSe and CdS nanoparticles, *J. Sol. State Chem.* 172 (2003) 102–110, ISSN: 0022-4596

Lindsjö, M. (2005). *On The Nature of Main-Group Polycations - An Odyssey*, PhD Thesis, KTH Chemical Science and Engineering, Stockholm 2005

Merz, W.J. & Nitsche, R. (1964). Ferroelectricity in SbSI and other compounds of the group 15-16-17 elements, *Izvest. Akad. Nauk SSSR Ser. Fiz.* 28 (1964) 681-682, ISSN: 0367-6765

Mizukoshi, Y.; Nakamura, H.; Bandow, H.; Maeda, Y. & Nagata, Y. (1999). Sonolysis of organic liquid: effect of vapour pressure and evaporation rate, *Ultrason. Sonochem.* 6 (1999) 203-209, ISSN: 1350-4177

Molnar, B.; Johannes, R. & Haas, W. (1965). Properties of single-crystal SbSI, *Bull. Am. Phys. Soc.* 10 (1965) 109, ISSN: 0003-0503

Moulder, J. F.; Stickle, W. F.; Sobol, P. E. & Bomben, K. D. (1995), *Handbook of X-ray Photoelectron Spectroscopy*, Physical Electronics, Inc., ISBN-13 9780964812413, Chanhassen

Nakonechnyi, Yu.S.; Gorvat, A.A.; Lyakhovitskaya, V.A.; Zadorozhnaya, L.A. & Chapur, D.V. (1979). Low-temperature dielectric anomalies and domain structure in SbSl crystals, *Kristallografiya* 24, no.4 (1979) 793-797, ISSN: 0023-4761

Nitsche, R. & Merz, W. J. (1960). Photoconduction in ternary V-VI-VII compounds, J. Phys. Chem. Solids 13 (1960) 154-155, ISSN: 0022-3697

Nitsche, R.; Roetschi, H. & Wild, P. (1964). New ferroelectric V-VI-VII compounds of the SbSI type, *Appl. Phys. Letters* 4 (1964) 210-211, ISSN: 0003-6951

Nowak, M.; Bober, Ł.; Borkowski, B.; Szperlich, P.; Nowrot, A.; Kępińska, M.; Stróż, D. & Sozańska, M. (2010a). Quantum efficiency coefficient for photogeneration of carriers in SbSI nanowires, send to *J. Phys.: Condens. Matter* (2010), ISSN: 0953-8984

Nowak, M.; Kauch, B. & Szperlich, P. (2009a). Determination of energy band gap of nanocrystalline SbSI using diffuse reflectance spectroscopy, *Rev. Sci. Instrum.* 80 (2009) 046107-1-3, ISSN: 0034-6748

Nowak, M.; Kauch, B.; Szperlich, P.; Jesionek, M.; Kępińska, M.; Bober, Ł.; Szala, J.; Moskal, G.; Rzychoń, T. & Stróż D.; (2009b). Sonochemical Preparation of SbSeI Gel, *Ultrason. Sonochem.* 16 (2009) 546–551, ISSN: 1350-4177

Nowak, M.; Kauch, B.; Szperlich, P.; Stróż, D.; Szala, J.; Rzychoń, T.; Bober, Ł.; Toroń, B. & Nowrot A. (2010b), Sonochemical preparation of $SbS_{1-x}Se_xI$ nanowires, *Ultrason. Sonochem.* 17 (2010) 487–493, ISSN: 1350-4177

Nowak, M.; Jesionek, M.; Szperlich, P.; Szala, J.; Rzychoń, T. & Stróż, D. (2009c). Sonochemical growth of antimony sulfoiodide in multiwalled carbon nanotube, *Ultrason. Sonochem.* 16 (2009) 800-804, ISSN: 1350-4177

Nowak, M.; Mroczek, P.; Duka, P.; Kidawa, A.; Szperlich, P.; Grabowski, A.; Szala, J. & Moskal, G. (2009d). Using of textured policrystalline SbSI in actuators, *Sensors and Actuators* A 150 (2009) 251–256, ISSN: 0924-4247

Nowak, M. & Szperlich P. (2010). Temperature dependence of energy bandgap and spontaneous polarization of SbSI nanowires, send to *Thin Solid Films* (2010), ISSN: 0040-6090

Nowak M.; Szperlich, P.; Bober, Ł.; Stróż, D.; Jesionek, M.; Nowrot, A.; Mistewicz, K.; Starczewska, A. & Kępińska, M. (2010c). Mechanism of sonochemical synthesis of SbSI nanowires, send to *Ultrason. Sonochem.* (2010), ISSN: 1350-4177

Nowak, M.; Szperlich, P.; Bober, Ł.; Szala, J.; Moskal, G. & Stróż, D. (2008). Sonochemical Preparation of SbSI Gel, *Ultrason. Sonochem.* 15 (2008) 709–716, ISSN: 1350-4177

Nowak, M.; Szperlich, P.; Kidawa, A.; Kępińska, M.; Gorczycki, P. & Kauch, B. (2003). Optical and photoelectrical properties of SbSI, *Proc. SPIE* 5136 (2003) 172-177, ISSN: 0277-786X

Nowak, M.; Szperlich, P.; Talik, E.; Szala, J.; Rzychoń, T.; Stróż, D.; Nowrot, A. & Solecka, B. (2010d). Sonochemical preparation of antimony subiodide, *Ultrason. Sonochem.* 17 (2010d) 219-227, ISSN: 1350-4177

Nowak, M.; Talik, E.; Szperlich, P. & Stróż, D. (2009e). XPS analysis of sonochemically prepared SbSI ethanogel, *Appl. Surf. Sci.* 255 (2009e) 7689–7694, ISSN: 0169-4332

Park, S-A.; Kim, M-Y.; Lim, J-Y.; Park, B-S.; Koh, J-D. & Kim, W-T. (1995). Optical Properties of Undoped and V-Doped VA-VIA-VIIA Single Crystals, *Phys. Stat. Sol. B* 187 (1995) 253-260, ISSN: 0370-1972

Park, S.-A.; Yun, S.-H.; Kim, W.-T.; Choe, S.-H. & Kwun, S.-I.(1990). Optical properties of $SbS_{1-x}Se_xI$ and $SbS_{1-x}Se_xI$:Co single crystals, *New Physics* 30 (1990) 763-767, ISSN: 0374-4914

Philips-Invernizzi, B.; Dupont, D. & Caze, C. (2001). Bibliographical review for reflectance of diffusing media, *Opt. Eng.* 40 (2001) 1082-1092, ISSN: 0091-3286

Pipinys, P.; Lapeika, V. & Audzijonis, A. (2004). Electronic Conduction Model in Ferroelectrics Based on Phonon-Assisted Tunneling Emission, *Ferroelectrics* 313 (2004) 91-98, ISSN: 0015-0193

Popik, Yu. V. & Betsa, V.V. (1988). Electronic state of the surface and local levels in SbSI single crystals, *Fizika Tverdogo Tela* 30, no.5 (1988) 1282-1288, ISSN: 0367-3294

Popolitov, V. I. & Litvin, B. N. (1970). Synthesis of ternary chalcohalides $A^VB^{VI}C^{VII}$ single crystals, In: *Investigations of Crystallization Processes under Hydrothermal Conditions*, Lobachev, A. N. (Ed.), 55-68, Nauka, Moscow

Popolitov, V. I. ; Litvin, B. N.; Lobachev, A. N.; Yurin, V.A. & Agal'tsov, L.A. (1969). Hydrothermal synthesis and ferroelectric properties of SbSI crystals, *Izvest. Akad. Nauk SSSR Ser. Fiz.* 33, No 2 (1969) 341-343, ISSN: 0367-6765

Pouga, G. D.; Pouga, P. P.; Maksimetz, V. V.; Boretz, A. N.; Hroshik, I. I.; Bercha, D. M. & Chepur, D. V. (1973). Optical and dielectric properties of semiconductor-ferroelectrics $SbS_xSe_{1-x}I$, *Ferroelectrics* 6, no.1-2 (1973) 111-113, ISSN: 0015-0193

Rittenmyer, K.M.; Alexandrakis, G.C. & Dubbelday, P.S. (1988). Detection of fluid velocity and hydroacoustic particle velocity using a temperature autostabilized nonlinear dielectric element (TANDEL). *J. Acoust. Am. Soc.* 84, no.6 (1988) 2002-2006, ISSN: 0001-4966

Samara, G.A. (1975). Effects of pressure on the dielectric properties of and the vanishing of the ferroelectricity in SbSI, *Ferroelectrics 9*, no.3-4 (1975) 209-219, ISSN: 0015-0193

Scott, J.F. (2006).Nanoferroelectrics: statics and dynamics, *J. Phys.: Condens. Matter* 18 (2006) R361–R386, ISSN: 0953-8984

Spitsyna, V. D.; Lyakhovitskaya, V. A. & Belyaev, L. M. (1975). Isomorphous replacement in SbSI, *Kristallografiya* 20 (4) (1975) 840-842, ISSN: 0023-4761

Starczewska, A.; Wrzalik, R.; Nowak, M.; Szperlich, P.; Bober, Ł.; Szala, J.; Stróż, D. & Czechowicz, D. (2008). Infrared spectroscopy of ferroelectric nanowires of antimony sulfoiodide, *Infrared Physics & Technology* 51 (2008) 307-315, ISSN: 1350-4495

Starczewska, A.; Wrzalik, R.; Nowak, M.; Szperlich, P.; Jesionek, M.; Moskal, G.; Rzychoń, T.; Szala, J.; Stróż, D. & Maślanka P. (2009). Influence of the solvent on ultrasonically produced SbSI nanowires, *Ultrason. Sonochem.* 16 (2009) 537–545, ISSN: 1350-4177

Suslick, K.S.; Hammerton, D.A. & Cline, Jr. R.E. (1986), The Sonochemical Hot Spot, *J. Am. Chem. Soc.* 108 (1986) 5641-5642, ISSN: 0002-7863

Szałajko, M. & Nowak, M. (2007). Quantum efficiency coefficient for photogeneration of carriers in gallium sulphide single crystals, J. Phys.: Condens. Matter 19 (2007) 196210, ISSN: 0953-8984

Szperlich, P.; Nowak, M.; Bober, Ł.; Szala, J. & Stróż, D. (2009). Ferroelectric properties of ultrasonochemically prepared SbSI ethanogel, *Ultrason. Sonochem.* 16 (2009) 398–401, ISSN: 1350-4177

Toyoda, K. (1986). Electrical properties of SbSI crystals in the vicinity of the ferroelectric Curie point, *Ferroelectrics* 69, no.3-4 (1986) 201-215, ISSN: 0015-0193

Toyoda, K. & Ishikava, J. (1970). Transport phenomena in SbSI, *J. Phys. Soc. Japan* 28 Suppl. (1970) 451-453, ISSN: 0031-9015

Turjanica, I. D.; Nejezchleb, K. & Horak, J. (1969). Some optical, photoelectric and ferroelectric properies of mixed crystals $SbS_xSe_{1-x}I$, *Czech. J. Phys. B* 18 (1969) 1465-1467, ISSN: 0011-4626

Voutsas, G. P. & Rentzeperis, P. J. (1982). The crystal structure of antimony selenoiodide, SbSeI, *Z. Kristallogr.* 161 (1982) 111-118, ISSN: 0044-2968

Voutsas, G. P. & Rentzeperis, P. J. (1986). The crystal structure of the quaternary compound $SbSe_{0.75}S_{0.25}I$, *Z. Kristallogr.* 178 (1986) 289-295, ISSN: 0044-2968

Voynarovych, I.M.; Gomonnai, A.V.; Solomon, A.M.; Azhniuk, Yu.M.; Kikineshi, A.A.; Pinzenik, V.P.; Kis-Varga, M.; Daroczy, L. & Lopushansky, V.V. (2003). Characterization of SbSI nanocrystals by electron microscopy, *J. Optoelectron. Adv. Mater.* 5, no 3 (2003) 713-718, ISSN: 1454-4164

Wang, C.R.; Tang, K.B.; Yang, Q.; Hai, B.; Shen, G.Z.; An, C.H.; Yu, W.C. & Qian, Y.T. (2001). Synthesis of novel SbSI nanorods by hydrothermal method, *Inorg. Chemi. Commun.* 4, no 7 (2001) 339-341, ISSN: 1387-7003

Wemple, S.H.; Kahng, D. & Braun, H.J. (1967). Surface barrier diodes on semiconducting $KTaO_3$, *J. Appl. Phys.* 38 (1967) 353-359, ISSN: 0021-8979

Xia, Y.; Yang, P.; Sun, Y.; Wu, Y.; Mayers, B.; Gates, B.; Yin, Y.; Kim, F. & Yan, H. (2003). One-Dimensional Nanostructures: Synthesis, Characterization, and Applications, *Adv. Mater.* 15 (2003) 353-389, ISSN: 0935-9648

Yuhuan Xu.; Cheng, C.-H.; Hui, Ye. & Mackenzie, J.D. (2001). Electro-optic effect in a nano-crystals doped glass, *Ferroelectrics* 259 (1-4) (2001) 259-68, ISSN: 0015-0193

Yuhuan, Xu.; Del Monte, F.; Mackenzie, J.D.; Namjoshi, K.; Muggli, P. & Joshi, C. (1999). Nanocomposite of semiconducting ferroelectric antimony sulphoiodide dot-doped glasses, *Ferroelectrics* 230 (1-4) (1999) 11-20, ISSN: 0015-0193

Yun, W. S., Urban, J. J.; Gu, Q. and Park, H. (2002). Ferroelectric Properties of Individual Barium Titanate Nanowires Investigated by Scanned Probe Microscopy, *Nano Letters* 2, no 5 (2002) 447-450, ISSN: 1530-6984

Zeinally, A.Kh.; Agasiev, A.A. & Efendiev, Sh.M. (1974). Direct 'allowed' transitions in SbSI, *Fiz. Tekhn. Poluprovod.* 8, no.1 (1974) 197-200, ISSN: 0015-3222

Zhdan, A.G. & Artobolevskaya, E.S. (1971). Properties of contacts between SbSI single crystals and some metals. Fizika Tverdogo Tela 13, no.4, (1971) 1242-1244

Žičkus, K.; Audzijonis, A.; Batarūnas, J. & Šileika, A. (1984). The fundamental absorption edge tail of ferroelectric SbSI. *Phys. Stat. Sol. B* 125, no.2 (1984) 645-651, ISSN: 0370-1972

14

Nanowires with Unimaginable Characteristics

Hui Li[1] and Fengwei Sun[2]
[1] *Key Laboratory for Liquid-Solid Structural Evolution and Processing of Materials, Ministry of Education, Shandong University, Jinan 250061,*
[2]*Physics Department, Ocean University of China, Qingdao, China*

1. Introduction

The understanding and functionalization of nanowires has become a topic of major interest to the international materials science and engineering communities due to their improved mechanical, electrical and thermodynamic properties. Molecularly perfect materials such as metallic nanowires have opened up opportunities for the design of nanometric electronic devices. Nanowires have a very large surface-area/volume ratio as compared to bulk materials, so their structures and properties can be quite different from those of bulk materials. Because of these unique properties, it is believed that nanowires will be utilized as the next-generation structural materials, biosensors, and as circuitry and interconnects in future nanoscale computers and micro-mechanical components.

This chapter presents the detailed analysis and explanation of how nanowires behave in the mechanical, electronical and thermal properties. In section 1, we focus on the mechanical properties and superplastic deformation of the 1-D carbon nanowires(CNWs). The objective of this work is to understand the tensile behavior of CNWs. And the key finding of this work is that the superplastic properties of CNWs and the stability of the elongated atomic chain emanating from CNWs may be due to the presence of a high strength multishell lattice structure that forms as an outcome of the plastically deforming nanowire(Li et al, 2008; Li et al, 2009). In section 2, we make some discussion on the electronical properties of metallic nanowires. A major issue in the use of them is the quality of electronic transmission, as obtaining the minimum resistance is critical to accessing their intrinsic electric properties. We use density function theory (DFT) to calculate Ni nanowires that are embedded in a carbon nanotube, and assess its applicability to the study of electronic transport along nanowires. These optimized stable structures of the ultrathin Ni nanowires in the carbon nanotubes are always multishell packs that are composed of coaxial cylindrical shells. The transmission of these nanostructures is dependent upon the geometric structures of nanowires and their size of diameters. Because of the quantum size effect, the current-voltage curves of the nanowires are clearly nonlinear, which does not follow an ohmic pattern(Li et al, 2007). In section 3, a microscopic description of the melting behavior of nanowire is simulated. We employed the well-fitted tight binding many body potential to investigate the heating process of the helical structure of Ni nanowires, which are obtained by the means of genetic algorithm (GA). Interestingly, we find that the melting of nanowire starts from the interior atoms which are different from bulk materials. When the slow heating rate is applied, the central atoms first move along the wire axis direction at a rather low temperature, while the helical outer shells are almost invariant. It is worth noting that the moving central atoms are not discrete but to build up a long monostrand atomic chain. As the temperature arises, the regular monostrand atomic chain begins to become deformed and finally, the atomic chain is broken to form a new type of cluster(Li et al, 2005). The last section presents our conclusion.

2. Mechanical properties of nanowires and their composites

2.1 Simulation method

The numerical simulations are carried out by means of the classical molecular dynamics method (Ribarsky et al, 1988) in which Newtonian equations of motion are solved numerically for a set of atoms that interact via Brenner's "second generation" empirical many-body bond order potential(REBO) (Brenner et al, 2002). Brenner's REBO potential energy has already been used to model the mechanical properties of carbon (Sinnott et al, 1988). But in the second-generation REBO potential, the switching function significantly influences the forces in the vicinity of the inflection point (Shenderova et al, 1999), so we modify the value of interaction cut-off. Huhtala et al. (Huhtala et al, 2004) used another simulation method that is adaptive intermolecular reactive empirical bond order (AIREBO) potential (Stuart et al, 2000) to study the mechanical load transfer between shells of multiwalled carbon nanotubes. Sammalkorpi et al. (Sammalkorpi et al, 2004) used the reactive analytical bond order potential model parametrized by Brenner but modified the cut-off value to avoid the overestimation of the force required to break a bond (Belytschko et al, 2002). These three methods are all indispensable extensions to the well-known reactive empirical bond order Brenner potential. The superplastic behavior of CNWs is simulated by solving the equations of motion using Gear's predictor–corrector algorithm (Gear, 1971). The axial tension of perfectly structured CNWs with different diameters and lengths is achieved by applying a rate of 10 and 80 m/s for CNWs, respectively. The end atoms are then moved outward along the axis by small steps and followed by a conjugate gradient minimization method while keeping the end atoms fixed (Liew et al, 2004). The simulation is carried out at approximately 1 K to avoid thermal effect. Similar research on the carbon tube has been done by Wang et al. (Wang et al, 2007) at the same temperature. Each time step used in this simulation is 1 fs, and the simulations are allowed to run for 500,000 time steps. Although many researchers have carried MD simulations of nanowires using schemes such as Nose-Hoover thermostat (Nose, 1984; Hoover, 1985), we note that under high rate at the time scale of the dynamic deformation of CNW, there is usually no effective mechanism for heat to be conducted. So we provide a more realistic condition of the dynamic deformation of CNWs, using no temperature controlling algorithms. CNWs are obtained by means of the Materials Studio(MS) software package. We adopted a molecular dynamics-based "simulated annealing"(SA) method to perform structural optimization to obtain the nanowire which is used to be stretched.

2.2 Results and discussion

2.2.1 Superplasticity of carbon nanowires

Figure 1 shows two snapshots of the CNWs at the initial state. The lengths of CNWs A1 and A2 are 47.4 Å and 50.5 Å, respectively. The CNWs are all multishell packs with each shell formed by rows of atoms that are helically wound upward side by side. CNW A1 features four-strand packing, and CNW A2 is composed of six-strand atomic rows.

Fig. 1. Morphologies of carbon nanowires

CNW A1 undergoes superplastic deformation during the simulation process. The surprising result is that the maximum tensile strain, increases to nearly 245% before tensile failure. This superplastic deformation is the result of the slippage of carbon atoms as well as the intense interaction between them. Figure 2 shows the morphological changes of CNW A1 at different strains. The CNW becomes thinner and even in the middle to form a single atomic chain at ε=0.246. Further tension leads another part to become a single chain, as shown in Fig. 2(a) at ε=0.95. Eventually, the middle part becomes a single chain. When it is stretched completely, the initial CNW becomes an individual atomic chain with a diameter of one carbon atom. The whole process can be displayed by movie and is found to be very similar to drawing a thread from a silk cocoon. After it breaks, the carbon atomic chain is found to recoil and displays elastic-plastic behavior, as shown in Fig. 2(f).

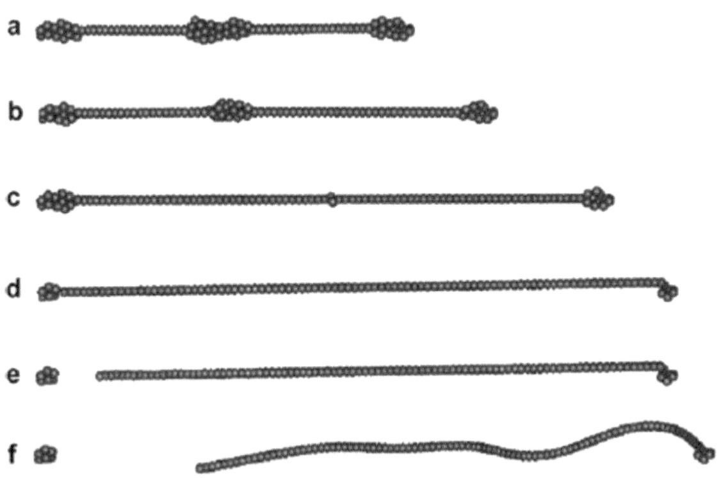

Fig. 2. There-dimensional views of morphological changes (axial deformation) for CNW (A1). (a) At ε=0.95 the two parts become a single chain. (b) At ε=1.42. (c) At ε=2.09 the middle part stretches to form a single chain and the stress is the largest (16.65Gpa). (d) At ε=2.45 the CNW is fully stretched. (e) At ε=2.457 the chain breaks. (f) After it breaks, the chain recoils.

In Fig. 3(a), the black curve indicates stress-strain relationship of the CNW A1. During the tensile straining, the stress-strain curve fluctuates corresponding to the fluctuation of the total potential-strain curve. This fluctuation mainly results from the slippage of carbon atoms from their spots. As one atom slips from one spot to another, the stress and the total potential spontaneously change from the peak value point to the valley point. Figure 3(a)

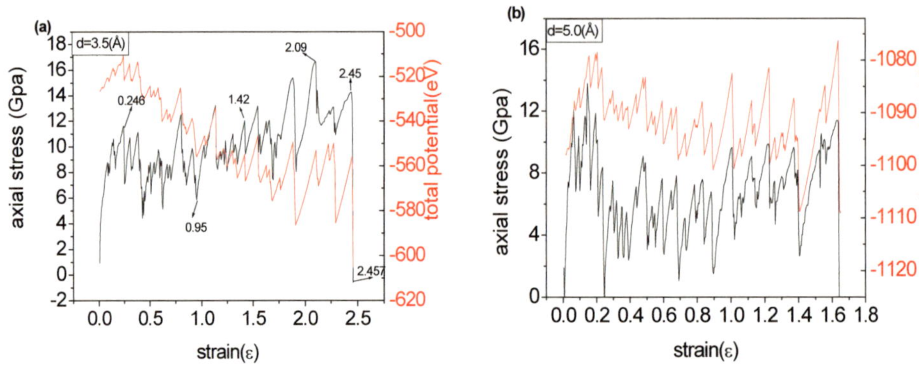

Fig. 3. (color online) Stress-strain and total potential-strain curves for CNWs with different diameters. (a)d=3.5Å(A1), (b)d=5.0Å(A2). The black curves correspond to stress-strain relationship, while the red curves correspond to total potential-strain relationship.

shows that the maximum stress for a CNW is 16.65 GPa. The red curves in Fig. 3 show the potential-strain relationship for CNWs with different diameters. When the strain increases, (as shown in Fig. 3b), the potential increases stair-like during the stretching process. But there is a little different trend for CNW with a smaller diameter in Fig. 3(a). Although the potential of the structure decreases with increasing the strain, it reflects the structure variation from complex sp3 hybridization to sp hybridization. At the initial stage, the structure (A1) was minimized to obtain the stable structure model. Some carbon atoms located in the center axis are surrounded by circumambient atoms. So the central carbon atoms will suffer strong interaction. During the stretching process, the single atom chain first forms from where C-C bond energy is weaker. The bond angle between carbon atoms at the joint is stretched to the degree of 180° to form the sp hybridization. As the stretching process continues, the number of carbon atoms surrounding the single atomic chain within the cut-off value becomes less, so the interaction between them becomes weak, in addition, sp hybridization is weaker than the sp3 hybridization. In order to explain clearly, we present the graphs which show the number of C atoms with different coordination numbers in Fig. 4.

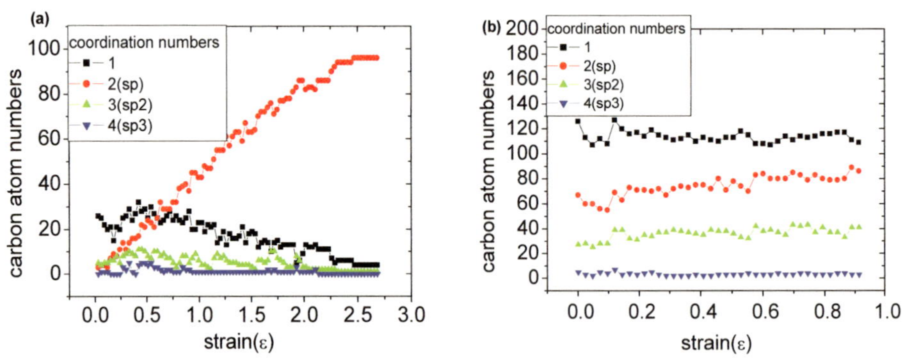

Fig. 4. (color online) The relationship between the number of C atoms with different coordination numbers (1,2,3,4) and the strain.(a) CNW with d=3.5 Å (A1) (b) d=8.5 Å.

As to the superplastic CNW A1, the number of C atoms forming sp hybridization increases as the stretching continues. And the number of sp2 and sp3 bonded atoms does not significantly increase during the pulling. The energy needed for CNWs (A1 and A2) to break are 8.17×10^{-10} nJ and 7.69×10^{-10} nJ, respectively, which are a little larger than the energy required to break a single C–C bond (5.76×10^{-10} nJ). As to CNWs with larger diameters, the number of C atoms forming sp hybridization does not vary remarkably as shown in Fig. 4b.

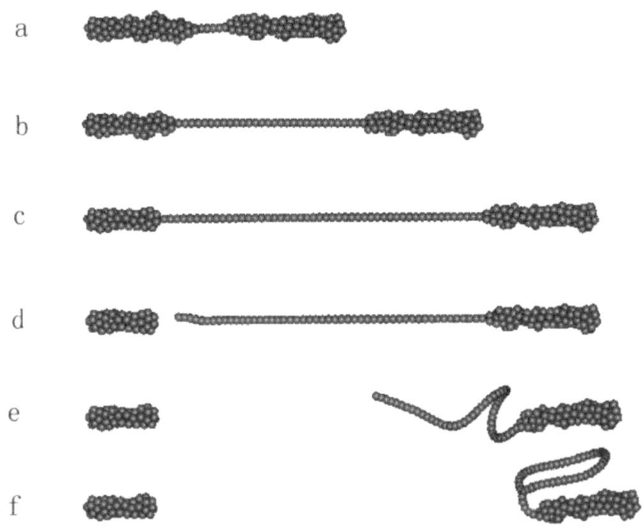

Fig. 5. Three-dimensional views of morphological changes (axial deformation) for CNW (A2) (a) At $\varepsilon=0.25$ the middle part forms a single chain. (b) At $\varepsilon=1.0$ the atomic chain stretches further. (c) At $\varepsilon=1.62$ the CNW has stretched to the maximum. (d) At $\varepsilon=1.63$ the atomic chain breaks. (e) The broken end recoils. (f) The broken end recoils, meeting with the single chain to form a loop.

Figure 5 shows CNW A2 at different strains. This is similar to the tensile process of CNW A1, but when CNW A2 breaks at $\varepsilon=1.63$ it appears to have not only superplasticity, but also elasticity. Figure 5 (e) and (f) clearly show that the broken end recoils and touches the single string to form a covalent bond loop. Figure 6 illustrates the maximum tensile strain of CNW as a function of the diameter of CNW. As the diameter increases from 3.5Å to 12.1Å, the maximum tensile strain decreases by nearly two thirds, i.e. from 245% to 82.3%, so the superplastic characteristics disappears while the diameter of CNW is larger than approximate 8Å. In contrast to CNW, maximum tensile strain of single walled carbon nanotube (SWCNT) with the diameter ranging from 6.26 to 13.56 Å is approximately 25~35%. Although the strength of SWCNT (Asaka et al, 2005; Ogata et al, 2003), multi-walled nanotubes (MWNTs) (Liew et al, 2004) and SWCNT bundle (Yu et al, 2000) are higher than CNW, but their maximum tensile strain are about 30%, 28%, 5.3%, respectively, much less than CNW.

2.2.2 Formation of nanobridge

In this section, we present an extensive set of simulation results for CNW encapsulated into carbon nanotube(CNT) to elucidate the mechanical properties of this material. Figure 7(a)

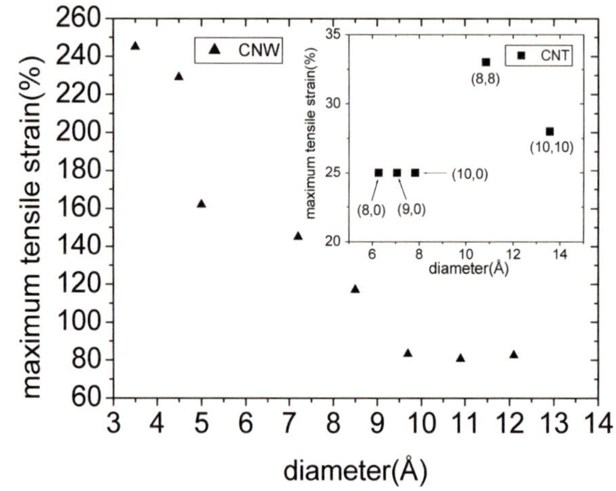

Fig. 6. The relation between maximum tensile strain and diameter of CNW. For comparison, the maximum tensile strains for SWCNT with different diameters are inserted.

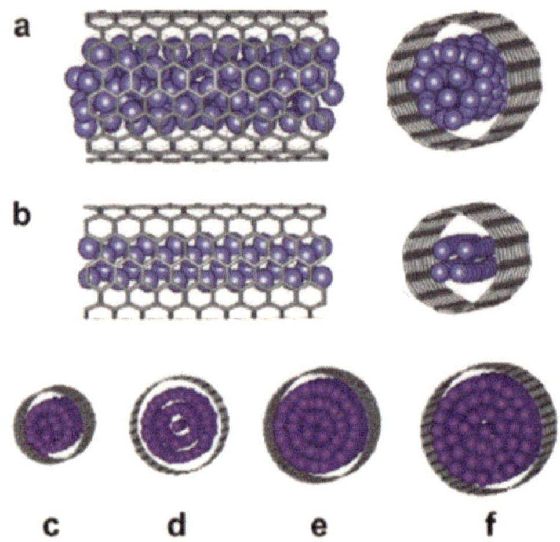

Fig. 7. (color online) Morphology of different CNTs encapsulating CNW. (a)CNW@CNT(9,9). (b) CNW@CNT(7,7). (c) CNW@CNT (20,0). (d) CNW@CNT (15,15). (e) CNW@CNT (30,0). (f) CNW@CNT (20,20).

shows the morphology of an 11-strand CNW@CNT(9,9) (i.e., an 11-strand nanowire including 102 atoms with a helical structure that is encapsulated in a CNT(9,9)). Its length and diameter are 21.9 Å and 12.22 Å, respectively. The atoms that constitute the CNW are colored blue for clarity.

In the initial stage, the average distance between the outer atoms of the CNW and the CNT wall is about 3.16 Å, so the interaction between them is the long-rang van der Waals potential using the Lennard-Jones potential (Lennard-Jones, 1924). The stretching process of an 11-strand CNW@CNT(9,9) is shown in Fig. 8.

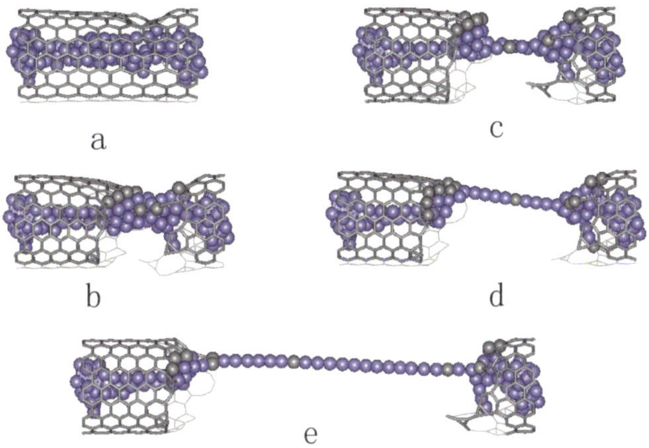

Fig. 8. (color online) Morphological changes of the 11-strand CNW@CNT(9,9) under elongation.(a) ε=0.22, the CNT begins to break and the CNW atoms attach to the surface of the CNT wall.(b) ε=0.51, the CNT fractures completely, and the CNT atoms and CNW atoms combine. (c) ε=0.91, the CNW begins to deform into a single chain. (d) ε=1.36, the single chain is stretched and the gray atom is composed of the initial CNT. (e) At ε=2.8, the CNW@CNT reaches its critical dimension.

There is a uniform increase in the C-C bond length, and the CNT atoms remain in their hexagonal structure. No defects are detected below the strain ε=0.22. As the strain reaches ε=0.228, the diameter of the tube apparently reduces and the necking effect (Troiani et al., 2003) is observed, as shown in Fig. 8(a). This is mainly because pentagon-heptagon dislocation (5|7) defect (Nardelli et al, 1998) dispersion occurs during the stretching process.

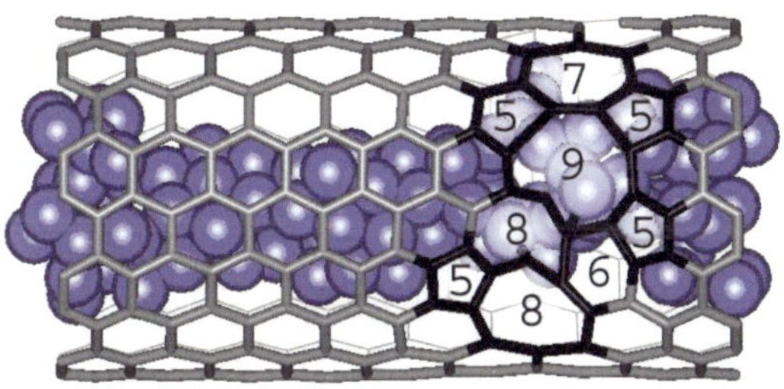

Fig. 9. (color online) Snapshot of the molecular dynamic simulation of the 11-strand CNW@CNT(9,9) at ε=0.228. The nucleation of the pentagon-heptagon dislocation can be seen, and much larger open rings form.

To show clearly the result, we present an atomic view of defect nucleation for CNW@CNT(9,9) in Fig. 9. As diameter of the CNT decreases, the interaction between the outer atoms of the CNW and those at the tube becomes stronger. And it results in forming a more complex hybridization structure; C-C bond can be rotated and some parts of the initial perfect hexagons are activated to form 5|7 defect during the elongation, then defect disperses all over the tube wall. With the combined effect of tension and rotation of C-C bond, some higher order rings such as octagon and nonagon are observed. The tube tension releases its excess strain via the spontaneous formation of topological defects. The closest distance between the atom of the CNW and one of the CNT is 1.41 Å at ε=0.38 (in general, sp3 hybridization occurs at the distance of 1.54 Å). Recent density functional-theory simulation (Lehtinen et al, 2003) indicated that carbon atoms would form covalent bonds with graphene sheets thus affecting shell sliding. Experiments (Strano et al., 2003) also provided evidence that C-C covalent bonds can exists on the surface of a perfect nanotube. As the strain continues to increase, the cylinder of the CNT breaks (as shown in Fig. 8b). Finally, the composite structure is found to be a single carbon atomic chain between the CNT's fragments. The bonds in the linear chain are found to be of a cumulene type (i.e., all of the bond lengths are nearly equivalent), as shown in Fig. 8 d and Fig. 8 e. The potential energy undergoes a critical point to override the energy barrier. When the CNW@CNT starts to form single chain, the potential energy curve reveals fluctuation. Furthermore, this nanobridge structure resembles the experimental result of pulling out a long carbon nanowire at the end of a carbon nanotube (Rinzler et al, 2003). And in theoretical aspect, Wang et al. has also used the same MD simulation method to obtain a long stable single carbon nanowire by pulling some corner atoms of a graphite layer (Wang et al, 2007). Both of them obtained the stable single chain structure employing different methods.

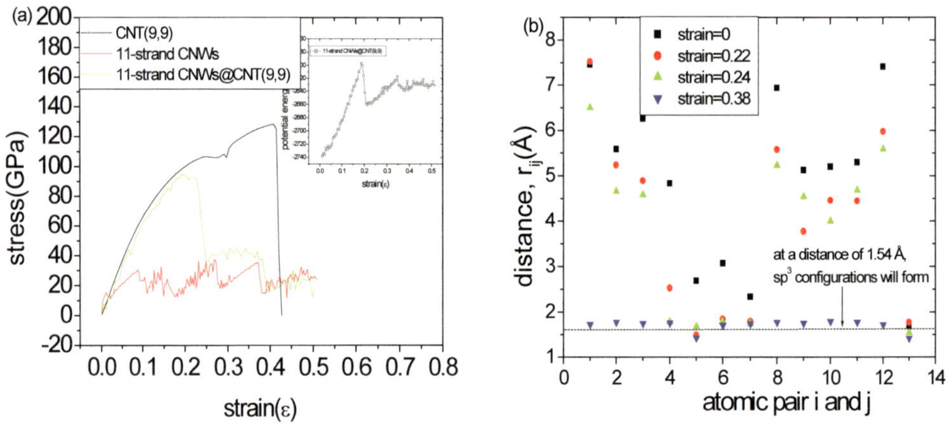

Fig. 10. (color online) (a) Stress-strain curves for the CNT(9,9), 11-strand CNW, and 11-strand CNW@CNT(9,9). Potential energy versus strain for the MD simulation of the stretching of the 11-strand CNW@CNT(9,9) is inserted. (b) Distances between selected atoms from the CNW and CNT at different strains.

Figure 10(a) shows the stress-strain curves for the individual CNT, CNW, and CNW@CNT at the same stretching rate. The critical stress for the CNTs is 127 Gpa, which is in good agreement with the experiment value (150Gpa) obtained by Demczyk (Demczyk et al, 2002). The critical stress of the CNW and CNW@CNT are 35Gpa and 94Gpa, respectively. The strength of the CNT is much larger than the composite nanostructure made up of CNW inserted into CNT. Insertion of the CNW into CNT does not increase the critical strength of the CNT, but actually weakens it. This is mainly because during the stretching process, the atoms of the CNW attach to the defect of the CNT, causing the bond to bend. As the strain increases, the 5|7 defect becomes a large "hole" that accelerates the fracturing of the CNT. The figure of how the potential energy for CNW@CNT(9,9) varies during the elongation is also inserted in Fig. 10(a). It rises steadily until a few C-C bonds break and simultaneously rearrange themselves to form a 5|7 defect. As the defect becomes larger, the potential energy decreases. At the same time, the atoms of CNW attach to the breaking fragment of the CNT to form a new strong covalent bond. As the strain increases, the fracture quickly propagates along the surface of the composite structure. The latter fluctuation of the curve is mainly the result of the gliding of the carbon atoms. As an atom slips from one spot to another, the total potential changes spontaneously from a peak to a valley. Figure 10(b) depicts the distance between a selected pair of atoms, one of which is from the CNT and the other is from the CNW. At ε=0.22, the potential reaches the maximum, and the distances of the four pairs of atoms reaches about 1.54 Å. Between ε=0.22 and ε=0.24, the CNT begins to break away from the circumference of the tube, and the potential energy falls steeply. At ε= 0.38, the distance between the 13 pairs of atoms approaches 1.54 Å, which lead to a complex hybridization that causes an increase in the potential energy.

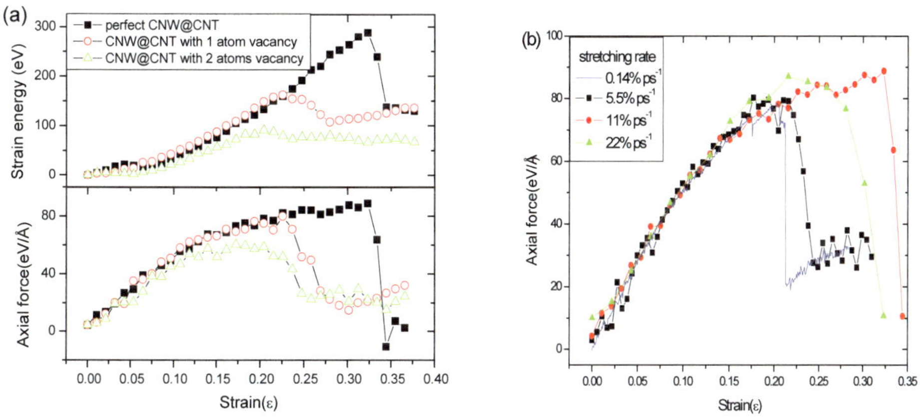

Fig. 11. (color online) (a) the upper one shows the relationship between strain energy and strain for perfect CNW@CNT and CNW@CNT with vacancy. The lower one shows the axial force as a function of strain for CNW@CNT as referred. (b) Axial forces as a function of strain under different stretching rates.

CNT exhibits mainly brittle behavior at high speeds, whereas the CNW@CNT shows apparent flexibility and plasticity. In contrast to our research, Marques et al. predicted that under high temperatures and in the presence of defects (Stone-Wales or vacancies) the tube would exhibit mainly plastic deformation, and the stretching tube would form a single wire composed of fourteen atoms. He (Marques et al, 2004) suggested that such defects (Stone-Wales or vacancies) play an important role in the plastic dynamics of the necking and thinning of a CNT. In order to show how the vacancy defect of CNT affects the stretching behavior of CNW@CNT, we present the variety of strain energy and axial force in the tensile process in Fig. 11(a). The solid square curve corresponds to CNW embedded into perfect CNT(9,9). The hollow circle and triangle curves correspond to CNW embedded into CNT with one atom and two atoms vacancy, respectively. From the upper one in Fig. 11(a) showing the strain energy as a function of strain, CNW embedded into perfect CNT needs higher energy to override the barrier than the other two types. Furthermore, the strain energy curve suffers a sharp drop after the critical point. As the atom vacancy increases, the energy barrier decreases and the curves vary smoothly after the critical point. The lower one in Fig. 11(a) also shows the CNW embedded into perfect CNT can suffer higher axial force than the other two types. Figure 11(b) shows the computed force-strain relationship of the CNW@CNT under different stretching rate ranging from 5.5%ps^{-1} to 22%ps^{-1}. The curves show nonlinear effects before the critical point. The critical forces are 79.2, 88.7 and 87.0 eV/Å under the rates 5.5%ps^{-1}, 11% ps^{-1} and 22% ps^{-1}, respectively. Furthermore, the critical strain does not decrease as the stretching rate increases. The critical strain under stretching rate 5.5% ps^{-1} and 11% ps^{-1} are 22.5%, 32.3%, respectively. This is mainly due to the thermal fluctuation effect under lower stretching rate. This interesting result is similar to the experimental result of stretching nanocrystalline copper (Lu et al, 2001). As the stretching rate increases, some C-C bonds between atoms stretch larger in a shorter time. This abrupt variety in the structure leads to fracture easily. There is usually a modest stretching rate under which the critical strain of this kind of nano scale materials reaches the maximum.

3. Electronic properties of nanowires

The researches on the electron transport through small metallic and semi-conductor nanowires have been a topic of interest over the past few decades. There has also been tremendous interest in the use of metallic nanowires as molecular electronic devices(Krans et al, 2002). Many theoretical studies of ultrathin nanowires or some straight-line uniform ultrathin nanowires with helical multishell structures have been conducted using the atomistic simulation of several metals, and shows other simple and transition metals have found no better linearity for nanowires than that displayed by gold (Sen et al, 2001). Single atomic chains have been prepared by molecular-beam epitaxy on ordered stepped surfaces that are distinguished by an inherent periodic one-dimensional pattern. This widely applicable technique permits the preparation of large-area nanostructures of outstanding quality. Another method that has been used to produce very thin wires and study their structure is the electron beam irradiation of a Au(100)-oriented thin film(Kondo et al, 2000). Several approaches have been used to study the conductance of metallic nanowires, showing conductance curves with plateaus and jumps, but displaying different profiles (Rodrigues et al, 2000). Garcia and Burk(Garcia et al, 1997; Burki et al, 1999) calculated the conductance of disordered nanocontacts and showed that electron scattering due to random impurities shifts nG_0 peaks to lower values and reduces their heights. A strong impurity scattering was predicted by Brandbyge et al., who showed that an electron scatter located at the center of a nanocontact almost washes out the $1G_0$ conductance peak (Brndbyge et al, 1997). Sakai(Enomoto et al, 2002) measured the quantized conductance of Au-Pd and Au-Ag alloy nanocontacts for a wide range of Pd and Ag concentrations and studied how the $1G_0$ conductance of Au changes with alloying. Despite all this progress, only few studies have been conducted on the atomic structures of metallic nanowires and their electrical transport properties. Thus, the objective of this chapter is to use atomistic simulations to systematically study the structures of Ni nanowires and explain the correlation between the electron conductance and atomic structures.

3.1 Simulation method

All the calculations were performed using MS molecular modeling software packages. These nanowires are firstly optimized by molecular dynamics (MD) based method such as the module of DISCOVER, Forcite or SA method to perform structural optimization. Then the low-energy configurations from the above MD simulation within the empirical potential were further optimized using density function theory (DFT) with the Perdew-Burke-Ernzerh exchange correlation function (Perdew et al, 1996) and an all-electron basis set of the double-numerical-plus-d-polarization type, as implemented in the DMOL package. In the DFT optimizations, both the atomic coordinates and 1-D super cell lengths along the wire axis were fully relaxed.

The current-voltage *(I-V)* characteristic and electron transmission probability were calculated using the nonequilibrium Green function (NEGF) formalism and density functional theory (DFT) which determined the *I-V* characteristics and transmission of nanowires sandwiched between the two metallic contacts, one of which could be a scanning probe. NEGF provides a powerful conceptual and computational framework for treating quantum transport in atomic-level materials, such as nanotubes, nanowires, and moleculars. The calculation details are given as follows:

The primary objective of this stage is to specify the structure of the nanowire. The nanowires and their geometry connected with gold atoms of the contact to form covalent bond. The nanowire-contact distance that we used is constant when different nanowires were connected to contacts. Gold (111) films were used as electrodes and the nanowires were perpendicularly sandwiched between the electrodes. The vertical distance between the end atoms of the nanowire and the gold contacts was set to 1.900 Å. The self-consistent-field approach is coupled with nonequilibrium Green's function (NEGF) formalism to describe electronic transport under an applied bias (Damle et al, 2002). The whole system is divided into a contact subspace and a nanowire subspace. This makes the simulation procedure simple to implement.

3.2 Electrical transport properties of Nickel nanowires

The optimized structures of Ni nanowires are shown in Fig. 13, where the number of inserted Ni atoms is 8, 16, 16, 24, and 30. In our simulations, an 8-atom stable single chain was obtained as Wire 1. Wire 2 formed in CNT(9,0) is an optimized double-chain structure that is composed of two parallel single chains in which the atoms are arranged in an interleaving fashion. Wire 3, another stable nanowire formed in CNT (9,0) is made up of two helical strand of atomic chains. Wire 4 features three-strand packing, and Wire 5 is a nonhelical shell structure. In general, the stable structures of the ultrathin Ni nanowires in the carbon nanotubes are multishell packs that are composed of coaxial cylindrical shells. In some cases, the nickel nanowires have a single-atom chain at the center. Each shell is formed by rows of atoms that are helically wound upwards side by side.

The transmission probability *(TE)* is dependent on the microstructure as shown in Fig. 14(a), for the *TE* of wire 2 with two parallel atomic chains is larger than Wire 3 with a helical structure. Transmission properties of the systems are determined by the electronic structure of the combined nanowire and electrode systems. The distance between the electrodes plays an important role in the current of the junction. In simple terms, the metal-nanowire-metal system can be viewed as a finite quantum well that sets up in between the metal surfaces. The direct tunneling between the electrodes, despite the possible electrode-induced gap states in the metal-air-metal junctions, is considered to be too small to contribute considerably to the current of the typical size molecular junctions. Interestingly, shown in Fig. 14(c), the shape of the curve of the single chain (wire 1) is markedly different from that of the chain that is embedded in the carbon nanotube. Ke and Baranger(Ke et al, 2006) investigated electrical transport in the nanotube-metal junctions of carbon tubes surrounded

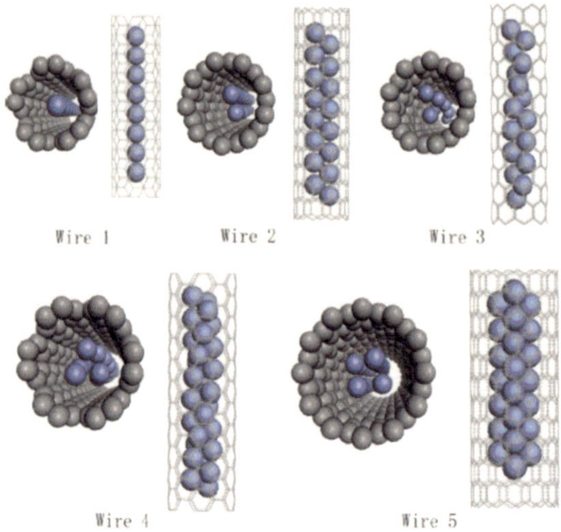

Fig. 13. (Color online) Images of the metallic nanowires with different diameters: Wire 1, single atomic chain; wire 2, double chain composed of two parallel atomic chains; wire 3, helical double atomic chain; wire 4, helical shell structure wound by three atomic strands; wire 5, shell structure composed of four parallel atomic chains.

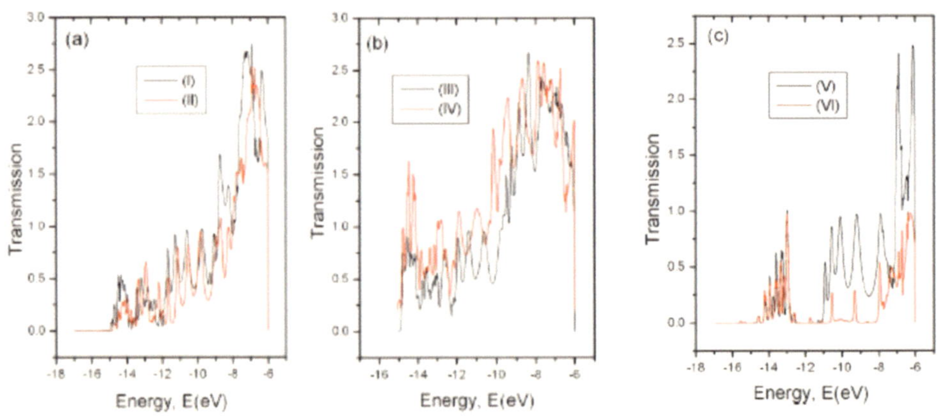

Fig. 14. (Color online) Transmission function *TE* for a series of nanowires and nanochains, with the transmission function of a chain embedded in a carbon tube for comparison: (I) curve for wire 2 (parallel double chains); (II) curve for wire 3 (helical double chains); (III) curve for wire 4 (three helical strands); (IV) curve for wire 5 (four parallel strands); (V) curve for wire 1 (single atomic chain); (VI) curve for the atomic chain embedded in a carbon nanotube.

by metal atoms. They found that the transport property of the single chain was much better than that of the carbon tube surrounded by the metal atoms. The coated shell around the chain would decrease the transmission property of single chain. In our calculations, the carbon tube can be viewed as the coated shell which makes the transmission decrease. To make sure of the coupling between the Ni atomic chain and the electrodes, the studied CNT is shorter than the Ni atomic chain. In our studies, the CNT can be looked upon as an adsorbate like in Bogozi's study. It is widely accepted that the adsorbate induced conductance change is due to the scattering of the conduction electrons by the adsorbates as the electrons impinge on the surface(Bogozi et al, 2001). The interaction between the CNT and the Ni atomic chain leads to rearrangement of the electrons in system which remarkably influences the transmission of the Ni atomic chain. Nearly the whole surface of the Ni atomic chain is covered by the CNT, which causes a sharp decrease in the transmission.

In Fig. 15, we plot the *I-V* characteristic curves that correspond to the different nanowires. Six *I-V* curves were obtained from different samples, many of which showed the same general, nonlinear, symmetric shape. An interesting point in Fig. 15 is the region around the zero voltage in the I-V curves. For the single atomic chain in Fig. 15(a), a clear flat was observed around the zero voltage, but as for others, the flat was not clear. The slope of the I-V curve for wire 5 is the largest among all the nanowires. Furthermore, the conductance increases with diameter of nanowire. Increasing the diameter of a nanowire therefore appears to improve the transmission area and thus the conductance. A recent experiment (Guo et al, 2003) also showed that a larger diameter yields a larger conductance, as is the case with bulk systems. The experimental results of Ohnishi and Kondo further indicated that the strand number of nanowire would play role in conductance of nanowires. They found that a double strand has twice the unit conductance of a single strand if the interaction between the two individual rows is not strong. Our calculation is consistent with the experimental results of Ohnishi and Kondo.

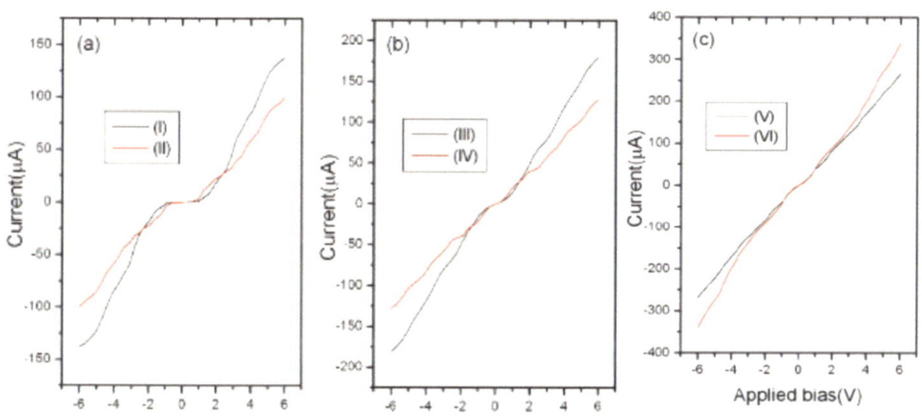

Fig. 15. (Color online) I-V characteristic curves obtained from our calculations: (I) curve for wire 1; (II) curve for wire 2; (III) curve for wire 3; (IV) curve for wire 4; (V) curve for wire 5; (VI) curve for the atomic chain embedded in a carbon nanotube.

It also can be seen in Fig. 15 that the current increases with the external bias. It is clear that the computed *I-V* curve does not follow an ohmic pattern. This is a characteristic feature of nanostructured materials. Due to the observed nonlinear I-V characteristic, there exists no true linear resistivity for all samples. Correia's research results further demonstrated the nonlinear character of *I-V* curves for metallic nanowires(Costa-Kramer et al, 1997), and he suggested that the most plausible explanation for this nonlinear behavior was connected with the Coulomb blockade phenomenon in small conducting systems. We believe that the nonlinear behavior is caused by the quantum size effect. When the size of a metallic nanowire or nanochain is comparable to the electron de Broglie wavelength, electrons in the nanowire and those transmitted through the nanowire manifest the quantum size effect. Under spatial confinement, the electron energies are also quantized, which leads to many effects. Electron transport in nanowires must be governed by these quantum effects and also by the enhanced boundary scattering. As electrons in a quantum wire are confined, they form quantized standing wave states along the nanowire. This kind of quantum size effect can cause the properties of nanowires and nanochains, such as the oscillation behavior with diameter in *I-V* curves, to differ significantly from those of bulk materials. The difference between the *I-V* curve of the single chain and that of the chain embedded in a carbon nanotube is also clearly noticeable, and the quantum effect between the nanochain and the carbon nanotube causes the nonlinear feature to become quite evident.

4. Melting behavior of 1-D Nickel nanowire

Nanoscale materials have unique properties different from crystal structures. Currently various types of nanodevices are being studied, and, in particular, metallic nanowires receive much attention in both industrial and academic fields because of their magic structure and conductance. Takayanagi's group succeeded in making suspended gold and platinum nanowires(Oshima et al, 2002) in ultrahigh vacuum. Many theoretical studies on ultrathin nanowires have been done using atomistic simulations for several metals, such as Ag (Finbow et al, 1997). Much work has been done for synthesization of nanowire by many new methods(Ono et al, 1997). Genetic algorithm study of the structure of nanowire has been carried out by our research group(Wang et al, 2001). A microscopic description of the melting behavior of nanowire is probably one of the most important problems in physical and material sciences. Simulations have been carried out for nanoscale-width platinum and silver wires. Melting of gold clusters has been investigated to study the thermal evolution of structural and dynamic properties (Cleveland et al, 1999). The thermal properties of ultrathin copper nano-bridges were investigated by Kang (Kang et al, 2002). Although these numerical studies account for the general feature of the melting transition, the microscopic process during the phase transition is still unclear. Experiments have not allowed us to directly observe the melting process. Molecular dynamics (MD) simulations have become a favorite tool for investigating the physics of nanowires theoretically. In this letter, we use MD simulation to address this issue.

4.1 Simulation methodology

We chose to employ the well-fitted tight-binding many body potential with proven ability to reliably model various static and dynamic properties of transition and noble metals. The configurations of nanowires are obtained by the genetic algorithm (GA) previously used by our group with one-dimensional periodical boundary condition along the wire axis direction. The MD time step is chosen as 2.5 fs. The MD simulations start from an initial temperature (400 K). The temperature gradually increases toward final temperature (1500 K) at two heating rates equal to 20 K/ fs and $1x10^{10}$ K/ s, respectively.

4.2 Results and discussion

Figure 16 shows the two kinds growth sequences of nickel nanowires obtained by GA optimization. The nickel nanowire that we studied is multishell structure, which is composed of coaxial cylindrical tubes. Each shell of nickel wire winds up helically by atomic rows. The surface of each shell exhibits a nearly triangular network. W3-1 and W3-2 have trigonal multishell packing. The thinnest wire W3-1 has a three-strand structure. W5-1 and W5-2 are not helical packing. W5-1 and W5-2 are composed of two-shell and three-shell centered pentagonal structures, respectively.

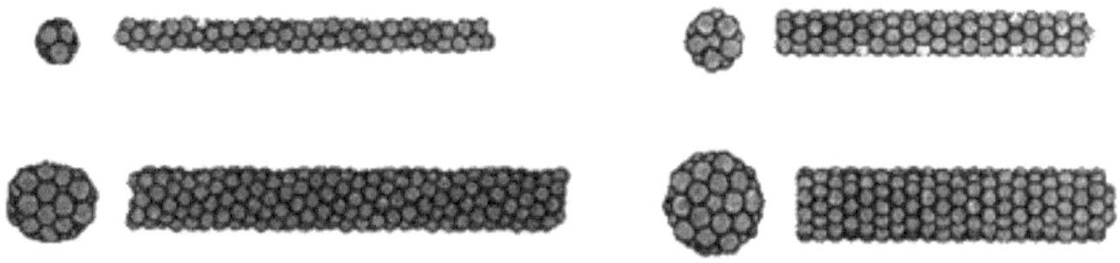

Fig. 16. Morphologies of structural growth sequences in selected nickel nanowires. The nanowire from the top to bottom on the left-hand side corresponds to W3-1, W3-2, and the nanowire from the top to bottom on the right-hand side corresponds to W5-1, W5-2, respectively.

Figure 17 gives us an overall physical picture of the melting process of nanowire. We find that the melting of nanowire starts from the interior atoms. Interestingly, when the slow heating rate is applied, the central atoms first move along the wire axis direction at a rather low temperature, while the helical outer shells are almost invariant. It is worth noting that the moving central atoms are not discrete but to build up a long monostrand atomic chain. The monostrand atomic chain is very stable. The first single atomic chain can exist in a wide temperature region (850–900 K). As the temperature arises, the regular monostrand atomic chain begins to become deformed and finally, the atomic chain is broken to form a new type of cluster.

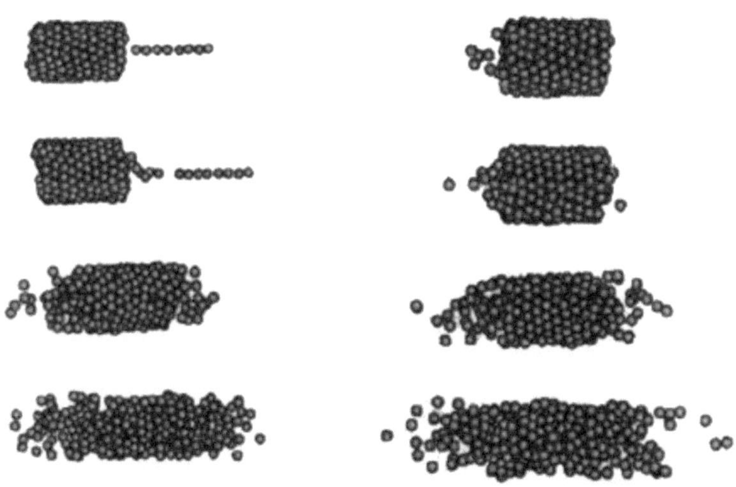

Fig. 17. Structural evolution of nickel nanowire with the temperature increase. The picture from the top to the bottom on the left-hand side corresponds to the 850, 900, 1000, 1050 K, and the picture from the top to the bottom on the right-hand side corresponds to 900, 950, 1050, 1100 K. The pictures on the left-hand side correspond to the heating rate 10^{10} K/s, thepictures on the right correspond to the heating rate 20 K/ s.

The atomic volume(V) and internal energy (E) are the common approaches for monitoring structural changes of the nanowire melting illustrated in Fig. 18. This figure exhibits three temperature regions, denoted by A, B, and C, in which the nanowire behaves differently. At the temperature region A, the atoms are located in well-defined lattice-site equilibrium positions, around which the atoms execute relatively small amplitude motions, with slight variations of interatomic distances, as long as structural changes do not occur. Upon heating the nanowire undergoes a transition from the temperature region A to region B in which atomic volume and internal energy increase very strongly, thus indicating that structural changes occur within the nanowire. With the temperature increase, the system undergoes a transition from region B to region C in which the atomic volume and internal energy is relatively high. It is worth mentioning that, when the slow heating rate is applied in this simulation, the melting temperature becomes low.

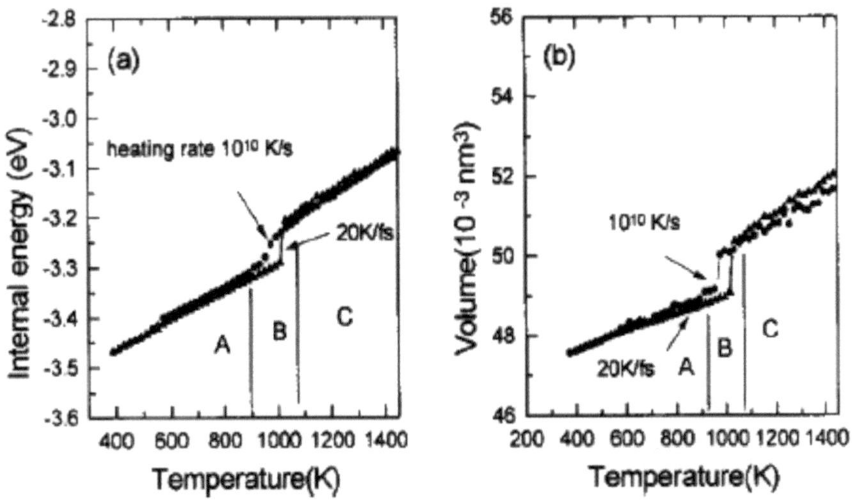

Fig. 18. The internal energy (a) and atomic volume (b) of the nanowire as a function of temperatures.

Figure 19 shows the pair correlation function (PCF) of nanowires at different temperatures. At low temperature, the crystalline structure is highly stable. With increasing temperature the peaks are broadened and lowered. Some peaks are nearly lost. At temperature T=900–1200 K, a diffusion dynamics is thermally activated. This is signaled by the corresponding pair correlation function. Finally, the crystal order is broken and the melting occurs. The shape of the PCF at the temperature 900–1200 K implies that coexistence of liquid and crystal is presented.

To further illustrate the melting phenomena and to distinguish the local symmetry of clusters in the melting process of nanowire, the common bond pair analysis previously used by us is adopted to characterize the local cluster's changes in the melting process of nanowire. In this technique, two atoms are said to be near neighbors if they are within a specified cutoff distance of each other. The first index denotes which peak of the PCF belongs to the pair under consideration; the second one counts the number of common nearest neighbors of that pair; the third one specifies the number of particles denoted by the second one; a fourth one sometimes is necessary to distinguish configurations having the

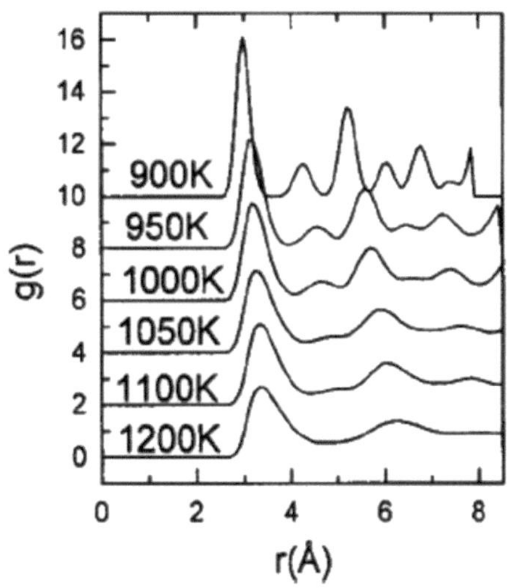

Fig. 19. The pair correlation functions at different temperatures in the melting process of nanowire.

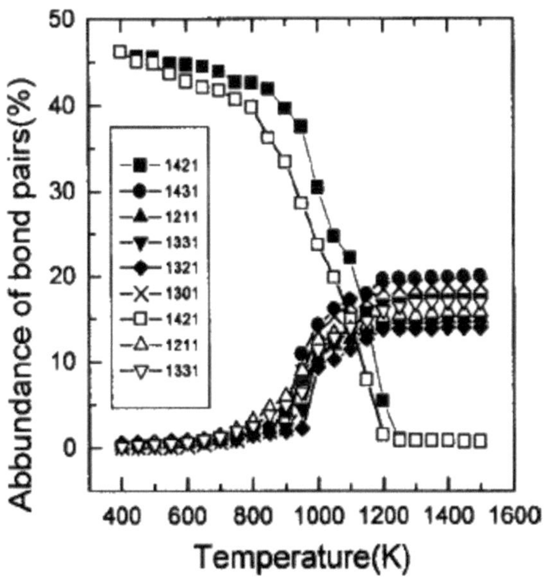

Fig. 20. Abundance of selected local clusters as a function of temperatures. Open symbol corresponds to the heating rate 10^{10} K/ s, solid symbol corresponds to the heating rate 20 K/ fs.

first three indices equal, yet topologically different. This method is able to distinguish various local structures between fcc, hcp, bcc, and the icosahedral structures. The fcc structure has the type 1421 bond pair, whereas the hcp crystal has the equal number of 1421 and 1422 bond pairs. The 1201 and 1311 bond pairs represent the rhombus symmetrical features of short-range order. Fig. 20 shows the evolution of the local cluster during the melting process of the nanowire. The increase in the abundance of the local clusters 1331, 1321, 1311, etc., and the decrease of the clusters 1421 indicate the occurrence of the melting of nanowire. The fcc structure in nanowire is presented through the existence of 1421 pairs, but not found in liquid from the melting of nanowire. The pairs 1421 are the main components in the nanowire. The overall melting point of nanowire can be defined as 1050 K from the sudden changes of local cluster curves. The local clusters in liquid from the nanowire melting are not decahedron, but rhombus symmetrical structures.

Our simulation result shows that as a result of melting process the amount of rhombus symmetric structure reduces in favor of the crystalline order. During the melting process, the local cluster 1421 in nanowire would decompose into other local clusters such as 1311 and 1321, etc. Local cluster diffusion leads to the occurrence of the melting. Cluster diffusion from equilibrium position is the main melting features of nanowire. Because there is frustration which comes from the difficulty in close packing with perfect cluster in flat three-dimensional space, therefore liquid cannot be composed of only one kind of cluster. The existence of various local clusters in liquid just meets the packing demand. Therefore a suitable proportion of local cluster is an important character for the nanowire melting. With the temperature increase, the local cluster begins to diffuse; this will lead to some new defects in nanowires. These defects would cause the decease of the average binding energy. The nanowires with a lot of defects should have lower thermal stability. Therefore, defect would accelerate the melting of the nanowire. Figure 20 shows that the evolution of the local cluster at two different heating rates is quite different. The transition temperature at slow heating rate is lower than that at fast heating rate.

It must be pointed that the 1421 represents the fcc microcrystal structural unit rather than integrals. 1431, 1331, 1321, 1301, and 1211 are also clusters which are ordered structures. Nickel nanowire contains a lot of 1421 microcrystal units. In this study, as the temperature increases, not only is diffusion of single atom observed, but also the diffusion of local clusters is found. Therefore, we think the nanowire melting is due to the diffusion of local clusters. In Fig. 20, the clusters 1421, 1431, 1331, 1321, etc., are, in fact, some leftover crystal units. The existence of these leftover crystal structural units implies that there are some leftover crystal units in liquid resulting from the nanowire melting. That is to say, the fact that some small crystal structural units moves from the nanowire causes the beginning of the nanowire melting. Melting behaviors of other nanowires with different diameters are also studied. It is noted that the melting point depends on the diameter size. In general, larger diameter would cause higher melting point. To explore the size dependence of nanowire melting temperature, we plot the overall melting temperature Tm versus wire diameter (D) in Fig. 21.

The T_m for the wires with the hexagonal growth sequence fit well to a linear dependence as $T_m = T_0 - \delta / D$, where T_0=1638K is the extrapolated melting temperature at the infinite limit, δ=857 K nm describes the linear dependence of T_m with wire diameter D (in units of nanometers). (When the slow heating rate is applied, T_0=1503 K, δ=782.57 K nm.) Such 1/D dependence of melting temperature for nanowire is similar to the well-known size relationship for metal nanoclusters. The heating rate has great influence on the melting temperatures of nanowires. The melting point at the slow heating rate is lower than that at the fast melting point. As compared with the linear size dependence for the hexagonal wires, almost belong to the same size relationship while the T_m of other nanowires deviate from such linear fit. These differences indicate that the atomic structures of nanowires play a significant role in determining the melting behavior of nanowires.

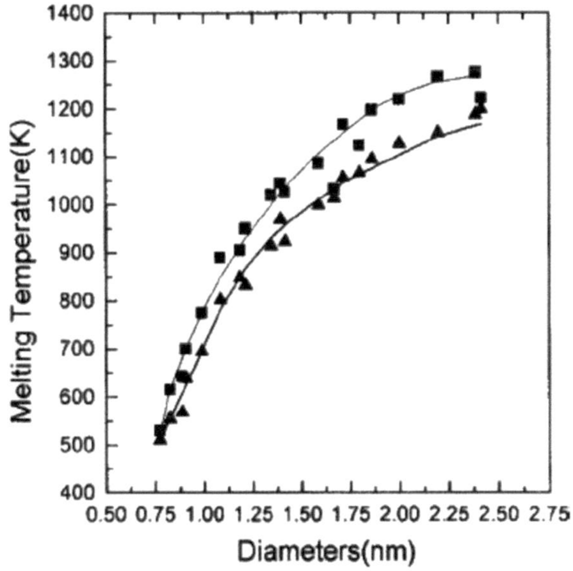

Fig. 21. Overall melting temperature T_m *(K)* vs nanowire diameters D (nm). The square symbol and the top line correspond to the heating rate 20 K/ fs, and the triangular symbol and the bottom line correspond to the heating rate 1×10^{10}K/s, respectively.

5. Conclusion

Our simulation results have shown three aspects of properties of nanowires which are as follows:

1. Simulation results indicate that the maximum tensile strain of carbon nanowire is up to 245% before tensile failure. The maximum stress for a CNW is 16.65 GPa which is lower than that of the carbon nanotubes. During the elongation process, a stable single atomic chain is obtained just like to draw a silk from a cocoon. The superplastic properties of carbon nanowires could prove useful in the fabrication of carbon nanocomposites.
2. This composite of CNW@CNT undergoes a large elongation in which a pentagon-heptagon dislocation (5|7) defect, octagonal defects and higher-order rings are observed. Insertion of the CNW into CNT increases the plastic elongation of a CNW@CNT under axial stretching. As the strain continues to increase, the composite structure is found to become a long single carbon atomic chain. Interestingly, insertion of the CNW into CNT does not increase the critical strength of the CNT but weaken. A large hole caused by (5|7) defect make the carbon tube fracture easily.
3. The electronic transport properties of these nanostructures have been studied, and we find that transmission is dependent on the geometric structures of nanowires and their size of diameters. Because of the quantum size effect, the I-V curves of the nanowires are clearly nonlinear, which does not follow an ohmic pattern.
4. Molecular dynamics simulation has been performed to study the melting process of nanowire by means of a tight binding model potential. We set up in the present work a general picture of the physical process which governs the melting phenomenon in nickel nanowire. The atomic volume, internal energy, the pair correlation function, and

the local clusters provide a strong and clear signature of structural changes as the melting occurs in nanowire. The thermal stability of nickel nanowires is dependent on the structures and size of the nanowire; the melting temperature of nickel nanowire is lower than the bulk's.

6. Acknowledgements

We would like to acknowledge the support from the National Natural Science Foundation of China [Grant Nos.50971081, 50831003, 50871062 and 50772061] and the National Basic Research Program of China [Grant No.2007CB613901]. This work is supported by a grant from the National Science Fund for Distinguished Young Scholars [Grant No. 50625101] and Scientific Research Foundation for Returned Scholars [Grant No. JIAO WAI SI LIU2007-1108], Ministry of Education of China. We also thank the support from the Natural Science Fund for Distinguished Young Scholars of Shandong [Grant No. JQ200817]. This work is also supported by the Natural Science Fund of Shandong Province (ZR2009FM043). This work was supported by the Ph. D. programs foundation of ministry of education of China (No. 20090131110025).

7. References

Li, H.; Sun, F.W.; Li, Y.F.; Liu, X.F. & Liew, K.M. (2008). Theoretical studies of the stretching behaviour of carbon nanowires and their superplasticity, Scripta Mater, Vol. 59, 479-482.

Li, H.; Sun, F.W.; Liew, K.M. & Liu, X.F. (2009) Stretching behavior of a carbon nanowire encapsulated in a carbon nanotube, Scripta Mater, Vol. 60, 129-132.

Li, H.; Zhang, X. Q.; Sun, F. W.; Li, Y. F.; Liew, K. M.; He, X. Q. (2007). Theoretical study of the electrical transport of nickel nanowires and a single atomic chain, J. Appl. Phys, Vol. 102, 013702.

Li, H.; Pederiva, F. & Wang G. H. (2005). How does the Nickel Nanowire Melt? Appl.Phys.Letts, Vol. 86(1), 011913.

Ribarsky, M.W. & Landman, U. (1988). Dynamical simulations of stress, strain, and finite deformations, Phys. Rev. B, Vol. 38, 9522-9537.

Brenner, D.W.; Shenderova, O.A.; Harrison, J.A.; Stuart, S.J.; Ni, B. & Sinnott, S.B. (2002). A second-generation reactive empirical bond order (REBO) potential energy expression for hydrocarbons, J. Phys.: Condens. Mater, Vol. 14, 783-802.

Sinnott, S.B.; Shenderova, O.A.; White, C.T. & Brenner, D.W. (1988). Mechanical Properties of Nanotubule Fibers and Composites Determined from Theoretical Calculations and Simulations, Carbon, Vol. 36, 1-9.

Shenderova, O.A. & Brenner, D.W. (2000). Atomistic modeling of the fracture of polycrystalline diamond, Phys. Rev. B, Vol. 61, 3877-3888.

Huhtala, M.; Krasheninnikov, A.V.; Aittoniemi, J.; Stuart, S. J.; Nordlund, K. & Kaski, K. (2004). Improved mechanical load transfer between shells of multiwalled carbon nanotubes, Phys. Rev. B, Vol. 70, 045404.

Stuart, S.J.; Tutein, A.B. & J.A. Harrison. (2000). A reactive potential for hydrocarbons with intermolecular interactions, J. Chem. Phys, Vol. 112, 6472.

Sammalkorpi, M.; Krasheninnikov, A.; Kuronen, A.; Nordlund, K. & Kaski, K. (2004). Mechanical properties of carbon nanotubes with vacancies and related defects, Phys. Rev. B, Vol. 70, 245416.

Belytschko, T.; Xiao, S. P.; Schatz, G.C.& R.S. Ruoff, (2002). Atomistic simulations of nanotube fracture, Phys. Rev. B, Vol. 65, 235430
Gear, C.W. (1971). Numerical Initial Value Problems in Ordinary Differential Equations, Prentice-Hall, New Jersey, USA.
Liew, K.M.; Wong, C.H.; He, X.Q.; Tan, M.J. & S.A. Meguid, (2004). Nanomechanics of single and multiwalled carbon nanotubes, Phys. Rev. B, Vol. 69, 115429.
Wang, Q.; Duan, W.H.; Liew, K.M. & X.Q. He, (2007). Local buckling of carbon nanotubes under bending, Appl. Phys.Lett. Vol. 90, 033110.
Nose, S. (1984). A molecular dynamics method for simulations in the canonical ensemble, Mol. Phys. Vol. 52, 255–268.
Hoover, W.G. (1985). Canonical dynamics: Equilibrium phase-space distributions, Phys. Rev. A, Vol. 31, 1695-1697.
Asaka, K.; Kizuka, T. (2005). Atomistic dynamics of deformation, fracture, and joining of individual single-walled carbon nanotubes, Phys. Rev. B, Vol. 72, 115431.
Ogata, S.; Shbutani, Y. (2003). Ideal tensile strength and band gap of single-walled carbon nanotubes, Phys. Rev. B, Vol. 68, 165409.
Liew, K.M.; He, X.Q. & C.H. Wong, (2004), On the study of elastic and plastic properties of multi-walled carbon nanotubes under axial tension using molecular dynamics simulation, Acta Mater. Vol. 52, 2521.
Yu, M. F.; Files, B.S. & R.S. Ruoff, (2000). Tensile loading of ropes of single wall carbon nanotubes and their mechanical properties, Phys. Rev. Lett. Vol. 84, 5552.
Lennard-Jones, L.E. (1924). On the Determination of Molecular Fields. I. From the Variation of the Viscosity of a Gas with Temperature, Proc. R. Soc. Lond. Ser. A ,Vol. 106, 441.
Troiani, H. E.; Miki-Yoshida, M.; Camacho-Bragado, G. A.; Marques, M. A.; Rubio, L. A.; Ascencio, J. A. & Jose-Yacaman, M. (2003). Direct Observation of the Mechanical Properties of Single-Walled Carbon Nanotubes and Their Junctions at the Atomic Level, Nano Lett. Vol. 3, 751-755.
Nardelli, M. B.; Yakobson, B. & Bernholc, J. (1998). Mechanism of strain release in carbon nanotubes, Phys.Rev. B, Vol. 57, R4277-R4280.
Lehtinen, P. O.; Foster, A. S.; Ayuela, A.; Krasheninnikov, A.; Nordlund, K. & Nieminen, R. M. (2003). Magnetic Properties and Diffusion of Adatoms on a Graphene Sheet, Phys. Rev. Lett. Vol. 91, 017202.
Strano, M.S. et al., (2003). Electronic Structure Control of Single-Walled Carbon Nanotube Functionalization, Science. Vol. 301, 1519-1522.
Rinzler, A. G.; Hafner, J. H. & Nikolaev, P. (1995). Unraveling nanotubes: field emission from an atomic wire, Science. Vol. 269, 1550-1553.
Wang, Y.; Ning, X. J.; Lin, Z. Z.; Li, P. & Zhuang, J. (2007). Preparation of long monatomic carbon chains: Molecular dynamics studies, Phys. Rev. B, Vol.76, 165423.
Demczyk, B. G.; Wang, Y. M.; Cumings, J.; Hetman, M.; Han, W.; Zettl, A. & R. O. Ritchie, (2002). Direct mechanical measurement of the tensile strength and elastic modulus of multiwalled carbon nanotubes, Mater. Sci. Eng. A, Vol. 334, 173-178.
Troiani, H.E.; Miki-Yoshida, M.; Camacho-Bragado, G. A.; Marques, M. A. L.; Jose-Yacaman, M. & Rubio, A. (2004). Direct Observation of the Mechanical Properties of Single-Walled Carbon Nanotubes and Their Junctions at the Atomic Level, Nano Lett. 4 811-815.
Lu, L.; Li, S.X. & Lu, K. (2001). An abnormal strain rate effect on tensile behavior in nanocrystalline copper, Scripta Mater. Vol. 45 1163-1169.

Krans, J. M.; Ruitenbeek, J. M. V; Fisun, V. V.; Yanson, J. K. & deJongh, L. J. (1995). The signature of conductance quantization in metallic point contacts, Nature, Vol. 375, 767-769.

Sen, P.; Ciraci, S.; Buldum, A. & I. P. Batra, (2001). Structure of aluminum atomic chains, Phys. Rev. B, Vol. 64, 195420.

Gambardella, P.; Blanc, M.; Brune, H.; Kuhnke, K. & K. Kern, (2000). One-dimensional metal chains on Pt vicinal surfaces, Phys. Rev. B, Vol. 61, 2254.

Kondo, Y.; Takayanagi, K. (2000). Synthesis and Characterization of Helical Multi-Shell Gold Nanowires , Science, Vol. 289, 606-608.

Rodrigues, V.; Fuhrer, T.; Ugarte1, D. (2000). Signature of Atomic Structure in the Quantum Conductance of Gold Nanowires, Phys. Rev. Lett, Vol. 85 (19), 4124-4127.

Garcia-Mochales, P.; Serena, P. A. (1997). Disorder as Origin of Residual Resistance in Nanowires, Phys. Rev. Lett, Vol. 79, 2316-2319.

Burki, J.; Stafford, C. A.; Zotos, X. & Baeriswyl, D. (1999). Cohesion and conductance of disordered metallic point contacts, Phys. Rev. B, Vol. 60, 5000-5008.

Brandbyge, M.; Yacobsen, K. W.; Norskov, J. K. (1997). Scattering and conductance quantization in three-dimensional metal nanocontacts, Phys. Rev. B, 55, 2637-2650.

Enomoto, A.; Kurokawa, S. & Sakai, A. (2002). Quantized conductance in Au-Pd and Au-Ag alloy nanocontacts, Phys. Rev. B, Vol. 65,125410.

Perdew, J. P.; Burke, K.; Ernzerhof, M. (1996). Generalized Gradient Approximation Made Simple, Phys. Rev. Lett, Vol. 77, 3865-3868.

Damle, P. S.; Ghosh, A. W. & Datta, S. (2002). First-principles analysis of molecular conduction using quantum chemistry software, Chem. Phys, Vol. 281,171-187.

Ke, S. H.; Yang, W. & H. U. Baranger, (2006). Nanotube-metal junctions: 2- and 3-terminal electrical transport, J. Chem. Phys, Vol. 124, 181102.

Bogozi, A.; Lam, O.; He, H.; Li, C.; Tao, N. J. L.; Nagahara, A.; Amlani, I. & Tsui, R. (2001). Molecular Adsorption onto Metallic Quantum Wires, J. Am. Chem. Soc, Vol. 123, 4585-4590.

Guo, A. J.; Wang, Q.; Lundstrom, M. & Dai, H. (2003). Ballistic Carbon Nanotube Transistors, Nature, Vol. 424, 654-657.

Costa-Kramer, J. L.; Garcia, N.; Garcia-Mochales, P.; Serena, P. A.; Marques, M. I. & Correia, A. (1997). Conductance quantization in nanowires formed between micro and macroscopic metallic electrodes, Phys. Rev. B, Vol. 55, 5416-5424.

Oshima, Y.; Koizumi, H.; Mouri, K.; Jirayama, H.; & Takayanagi, K. (2002). Evidence of a single-wall platinum nanotube, Phys.Rev. B, Vol. 65, 121401.

G. M. Finbow, R. M. Lynden-Bell, and I. R. McDonald, (1997). Atomistic simulation of the stretching of nanoscale wires, Mol. Phys, Vol. 92, 705-714.

Ono, T. Saitoh, H. & Esashi, M. (1997). Si nanowire growth with ultrahigh vacuum scanning tunneling microscopy, Appl. Phys. Lett, Vol. 70, 1852.

Wang, B. L.; Yin, S. Y.; Wang, G. H. & Zhao, J. J. Novel structures and properties of gold nanowires, (2001). Phys. Rev. Lett, Vol. 86, 2406

Cleveland, C. L.; Luedtke, W. D. & Landman, U. (1999). Melting of gold clusters, Phys. Rev. B, Vol. 60, 5065-5077.

Kang J. W. & Hwang, H. J. (2002). Thermal properties of ultra-thin copper nanobridges, Nanotechnology, Vol. 13, 503-509.

15

Mn–Fe Nanowires Towards Cell Labeling and Magnetic Resonance Imaging

Ken Cham-Fai Leung and Yi-Xiang J. Wang
Center of Novel Functional Molecules, Department of Chemistry and Department of Diagnostic Radiology and Organ Imaging, Prince of Wales Hospital The Chinese University of Hong Kong, Shatin, NT Hong Kong SAR, P. R. China

1. Introduction

In the past two decades, our understanding in biology, materials science and nanotechnology has expanded rapidly. The inevitable intersection of these three disciplines has set in motion the development of an emerging research area, nanobiotechnology or nanobiomedical science, which offers exciting and abundant opportunities for discovering new processes and phenomena. In particular, the advances in the synthesis and characterization of nanoscale materials allow scientists to understand and control the interactions between nanomaterials (e.g., nanowires, nanofibers, nanoparticles, nanobelts or nanoribbons, and nanotubes) and biological entities (e.g., nucleic acid, proteins, or cells) at molecular or cellular levels. These advances promise major achievements in the life sciences. By way of an example, the research on magnetic nanomaterials (Skomski, 2003) has attracted a lot of attention because of their numerous applications including magnetic separation of biomolecules (Nam et al., 2003), as biocompatible contrast agents for magnetic resonance imaging (MRI) (Wang et al., 2001; Berry & Curtis 2003; Pankhurst et al., 2003; Tartaj et al., 2003; Nitin et al., 2004; Lee et al., 2007), magnetic recording (Darques et al., 2009), spintronic devices and magnetic sensing (Thurn-Albrecht et al., 2000; Allwood et al., 2002; Redl et al., 2003). Relatively large, near-monodisperse spherical iron oxide (Fe_3O_4, magnetite) nanoparticles with average core diameters of 200–1000 nm are commercially available (e.g., Feridex), some of them are used as contrast agents for *in vivo* clinical MRI. These relatively large magnetite nanoparticles can be obtained by a structural refinement process of bulk materials or a recrystallization from small nanoparticles. In respect to biological applications, the particles' surface is usually coated with dextran as a biocompatible, protective layer. Biodegradable magnetic nanoparticles can be labeled intracellularly towards cellular and molecular imaging, and also used for targeted drug carrier together with controlled release capabilities (Lübbe et al., 2001), and radio frequency therapy of tumours via hyperthermia (Fortin et al., 2007). There are indications that the size, shape, structure, and the functional surfaces (e.g., amine, hydroxyl, poly(ethylene glycol), etc.), texture and porosity of many biocompatible nanomaterials are important parameters which influence the rate of cell attachment, internalization, uptake and vascular dynamics (Corot et al., 2006; Wilhelm & Gazeau 2008; Mitragotri & Lahann 2009; Park et al., 2009).

The properties of one-dimensional (1-D) nanowires toward biological systems are attracting increasing attention recently (Fang & Kelley 2009; Cohen-Karni et al., 2009). Generally, nanomaterials having elongated shapes and correspondingly increased surface area are more effective *in vivo* due to the geometrically enhanced multivalent interaction between ligands and receptors (Mitragotri & Lahann 2009). In comparison to the zero-dimensional nanoparticles, 1-D nanowires provide more parallel-aligned surface functionalities per molecular structure, which are suitable for multivalent molecular recognition on cell surfaces. Potential application of the magnetic nanowires may offer endothelium surface magnetic labeling because of their steric hindrance in passing into the extra-vascular space. Macrophage cells may take up the nanowires more rapidly than nanoparticles by their differences in mass. It is of current interest to investigate the cell labeling efficacy with 1-D nanomaterials, cytotoxicity, stability in different cell compartments, and fate of the materials. Recently, Park *et al.* reported 1-D magnetic nanoworms for *in vivo* tumor targeting (Park et al., 2009). Superparamagnetic iron oxide nanoworms that have a long dimension of ~70 nm and thickness of ~30 nm have been used for targeting xenograft tumors. The nanoworms exhibit superior *in vivo* tumor-targeting ability than nanoparticles (mean diameter ~30 nm). The authors conclude that the blood half-life of a targeting molecules-nanomaterial ensemble is a key consideration when selecting the appropriate ligand and nanoparticle chemistry for tumor targeting.

Although different magnetic nanowires have been fabricated by solution methods—a bottom-up approach (versus a top-down aprroach which produces nanostructures by refinement of bulk materials) (Xia et al., 1999; Stephanopoulos et al., 2005)—and characterized by various techniques (Whitney et al., 1993; Meier et al., 1996; Doudin et al., 1996; Ferré et al., 1997; Fert & Piraux 1999; García & Miltat 2002; Nielsch et al., 2002; Chen et al.(a), 2003; Chen et al.(b), 2003; Love et al., 2003; Ponhan & Maensiri 2009), however, the effective preparation, cytotoxicity, as well as cell labelling efficacy of different cell types using relatively rigid, long magnetic nanowires have been seldomly investigated. Therefore, we have recently investigated rigid 1-D magnetic nanostructures as effective contrast agents for MRI, and discovered the preparation of Mn-Fe oxide nanowires with amine-functional peripheries, which are formed by a self-assembling organization of their corresponding small $MnFe_2O_4$ nanoparticles – a process of assembling approach using cystamine as the linker. This approach, which utilizes chemicals and supramolecular driving forces to arrange small components into an ordered conformation, represents an effective and inexpensive way to achieve more complex and functional nanoarchitectures. Herein, the properties of Mn-Fe nanowires with lengths ranging from 400-1000 nm and widths ranging from 8-35 nm for MRI contrast and the potential of macrophage cell uptake are also reported (Leung et al., 2008; Leung et al., 2009).

2. Approach

2.1 Synthesis of the nanostructures

Spontaneous organization of small individual nanostructures into large and well-defined nanowires, represents (Tang et al., 2002; Grubbs 2007) a facile way to obtain useful nanomaterials for magnetic devices (Hangarter et al., 2007; Wu et al., 2007). The key to prepare our target 1-D Mn-Fe nanowires for cell labeling and MRI contrast could be achieved in such a way that small building blocks – $MnFe_2O_4$ nanoparticles which were prepared by co-precipitation at elevated temperature (Sousa et al., 2001; Aquino et al., 2002;

Deng et al., 2005), were rationally organized into Mn-Fe nanowires with larger lengths in relatively high yields. We have found out that suitable amount of linker – cystamine ($NH_2CH_2CH_2S–SCH_2CH_2NH_2$) could induce the organization of the as-synthesized $MnFe_2O_4$ nanoparticles into novel amine-functionalized Mn-Fe nanowires in good yields under basic condition and magnetic stirring for 24 hours. The precipitate was separated by centrifugation and washed with water/ethanol mixture to afford the 1-D Mn-Fe nanostructures.

2.2 Characterization of the nanostructures

The morphology of as-prepared $MnFe_2O_4$ nanoparticles was characterized by high-resolution TEM, revealing their morphologies with an average diameter of 5 nm (**Figure 1A**). One drop of sample in ethanol suspension was added to the holey carbon-coated

Fig. 1. High-resolution transmission electron microscopic images of (A) $MnFe_2O_4$ nanoparticles; (B) nanoneedles; (C) nanorods; and (D) nanowires.

copper grid and was allowed to evaporate to dryness. For the prepared Mn-Fe nanostructures, TEM images revealed that they were in substantial amounts (**Figures 1B-D**). The nanoneedles possessed (**Figure 1B**) an average length of 400 nm and width of 8 nm; while the nanorods possessed (**Figure 1C**) an average length of 800 nm and width of 30 nm. For the nanowires, they possessed (**Figure 1D**) an average length of 1 μm and width of 35 nm.

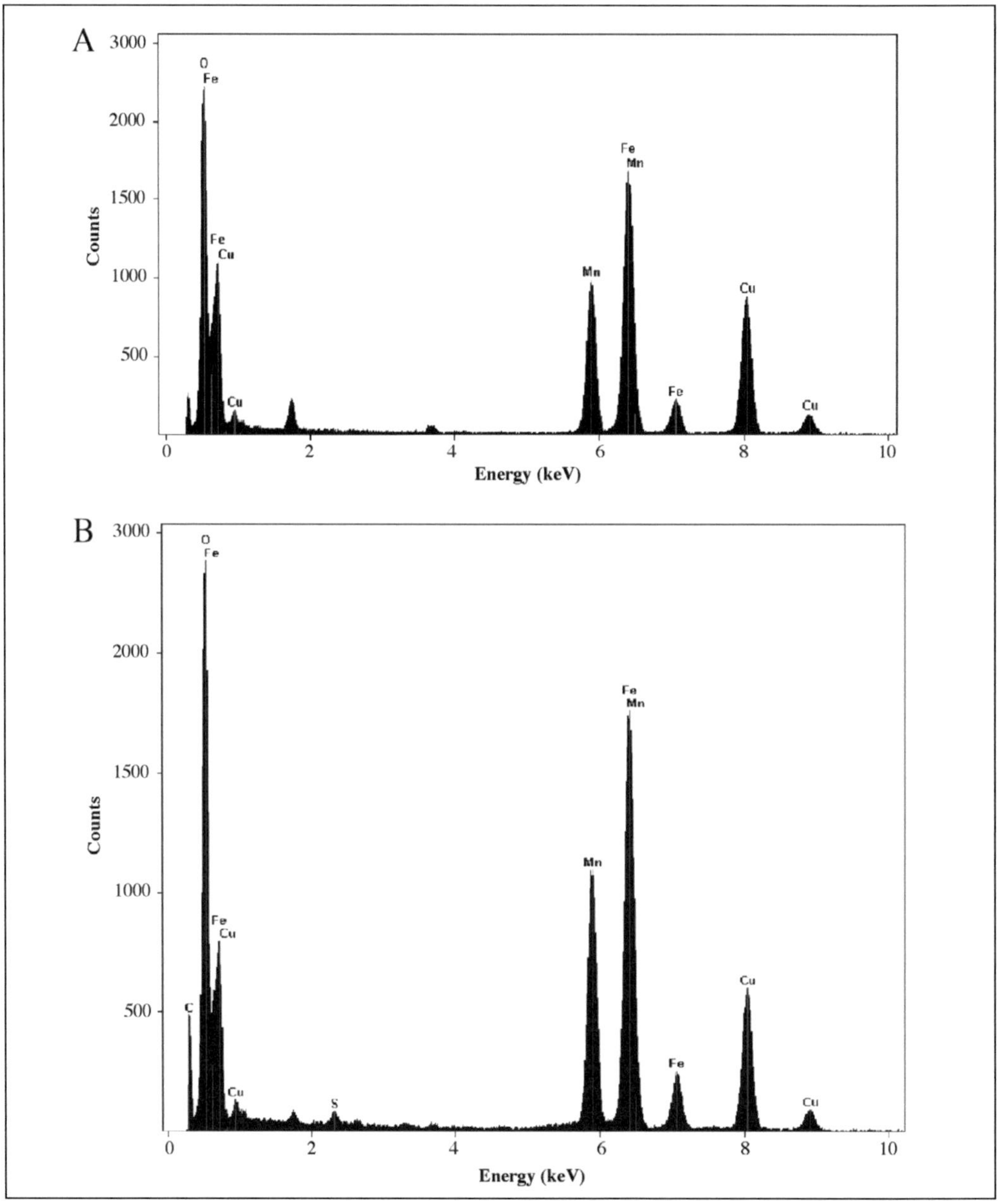

Fig. 2. Energy-dispersive X-ray spectroscopic analyses (on copper grid) of (A) $MnFe_2O_4$ nanoparticles and (B) Mn-Fe nanowires. The obseved signals for Cu originated from the copper grid.

In addition to the TEM characterization, inductively coupled plasma-optical emission spectroscopy (ICP-OES) and energy-dispersive X-ray (EDX) spectroscopy were employed for the determination of elemental contents (Mn, Fe, S, O and N) of the nanoarchitectures (**Table 1, Figure 2**). EDX measurements were performed by locating a region (~20 nm × 20 nm) with substantial amount of materials on copper grid without carbon coating. By comparing between the EDX spectra of the $MnFe_2O_4$ nanoparticles (**Figure 2A**) and nanowires (**Figure 2B**), the spectrum of the nanowires revealed additional signals of sulfur and nitrogen as well as enhanced signals of carbon, originating from the attached cystamine linker. The observed EDX signal of copper, which originates from the TEM sample grid, has been omitted in the calculations of the elemental contents present in our nanostructures. The X-ray diffraction (XRD) analysis revealed that the $MnFe_2O_4$ nanoparticles exhibited several peaks corresponding to the characteristic interplanar spacings 220, 311, 400, 511 and 400 of the spinel structure with 2θ 31.5, 35.0, 42.4, 56.2 and 61.7, respectively. These results are similar to the reported values in the literature (Aquino et al., 2002).
ICP-OES samples were dissolved in 2% HCl solution with a few drops of $SnCl_2$ solution. Iron absorption was observed at 238.20 nm while the manganese absorption was observed at 257.61 nm. Although the separate measurements (**Table 1**) by ICP-OES and EDX occurred with the errors that are less than 1%, there existed relatively large errors (0.3–48%) when comparing the results obtained in both methods. Generally, ICP-OES is regarded as an accurate method to determine the metal ion concentrations while EDX spectroscopy is an evaluation of the existence of elements in a specific area on sample grid surface. The ratio of Mn:Fe is approximately 1:2. From the $MnFe_2O_4$ nanoparticles to the nanowires, there was a trend in the quantitative ICP-OES measurements that both the Mn and Fe metal contents decrease slightly. Moreover, the sulfur and nitrogen contents originated from the cystamine linker appeared in the EDX measurement of the 1-D nanoarchitectures. These increased organic characteristics indicate that the amounts of linker play a crucial role in controlling the sizes of the nanostructures.

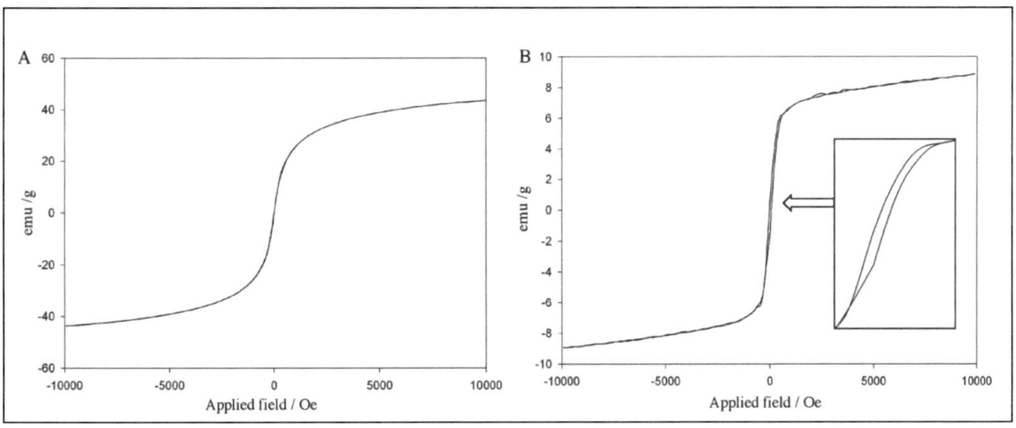

Fig. 3. VSM spectra of (A) $MnFe_2O_4$ nanoparticles and (B) Mn-Fe nanowires showing the difference in magnetic response.

The magnetic properties of the nanomaterials have been investigated using a vibrating sample magnetometer (VSM). The VSM spectra (*M-H*) of magnetization (emu/g) versus magnetic field (Oe) are examined. **Figure 3A** shows a hysteresis loop of typical $MnFe_2O_4$ nanoparticles measured by sweeping the external field between –10000 to 10000 Oe at room temperature. The magnetization curve shows no remanemce or coercivity at room temperature, demonstrating the superparamagnetic character. Superparamagnetism is the responsiveness to an applied magnetic field without retaining any magnetism after removal of the applied magnetic field. The saturated magnetization value of $MnFe_2O_4$ nanoparticles is determined to be 43.7 emu/g. On the other hand, **Figure 3B** shows a hysteresis loop of Mn-Fe nanowires measured by sweeping the external field between -10000 to 10000 Oe at room temperature. The magnetization curve shows that the coercivity and remanence are small at room temperature (**Figure 3B** inset). The saturated magnetization value of Mn-Fe nanowires is determined to be 9.0 emu/g. The decreased saturated magnetization value is in part due to the increasing thickness of organic coating layer of the nanowires. Other reasons are the presence of a magnetic dead or anti-ferromagnetic organic layer on the nanowire's periphery.

For high-density information storage, the superparamagnetic relaxation of magnetization direction in tiny magnetic data bits has to be avoided in order to keep the digital data stored for a desired period of time. For biomedical applications such as magnetic resonance imaging, hyperthermia, drug delivery, and catalysis, in contrast, superparamagnetic property of materials would be essential from which the materials do not retain any magnetization in the absence of an externally applied magnetic field. In our system, soft magnetized nanowires have been employed as contrast agents for magnetic resonance imaging.

Mn-Fe structure	average diameter / nm	average length / nm	ICP-OES		EDX			
			Mn%	Fe%	Mn%	Fe%	O%	N%
nanoparticle	5	-	24.7 ± 0.1	40.6 ± 0.2	15.4 ± 0.2	40.7 ± 0.3	43.9 ± 0.4	-
nanoneedle	8	400	20.4 ± 0.1	33.2 ± 0.2	10.6 ± 0.1	52.2 ± 0.3	32.9 ± 0.3	2.0 ± 0.1
nanorod	30	800	19.7 ± 0.1	32.3 ± 0.2	20.0 ± 0.3	38.9 ± 0.4	36.5 ± 0.5	4.2 ± 0.3
nanowire	35	1000	17.1 ± 0.1	28.2 ± 0.1	23.2 ± 0.2	39.9 ± 0.2	31.4 ± 0.3	5.5 ± 0.3

Table 1. Summarized results of ICP-OES and EDX measurements

2.3 Magnetic resonance imaging

For magnetic resonance imaging (MRI), the capability of nanoarchitectures to influence the *T2* relaxation time was studied using a clinical 1.5 T whole-body MR system (Sonata, Siemens Symphony, Erlangen, Germany) with a standard human knee coil radio frequency coil for excitation and signal reception. Nanoarchitectures were diluted in distilled water at nanowire concentrations of 100, 10, 1 and 0.1 μg/mL. For MR measurements, the nanomaterial suspensions (1 mL) were filled in 2 mL Eppendorf vials each. These vials were placed in a water bath. Sonication was applied for 10 min prior to MRI using at 35 kHz in water at ambient temperature. The imaging sequence was a standard Carr-Purcell-Meiboom-Gill pulse sequence with the following parameters: TR = 2000 milisecond, TE range 30-960 milisecond, 32 echoes, FOV = 134 × 67 mm^2, matrix = 128 × 64, slice thickness 5 mm, NEX = 3. *T2* relaxation times were calculated by a linear fit of the logarithmic ROI signal amplitudes versus TE.

When a cylindrical magnetic nanorod segment is disk-shaped, its magnetization axis lies perpendicular to the rod axis. It also follows that if the segment is longer than its width, its axis of magnetization is parallel to the nanorod axis (Ferré et al., 1997). Therefore, the magnetic properties of nanostructures can be tuned over a wide range by tuning the aspect ratio of the magnetic block and its composition (Fert & Piraux 1999). Experimentally, the *T*2 relaxivities (r2) for nanoneedles, nanorods, and nanowires were determined to be 20.81 ± 0.58, 8.10 ± 0.31, and 6.62 ± 0.42 $mM^{-1}sec^{-1}$, respectively (**Figure 4**). These values are at the lower end of the relaxivity of other iron oxide nanoparticle derivatives, such as VSOP-C184 has a r2 of 33 $mM^{-1}sec^{-1}$ and SHU-555C (Supravist) has a r2 of 38 $mM^{-1}sec^{-1}$ (Wang et al., 2001; Nitin et al., 2004; Lee et al., 2007; Lu et al., 2007; Lee et al., 2008). It is interesting to notice that though the iron concentration of nanoneedles is not very different to the other nanostructures, it has significantly higher MR relaxivity than the other nanostructures. It is known that MR relaxivity depends in part on the crystal lattice of the iron oxide. Our results suggest that the nanoneedles with higher MR relaxivity are majorly composed of iron oxide crystal lattice similar to those of the nanoparticles (Wang et al., 2001). The development of larger 1-D nanostructures leaded to lower MR relaxivities. Despite the slightly weaker MR relaxivity of the nanostructures developed in the current study, it is expected that the unique shape of these nanostructures can find applications for cell labeling, as drug carrier, magnetic separation of labeled cells and other biological entities, and also in the application of magnetic targeting (Alexiou et al., 2000; Pankhurst et al., 2003; Tartaj et al., 2003; Berry & Curtis 2003). It may be possible to further modify the systems with the aim to increase their relaxivities. It has been reported that the doping of Mn into Fe_3O_4 can also increase the MR

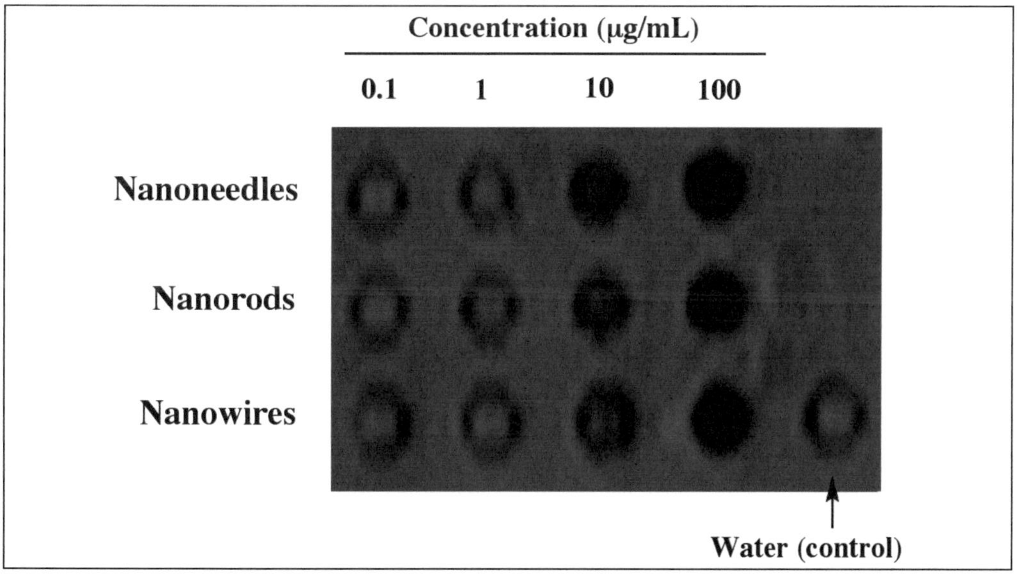

Fig. 4. Spin echo MR image of the Mn-Fe nanostructure suspensions obtained at 1.5 Tesla with time of repetition (TR) 2000 millisecond and time of echo (TE) of 960 millisecond. All the Mn-Fe nanostructures with concentration of 100 μg/mL decrease the MRI signal to background signal (dark contrast). Mn-Fe nanostructures with concentration of 10 μg/mL decrease the MRI signal substantially, though less so with nanorods. Compared to the water signal, contrast due to 1 μg/mL nanostructures is just detectable.

relaxivity as compared with the pure iron oxide counterpart (Lee et al., 2007). For *in vivo* tissue contrast, the weaker MR relaxivity can be compensated by using *T2** MRI sequence and higher magnetic field such as 3 Tesla or 7 Tesla.

The nanowires were incubated with RAW264.7 cells followed by Prussian blue staining. The results indicated that the Mn-Fe nanowires were effectively incorporated into RAW264.7 cells without the addition of any transfecting agent (e.g., liposomes) (**Figure 5**). **Figures 5B–5D** reveal the optical microscopic images of the RAW264.7 cells incubated with nanowires for 2 hours at the concentration of 10, 50, 100 μg/mL respectively. As increasing

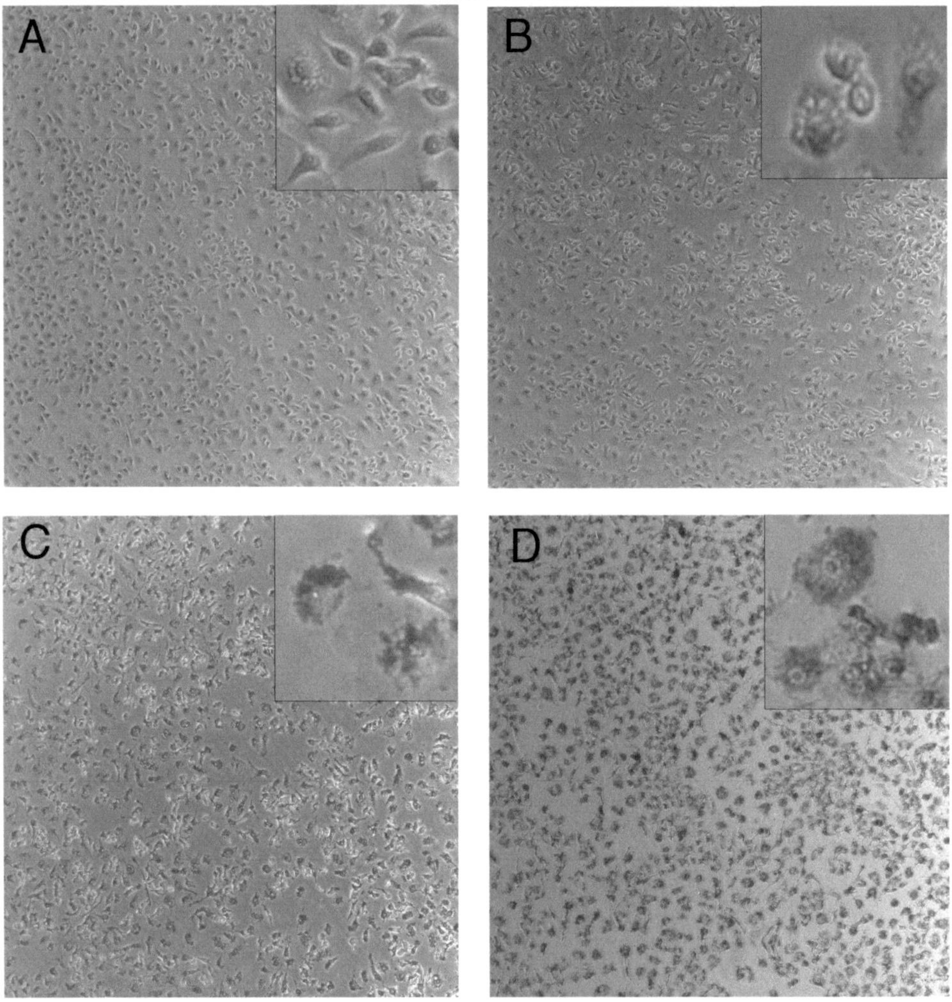

Fig. 5. Optical microscope images of the Raw264.7 cells [Original magnification: 100×]: (A) Control cells without the labeling of nanowires. (B)–(D) Cells incubated with nanowires for 2 hours at the concentration of 10, 50, 100 μg/mL, respectively. Prussian blue staining was carried out. As increasing the labeling concentration with nanowires, increased amounts of the nanowire uptake are observed as shown in the blue-stained part inside the cells. For (C) and (D), the cell labeling efficacies with the nanowires are approximately 100%.

the labeling concentration with nanowires, increased amounts of the nanowire uptake were observed as in the blue-stained part inside the cells. At the nanowire concentrations of 50 and 100 μg/mL (**Figures 5C and 5D**), the cell labeling efficacies with the nanowires were approximately 100%.

TEM analysis was also performed to confirm the location of nanowires inside the RAW264.7 cells. For the TEM analysis, specimens were fixed in 2.5% glutaraldehyde and 2% paraformaldehyde in 0.1 M cacodylate buffer (pH 7.2) containing 0.05% $CaCl_2$ for 2 hours at room temperature, followed by post-fixation in 2% osmium tetroxide in the same buffer solution for 2 hours at room temperature. After dehydration and embedded in Spurr's resin, ultra-thin sections (80 nm) were cut before examined under a FEI/Philips Tecnai 12 TEM operated at 80 kV. TEM results demonstrated that the nanowires were located within the lyzosome and cell vesicles (**Figure 6**). No obvious sub-cellular superstructure injury and cell apoptosis was observed. However, modest amount of the nanoparticles were observed inside the lyzosome and cell vesicles, a result which indicates that the nanostructures might be susceptible to enzymatic degradation in the slightly acidic lyzosome environment. It is also expected that some nanowires incorporated in the cells were fragmented or partially cut during the preparation of the ultra-thin section for TEM.

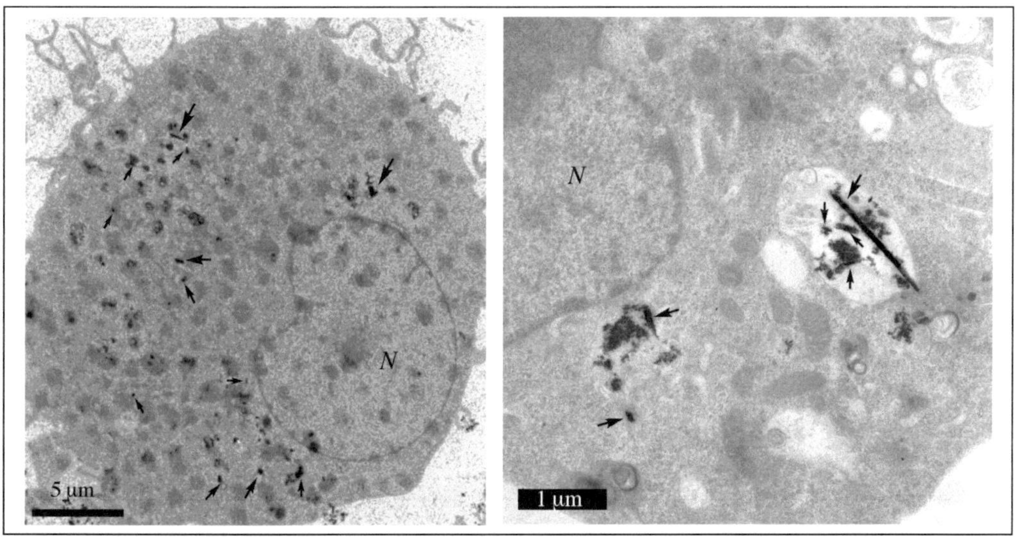

Fig. 6. TEM images of the nanowires-labeled RAW264.7cells. The arrow denotes some of the nanowires present in the lysosomes and vesicles.

It has been demonstrated that certain biocompatible synthetic materials with amine functional peripheries could enhance the degree of cell adhesion and transfection (Pollard et al., 1998; Wang et al., 2009). As a result, our reported Mn-Fe nanostructures which contain free amines would be beneficial not only to the macrophage cells but also can extend their labeling efficacy to other cell lines.

To assess the biocompatibility of the nanostructures, after RAW264.7 cells incubated with 0.1, 1, 10, 50, 100 μg/mL nanowires for 2 hours, Trypan blue exclusion assay (Sigma T6146) was performed to assess the viability of the cells. To assess the cell proliferation potential post labeling, RAW264.7 cells were cultured in 96-well plate at the density of 2500 cell/well with DMEM including 10% FCS. After labeling cells with 0.1, 1, 10, 50 or 100 μg/mL nanowires for 2 hours, nanowires were removed from the plate and PBS was used to rinse the residual nanostructures. Fresh DMEM including 10% FCS was added again for normal cell growth. 3-[4,5-Dimethylthiazol-2-yl]-2,5-diphenyl tetrazolium bromide (MTT) assay was performed to detect the proliferation of the nanowire-labeled RAW264.7 cells.

The Trypan blue exclusion assay results and MTT assay results are shown in **Figure 7**. RAW264.7 cell viability was not significantly affected from the labeling concentration of 0.1 up to 50 μg/mL with nanowires, and no apparent growth inhibition of RAW264.7 cells was observed after labeling up to 50 μg/mL of nanowires (inclusive). These results revealed the satisfactory safety profiles of these nanostructures.

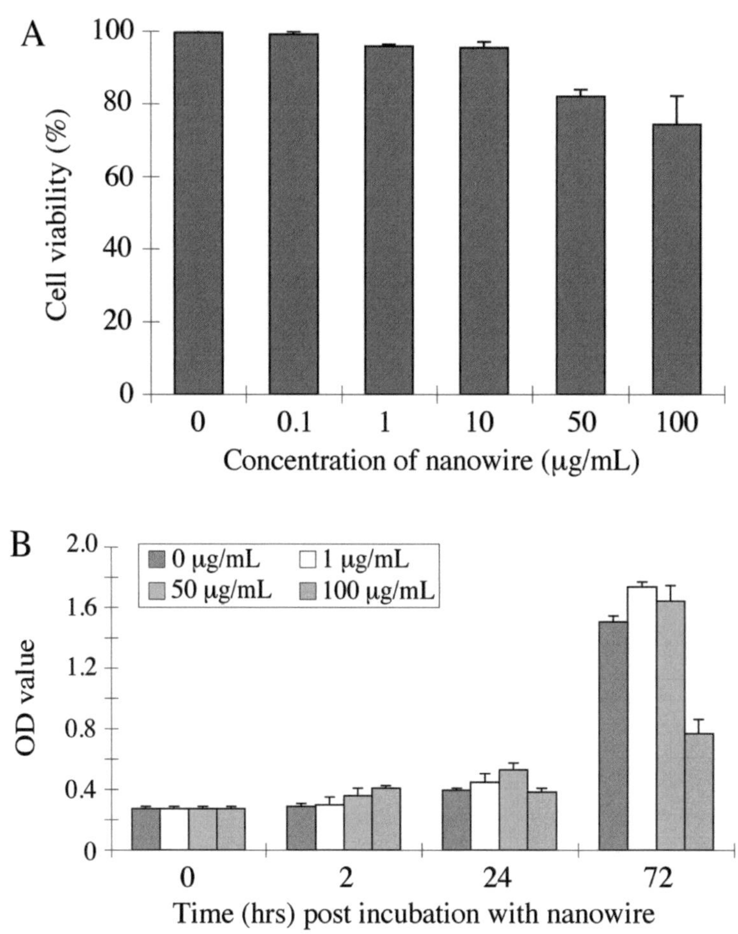

Fig. 7. The Trypan blue exclusion assay and MTT assay results. RAW264.7 cells were incubated with 0.1, 1, 10, 50, 100 μg/mL of nanowires for 2 hours. Satisfactory biocompatibility is shown from the concentration of 0.1 to 50 μg/mL. At 100 μg/mL, cell death and cell proliferation inhibition was observed.

3. Conclusion

In summary, the manganese-doped iron oxide nanoparticles are capable to organize into one-dimensional magnetic Mn-Fe nanostructures of different sizes with cystamine. The use of these materials which are employed as the cell-labeling agent for magnetic resonance imaging (MRI), has been explored. All nanoarchitectures demonstrate remarkable magnetic resonance *T*2 relaxivities. The cell viability test and growth curve reveal the satisfactory safety profiles of these nanostructures. This method represents an efficient and facile way to prepare novel functional nanoarchitectures for potential use in magnetic separation of labeled cells and other biological entities, magnetic resonance imaging, carriers of therapeutic molecules and magnetic targeting, and radio frequency therapy of tumours via hyperthermia (Alexiou et al., 2000; Berry & Curtis 2003; Pankhurst et al., 2003; Tartaj et al., 2003). Eventually, nanoarchitectures of specific sizes can be potentially obtained by this method and used as MRI contrast agents. The labeling efficacies of these nanostructures can be investigated with different cell types such as inflammation cells and liver cells (Du et al., 2007), etc. The comparative study on the use of both magnetic nanoparticles and nanowires as effective, biocompatible MRI contrast agents for different cell lines should be further investigated.

4. References

Alexiou, C.; Arnold, W.; Klein, R. J.; Parak, F. G.; Hulin, P.; Bergemann, C.; Erhardt, W.; Wagenpfeil, S. & Lübbe, A. S. (2000). Locoregional cancer treatment with magnetic drug targeting. *Cancer Res.*, 60, 6641-6648

Allwood, D. A.; Xiong, G.; Cooke, M. D.; Faulkner, C. C.; Atkinson, D.; Vernier, N. & Cowburn, R. P. (2003). Submicrometer ferromagnetic NOT gate and shift register. *Science*, 296, 2003-2006

Aquino, R.; Tourinho, F. A.; Itri, R.; Lara, M. C. F. L. & Depeyrot, J. (2002). Size control of $MnFe_2O_4$ nanoparticles in electric double layered magnetic fluid synthesis. *J. Mag. Mag. Mater.*, 252, 23-25

Berry, C. C. & Curtis, A. S. G. (2003). Functionalisation of magnetic nanoparticles for applications in biomedicine. *J. Phys. D: Appl. Phys.*, 36, R198-R206

Chen, M.; Searson, P. C. & Chien, C. L. (2003). Micromagnetic behavior of electrodeposited Ni/Cu multilayer nanowires. *J. Appl. Phys.*, 93, 8253-8255

Chen, M.; Sun, L.; Bonevich, J. E.; Reich, D. H.; Chien, C. L. & Searson, P. C. (2003). Tuning the response of magnetic suspensions. *Appl. Phys. Lett.*, 82, 3310-3312

Cohen-Karni, T.; Timko, B. P.; Weiss, L. E. & Lieber, C. M. (2009). Flexible electrical recording from cells using nanowire transistor arrays. *Proc. Natl. Acad. Sci. USA*, 106, 7309-7313

Corot, C.; Robert, P.; Idée, J. M. & Port, M. (2006). Recent advances in iron oxide nanocrystal technology for medical Imaging. *Adv. Drug Deliv. Rev.*, 58, 1471-1504

Darques, M.; Spiegel, J.; De la Torre Medina, J.; Huynen, I. & Piraux, L. (2009). Ferromagnetic nanowire-loaded membranes for microwave electronics. *J. Magn. Magn. Mater.*, 321, 2055-2065

Deng, H.; Li, X.; Peng, Q.; Wang, X.; Chen, J. & Li, Y. (2005). Monodispersed magnetic single-crystal ferrite microspheres. *Angew. Chem., Int. Ed.*, 44, 2782-2785

Doudin, B.; Blondel, A. & Ansermet, J.-Ph. (1996). Arrays of multilayered nanowires. *J. Appl. Phys.*, 79, 6090-6094

Du, L.; Chen, J.; Qi, Y.; Li, D.; Yuan, C.; Lin, M. C.; Yew, D. T.; Kung, H.-F.; Yu, J. C. & Lai, L. (2007). Preparation and biomedical application of a non-polymer coated superparamagnetic nanoparticle. *Int. J. Nanomed.*, 2, 805-812

Fang Z. & Kelley, S. O. (2009). Direct electrocatalytic mRNA detection using PNA-nanowire sensors. *Anal. Chem.*, 81, 612-617.

Ferré, R.; Ounadjela, K.; George, J. M.; Piraux, L. & Dubois, S. (1997). Magnetization processes in nickel and cobalt electrodeposited nanowires. *Phys. Rev. B*, 56, 14066-14075

Fert, A. & Piraux, L. (1999). Magnetic nanowires. *J. Magn. Magn. Mater.*, 200, 338-358

Fortin, J. P.; Wilhelm, C. Servais, J.; Ménager, C.; Bacri, J. C. & Gazeau, F. (2007). Size-sorted anionic iron oxide nanomagnets as colloidal mediators for magnetic hyperthermia. *J. Am. Chem. Soc.*, 129, 2628-2635

García, J. M. & Miltat, A. T. J. (2002). MFM imaging of nanowires and elongated patterned elements. *J. Magn. Magn. Mater.*, 249, 163-169

Grubbs, R. B. (2007). Solvent-tuned structures. *Nat. Mater.*, 6, 553-555

Hangarter, C. M.; Rheem, Y.; Yoo, B.; Yang, E.-H. & Myung, N. V. (2007). Hierarchical magnetic assembly of nanowires. *Nanotechnology*, 18, article number 205305

Lee, J.-H.; Huh, Y. M.; Jun, Y.-W.; Seo, J.-W.; Jang, J.-T.; Song, H.-T.; Kim, S.; Cho, E.-J.; Yoon, H.-G.; Suh, J.-S. & Cheon, J. (2007). Artificially engineered magnetic nanoparticles for ultra-sensitive molecular imaging. *Nat. Med.*, 13, 95-99

Lee, H.-Y.; Lee, S.-H.; Xu, C.; Xie, J.; Lee, J.-H.; Wu, B.; Koh, A. L.; Wang, X.; Sinclair, R.; Wang, S. X.; Nishimura, D. G.; Biswal, S.; Sun, S.; Cho, S. H. & Chen, X. (2008). Synthesis and characterization of PVP-coated large core iron oxide nanoparticles as an MRI contrast agent. *Nanotechnology*, 19, 165101-165106

Leung, K. C.-F.; Wang, Y.-X.; Wang, H.-H. & Chak, C.-P. (2008). Novel one-dimenstional Mn-Fe oxide nanowires for cell labeling and magnetic resonance imaging. *IEEE Proceedings of the 5th International Conference on Information Technology and Application in Biomedicine*, 193-195, Shenzhen, China, May 30-31, 2008, IEEE

Leung, K. C.-F.; Wang, Y.-X. J.; Wang, H.-H.; Xuan, S.; Chak, C.-P. & Cheng, C. H. K. (2009). Biological and magnetic contrast evaluation of shape-selective Mn-Fe nanowires. *IEEE Trans. Nanobiosci.* 8, 192-198.

Love, J. C.; Urbach, A. R.; Prentiss, M. G. & Whitesides, G. M. (2003). Three-dimensional self-assembly of metallic rods with submicron diameters using magnetic interactions. *J. Am. Chem. Soc.*, 125, 12696-12697

Lu, C.-W.; Hung, Y.; Hsiao, J.-K.; Yao, M.; Chung, T.-H.; Lin, Y. S.; Wu, S.-H.; Hsu, S.-C.; Liu, H.-M.; Mou, C.-Y.; Yang, C.-S.; Huang, D.-M. & Chen, Y.-C. (2007). Bifunctional magnetic silica nanoparticles for highly efficient human stem cell labelling. *Nano Lett.*, 7, 149-154

Lübbe, A. S.; Alexiou, C. & Bergemann, C. (2001). Clinical applications of magnetic drug targeting. *J. Surg. Res.*, 95, 200-206

Meier, J.; Doudin, B. & Ansermet, J.-Ph. (1996). Magnetic properties of nanosized wires. *J. Appl. Phys.*, 79, 6010-6012

Mitragotri, S. & Lahann, J. (2009). Physical approaches to biomaterial design. *Nat. Mater.*, 8, 15-23

Nam, J.-M.; Thaxton, C. S. & Mirkin, C. A. (2003). Nanoparticle-based bio-bar codes for the ultrasensitive detection of proteins. *Science*, 301, 1884-1886

Nielsch, K.; Hertel, R.; Wehrspohn, R. B.; Barthel, J.; Kirschner, J.; Gösele, U.; Fischer, S. F. & Kronmüller, H. (2002). Switching behavior of single nanowires inside dense nickel nanowire arrays. *IEEE Trans. Magn.*, 38, 2571-2573

Nitin, N.; LaConte, L. E. W.; Zurkiya, O.; Hu, X. & Bao G. (2004). Functionalization and peptide-based delivery of magnetic nanoparticles as an intracellular MRI contrast agent. *J. Biol. Inorg. Chem.*, 9, 706-712

Pankhurst, Q. A.; Connolly, J.; Jones, S. K. & Dobson, J. (2003). Applications of magnetic nanoparticles in biomedicine. *J. Phys. D: Appl. Phys.*, 36, R167-R181

Park, J.-H.; von Maltzahn, G.; Zhang, L.; Derfus, A. M.; Simberg, D.; Harris, T. J.; Ruoslahti, E.; Bhatia, S. N. & Sailor, M. J. (2009). Systematic surface engineering of magnetic nanoworms for in vivo tumor targeting. *Small*, 5, 694-700

Pollard, H.; Remy, J.-S.; Loussouarn, G.; Demolombe, S.; Behr, J.-P. & Escande, D. (1998). Polyethylenimine but not cationic lipids promotes transgene delivery to the nucleus in mammalian cells. *J. Biol. Chem.*, 273, 7507-7511

Ponhan, W. & Maensiri, S. (2009). Fabrication and magnetic properties of electrospun copper ferrite ($CuFe_2O_4$) nanofibers. *Solid State Sci.*, 11, 479-484

Redl, F. X.; Cho, K. S.; Murray, C. B. & O'Brien, S. (2003). Three-dimensional binary superlattices of magnetic nanocrystals and semiconductor quantum dots. *Nature*, 423, 968-971

Skomski, R. (2003). Nanomagnetics. *J. Phys. Condens. Matt.*, 15, R841-R896

Sousa, M. H.; Tourinho, F. A.; Depeyrot, J.; da Silva, G. J. & Lara, M. C. F. L. (2001). New electric double-layered magnetic fluids based on copper, nickel, and zinc ferrite nanostructures. *J. Phys. Chem. B*, 105, 1168-1175

Stephanopoulos, N.; Solis, E. O. P. & Stephanopoulos, G. (2005). Nanoscale process systems engineering: toward molecular factories, synthetic cells, and adaptive devices. *AIChE J.*, 51, 1858-1869

Tang, Z.; Kotov, N. A. & Giersig, M. (2002). Spontaneous organization of single CdTe nanoparticles into luminescent nanowires. *Science*, 297, 237-240

Tartaj, P.; del Puerto Morales, M.; Veintemillas-Verdaguer, S.; Teresita González-Carreño, T. & Serna, C. J. (2003). The preparation of magnetic nanoparticles for applications in biomedicine. *J. Phys. D: Appl. Phys.*, 36, R182-R197

Thurn-Albrecht, T.; Schotter, J. ; Kastle, G. A. ; EmLey, N. ; Shibauchi, T. ; Krusin-Elbaum, L.; Guarini, K.; Black, C. T.; Tuominen, M. T. & Russell, T. P. (2000). Ultrahigh-density nanowire arrays grown in self-assembled diblock copolymer templates. *Science*, 290, 2126-2129

Wang, H.-H.; Wang, Y.-X. J.; Leung, K. C.-F.; Au, D. W. T.; Xuan, S.; Chak, C.-P.; Lee, S. K. M.; Sheng, H.; Zhang, G.; Qin, L.; Griffith, J. F. & Ahuja, A. T. (2009). Durable mesenchymal stem cell labelling using polyhedral superparamagnetic iron oxide nanoparticles. *Chem. Eur. J.*, 15, 12417-12425.

Wang, Y.-X. J.; Hussain, S. M. & Krestin, G. P. (2001). Superparamagnetic iron oxide contrast agents: physicochemical characteristics and applications in MR imaging. *Eur. Radiology*, 11, 2319-2331

Whitney, T. M.; Jiang, J. S.; Searson, P. C. & Chien, C. L. (1993). Fabrication and magnetic properties of arrays of metallic nanowires. *Science*, 261, 1316-1319

Wilhelm, C. & Gazeau, F. (2008). Universal cell labelling with anionic magnetic nanoparticles. *Biomaterials*, 29, 3161-3174

Wu, H.; Zhang, R.; Liu, X.; Lin, D. & Pan, W. (2007). Electrospinning of Fe, Co, and Ni nanofibers: synthesis, assembly, and magnetic properties. *Chem. Mater.*, 19, 3506-3511

Xia, Y.; Rogers, J. A.; Paul, K. E. & Whitesides, G. M. (1999). Unconventional methods for fabricating and patterning nanostructures. *Chem. Rev.*, 99, 1823-1848

16

pH Dependent Hydrothermal Synthesis and Photoluminescence of Gd_2O_3:Eu Nanostructures

Kyung-Hee Lee, Yun-Jeong Bae and Song-Ho Byeon*
Department of Applied Chemistry, Kyung Hee University
Republic of Korea

1. Introduction

The size and shape of nanocrystals (nanospheres, nanorods, nanotubes, and nanowires) are crucial parameters for controlling nanocrystal properties such as electrical transport, optical, and magnetic properties. (Hu al., 1999; Kazes et al., 2002) In particular, one-dimensional structures are the fundamental units for anisotropic shape control, presenting general shape and assembly control strategies for more complex structures. The most exciting and rapidly expanding field would be carbon nanotubes from graphite by the rolling of graphen sheets.(Ebbesen & Ajayan, 1992; Iijima, 1991; Iijima & Ichihashi, 1993; Rao et al., 2004) The MX_2 layers of metal dichalcogenides (M = Mo, W and X = S, Se) are analogous to the single graphene sheets, being able to roll into curved structures. Considerable progress has been made in the synthesis of the fullerene and nanotube structures of MX_2.(Parilla et al., 1999; Rao & Nath, 2003; Tenne, 1995) Nanorods (or nanoneedles) and nanowires have also been extensively explored because the functional nanomaterials with a restricted dimension offer the opportunities for investigating the influence of shape and dimensionality on the physical properties.(El-sayed, 2001; Xia et al., 2003) For instance, ZnO nanowire sensitized by narrow band-gap materials have been studied for use in photovoltaic device applications due to their facile synthesis and excellent optical properties.(Birkmire & Eser, 1997; Law et al., 2005; Levy-Clement et al., 2005; Leschkies et al., 2007) The selective synthesis of such structures can be achieved by a morphology control in the crystallization process including nucleation and growth. Various synthetic methods have been developed for many important materials.

The interest in studying lanthanide oxides and hydroxides stems from their potential applications including dielectric materials for multilayered capacitors, luminescent lamps and displays, solid-laser devices, optoelectronic data storages, waveguides, and heterogeneous catalysts.(Cutif et al., 2004) Recently, lanthanide-doped oxide nanoparticles are of special interests as potential materials for an important new class of nanophosphors. For instance, when the oxide nanophosphors are applied for a fluorescent labeling, there are several advantages such as sharp emission spectra, long life times, and high resistance against photobleaching in comparison with conventional organic fluorophores and quantum dots.(Beaurepaire et al., 2004; Louis et al., 2005) Many lanthanide-doped oxides generate visible light in fluorescent lamps and emissive displays. Excitation of photoluminescent phosphors takes place using ultraviolet (UV) photons generated by a discharge in fluorescent lamp or plasma display panel (PDP). Cathodoluminescence, thermoluminescence, and electroluminescence of phosphors are also important for cathode-ray tube (CRT), field emission display (FED), radiation detector, and electroluminescence display (ELD) applications.(Yen et al., 2007) An improved performance of displays and lamps requires high quality of phosphors for sufficient brightness and long-term stability. In practice, a densely packed layer of small size particles can improve aging problems. On the other hand, when the particle is smaller than a critical value, the luminescence efficiency decreases because of increased light reabsorption and the luminescence quenching by the surface layer. Thus, high-concentration of surface defects and microstrains could greatly reduce the total luminescent intensity of nanosized luminescent materials. The concentration quenching ranges of activators are also different from those of corresponding bulk materials.(Duan et al., 2000; Zhang et al., 2003) Therefore, the morphology and size, the stoichiometry and composition, and the surface characteristics must be controlled in order to achieve the desirable objectives of improved nanophosphors.

As one of lanthanide oxides and hydroxides, Gd_2O_3 nanopaticle is a promising host matrix for multiphoton and up-conversion excitation.(Guo et al., 2004; Hirai & Orikoshi, 2004; Zhou et al., 2003) The gadolinium oxide doped with Eu^{3+} (Gd_2O_3:Eu) exhibits paramagnetic behavior (S = 7/2) as well as strong UV and cathode-ray excited luminescences, which are useful in biological fluorescent label, contrast agent, and display applications.(Blasse & Grabmaier, 1994; Goldys et al., 2006; Nichkova et al., 2005) Very recently, the magnetic resonance relaxation property of the colloidal solution of layered gadolinium hydroxide exhibited the sufficient positive contrast effect for magnetic resonance imaging (MRI).(Lee et al., 2009) In addition, Gd_2O_3:Eu is a very efficient X-ray and thermoluminescent phosphor.(Rossner & Grabmaier, 1991; Rossner, 1993) Diverse preparation methods have been developed to reduce the reaction temperature and achieve a small particle size of high quality Gd_2O_3:Eu phosphors.(Erdei et al., 1995; Ravichandran et al., 1997; Shea et al., 1996; Yan et al., 1987)

In an attempt to produce various inorganic materials in the form of isotropic or anisotropic nanostructures, the solvothermal reaction has widely been adopted due to its simplicity, high efficiency, and low cost. The hydrothermal synthetic routes to the nanostructures of lanthanide hydroxides are well introduced in the literatures. The hydrothermal treatments for colloidal precipitates of lanthanide hydroxides resulted in diverse nanostructures such as nanospheres, nanorods, nanowires, nanoplates, nanotubes, and nanobelts.(Wang et al., 2003; Wang & Li, 2002; Wang & Li, 2003) It is noted that the hydrothermal reaction is generally sensitive to the temperature, pH, or aging conditions. Thus, when we apply the hydrothermal technique to synthesize $Gd(OH)_3$:Eu as a precursor for Gd_2O_3:Eu nanophosphor, a strong pH dependence is observed in particle shape. In this chapter, we introduce a selectively controlled low-temperature hydrothermal synthesis of $Gd(OH)_3$:Eu phosphor at different pHs and subsequent dehydration into Gd_2O_3:Eu in the form of nanorods with different aspect ratios, nanowires, nanospheres, and nanotubes.(Bae et al., 2009; Lee et al., 2009) Because the concentration quenching behavior has not yet been well investigated for the phosphor nanowires, a better understanding is required for the parameters affecting the radiative or nonradiative relaxation behaviors in nanowire structures. Therefore, the growth behavior and photoluminescence property of high quality nanowires of Gd_2O_3:Eu phosphor are also described here. $Gd_{2-x}Eu_xO_3$ solid solution is examined to get an insight for the relationship between the quenching concentration of activator (Eu^{3+}) and the optimum photoluminescence characteristics of nanowires with high aspect ratio.

2. Hydrothermal synthesis of Gd_2O_3:Eu phosphor

2.1 pH dependent hydrothermal synthesis

Several shapes (i.e. nanospheres, nanorods, nanotubes, and nanowires) of $Gd(OH)_3$:Eu nanoparticles are synthesized at different pHs by using the hydrothermal method. Thus, the shape selective synthesis of $Gd(OH)_3$:Eu from nanorods with considerably different aspect ratios to nanowires, nanospheres, and nanotubes can be successively achieved with increasing the pH of initial solution for hydrothermal reaction from about 6 to 14. In a typical synthesis, the stoichiometric amounts of Gd_2O_3 and Eu_2O_3 are dissolved in HNO_3 solution. After clear solution is formed by uniform stirring, aqueous KOH solution is added until the pH of solution is adjusted to be in the range of 6 ~ 14 for the formation of colloidal hydroxide precipitates. For the hydrothermal growth of $Gd(OH)_3$:Eu particles, the resulting colloidal mixture is put into a Teflon-lined stainless steel autoclave at room temperature. The autoclave is then sealed and maintained at 120 - 180°C for several hours. In general, the solution is continuously stirred during the hydrothermal treatment. But the reaction can be performed without stirring for aging. After the reaction is completed, the solid product $Gd(OH)_3$:Eu is collected by filtration, washed with distilled water, and dried. Subsequent dehydration of $Gd(OH)_3$:Eu by heating at 500°C for several hours yields Gd_2O_3:Eu oxide. To maintain the morphology of $Gd(OH)_3$:Eu, the heating rate is controlled at slower than 3 °C /min. In order to compare the crystallization, morphology change, and luminescence behaviors at different temperatures, the additional heat treatments for all Gd_2O_3:Eu powders are successively carried out at 700 and 800 °C for several hours.

It is noted that the precipitates showing a plate-type morphology are obtained after hydrothermal treatment at pH < 7, such products show an X-ray powder diffraction (XRD) pattern quite different from that of $Gd(OH)_3$. Recently, it was reported that $Gd_2(OH)_5(NO_3)\,nH_2O$ of layered rare-earth hydroxide structure is obtained in the range of pH = 6 ~ 7.(Lee & Byeon, 2009) Because large capacity and high affinity for the ion-exchange reaction, this compound can be used as host materials for a wide range of active molecules. Fig. 1 shows typical XRD patterns of the as-synthesized $Gd_2(OH)_5(NO_3)\,nH_2O$:Eu at pH ~ 7, $Gd(OH)_3$:Eu from hydrothermal process at pH ~ 8 (nanorods; see section 3.1) and ~ 11 (nanowires; see section 3.1), and Gd_2O_3:Eu obtained after heat treatment at 500 °C. A series of strong (00*l*) reflections observed in the XRD pattern of $Gd_2(OH)_5(NO_3)\,nH_2O$:Eu (Fig. 1a) is characteristic of a layered phase and corresponding to an interlayer separation of ~ 8.5 Å. In contrast, all the reflections for $Gd(OH)_3$:Eu and Gd_2O_3:Eu can be indexed to the hexagonal (space group $P6_3/m$; Figs. 1b and 1c) and the cubic (space group *I*a3; Fig.1d) phases respectively.(JCPDS card 83-2037; Buijs et al., 1987) While $Gd(OH)_3$:Eu obtained at pH ~ 8 is well crystallized, the increase of solution pH to ~ 11 and ~ 13 results in a relatively weak and broad intensity patterns, indicating a poorer crystallinity. The full-widths at half-maximum (FWHM) of the (110), (101), (201) reflections of $Gd(OH)_3$:Eu obtained at pH ~ 8 are smaller than one half in comparison with those obtained at pH ~ 11. Furthermore, it is of particular interest that the relative intensities of (110) and (101) diffractions for $Gd(OH)_3$:Eu are significantly different depending on the solution pH for hydrothermal synthesis. The intensity ratios between (110) and (101) reflections of the hydroxide nanorod is around 0.5 (Fig. 1b). Similar XRD patterns are observed for $Gd(OH)_3$:Eu oxides which are prepared at different pHs other than ~ 11 and the (110)/(101) intensity ratios are typically in the range of 0.5 ~ 1.0. These observations are consistent with the previously reported results for

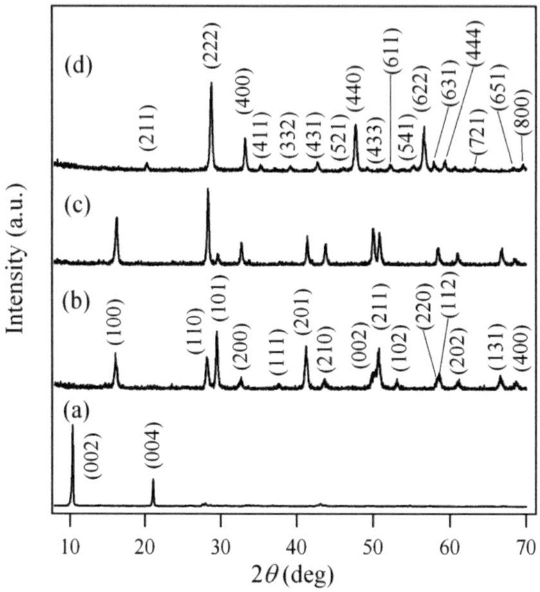

Fig. 1. Powder X-ray diffraction patterns of $Gd(OH)_5(NO_3)\ nH_2O$:Eu and $Gd(OH)_3$:Eu prepared by hydrothermal reaction at 180 °C and pH ~ (a) 7, (b) 13 (nanorods), and (c) 11 (nanowires) and (d) Gd_2O_3:Eu obtained after dehydration at 500 °C.

$Gd(OH)_3$.(Wang & Li, 2002; Wang et al., 2003; Louis et al., 2003) In contrast, such intensity ratio dramatically increased up to ~ 8 in the XRD pattern of the hydroxide prepared at pH ~ 11 (Fig. 1c). This would suggest that the (110) orientation is strongly preferred for $Gd(OH)_3$:Eu prepared at around this pH range. The comparison of the XRD patterns and morphology changes of $Gd(OH)_3$:Eu (see section 3.1) indicates that the growth direction of $Gd(OH)_3$:Eu from nanorods to nanowires is parallel to the (110) planes. Some more details are discribed in the next section.

2.2 Synthesis of high quality nanowires of Gd_2O_3:Eu at constant pH

The careful adjustment of pH for the hydrothermal reaction is required to induce well developed nanowire structure with sufficiently high aspect ratio. In contrast to the synthesis of lanthanide phosphate nanowires at pH = 1 – 2 under the hydrothermal condition,(Fang et al., 2003) the optimal pH range for the growth of $Gd(OH)_3$:Eu nanowires is close to around 11. Nanowires of $Gd_{1-x/2}Eu_{x/2}(OH)_3$ solid solution (x = 0.08, 0.12, 0.16, 0.20, and 0.24) are accordingly synthesized at pH ~ 11 by the procedure similar to section 2.1. The high quality bulk powder of Gd_2O_3:Eu phosphor can be prepared according to the citrate route.(for example, see Byeon et al., 2002)

It is well known that, besides pH, the concentration of solution can also strongly affect the transport behavior of constituting ions and the growth behavior of particles in the solvothermal synthesis. Thus, the highly anisotropic growth to the more uniform nanowires with several tens of micrometers in length is achieved by maintaining the higher concentration of the precursors dissolved in the initial HNO_3 solution for hydrothermal synthesis. The temperature is another factor influencing the growth of $Gd(OH)_3$:Eu nanowires. For instance, no nanowire is induced below 120 °C. The nanorods with relatively high aspect ratios are obtained at 120 – 160 °C under the same pH condition. The growth of nanowires longer than 10 micrometers can be achieved by raising the reaction temperature above 160 °C. Fig. 2 compares the XRD pattern of $Gd(OH)_3$:Eu bulk powder with those of the as-synthesized hexagonal $Gd(OH)_3$:Eu nanowires from hydrothermal process at pH ~ 11 and the cubic Gd_2O_3:Eu nanowires obtained after heat treatment at 500 °C. Similarly to the case in section 2.1, the intensity ratio between (110) and (101) reflections for the hydrothermally synthesized $Gd(OH)_3$:Eu nanowires are significantly different from those of the bulk powder; such relative intensity ratio of the bulk hydroxide is close to 1.0 (Fig. 2a), which is significantly smaller than that (~8) of the hydroxide nanowire (Fig. 2b). Single crystalline nanowires of hydroxides and oxides have been extensively studied.(Sharma & Sunkara, 2002; Tang et al., 2005; Singh et al., 2007) Their formations are related to the fact that the growth rate along one crystallographic direction is significantly faster than along the other directions. It has been reported that the growth direction of $Gd(OH)_3$ to nanorods is parallel to the (110) planes of the hexagonal unit cell.(Du & van Tendeloo, 2005) High resolution transmission electron microscope (HRTEM) images of $Gd(OH)_3$:Eu nanorod and nanowire are compared in Fig. 3. The fine fringes demonstrate that the as-synthesized $Gd(OH)_3$:Eu nanorods and nanowires are single crystalline. The spacing between fringes along both the rod and wire axes is about 0.32 nm which is close to the interplanar spacing of the (110) plane. This agreement confirms that the exceptionally high (110)/(101) intensity ratio of nanowires (Fig. 2b) is attributed to the preferential growth of nanowire parallel to the (110) planes. Although $Gd(OH)_3$ nanorods with relatively high aspect ratios have been described as nanowires in the literatures, true nanowires would show highly increased (110)/(101) intensity ratios in their XRD patterns.

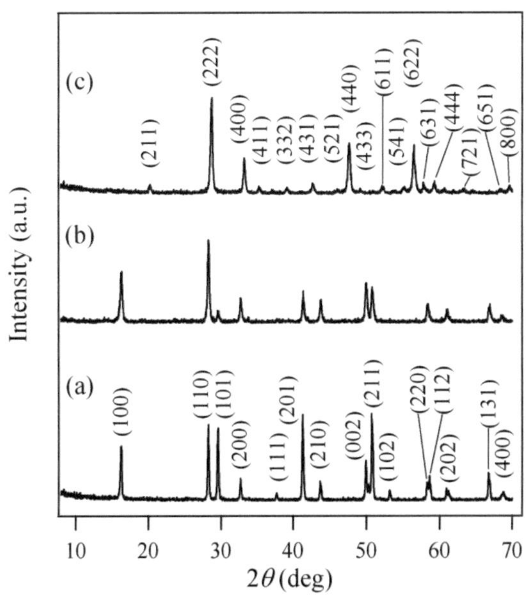

Fig. 2. Powder X-ray diffraction patterns of (a) the bulk $Gd(OH)_3$:Eu powder, (b) as-synthesized $Gd(OH)_3$:Eu nanowires from hydrothermal reaction, and (c) Gd_2O_3:Eu nanowires obtained after dehydration at 500 °C.

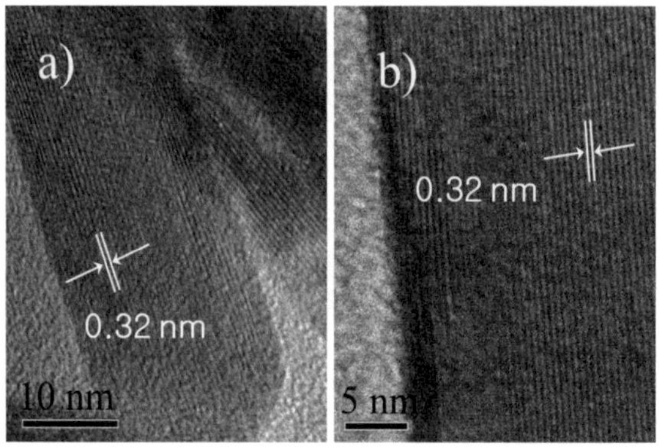

Fig. 3. HRTEM photographs of $Gd(OH)_3$:Eu prepared by hydrothermal reaction at 180 °C. pH ~ (a) 13 (nanorods) and (b) 11 (nanowires).

3. Morphological aspects of Gd_2O_3:Eu nanophosphor

3.1 pH dependent shape evolution of $Gd(OH)_3$:Eu and Gd_2O_3:Eu

The selective synthesis of diverse nanomaterials is challenging for understanding physical properties derived from well difined shape dimensionality. For example, the anisotropic growth of lanthanide orthophosphates ($LnPO_4$ where Ln = lanthanides) nanostructure can be enhanced by kinetically controlled hydrothermal processes using carefully selected chelating ligands.(Yan et al., 2005) Among many parameters affecting the solvothermal synthesis, the adjustment of pH was found to play a key role in selectively controlling the morphology of $Gd(OH)_3$:Eu nanostructures. In particular, KOH would influence on the nucleation and anisotropic growth of particles. For instance, the layered gadolinium hydroxynitrate $Gd_2(OH)_5(NO_3)\cdot nH_2O$:Eu is obtained at pH = 6 - 7. The field emission scanning electron microscopy (FE-SEM) image (Fig. 4a) of $Gd_2(OH)_5(NO_3)\cdot nH_2O$:Eu obtained at pH ~ 7 shows that they comprise a plate-like microcrystalline powders. Because of very weak intensity and insufficient number of non-(00*l*) reflections observed in XRD patterns of $Gd_2(OH)_5(NO_3)\cdot nH_2O$:Eu (Fig. 1a), the arrangement mode within the *ab* plane is measured by the selected area electron diffraction (SAED). As shown in Fig. 4b, the clear spots observed in the reciprocal lattice, which are rectangularly arranged, confirm an order within the *ab* plane of crystals (a ~ 7.3 Å, b ~ 12.9 Å).

Fig. 4. (a) FE-SEM photograph and (b) selected area elctron diffraction patterns of $Gd_2(OH)_5(NO_3)\cdot nH_2O$:Eu prepared by hydrothermal reaction at 180 °C, pH ~ 7.

On the contrary, at pH > 8, $Gd(OH)_3$:Eu particles are produced in the form of nanorods with variable aspect ratios, nanowires, or the mixture of nanosheets, nanotubes, and nanorods, depending on the solution pH. Fig. 5 shows FE-SEM images of $Gd(OH)_3$:Eu synthesized at pH ~ 8 and Gd_2O_3:Eu obtained by subsequent heat treatment at 500 and 800 °C. It can be seen in Fig. 5a that the as-synthesized $Gd(OH)_3$:Eu are composed of highly uniform nanorods. The aspect ratio of nanorods is tunable by controlling the experimental conditions of pH, temperature, and concentration. After dehydrating this hydroxide to Gd_2O_3:Eu at high temperatures up to 800 °C, the nanorod shape is completely retained as shown in Figs. 5b and 5c. It has already been reported that the nanorod shape of $Gd(OH)_3$:Eu is maintained after the dehydration into Gd_2O_3:Eu.(Chang et al., 2005) Their average diameter and length are around 200 nm and 1 - 1.5 μm, respectively. The morphology of $Gd(OH)_3$:Eu synthesized at pH ~ 9 is shown in Fig. 6a. This powder also consists of uniform rod-like particles. Compared with those obtained at pH ~ 8, the diameter of $Gd(OH)_3$:Eu nanorods is smaller but the length is longer when they are prepared at pH ~ 9.0. The nanorod structure is not collapsed after thermal dehydration (Figs. 6b and 6c). Their average diameter and length are around 150 nm and 1.5 - 2 μm, respectively.

During hydrothermal synthesis, it is revealed that the higher the pH of starting solution is, the smaller diameter of $Gd(OH)_3$:Eu nanorods is induced. In contrast, the average length of the nanorods is increased so that the aspect ratio of $Gd(OH)_3$:Eu nanorods becomes larger with the increase of solution pH at the same concentration and reaction temperature. Fig. 7 shows typical images of the as-synthesized $Gd(OH)_3$:Eu at pH ~ 10 and its dehydrated oxide

Fig. 5. FE-SEM photographs of (a) $Gd(OH)_3$:Eu prepared by hydrothermal reaction at 180 °C, pH ~ 8 and Gd_2O_3:Eu obtained after dehydration at (b) 500 °C and (d) 800 °C.

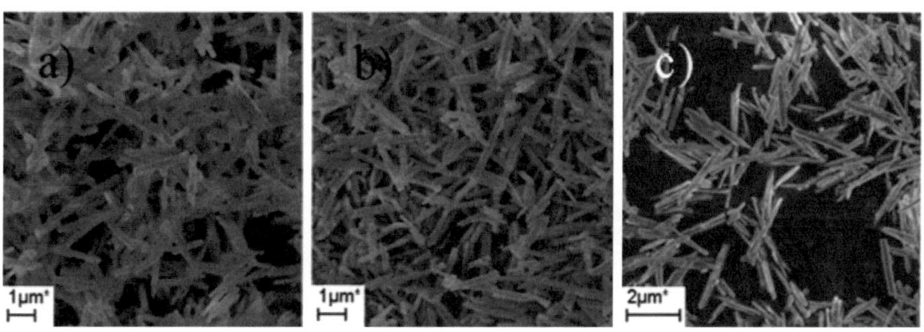

Fig. 6. FE-SEM photographs of (a) $Gd(OH)_3$:Eu prepared by hydrothermal reaction at 180 °C, pH ~ 9 and Gd_2O_3:Eu obtained after dehydration at (b) 500 °C and (d) 800 °C.

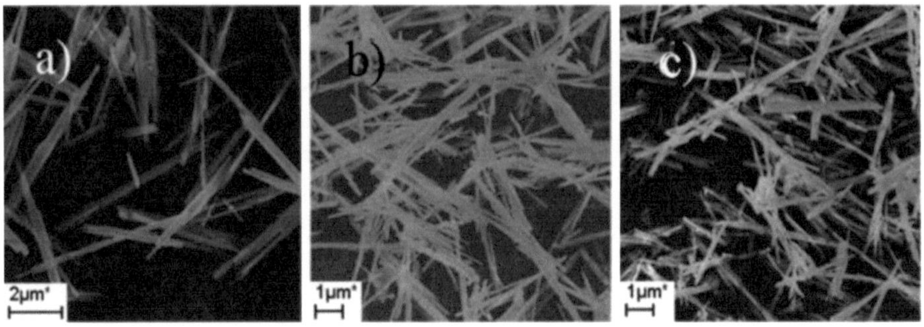

Fig. 7. FE-SEM photographs of (a) $Gd(OH)_3$:Eu prepared by hydrothermal reaction at 180 °C, pH ~ 10 and Gd_2O_3:Eu obtained after dehydration at (b) 500 °C and (d) 800 °C.

Gd_2O_3:Eu. Both compounds exhibit the nanorod structures with average diameter of about 100 nm and length of 3 - 4 μm. The maximal aspect ratios of $Gd(OH)_3$:Eu and Gd_2O_3:Eu particles are achieved in the higher pH range. The typical FE-SEM and TEM images of $Gd(OH)_3$:Eu synthesized at pH ~ 11 are shown in Figs. 8a and 8b, respectively. As can be seen in these figures, the as-synthesized $Gd(OH)_3$:Eu is composed of uniform nanowires whose lengths are close to several tens of micrometers. The images shown in Fig. 8b with higher magnification suggests that the diameters of nanowires are in the range of 20 - 30 nm. Figs. 8c and 8d display the FE-SEM of Gd_2O_3:Eu nanowires converted by thermal treatment of $Gd(OH)_3$:Eu nanowires at 500 °C and 800 °C, respectively. The temperature and activator (Eu^{3+}) concentration dependence of Gd_2O_3:Eu nanowires are additionally described in next section (3.2.). Further addition of KOH solution to adjust the pH to higher than 11 significantly reduces the aspect ratio of $Gd(OH)_3$:Eu particles to produce essentially

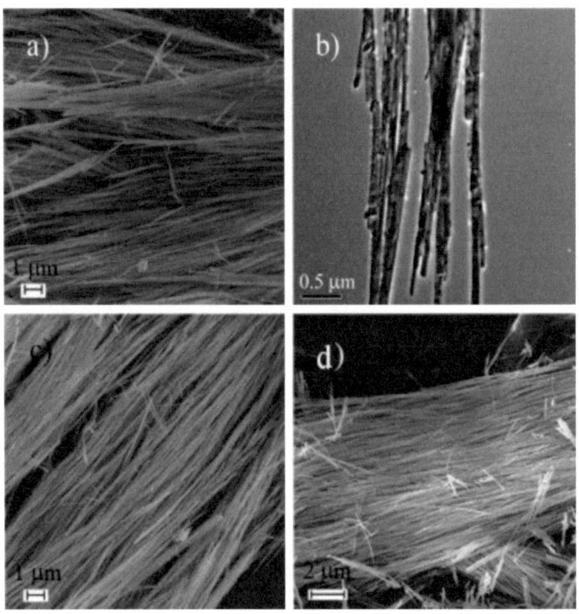

Fig. 8. FE-SEM and TEM photographs of (a and b) $Gd(OH)_3$:Eu prepared by hydrothermal reaction at 180 °C, pH ~ 11 and Gd_2O_3:Eu obtained after dehydration at (c) 500 °C and (d) 800 °C.

nanorods. The XRD pattern of this hydroxide is quite similar to Fig. 1b, the (110)/(101) intensity ratio being close to 0.7. It was proposed that the high OH^- ion concentration is preferable for the high aspect ratio but greatly reduce the ionic motion for the one-dimensional growth.(Wang & Li, 2002) This would imply that an optimal pH condition is required for the growth of true nanowires with high aspect ratio. Fig. 9a shows a FE-SEM image of the as-synthesized $Gd(OH)_3$:Eu powder by hydrothermal reaction at pH ~ 12. Although the particle shapes are uniform and wire-like, the aspect ratio is significantly decreased in comparison with that of true nanowires (Fig. 8) prepared at pH ~ 11. The morphology images of Gd_2O_3:Eu also exhibit the nanorod shapes (Figs. 9b and 9c), indicating the maintenance of morphology. The diameter and length of rigid nanorods are in the range of 20 - 30 nm and 2 - 3 μm, respectively.

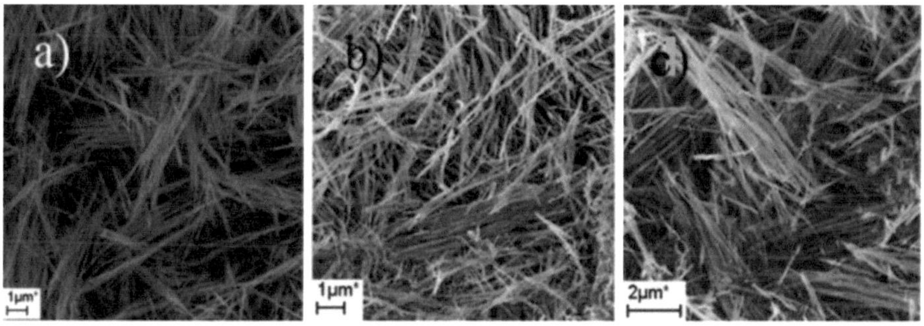

Fig. 9. FE-SEM photographs of (a) $Gd(OH)_3$:Eu prepared by hydrothermal reaction at 180 °C, pH ~ 12 and Gd_2O_3:Eu obtained after dehydration at (b) 500 °C and (d) 800 °C.

Compared with those prepared at lower pH ranges, the morphology of $Gd(OH)_3$:Eu obtained at pH ~ 13 strongly depends on the reaction temperature; the formation of nanowires is not induced but instead both nanorod and nanotube structures seems in competition in this pH range. The formation of nanorods with low aspect ratio is preferred at higher than 160 °C while the nanotubes are mainly obtained at lower than 140 °C. When the hydrothermal reaction is carried out with the initial solution of pH ~ 13, the $Gd(OH)_3$:Eu crystallites synthesized at 180 °C display the uniform morphology of nanorods with 20 - 30 nm in diameter and 200 - 300 nm in length (Fig. 10a). In contrast to the observations in other nanorods prepared at lower pHs, the morphology of $Gd(OH)_3$:Eu nanorods obtained at this pH range was not retained after the thermal transformation into Gd_2O_3:Eu. Figs. 10b and 10c are the FE-SEM and TEM images of Gd_2O_3:Eu obtained from heat treatment of $Gd(OH)_3$:Eu at 500 °C. Relatively regular Gd_2O_3:Eu particles are all quasi-spherical and the average particle size is close to 30 - 60 nm. Observation of the fine fringes supports a formation of regular crystalline lattice. On the other hand, the mixture of nanorods, nanotubes, and nanosheets is produced when $Gd(OH)_3$:Eu is synthesized at 120 °C. Fig. 10d shows that the nanotubes have outer diameters less than 30 nm and lengths of 150 - 200 nm. The nanorods have the aspect ratio smaller than that prepared at 180 °C. Considering that the nanosheets are also observed, the formation of sheet-structure is likely followed by the formation of nanotubes at a lower temperature, which in turn grows to more stable nanorods at a higher temperature. The optimal condition to synthesize uniform nanotubes of $Gd(OH)_3$:Eu is strongly dependent on the concentration of KOH and temperature. Similarly to the nanorods prepared at 180 °C, no shape is retained in the mixture of nanosheets, nanotubes, and nanorods obtained at 120 °C but the spherical nanoparticles with size of 10 - 40 nm are mainly obtained after dehydration into Gd_2O_3:Eu (Fig. 10e). The observed fine fringes are associated with the regular crystalline lattice (Fig. 10f). The spacings between fringes, 0.32, 0.27, and 0.20 nm correspond to the interplanar spacing of the (220), (400), and (440) plane of cubic cell, respectively. The origin of collapse of this nanorod shape is not straightforward. If we consider that the nanosheet, nanorod, and nanotube morphologies are in competition, such collapse would be attributed to a metastable nanorod structure. Many different strategies for the synthesis and characterization of inorganic nanotubes have been reported.(Rao & Nath, 2003) In the majority of the cases, the nanotube structures are induced by a rolling of single sheets from the layered lattices. Some oxide nanotubes have been synthesized by employing a hydrothermal technique. For instance, the hydrothermal synthesis of single-crystalline α-Fe_2O_3 nanotubes has been accomplished.(Jia et al., 2005)

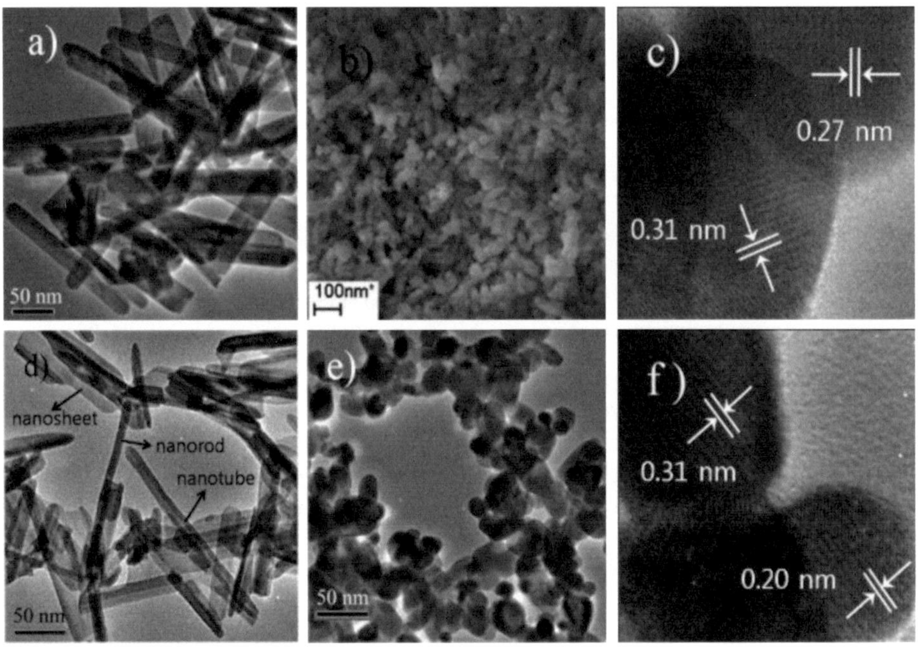

Fig. 10. FE-SEM and TEM photographs of (a) $Gd(OH)_3$:Eu prepared by hydrothermal reaction at 180 °C, pH = 12.9 and (b and c) Gd_2O_3:Eu obtained after dehydration at 500 °C. TEM photographs of (d) $Gd(OH)_3$:Eu prepared at 120 °C, pH ~ 13 and (e and f) Gd_2O_3:Eu obtained after dehydration at 500 °C.

When the direct synthesis of oxide nanotubes is difficult, a precursor prepared by hydrothermal reaction can be used under the appropriate conditions. Highly crystalline TiO_2 nanotubes were synthesized by hydrogen peroxide treatment of low crystalline TiO_2 nanotubes prepared by hydrothermal methods.(Khan et al., 2006) In particular, CeO_2 nanotubes have been prepared by the controlled annealing of the as-synthesized $Ce(OH)_3$ nanotubes from hydrothermal synthesis.(Tang et al., 2005) A similar result was expected for a thermal dehydration of $Gd(OH)_3$:Eu nanotubes to Gd_2O_3:Eu nanotubes. Unfortunately, $Gd(OH)_3$:Eu nanotubes are collapsed to yield spherical nanoparticles, no Gd_2O_3:Eu nanotubes being obtained after dehydration at 500 °C. The nanotubes of cubic Gd_2O_3:Eu, which is not a lamella structure, will require numerous defects and twin orientation relationships, which is energetically unfavorable, during rearrangement for the structural transformation. The difference in strain and curvature between the outer and inner surfaces of nanotubes will induce a different contraction tensions around defects and twins. This torsion would consequently lead to a collapse of tube structure. The hydrothermal reaction at pH ~ 14 results in $Gd(OH)_3$:Eu nanorods with low aspect ratios. This hydroxide is not well crystallized and the rod shape is not completely retained after thermal dehydration into Gd_2O_3:Eu. Instead, the mixture of nanospheres andnanorods is obtained.

3.2 The growth behavior of Gd_2O_3:Eu nanowire

Oxide nanowires can be synthesized by diverse methods. Single crystalline cubic spinel $LiMn_2O_4$ nanowires were prepared by using $Na_{0.44}MnO_2$ nanowires as a self-template.(Hosono et al., 2009) $ZnSnO_3$ nanowires were prepared by using D-fructose as a molecule template.(Fang et al., 2009) Non-catalytic resistive heating of pure metal wires or foils at ambient conditions was developed to grow the nanowires of metal oxides.(Rackauskas et al., 2009) A vapor–solid route was employed for the catalyzed synthesis of vertically aligned V_2O_5 nanowire arrays with tunable lengths and substrate coverage.(Velazquez & Banerjee, 2009) The hydrothermal synthesis also provides an effective route to fabricate the oxide nanowires. Single-crystalline tetragonal perovskite-type PZT oxide nanowires has been synthesized by applying a polymer-assisted hydrothermal method.(Xu et al., 2005) Uniform single-crystalline $KNbO_3$ oxide nanowires have also been obtained by employing the hydrothermal route.(Magrez et al., 2006) However, a direct hydrothermal synthesis of Gd_2O_3:Eu oxide nanowires is difficult because of high stability of the hydroxide form at high pH conditions. Instead, Gd_2O_3:Eu nanowires can be obtained via dehydration of hydrothermally synthesized hydroxide forms as described in section 3.1. Fig. 11a shows again the FE-SEM image of Gd_2O_3:Eu oxide of several tens of micrometers in length. Fig. 11b with higher magnification suggests that the diameters of nanowires are in the range of 10 ~ 30 nm. It is generally expected that the highly anisotropic shapes of nanoparticles would collapse when they transform into a different structure of phase by heat treatment.(Liang & Li, 2003) In this respect, it is of interest that the nanowire shape of $Gd(OH)_3$:Eu are not broken after the transformation into Gd_2O_3:Eu structure by heat treatment. The structural transformation from hexagonal $Gd(OH)_3$ to cubic Gd_2O_3 proceeds through the formation of intermediate monoclinic GdOOH. If we consider that the single-crystalline character of $Gd(OH)_3$ nanorods is retained in GdOOH,(Chang et al., 2006) a sequential transformation from hexagonal to monoclinic and finally to cubic structure, rather than an abrupt transformation, could be responsible for the maintenance of wire-shapes in Gd_2O_3:Eu. The hydrothermal method for the synthesis of $Gd(OH)_3$:Eu nanowires consequently provides a selective route to the nanowires of red-emitting Gd_2O_3:Eu oxide nanowires. Despite the maintenance of anisotropic particle shape, however, the structural transformation of nanowires from hexagonal to cubic symmetry will require numerous defects associated with the presence of microstrains. HRTEM image of Gd_2O_3:Eu nanowires shown in Fig. 11c supports this feature. Besides the fine fringes spaced by about 0.27 nm, which is close to the interplanar spacing of the (400) plane, the other weak fringes and wave-like images likely attributed to the stacking faults are observed as indicated by dotted line in the circle of this figure. The concentration of Eu^{3+} also influences on the formation of $Gd_{1-x/2}Eu_{x/2}(OH)_3$ and consequently $Gd_{2-x}Eu_xO_3$ nanowires. SEM images of nanowires are

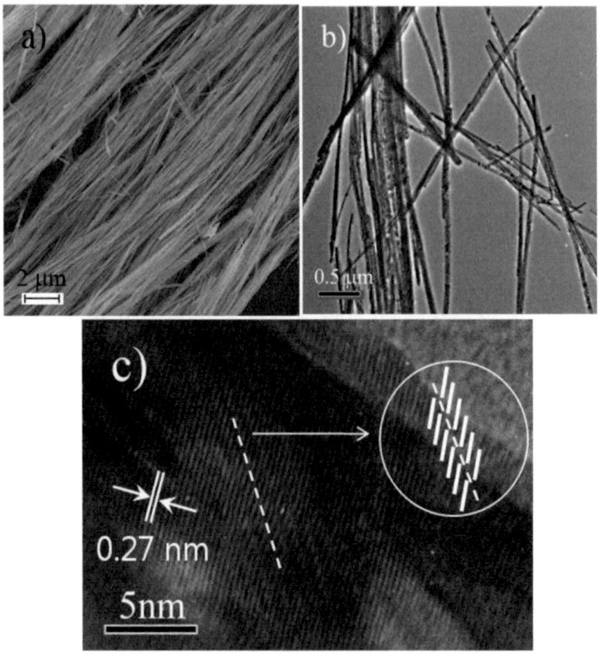

Fig. 11. (a) FE-SEM, (b) TEM, and (c) HRTEM photographs of Gd_2O_3:Eu nanowires (A stacking fault is represented in the circle.) obtained after dehydration at 500 °C.

compared as a function of Eu^{3+} concentration(x) and heating temperature in Fig. 12. As shown in these images, when the concentration of doped Eu^{3+} increases from 0.08 up to ~ 0.20, the morphology of Gd_2O_3:Eu nanowire tends to be improved. Further increase of doping concentration results in an abrupt decrease in aspect ratio and then nanorods are obtained. The nanowire morphology of Gd_2O_3:Eu is maintained even after heat treatment up to 700 and 800 °C. Partial collapse of nanowires to the irregular nanorods is induced at higher than 900 °C.

4. Photoluminescence spectra of Gd_2O_3:Eu nanophosphors

4.1 Correlation between photoluminescence and aspect ratio of Gd_2O_3:Eu nanoparticles

The optical characteristics and performances of nanometer-sized phosphor materials are generally dependent on their crystal structures, size, and morphologies. For instance, a difference of about 10 nm in the charge-transfer band position is observed between Gd_2O_3:Eu nanorods and microrods despite the same composition.(Chang et al., 2005) The correlation between morphology and photoluminescence (PL) intensity of Gd_2O_3:Eu phosphor provides an insight for a particle shape of the oxide nanophosphor with high PL efficiency. Indeed, the photoemission spectra of Gd_2O_3:Eu with different aspect ratios support that a systematic difference in PL intensity is induced as a function of the particle morphology. Fig. 13a compares the PL emission spectra of Gd_2O_3:Eu phosphors as a function of the pH value at which corresponding hydroxide precursors are synthesized. For comparison, the emission intensity measured from a commercial Y_2O_3:Eu is also plotted as a reference in the figure. The observed several emission bands are associated with the $^5D_0$ - ^7F_J

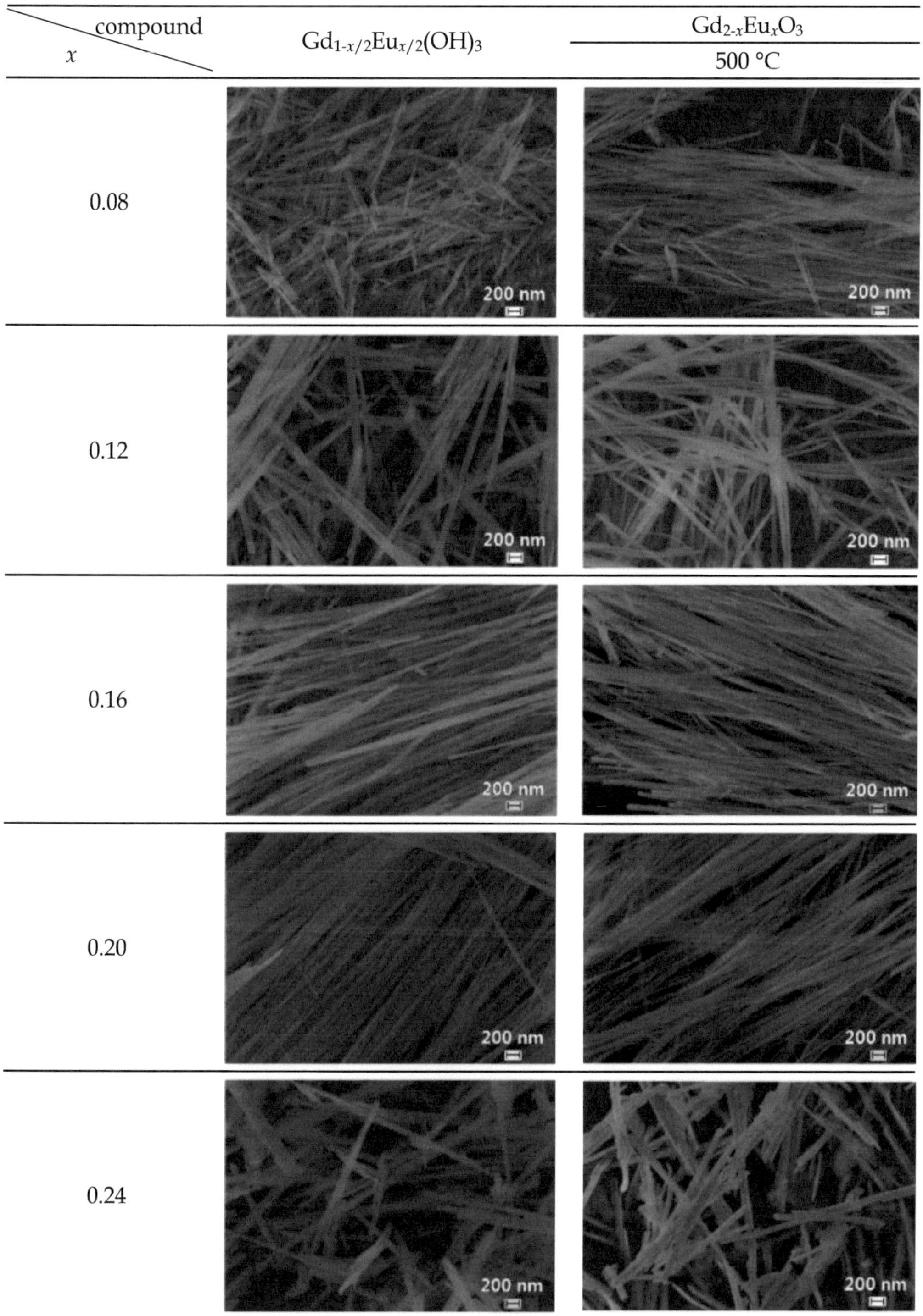

Fig. 12. Morphology evolution of $Gd_{1-x/2}Eu_{x/2}(OH)_3$ and $Gd_{2-x}Eu_xO_3$ nanowires as a function of Eu^{3+} concentration and temperature.

compound / x	$Gd_{2-x}Eu_xO_3$ 600 °C	$Gd_{2-x}Eu_xO_3$ 800 °C
0.08	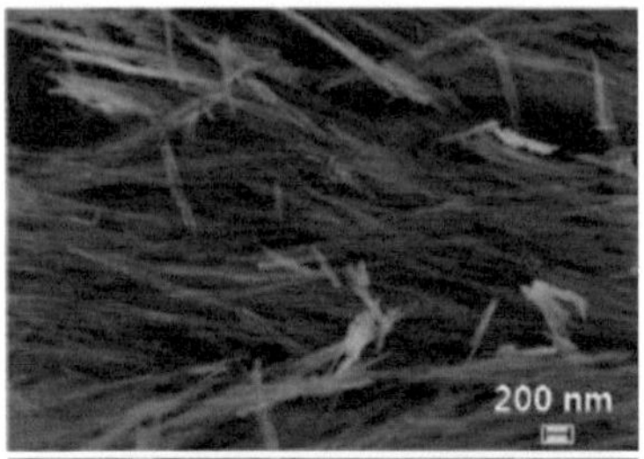	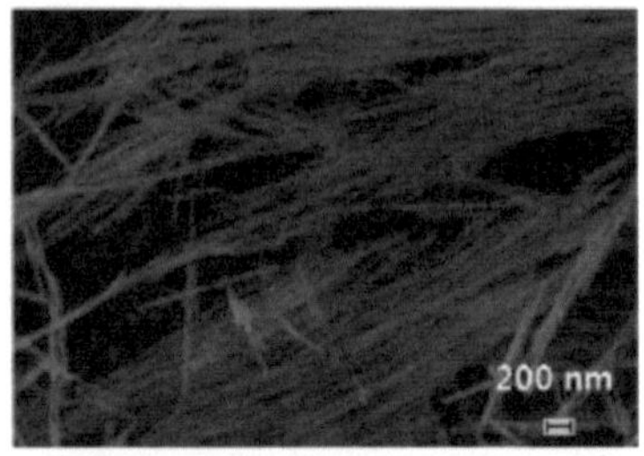
0.12		
0.16		
0.20	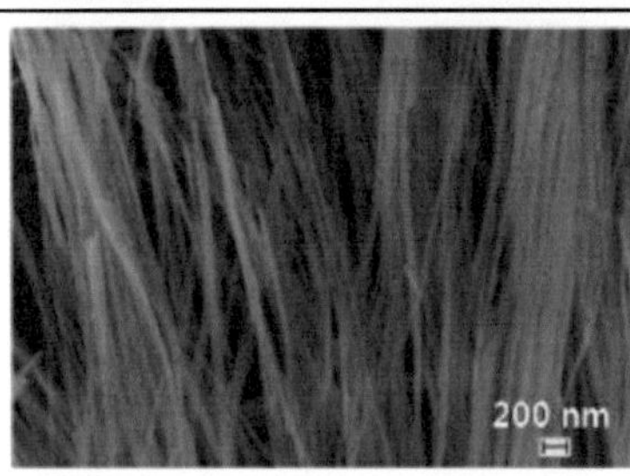	
0.24		

Fig. 12. (continued)

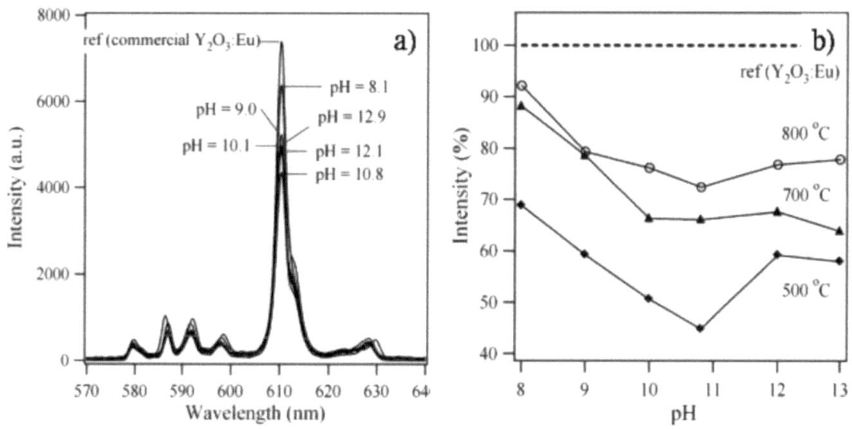

Fig. 13. (a) PL emission spectra of Gd_2O_3:Eu obtained after dehydration of $Gd(OH)_3$:Eu at 800 °C. PL emission spectrum of commercial Y_2O_3:Eu is also compared as a reference. (b) Comparison of the relative PL emission intensity of Gd_2O_3:Eu phosphors obtained after heat treatment for 5 h as a function of the solution pH for the hydrothermal synthesis. All intensities were measured as values relative to that of commercial Y_2O_3:Eu.

(J = 1 ~ 5) transitions of the Eu^{3+} ion, where the most intense emission at 610 nm is assigned to the $^5D_0$ - $^7F_2$ transition.(Brecher et al., 1967) As shown in this figure, Gd_2O_3:Eu oxide from the hydroxide prepared at pH ~ 8 exhibits strong emission whose intensity is close to 90 % in comparison with that of commercially available Y_2O_3:Eu. The commercial Y_2O_3: Eu phosphor is generally sintered at temperature above 1300 °C. An adoption of hydrolysis technique using urea for the synthesis of Y_2O_3: Eu also requires the firing temperature of 1150 - 1400 °C to achieve an optimum luminescent property.(Matijevic & Hsu, 1987; Jiang et al., 1998; Jing et al., 1999) It is accordingly notable that the emission intensity of Gd_2O_3:Eu nanorods obtained at 800 °C is comparable with that of commercial Y_2O_3: Eu. In Fig. 13b, the pH dependent PL intensities of Gd_2O_3:Eu phosphors are summarized as a function of the dehydration temperature of corresponding hydroxide precursors. The emission intensity of Gd_2O_3:Eu exhibits a minimum value when the pH for hydrothermal synthesis of the hydroxide increases. The aspect ratio of $Gd(OH)_3$:Eu (or Gd_2O_3:Eu) is enhanced with increasing the solution pH from ~ 8 to ~ 11 which is the optimal value for the formation of nanowires. In contrast, the PL emission intensity of Gd_2O_3:Eu is monotonically reduced with increase of the aspect ratio in this pH range. The lowest emission is observed with the nanowires (pH ~ 11) of the highest aspect ratio. The intensity of Gd_2O_3:Eu nanowires obtained after dehydration at 800 °C is close to 70 % in comparison with that of commercial Y_2O_3:Eu. When pH > 11, the aspect ratio of Gd_2O_3:Eu nanorods is decreased again and then the PL emission intensity is enhanced.

Comparing the PL behavior as a function of the particle morphology, the emission intensity of Gd_2O_3:Eu decreases when the aspect ratio becomes larger. One of the origins for such a correlation could be a difference in the surface area of particles. An important source of luminescence quenching in nanosized particles is the surface, where the coordination of the atoms differs from that in the bulk.(Counio et al., 1998) As a result of higher aspect ratios, the enlarged surface area of crystallites would result in an increase of surface defects which can be the nonradiative recombination centers (see section 4.2). Thus, the high concentration of stacking faults and twins in the surface of nanowires can be an origin for a decrease of the emission intensity observed in higher aspect-ratio-particles.

4.2 Photoluminescence and quenching concentration of Gd_2O_3:Eu nanowires

Few researchs have been addressed to the photoluminescence and cathodoluminescence of oxide phosphor nanowires. Tin oxide nanowires grown by vapor-liquid-solid process showed the strong UV emission at low temperatures related to the near-band-edge transition.(Chen et al., 2009) Ga_2O_3-SnO_2 nanowires of heterostructure showed a sharp transition region with emission bands from green to red cathodoluminescences.(Maximenko et al., 2009) In Fig. 14A, the photoluminescence (PL) emission spectra of Gd_2O_3:Eu nanowires are compared as a function of heating temperature. As shown in this figure, the emission intensity of Gd_2O_3:Eu nanowires measured at 610 nm is enhanced with increasing the dehydration temperature. When obtained even after dehydration at 800 °C, however, the nanowire phosphor exhibits about 60 % of emission intensity in comparison with that of commercial Y_2O_3:Eu. As pointed out in section 4.1, the high concentration of defects such as stacking faults in the surface of Gd_2O_3:Eu nanowires (Fig. 11c) would act as the nonradiative recombination centers and accordingly be responsible for the low PL emission intensity. The large surface of nanoparticles provides efficient quenching centers for the deexcitation via traps. The density of phonons of the nanoparticles, which is crucial for the resonant energy transfer, is also much lower than that of normal crystal. All of these effects result in a decreased probability of energy transfer among activator ions. As a consequence, higher activator concentrations have been generally observed for the optimized radiative recombination in nanosized particles than in normal materials.(Duan et al., 2000; Zhang et al., 2003) While an increase of the activator concentration improves the probability of the energy transfers to the activator ions and therefore radiative recombination, the probability

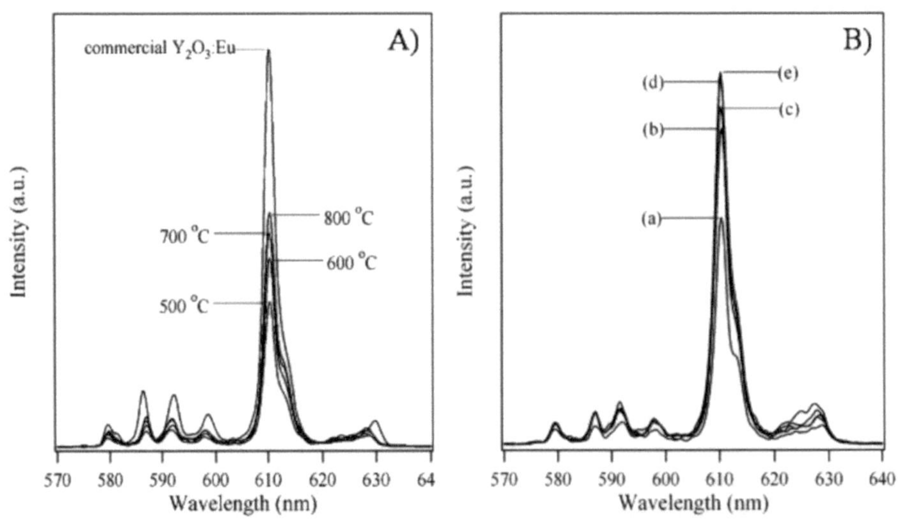

Fig. 14. (A) PL emission spectra (λ_{ex} = 254 nm) of Gd_2O_3:Eu nanowires as a function of heating temperature. All intensities were measured as values relative to that of commercial Y_2O_3:Eu. (B) Comparison of PL emission spectra (λ_{ex} = 254 nm) of $Gd_{2-x}Eu_xO_3$ nanowires as a function of the Eu^{3+} concentration; x = 0.08 (a), 0.12 (b), 0.16 (c), 0.20 (d), and 0.24 (e).

of energy transfer between activator ions also increases to induce a concentration quenching. Because there are no intermediate energy levels between the excited $^5D_0$ state and the ^7F_J states of Eu^{3+} to act as a bridge for cross-relaxation, the concentration quenching effect is essentially ascribed to the possible nonradiative transfer between neighboring Eu^{3+} ions. (Zhang et al., 1998; Dhanaraj et al., 2001) The quenching concentration of Eu^{3+} activators for $Gd_{2-x}Eu_xO_3$ nanowires is determined by measuring the PL intensity in the range of $0.08 \le x \le 0.24$. The relationship between photoemission intensity of $^5D_0$ - $^7F_2$ at 610 nm and Eu^{3+} activator concentration (x) for the $Gd_{2-x}Eu_xO_3$ nanowires is shown in Fig. 13B. The quenching concentration for nanowires is in the range $0.16 < x < 0.20$, which is much higher value in comparison with that ($x < 0.08 \sim 0.10$) for the bulk Gd_2O_3:Eu powder. Although the PL intensity of $x = 0.24$ member is comparable to that of $x = 0.20$ member, the nanowire shape is no longer observed and instead the nanospheres or nanorods with low aspect ratios is formed after high temperature annealing (Fig. 12).

5. Conclusion

Gadolinium oxide (Gd_2O_3) with a cubic structure is used as an efficient host matrix for trivalent rare earth ions for the fabrication of nanocrystalline phosphor materials. Because of the high refractive index, the large band gap (5.4 eV), the high resistivity and the high relative permittivity, gadolinium oxide is a promising material for high-k gate dielectrics, waveguides, and high resolution X-ray image detectors. Gd^{3+} is also a known contrast agent for magnetic resonance imaging (MRI), and thus the rare-earth doped Gd_2O_3 labels can be used for dual, fluorescence and MR imaging applications. In this chapter, we described that various single crystalline nanostructures (nanorods, nanowires, nanotubes, nanospheres) of red-emitting Gd_2O_3:Eu phosphor can be prepared by selective hydrothermal synthesis of $Gd(OH)_3$:Eu at different pHs and subsequent dehydration at high temperatures. The aspect ratios of phosphor particles are tunable by simply adjusting the pH of the initial solution for hydrothermal synthesis of $Gd(OH)_3$:Eu. In particular, the nanowires of $Gd(OH)_3$:Eu can be selectively prepared at pH ~ 11. Highly uniform nanowires of 20 ~ 30 nm in diameter can grow up to several tens of micrometers in length. The shape of nanowire obtained under hydrothermal pressures are retained after the structural transformation from hexagonal $Gd(OH)_3$:Eu to cubic Gd_2O_3:Eu at high temperatures. Therefore, the selective control of $Gd(OH)_3$:Eu morphology provides a strategy for the selective control of one-dimensional oxide nano-phosphor Gd_2O_3:Eu. This method for the synthesis of Gd_2O_3:Eu nanowires is quite simple and facile. No catalyst is required to serve as the energetically favorable site for the absorption of reactants. No template is added to direct the growth of nanowires. The relative emission intensity of Gd_2O_3:Eu is reduced with increasing the aspect ratio of nanoparticles and the quenching concentration of activators is significantly increased in Gd_2O_3:Eu nanowires ($x = 0.20$) compared with that of the bulk powder ($x = 0.08 \sim 0.10$). Thus, the low luminescence efficiency of Gd_2O_3:Eu nanowires can be highly improved by doping more Eu^{3+} into the host lattice. This compensation of luminescence efficiency would be of great benefit to the practical use of Gd_2O_3:Eu phosphor nanowires.

6. Acknowledgment

The authors acknowledge the grant R01-2008-000-10442-0 from the Korea Science and Engineering Foundation.

7. References

Bae, Y.-J.; Lee, K.-H. & Byeon, S.-H. (2009). Synthesis and Eu^{3+} concentration-dependent photoluminescence of $Gd_{2-x}Eu_xO_3$ nanowires. *J. Lumin.*, 129, 1, (2009) 81-85, ISSN 0022-2313

Beaurepaire, E.; Buissette, V.; Sauviat, M.-P.; Mercuri, A.; Martin, J.-L.; Lahlil, K., Giaume, D.; Huignard, A. Gacoin, T.; Boilot, J.-P. & Alexandrou, A. (2004). Functionalized fluorescent oxide nanoparticles: artificial toxins for sodium channel targeting and imaging at the single-molecule level. *Nano Lett.*, 4, 11, (2004) 2079-2083, ISSN 1530-6984

Birkmire, R.W. & Eser, E. (1997). Polycrystalline thin film solar cells: present status and future potential. *Annu. Rev. Mater. Sci.*, 27, (1997) 625–653, ISSN 0084-6600

Blasse, G. & Grabmaier, B.C. (1994). *Luminescent Materials*, Springer, ISBN 3-540-58019-0, Berlin

Brecher, C.; Samelson, H.; Lempicki, A.; Riley, R. & Peters, T. (1967). Polarized spectra and crystal-field parameters of Eu^{+3} in YVO_4. *Phys. Rev.*, 155, 2, (1967) 178-187, ISSN 0031-899X

Buijs, M.; Meyerink, A. & Blasse, G. (1987). Energy transfer between Eu^{3+} ions in a lattice with two different crystallographic sites: Y_2O_3:Eu^{3+}, Gd_2O_3:Eu^{3+} and Eu_2O_3. *J. Lumin.*, 37,1, (1987) 9-20 , ISSN 0022-2313

Byeon, S.-H.; Ko, M.-G.; Park, J.-C. & Kim, D.-K. (2002). Low-temperature crystallization and highly enhanced photoluminescence of $Gd_{2-x}Y_xO_3$:Eu^{3+} by Li doping. *Chem. Mater.*, 14, 2, (2002) 603-608, ISSN 0897-4756

Chang, C.; Kimura, F.; Kimura, T. & Wada, H. (2005). Preparation and characterization of rod-like Eu:Gd_2O_3 phosphor through a hydrothermal routine. *Mater. Lett.*, 59, 8-9, (2005) 1037-1041, ISSN 0167-577X

Chang, C.; Zhang, Q.; Mao, D. (2006). The hydrothermal preparation, crystal structure and photoluminescent properties of GdOOH nanorods. *Nanotech.*, 17, 8, (2006) 1981-1985 , ISSN 0957-4484

Chen, R.; Xing, G.Z.; Gao, J.; Zhang, Z.; Wu, T. & Sun, H.D. Characteristics of ultraviolet photoluminescence from high quality tin oxide nanowires. *Appl. Phys. Lett.*, 95, (2009) 061908, ISSN 0003-6951

Counio, G.; Gacoin, T. & Boilot, J.-P. (1998). Synthesis and photoluminescence of $Cd_{1-x}Mn_xS$ ($x = 5\%$) nanocrystals. *J. Phys. Chem. B*, 102, 27, (1998) 5257-5260, ISSN 1089-5647

Cuif, J.P.; Rohart, E.; Macaudiere, P.; Bauregard, C.; Suda, E.; Pacaud, B.; Imanaka, N.; Masui, T. & Tamura, S. (2004). *Binary Rare Earth Oxides*, Kluwer Academic Publishers, ISBN 9781402025688, Dordrecht

Dhanaraj, J.; Jagannathan, R.; Kutty, T.R.N. & Lu, C.H. (2001). Photoluminescence characteristics of Y_2O_3:Eu^{3+} nanophosphors pPrepared using sol-gel thermolysis. *J. Phys. Chem. B*, 105, 45, (2001) 11098-11105, ISSN 1089-5647

Du, G. & Van Tendeloo, G. (2005). Preparation and structure analysis of $Gd(OH)_3$ nanorods. *Nanotechnology*, 16, 4, (2005) 595-597, ISSN 0957-4484

Duan, C.K.; Yin, M.; Yan, K. & Reid, M. (2000). Surface and size effects and energy transfer phenomenon on the luminescence of nanocrystalline X_1–Y_2SiO_5:Eu^{3+}. *J. Alloys Compd.*, 303-305, (2000) 371-375, ISSN 0925-8388

Ebbesen, T.W. & Ajayan, P.M. (1992). Large-scale synthesis of carbon nanotubes. *Nature*, 358, 6383, (1992) 220-222, ISSN 0028-0836

El-sayed, M. (2001). Some interesting properties of metals confined in time and nanometer space of different shapes. *Acc. Chem. Res.*, 34, 4, (2001) 257-264, ISSN 0001-4842

Erdei, S.; Roy, R.; Harshe, G.; Juwhari, S.; Agrawal, H. D.; Ainger, F.W. & White, W.B. (1995). The effect of powder preparation processes on the luminescent properties of yttrium oxide based phosphor materials. *Mater. Res. Bull.*, 30, 6, (1995) 745-753, ISSN 0025-5408

Fang, C.; Geng, B.; Liu, J. & Zhan, F. (2009). D-fructose molecule template route to ultra-thin $ZnSnO_3$ nanowire architectures and their application as efficient photocatalyst. *Chem. Commun.*, 2009, 2350–2352, ISSN 1359-7345

Fang, Y.-P.; Xu, A.-W.; Song, R.-Q.; Zhang, H.-X.; You, L.-P.; Yu, J.C. & Liu. H.-Q. (2003). Systematic synthesis and characterization of single-crystal lanthanide orthophosphate nanowires. *J. Am. Chem. Soc.*, 125, (2003) 16025-16034, ISSN 0002-7863

Goldys, E.M.; D.-Tomsia, K.; Jinjun, S.; Dosev, D.; Kennedy, I.M.; Yatsunenko, S. & Godlewski, M. (2006). Optical characterization of Eu-doped and undoped Gd_2O_3 nanoparticles synthesized by the hydrogen flame pyrolysis method. *J. Am. Chem. Soc.*, 128, 45, (2006) 14498-14505, ISSN 0002-7863

Guo, H.; Dong, N.; Yin, M.; Zhang, W.; Lou, L. & Xia, S. (2004). Visible upconversion in rare earth ion-doped Gd_2O_3 nanocrystals. *J. Phys. Chem. B*, 108, 50, (2004) 19205-19209, ISSN 1520-5207

Hirai, T. & Orikoshi, T. (2004). Preparation of Gd_2O_3:Yb,Er and Gd_2O_2S:Yb,Er infrared-to-visible conversion phosphor ultrafine particles using an emulsion liquid membrane system. *J. Colloid Interface Sci.*, 269, 1, (2004) 103-108, ISSN 0021-9797

Hosono, E.; Kudo, T.; Honma, I.; Matsuda, H. & Zhou, H. (2009). Synthesis of single crystalline spinel $LiMn_2O_4$ nanowires for a lithium ion battery with high power density. *Nano Lett.*, 9, 3, (2009) 1045-1051, ISSN 1530-6984

Hu, J.; Odom, T. W. & Lieber, C.M. (1999). Chemistry and physics in one dimension: synthesis and properties of nanowires and nanotubes. *Acc. Chem. Res.*, 32, 5, (1999) 435-445 , ISSN 0001-4842

Iijima, S. (1991). Helical microtubules of graphitic carbon. *Nature*, 354, 6348, (1991) 56-58 , ISSN 0028-0836

Iijima, S. & Ichihashi, T. (1993). Single-shell carbon nanotubes of 1-nm diameter. *Nature*, 363, 6430, (1993) 603-605, ISSN 0028-0836

Jia, C.-J.; Sun, L.-D.; Yan, Z.-G.; You, L.-P.; Luo, F.; Han, X.-D.; Pang, Y.-C.; Zhang, Z. & Yan, C.H. (2005). Single-crystalline iron oxide nanotubes. *Angew. Chem. Int. Ed.*, 44, 28, (2005) 4328-4333, ISSN 0570-0833

Jiang, Y.D.; Wang, Z.L.; Zhang, F.; Paris, H.G. & Summers, C.J. (1998). Synthesis and characterization of Y_2O_3: Eu^{3+} powder phosphor by a hydrolysis technique. *J. Mater. Res.*, 13, 10, (1998) 2950-2955, ISSN 0884-2914

Jing, X.; Ireland, T.; Gibbons, C.; Barber, D.J.; Silver, J.; Vecht, A.; Fern, G.; Trowga, P. & Morton, D.C. (1999). Control of Y_2O_3:Eu spherical particle phosphor size, assembly properties, and performance for FED and HDTV. *J. Electrochem. Soc.*, 146, 12, (1999) 4654-4658, ISSN 0013-4651

Kazes, M.; Lewis, D.Y.; Ebenstein, Y.; Mokari, T. & Banin, U. (2002). Lasing from semiconductor guantum rods in a cylindrical Microcavity. *Adv. Mater.*, 14, 4, (2002) 317-321, ISSN 0935-9648

Khan, M.A.; Jung, H.-T. & Yang, O.-B. (2006). Synthesis and characterization of ultrathin crystalline TiO_2 nanotubes. *J. Phys. Chem. B*, 110, 13, (2006) 6626-6630, ISSN 1520-5207

Law, M.; Greene, L.E.; Johnson, J.C.; Saykally, R. & Yang, P. (2005). Nanowire dye-sensitized solar cells. *Nat. Mater.*, 4, (2005) 455–459, ISSN 1476-1122

Lee, B.-I.; Lee, K.S.; Lee, J.H.; Lee, I.S. & Byeon, S.-H. (2009). Synthesis of colloidal aqueous suspensions of a layered gadolinium hydroxide: a potential MRI contrast agent. *Dalton Trans.*, (2009) 2490-2495, ISSN 1477-9226

Lee, K.-H.; Bae, Y.-J. & Byeon, S.-H. (2009). Nanostructures and photoluminescence properties of Gd_2O_3:Eu red-phosphor prepared via hydrothermal route. *Bull. Kor. Chem. Soc.*, 29, 11, (2008) 2161-2168, ISSN 0253-2964

Lee, K.-H. & Byeon, S.-H. (2009). Extended members of layered rare-earth hydroxide family, $RE_2(OH)_5NO_3 \cdot nH_2O$ (RE = Sm, Eu, and Gd): synthesis and anion-exchange behavior. *Eur. J. Inorg. Chem.*, 2009, 8, (2009) 929-936, ISSN 1434-1948

Leschkies, K.S.; Divakar, R.; Basu, J.; Enache-Pommer, E.; Boercker, J.E.; Carter, C.B.; Kortshagen, U.R.; Norris, D.J. & Aydil, E.S. (2007). Photosensitization of ZnO nanowires with CdSe quantum dots for photovoltaic devices. *Nano Lett.*, 7, (2007) 1793–1798, ISSN 1530-6984

Levy-Clement, C.; Tena-Zaera, R.; Ryan, M.A.; Katty, A. & Hodes, G. (2005). CdSe-sensitized p-CuSCN/nanowire n-ZnO heterojunctions. *Adv. Mater.*, 17, (2005) 1512–1515, ISSN 0935-9648

Liang, J.H. & Li, Y.D. (2003). Synthesis and characterization of $Ni(OH)_2$ single-crystal nanorods. *Chem. Lett.*, 32, 12, (2003) 1126-1127, ISSN 1348-0715

Louis, C.; Bazzi, R.; Flores, M.A.; Zheng, W.; Lebbou, K.; Tillement, O.; Mercier, B.; Dujardin, C. & Perriat, P. (2003). Synthesis and characterization of $Gd_2O_3:Eu^{3+}$ phosphor nanoparticles by a sol-lyophilization technique. *J. Solid State Chem.*, 173, 2, (2003) 335-341, ISSN 0022-4596

Louis, C.; Bazzi, R.; Marquette, C.A.; Bridot, J.L.; Roux, S.; Ledoux, G.; Mercier, B.; Blum, L.; Perriat, P. & Tillement, O. (2005). Nanosized hybrid particles with double luminescence for biological labeling. *Chem. Mater.*, 17, 7, (2005) 1673-1682, ISSN 0897-4756

Magrez, A.; Vasco, E.; Seo, J.W.; Dieker, C.; Setter, N. & Forro, L. (2006). Growth of single-crystalline $KNbO_3$ nanostructures. *J. Phys. Chem. B*, 110, 1, (2006) 58-61, ISSN 1520-5207

Matijevic, E. & Hsu, W.P. (1987). Preparation and properties of monodispersed colloidal particles of lanthanide compounds: I. gadolinium, europium, terbium, samarium, and cerium(III). *J. Colloid Interface Sci.*, 118, 2, (1987) 506-523, ISSN 0021-9797

Maximenko, S. I.; Mazeina, L.; Picard, Y.N.; Freitas, J.A.; Bermudez, Jr., V.M. & Prokes, S. M. Cathodoluminescence studies of the inhomogeneities in Sn-doped Ga_2O_3 nanowires. *Nano Lett.*, 9, 9, (2009) 3245-3251, ISSN 1530-6984

Nichkova, M.; Dosev, D.; Gee, S.J.; Hammock, B.D. & Kennedy, I.M. (2005). Microarray immunoassay for phenoxybenzoic acid using polymer encapsulated $Eu:Gd_2O_3$ nanoparticles as fluorescent labels. *Anal. Chem.*, 77, 21, (2005) 6864-6873, ISSN 0003-2700

Parilla, P.A.; Dillon, A.C.; Jones, K.M.; Riker, G.; Schulz, D.L.; Ginley, D.S. & Heben, M.J. (1999). The first true inorganic fullerenes? *Nature*, 397, 6715, (1999) 114, ISSN 0028-0836

Rackauskas, S.; Nasibulin, A.G.; Jiang, H.; Tian, Y.; Kleshch, V.I.; Sainio, J.; Obraztsova, E. D.; Bokova, S.N.; Obraztsov, A.N. & Kauppinen, E.I. (2009). A novel method for metal oxide nanowire synthesis. *Nanotech.*, 20, (2009) 165603(8pp), ISSN 0957-4484

Rao, C.N.R. & Nath, M. (2003). Inorganic nanotubes. *J. Chem. Soc., Dalton Trans.*, 1, (2003) 1-24, ISSN 1477-9226

Rao, C.N.R.; Mueller, A. & Cheetham, A.K. (2004). *The Chemistry of Nanomaterials*; Wiley-VCH: Weinheim, ISBN 3527306862, Germany

Ravichandran, D.; Roy, R. & White, W.B. (1997). Synthesis and characterization of sol-gel derived hexa-aluminate phosphors. *J. Mater. Res.*, 12, 3, (1997) 819-824, ISSN 0884-2914

Rossner, W. & Grabmaier, B.C. (1991). Phosphors for X-ray detectors in computed tomography. *J. Lumin.*, 48/49, 1, (1991) 29-36, ISSN 0022-2313

Rossner, W. (1993). The conversion of high energy radiation to visible light by luminescent ceramics. *IEEE Trans. Nucl. Sci.*, 40, 4, (1993) 376-379, ISSN 0018-9499

Sharma, S. & Sunkara, M.K. (2002). Direct synthesis of gallium oxide tubes, nanowires, and nanopaintbrushes. *J. Am. Chem. Soc.*, 124, 41, (2002) 12288-12293, ISSN 0002-7863

Shea, L.E.; McKittrick, J. & Lopez, O.A. (1996). Synthesis of red-emitting, small particle size luminescent oxides using an optimized combustion process. *J. Am. Ceram. Soc.*, 79, 12, (1996) 3257-3265, ISSN 0002-7820

Singh, D.P.; Neti, N. R.; Sinha, A.S.K. & Srivastava, O.N. (2007). Growth of different nanostructures of Cu_2O (nanothreads, nanowires, and nanocubes) by simple electrolysis based oxidation of copper. *J. Phys. Chem. C.*, 111, 4, (2007) 1638-1645, ISSN 1932-7455

Tang, B.; Zhou, L.; Ge, J.; Niu, J. & Shi, Z. (2005). Hydrothermal synthesis of ultralong and single-crystalline $Cd(OH)_2$ nanowires using alkali salts as mineralizers. *Inorg. Chem.*, 44, 8, (2005) 2568-2569, ISSN 0020-1669

Tang, C.; Bando, Y.; Liu, B. & Goldberg, D. (2005). Cerium oxide nanotubes prepared from cerium hydroxide nanotubes. *Adv. Mater.*, 17, 24, (2005) 3005-3009, ISSN 0935-9648

Tenne, R. (1995). Doped and heteroatom-containing fullerene-like structures and nanotubes. *Adv. Mater.*, 7, 12, (1995) 965-995, ISSN 0935-9648

Velazquez, J.M. & Banerjee, S. (2009). Catalytic growth of single-crystalline V_2O_5 nanowire arrays. *Small*, 5, 9, (2009) 1025–1029, ISSN 1613-6810

Wang, X. & Li, Y. (2002). Synthesis and characterization of lanthanide hydroxide single-crystal nanowires. *Angew. Chem. Int. Ed.*, 41, 24, (2002) 4790-4793, ISSN 0570-0833

Wang, X.; Sun, X; Yu, D.; Zou, B & Li, Y. (2003). Rare earth compound nanotubes. *Adv. Mater.*, 15, 17, (2003) 1442-1445, ISSN 0935-9648

Wang, X. & Li, Y. (2003). Rare-earth-compound nanowires, nanotubes, and fullerene-like nanoparticles: synthesis, characterization, and properties. *Chem. Eur. J.*, 9, (2003) 5627-5635, ISSN 0947-6539

Xia, Y.N.; Yang, P.D.; Sun, Y.G.; Wu, Y.Y.; Mayers, B.; Gates, B.; Yin, Y.D.; Kim, F. & Yan, Y.Q. (2003). One-dimensional nanostructures: synthesis, characterization, and applications. *Adv. Mater.*, 15, 5, (2003) 353-389, ISSN 0935-9648

Xu, G.; Ren, Z.; Du, P.; Weng, W.; Shen, G. & Han, G. (2005). Polymer-assisted hydrothermal synthesis of single-crystalline tetragonal perovskite $PbZr_{0.52}Ti_{0.48}O_3$ nanowires. *Adv. Mater.*, 17, 7, (2005) 907-910, ISSN 0935-9648

Yan, M.F.; Huo, T.C. D. & Ling, H.C. (1987). Preparation of $Y_3Al_5O_{12}$-based phosphor powders. *J. Electrochem. Soc.*, 134, 2, (1987) 493-498, ISSN 0013-4651

Yan, R.; Sun, X.; Wang, X., Peng, Q. & Li, Y. (2005). Crystal structures, anisotropic growth, and optical properties: Controlled synthesis of lanthanide orthophosphate one-dimensional nanomaterials. *Chem. Eur. J.*, 11, (2005) 2183-2195, ISSN 0947-6539

Yen, W.M.; Shionoya, S. & Yamamoto, H. (2007). *Phosphor Handbook*, CRC Press, ISBN 0-8493-3564-8, New York

Zhang, W.P.; Xie, P.B.; Duan, C.K.; Yan, K.; Yin, M.; Lou, L.R.; Xia, S.D. & Krupa, J.C. (1998). Preparation and size effect on concentration quenching of nanocrystalline Y_2SiO_5:Eu. *Chem. Phys. Lett.*, 292, 1-2, (1998) 133-136, ISSN 0009-2614

Zhang, W.W.; Zhang, W.P.; Xie, P.B.; Yin, M.; Chen, H.T.; Jing, L.; Zhang, Y.S.; Lou, L.R. & Xia, S.D. (2003). Optical properties of nanocrystalline Y_2O_3:Eu depending on its odd structure. *J. Colloid Interface Sci.*, 262, 2, (2003) 588-593, ISSN 0021-9797

Zhou, Y.; Lin, J. & Wang, S. (2003). Energy transfer and upconversion luminescence properties of Y_2O_3:Sm and Gd_2O_3:Sm phosphors. *J. Solid State Chem.*, 171, 1-2, (2003) 391-395, ISSN 0022-4596

17

Transition Metal-Doped ZnO Nanowires: En Route Towards Multi-colour Light Sensing and Emission Applications

N. Kouklin, M. Omari and A. Gupta
Department of Electrical Engineering and Computer Science,
University of Wisconsin-Milwaukee,
USA

1. Introduction

Opto-electronics is one of the largest and fastest evolving market segments, with revenues expected to surpass $1.2 trillion by 2017 (Optoelectronics Industry, 2007). Continuous improvements in the size, detection limit, reliability and spectral sensitivity of existing solid-state light sensors and image detectors, which remain a key component of almost every opto-electronic system and circuit, drive the field to new heights every year. Compared with other components, light sources and detectors recently have shown the most significant revenue gains, whereas highly-miniaturized and low-power solid-state photodetectors operating in extended, UV-visible spectral ranges will continue to have an important role, as they are expected to be ubiquitously used in many electronic systems ranging from high-capacity information storage to biochemical sensing, chemical and biological analysis, and astronomy.

Compound	Crystal structure; lattice constants, *nm*; symmetry group	E_g, eV	μ_n , μ_p cm^2/Vs	m_n / m_p
Cu_2O	Cubic; a = 0.42 ; O^4_h	2.17	~0.1 , ~100	~1.7
ZnO	Hex.; a = 0.32, c= 0.59 ; C^4_{6v}	3.35	~180-1000, -	~0.45
In_2O_3	Cubic; a = 1.01	2.8	-	~0.36
SnO_2	Cubic; a = 0.47 ; O^5_h	3.54	~260, ~300	-
TiO_2	Tetr.; a = 0.45, c = 0.29 ; D^{14}_{4h}	3.0	~0.16-0.57, -	-

Table 1. Structural and electronic characteristics of several key metal-oxide semiconductors; E_g is band gap energy, μ_n, μ_p are electron and hole mobilities, m_n /m_p is electron-hole effective mass ratio.

While many commercial UV-visible detectors have been realized based on the matured Si p-n junction technology, there are several limitations related to the use of Si materials for UV sensing, among which are aging effects and cooling requirements to subdue dark current. Metal oxides can easily help overcome both limitations as their band gap (E_g) is significantly larger compared with that of cSi (~ 1.1 eV), Table 1. Given the great thermal conductivity and strength of the chemical bonds of wide-band gap semiconductors, photodetectors engineered based on most of metal-oxides can withstand highly elevated T and hostile environments. Despite the fact that carrier mobilities are much smaller in cases of metal oxides compared with those of many conventional semiconductors, owing to the drastically increased dielectric strength of metal oxides, much larger carrier velocities can be readily achieved by applying much higher electric fields/device biases. The property remains critical for reducing the photogenerated carrier transit time and thus improving gain characteristics of the photodetectors.

Advances in nanofabrication methodologies now allow growing many metal-oxide nanowires, including in ZnO as dislocation-free, highly faceted single crystals, which according to the above-made discussion show significant promise for diverse nano-photodetector device as well as light-emission applications. At the same time, compared to cSi, ZnO nanowires typically exhibit a very limited intrinsic sensitivity to visible and infrared radiation, whereas the sensitivity to short-wavelength photons also tends to be reduced as a result of their small diameter and relative increase in the number of the surface defect states. In this chapter, we provide a brief review on the progress in engineering high-responsivity ZnO nanowire photodetectors operating in an extended, i.e., ultraviolet-visible spectral range. Particular attention is paid to the use and role of transition metal dopants to enhance the light sensitivity of ZnO nanowires/nanorods grown by seeded vapor-transport methods. Detailed consideration is given to several key aspects pertaining to the transport, photoconduction, and time-response characteristics of two-terminal metal-semiconductor-metal nano-photodetectors operating at both pre-avalanche and avalanche regimes. This review might be important to scientists working in with high-sensitivity and multispectral oxide-based nano-photodetectors, optical switches, and sensors.

2. Sensing light with ZnO nanowires

The relentless drive towards down-scaling traditional electronics has enabled a new class of devices that operate faster, consume less power, and cost less. This has catalyzed new fabrication approaches to push beyond the limits of conventional lithography. A variety of high-quality, one-dimensional electronic materials have been all synthesized by techniques other than photolithography. Such materials include nanotubes, nanobelts and nanowires in both conventional and non-conventional semiconductors, such as metal oxides. (Chik et al. 2004; Lee et al. 2004; Sen et al. 2007; Gupta et al. 2008; Wischmeie et al. 2008) Among different metal-oxides, wurtzite ZnO, a II-IV compound, has a direct-band gap of ~3.3 eV (at room temperature) and one of the largest exciton binding energies of ~ 60meV, which makes this semiconductor particularly attractive for many advanced short-wavelength device applications, including but not limited to, ultra-small photodetectors, optical switches, and light-emitting diodes.

The UV-light sensitivity of ZnO nanowires produced by various routes including vapor transport methods has been commonly verified and reported. (Fan et al. 2004; Fan et al. 2006; Yong 2006) The group of P. Yang was among the first to report a marked drop in the resistance of multi-terminal ZnO metal-semiconductor-metal detectors exposed to a UV light (~ 365 nm). (Kind et al. 2002) The device rise and decay times were typically better than ~1 s. No change in the current was registered when the devices were exposed to visible radiation (532 nm light). P. Yang's group also found that, in contrast with a conventional semiconductor-based sensor, the photocurrent changed nonlinearly with illumination intensity, presumably stemming from a complex interplay of electron-hole generation/recombination such as trapping/detrapping processes. Despite an overall good UV photoconduction response, the UV absorption efficiency of most ZnO nanowire-based detectors is to remain small on the whole, especially when the diameter of the nanowire is reduced below $\alpha^{-1} \sim 150$ nm (α stands for the intrinsic absorption coefficient of ZnO). At the same time, a relative increase in the number of surface defect states (Shalish et al. 2004) is known to further reduce the device absorption cross-section as defects provide the sites for fast, non-radiative recombinations. The effect becomes especially apparent as the nanowire diameter approaches the limit of tens of nm.

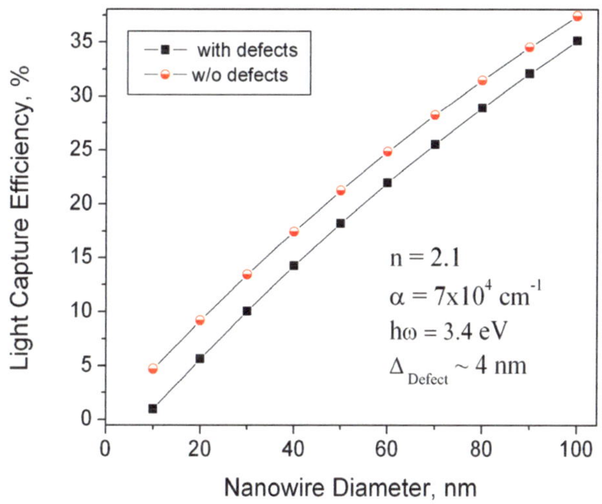

Fig. 1. Plots light capture efficiency as a function of nanowire diameter with and without a defect layer. The light polarization is assumed to be along the nanowire main axis. Reproduced with permission, J. Elect. Mat. 38 (2009).

Figure 1 plots the light absorption efficiency of the ZnO nanowire as a function of its diameter, which has been calculated by taking into account the effects of light reflection and absorption. It is apparent that the presence of a ~4 nm dead (i.e., surface defect) layer effectively decreases the interband absorption by a factor of five, i.e., down to ~ 1% in the nanowires with diameter of ~ 10 nm. While an increase in the spatial confinement of electrons and holes could generally improve the strength of many optical transitions (and thus the absorption in case of ZnO nanowires) owing to a stronger electron-hole wavefunction overlap, the quantum confinement effects are expected to be generally weak as most of the nanowires produced by vapor transport methods have diameters in the range of tens to hundreds of nm, thereby significantly exceeding the exciton Bohr radius in this material system (~2 nm). Therefore, the size quantization effects are unlikely to have any profound influence on the absorption characteristics of ZnO nanowires.
In the case of stoichiometric ZnO, the $4s^2$ Zn orbital is empty and the $2p^6$ O orbital is full, making ZnO an intrinsic semiconductor. However, in the case of oxygen-deficient ZnO, the excess $4s^2$ electrons can be easily ionized (~0.05eV), which explains why conductivity of

most of as-grown ZnO nanowires is of n-type. As a result, for any solid-state ZnO nanowire-based optical sensors that rely on band-to-band electronic transitions as the primary photocarrier generation mechanism, minimizing the undesired n-type doping effects becomes critical in order to engineer photosensors with improved photosensing characteristics and with a large on/off ratio. (Sen et al. 2007)

In the case of Ohmic contacts, the photoresponse of two-terminal, nanowire-based photodetectors can be calculated by invoking classical considerations, with the collected photocurrent having the following simple expression (Sze 2002):

$$I_{photo} = q\,(\eta \frac{P_{opt}}{h\nu})\,(\frac{\tau v_{drift}}{L}) = qGK\,,$$

where q is the electron charge, η is quantum efficiency, P_{opt} is incident optical power and hν is photon energy. In this formula, $G = \eta \frac{P_{opt}}{h\nu}$ is the number of free carriers generated per unit time and $\gamma = \frac{\tau v_{drift}}{L}$ is the device gain defined as a ratio of the rate at which free carrier is collected by the contacts to the rate at which it is photogenerated within the device. From the above formula, it is evident that in order to achieve optimum gain/photocarrier collection efficiency, the excess free-carrier lifetime, τ, and its drift velocity, v_{drift}, must be increased while the device size / carrier travel time must be reduced. At the same time, in contrast with bulk devices, many more free holes are to undergo annihilation on the molecular oxygen that is chemisorbed on the surface of the nanowires—the process that can significantly reduce the rate of bi-molecular recombination of free carriers, especially under high-injection conditions, i.e., when photogenerated excess carrier density $\Delta n >> n_0$, n_0 is equilibrium majority carrier density. As a result, the lifetime of the photogenerated excess carriers, τ, is significantly augmented in ZnO nanowires, making these nanostructures significantly more sensitive to UV optical radiation than their bulk counterparts.

These effects have been held responsible for high gain of ~ 10^8 attained in ZnO nanowire-based UV photodetectors (Soci et al. 2007). For comparison, a gain of ~ 10^5-10^6 is typically considered as an upper-limit in bulk and thin film photoconductor devices. Yet, the main drawback of increased carrier lifetime is slow device response time, also known as persistent photoconductivity. This limits the ability of the device to quickly switch from photoconduction to dark state, which is required in many modern opto-electronic device applications.

While the problem of reduced light absorption by ZnO nanowires can be partially mitigated by increased device gain, most of the metal oxides have another hefty drawback - a very limited sensitivity in the visible-infrared spectral range inherent with their large band gap, Table 1. While the increased visible-blindness represents a critical advantage for engineering visible-blind UV detectors the number of useful device technologies that can be realized with ZnO nanowires remains low, making it difficult for this nanomaterial to compete with commercially established Si-based photodetectors on the cost basis.

A controlled introduction of optically-active defect states within the band gap represents one of the most cost-efficient solutions to improve the photoresponse of ZnO in the sub-band gap optical range. Lin et al., for instance, reported a significant increase in the white light (400-800 nm) sensitivity of ZnO nanowires doped to only ~1% of Al , with a device on/off current ratio approaching as high as ~ 20, whereas for undoped devices the ratio was only ~ 1.06 (Lin et al. 2007). The effect is attributed to an optical excitation of free electrons from the Al-defect levels positioned ~ 80 meV below the conduction band. The photoconduction response of ZnO can be also improved by engineering heterojunctions with other semiconductors or high electronegativity metals such as Pt, Pd, etc., which are known to form Schottky-type junctions at the metal-ZnO interface (Jeong et al. 2003; Heo et al. 2004; Keem et al. 2004; Yong 2006). However, the results of more recent studies performed by us indicate additional possibilities of achieving strong sub-band-gap sensitivity based on avalanche-type photocarrier multiplication effects by intentionally introducing transition metal (TM) centers such as Cu within the core regions of ZnO. This has been found to substantially enhance the light sensitivity of ZnO nanowires in multiple spectral domains, which forms the topic of our next discussion.

3. Fabricating highly crystalline ZnO nanowires doped with Cu.

Role of TM-impurities.

Advances in nanofabrication now allow growing ZnO nanowires as dislocation-free, highly faceted single crystals (Chik et al. 2004), Consequently, such nanowires possess highly versatile and often refined electrical and optical properties. Following the previous discussion, achieving *p*-type doping in ZnO is critical not only from the standpoint of realizing *p-n* junction-based nanodevices such as electrically driven UV light-emitting diodes or detectors, but also improving the photosensitivity characteristics due to the charge compensation effect. Apart from spintronic applications, doping ZnO with TM impurities has a two-fold advantage. First, previous reports indicate that many TM dopants, including Cu, form deep acceptor states within the band gap of II-VI compounds (Dingle 1969; Heitz et al. 1992), including ZnS, CdS, and ZnO; thus these dopants can partially compensate/offset the n-type doping effects. Second, these centers have been found to act as a source of optically-induced charge transfer transitions and photocarrier multiplication effects, which render ZnO nanowires extremely sensitive to both UV and visible radiation and thereby help extend the optical sensitivity range of this metal-oxide.

Theoretical investigations of the electronic configurations of different substitutional TM impurities in groups II-VI and III-V have been undertaken within the framework of local density functional calculations (Gemma 1984), with the results confirming that most of TM dopants, including Fe, Ni and Cu, exhibit a highly rich transitional behavior and form deep multiplet states within the band gap of ZnO. Most of the doubly (TM^{2+}) and triply charged states (TM^{3+}) were also found to be electronically stable in this semiconductor system. At the same time, only Cu has been predicted to exist in a highly stable singly-ionized (TM^{+}) state in ZnO. Due to the strong ionicity of ZnO, most of TM impurities exhibit a strong 3d electron character (Dietz et al. 1963). The formation of multiplet states within the band gap of ZnO results from the splitting of the TM state into two E and T_2 symmetry states by the internal field of the crystal. In the case of ZnO, using a hole representation, Cu is predicted to exist in three primary electronic configurations/states: a ground state (Cu^{2+}, $^2T_2$) and two excited states (Cu^{2+}, ^2E) and (Cu^{+},h)) (Dahan et al. 1998). According to IR-absorption studies, the lowest energy transitions, i.e., involving Cu^{2+} ($^2T_2$-^2E) states, recently have been confirmed to be infrared-active, with a quantum efficiency approaching ~ 30% (Broser et al. 1994)

VLS Fabrication and Structural Characteristics of Cu:ZnO nanowires.

Several pathways have been developed and used to produce crystalline nanowires in different materials systems and in large quantities, including physical vapor deposition (Huang et al. 2001, Kong et al. 2001; Omari et al. 2008), electrochemical deposition (Kouklin et al. 2001; Kouklin et al. 2002; Wang et al. 2002), metalorganic chemical vapor deposition (Li et al. 2001), and Vapor-Liquid-Solid (VLS) growth methods (Wu and Liu 2002). The VLS pathway provides many technological advantages such as fast growth rates, low probability/ number of the extended defects such as dislocations, the possibility of diameter, and length control of the resultant nanowires. The VLS pathway also has been proven successful and previously used to produce nanowires in a variety of metal oxides, including but not limited to TiO_2, CdO, Cu_2O and ZnO (Huang et al. 2001; Wang et al. 2002; H. Chik et al. 2004; Keem et al. 2004; Fan et al. 2006; Zhao et al. 2007; Wischmeier et al. 2008).

The ability to control the nanowire diameter, density, and location is inherently provided with the use of proper diameter catalyst islands, as well as their precise positioning on the growth substrate. The latter also remains one of the key prerequisites to obtaining nanowires with identical electronic, optical, chemical, and mechanical characteristics (Chik et al. 2004; Kouklin and Sen et al. 2008; Sen and Kouklin, 2008). Finally, the VLS method also provides a simple means of incorporating TM impurities into ZnO nanowires either by time-controlled or the composition-tailored addition of TM precursors to the target (Zhou et al. 2004; Goris et al. 2008). Several TM impurities, including Co, Cu, Fe, and Ni, have been successfully incorporated within the crystalline core of ZnO nanowires (Zhou et al. 2004; Wang et al. 2007; Zhao et al. 2007; Kouklin 2008; Lin et al. 2009). S. Zhou, for example, produced wool-like films of ~65 nm diameter ZnO nanowires doped to 1-5 % of Cu by using intermixed CuI- ZnI_2 powders as the growth precursors. The results of x-ray diffraction and transmission electron microscopy (TEM), selected area electron diffraction, and high-resolution TEM microscopy suggest that the nanowires are single crystals, with the fast-growth direction corresponding with (101).

Our research findings also show that other precursors, particularly high purity powders of Cu, Zn, and graphite mixed in the proportion of Zn:Cu:C=3:1:1 can be also used to produce crystalline Cu:ZnO nanowires. In this case, the growth of the nanowires was achieved at ~800 ^{0}C in an Argon atmosphere with an Argon flow rate maintained at 0.12 L/min. Our routine SEM and TEM investigations indicate that the growth mechanism remains self-catalytic VLS, with the average growth rates of 1-2 microns per minute. Figure 2, left, shows the detailed SEM image of the nanowire tip. The presence of the metal nanoparticle at the apex of the nanowire is an attribute of self-catalytic growth. This pathway obviates the need to introduce metal catalyst (typically Au), which should help lower the fabrication cost and avoid potential problems with cross-contamination/uncontrolled doping.

In parallel, the results of Energy Dispersive Spectroscopy (EDS) have confirmed the presence of Cu traces in the samples, Figure 2, right, and agree with the Photoluminescence (PL) measurement results, which are discussed below. According to previous EPR studies, interstitial and substitutional Cu remain a most common defect in bulk ZnO crystals unintentionally/intentionally contaminated with Cu (Dietz et al. 1963; Broser and M.Schultz 1969). In the case of in-situ Cu doping, i.e. during VLS growth, incorporating Cu can be achieved by two separate routes: (1) diffusion through sidewalls and (2) direct deposition via a solid-liquid interface. It is likely that the diffusion-based incorporation will prevail at a

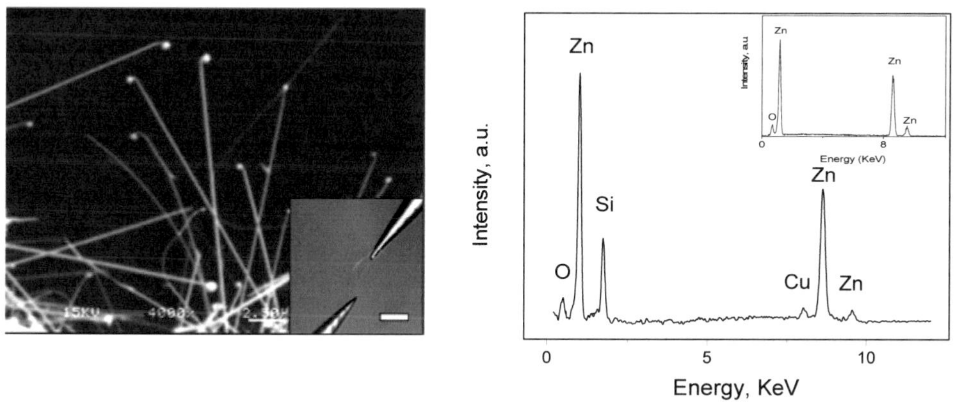

Fig. 2. (Left) A low-resolution side-view SEM image of Cu:ZnO nanowires grown by VLS on top of (111) Si substrate using Zn+Cu+C (graphite) mixture as a target; the inset shows an optical microscope image of ZnO nanowire suspended across the gap between two-terminals, scale bar 50 µm. (Right) EDS spectrum of Cu:ZnO nanowires; the inset shows an EDS spectrum of the ZnO nanowires grown away from the Cu target. Reproduced with permission, Adv. Mat 20 (2008).

very high T, similar to what has been observed in bulk and thin films of ZnO (Zhou et al. 2004). Instead, a direct incorporation pathway (i.e., via the solid-liquid interface) is likely to dominate at low T regime, as well as during growth initialization and termination stages. Due to an overall low melting point of metallic Zn (419 °C) compared with that of Cu (1100 °C), Zn droplets are expected to form first on the growth substrates while becoming saturated with Cu as the temperature is set to increase. According to a binary phase diagram of Cu-Zn alloy, at any growth stage the maximum Cu content is limited to 25% for the growth temperatures of 700-800 °C (Okamoto 2000). This makes the droplet stay saturated with Zn, thereby yielding the growth of ZnO rather than CuO nanowires. Given a close similarity in ionic radii of Cu^{2+} and Zn^{2+}, Cu is expected to precipitate mostly as substitutional or interstitial dopants in case of ZnO nanowires.

4. Light emission and photoconduction response of Cu:ZnO nanowires.

PL and PLE studies.

Light emission from Cu:ZnO nanomaterials can be categorized similarly to that from bulk ZnO, with the emission highly dependent on the nature of the physical mechanisms that control radiative recombinations, including free and donor/acceptor bound excitons, both shallow donor-acceptor pair recombination, and deep defect associated emission (DAP-emission), effectively covering the UV-IR optical range. Figure 3, left, lists some of emission lines that were experimentally obtained from ZnO samples. A typical near band-edge RT PL spectra collected from Cu:ZnO nanowires at room temperature is presented in Figure 3, right. The emission is dominated by two peaks centered at ~ 382 nm and ~493 nm, which are attributed to free exciton and blue-green (BG) emission bands, respectively. The peak positions of both bands were further found to be slightly red shifted by ~ 7 nm and ~ 3 nm with respect to their known positions in undoped ZnO nanowires. At the same time, an increase of ~2 orders of magnitude in the ratio of BG- to-excitonic intensities for Cu:ZnO samples is observed. An appearance of high-intensity BG band accompanied by quenching of the excitonic emission is a typical attribute of Cu:ZnO samples, i.e., similar trends were reported for thin film and nanowire samples in ZnO containing Cu impurities (Garces et al. 2002; Xu et al. 2004).

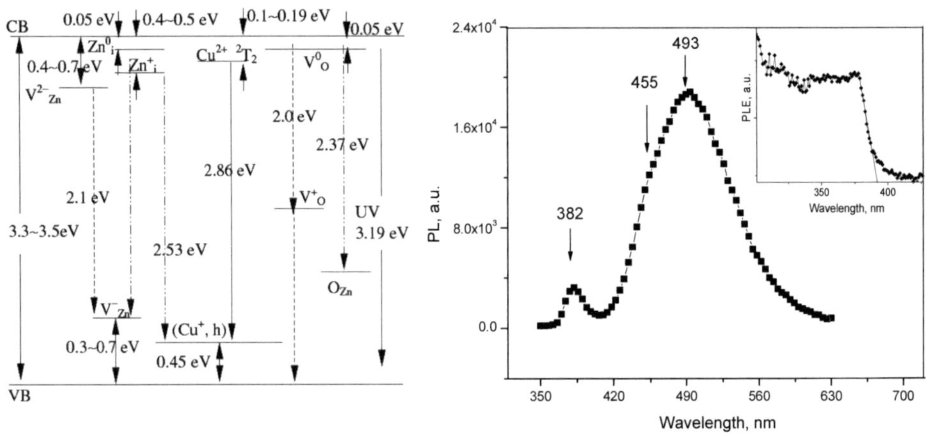

Fig. 3. (Left) Energy diagram of different defect states within the band-gap of ZnO and related optical transitions, reproduced with permission, Nanotechnology, **15** (2004); (right) room temperature micro-PL spectrum of Cu:ZnO nanowires, excited with a 325 nm light source. The inset displays the PLE spectrum with the emission monitored at 493 nm. Reproduced with permission, Adv. Mat **20**, (2008).

The DAP recombination involving Zn^+ and Cu^+ states is known to yield BG emission with a peak centered around ~493 nm (~2.51 eV) (Garces et al. 2002). On the other hand, a weak shoulder peak centered at ~455 nm (~ 2.72 eV) has been observed and assigned to Cu^{2+}-Cu^+ transitions, in which the hole remains localized on the Cu^+ center (Hausmann et al. 1979; Robbin et al. 1981).

Figure 3 right the inset, further shows PL-excitation (PLE) spectra obtained on Cu:ZnO nanowires at room temperature by monitoring the PL emission at ~493 nm. A step-like shape of PLE spectra can be observed and attributed to a strong increase in the light absorption by nanowires at the wavelength of ~ 390 nm. The extrapolated intercept on the $\hbar\omega$ -axis yields an effective band-gap of ~3.17 eV, which is again slightly red shifted (by ~ 50 meV) compared with that of undoped ZnO nanowire samples. No light emission could be obtained for photons with energies < E_{gap} in accordance with the previous works on Cu:ZnO materials. (Heitz et al. 1992)

Photoconduction in preavalanche regime.

The UV-photocurrent-voltage characteristics of Cu:ZnO metal-semiconductor-metal photodetectors engineered by precisely placing nanowires across the gap between two metal contacts were investigated in the low-intermediate voltage regime. For this, the central part of the nanowire devices was excited with a UV light of the wavelengths: 254, 302, and 365 nm. In the dark, the IV characteristics were typically found to be linear,

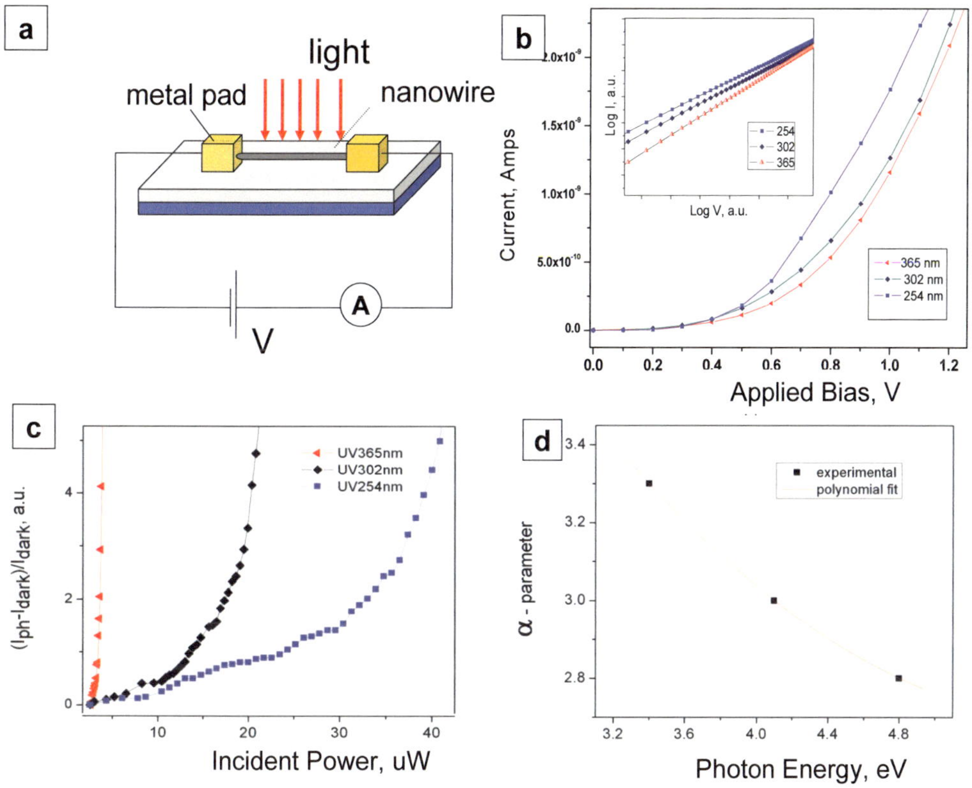

Fig. 4. (a) Schematics of Cu:ZnO-nanowire photodetector (b) Photo-I-V characteristics of Cu:ZnO photodetector obtained for applied biases up to ~1.2 V for excitation wavelengths of 254, 302, and 365 nm and power densities of ~5-10 mW/cm². The inset shows a log-log plot of UV photo-IVs plotted for an intermediate part of applied biases yielding power-law dependences of $I \sim V^{3.3}$, $I \sim V^{3.0}$ and $I \sim V^{2.8}$ for wavelengths of 365 nm, 302 nm, and 254 nm, respectively. (c) $(I_{ph}-I_{dark})/I_{dark}$ ratio as a function of excitation powers for different excitation wavelengths of 254, 302 and 365 nm. (d) α-parameter as a function of photon energy, fitted with a polynomial function $\alpha \cong 6.2 - 1.2x + 0.1x^2$. Reproduced with permission, J. Elect. Mat. 38 (2009).

whereas the dark currents were very small, i.e., approaching only few pA at applied bias of a few volts. In contrast, in the case of UV illumination, the photocurrent in the devices showed a strong increase whereas the conductance also changed with the applied bias. A shift from linear to nonlinear photo-IVs is attributed to a shift in the role of interface and bulk-limited currents invoked by the illumination. According to the results of more detailed investigations conducted by some of us (Kouklin 2008), the response at low biases (i.e., < ~0.5 V) is controlled by a Schottky barrier formed at the interface. On the other hand, at the intermediate bias regime, i.e., for bias falling in the range of ~ 0.5V-4.5 V, the response is power-law dependent as being trap-limited (Figure 4). In this regime, the photocurrent is found to change with voltage as $I \sim V^{3.3}$, $I \sim V^{3.0}$ and $I \sim V^{2.8}$ for excitation wavelengths of 365, 302, and 254 nm, respectively.

It should be mentioned that the shift of the transport from the Schottky to power-law dependent regime is generally expected in our devices as an increase in the electric field strength is to be accompanied by a reduction in the width/height of the Schottky barrier. This, in turn, helps the photogenerated carriers to tunnel through the barrier, which lowers the contact resistance and the voltage drop across the contact, with the transport thus becoming bulk-limited. Likewise, the barrier height associated with the shallow impurities/traps is to be also reduced, implying that the current should now be controlled by the trapping/detrapping of carriers on deep Cu states. It is also known that deep traps that are only distributed non-uniformly within the bandgap yield power-law dependent IV curves. For the trap density that changes exponentially within the band gap, i.e., when $n_t \propto e^{-E_t/kT_c}$, where T_c is some characteristic temperature reflecting the steepness of the trap distribution and E_t is the energy of the trap measured from the top of the conduction band, the device current is to change nonlinearly with applied bias given by the following equation $I \sim V^{\alpha+1}$, where $\alpha = T_c / T$ for T_c>T (Rose 1955). Experimentally, the power-law dependent IVs only have been obtained under UV illumination with α typically falling in the range of ~1.8 - 2.3. The result confirms the existence of optically-activated, exponentially-distributed deep-trap states in Cu-doped ZnO nanowires.

A dependence of the exponent on excitation wavelength is also observed to be polynomial, Figure 4d. The dependence of α on the excitation energy/wavelength is invoked by a shift in role different defect states play behind photocarrier transport. More specifically, as the excitation energy increases, the role of upper lying states behind trapping/detrapping weakens, which should yield smaller values of T_c and thus α, in accordance with the experimental findings. An interesting aspect is that this trend remains valid also for photon energy well exceeding E_g, which we attribute to the fact that some excitation processes are likely to originate on surface defect states positioned well outside the band gap.

According to intensity-dependent photocurrent measurements, the photocurrent is also found to change non-linearly with excitation powers in case of all UV wavelengths, Figure 4c (Kind et al. 2002). Finally, the on/off photocurrent measurements have been carried in this bias regime. The photocurrent changes exponentially with time, with the rise and decay time constants approaching ~10 and 39 s, respectively. The slow response is associated with overall slow trapping/detrapping processes that dominate the photoconduction response in the intermediate bias regime.

Photoconduction in avalanche, photocarrier-multiplication regime.

At high applied biases of ~4.5 V and above, a drastic shift in the photo IVs is observed, with the current becoming an exponential function of the applied bias, as seen in Figure 5, right. At a bias of 10 V, the device UV on/off factor, η has been seen to approach ~10^4. At the same time, undoped ZnO nanowire sensors measured similarly have yielded η ~ 7 at best at a bias of ~1.7 V, as shown in Figure 5, left, the inset. Likewise, and more importantly, a very strong decrease in the device resistance was also observed in the Cu:ZnO samples irradiated with a white light produced by an incandescent light source with a cut-off wavelength of ~440 nm. Under white-light excitation and bias exceeding ~4.5 eV the photocurrent also exhibited a strong, exponential-like increase with bias as in the case of UV illumination. At a bias of ~10 V, the obtained η was similarly large and approaching ~10^4. The responsivity of photosensor engineered using a nanowire with a length of ~100 μm and diameter of ~100 nm was further measured to be ~ 500 A/W for both white and UV lights, which significantly exceeds the sensitivity levels reported previously for ZnO nanowires for sub-band-gap photons. (Fan et al. 2004; Lin et al. 2007)

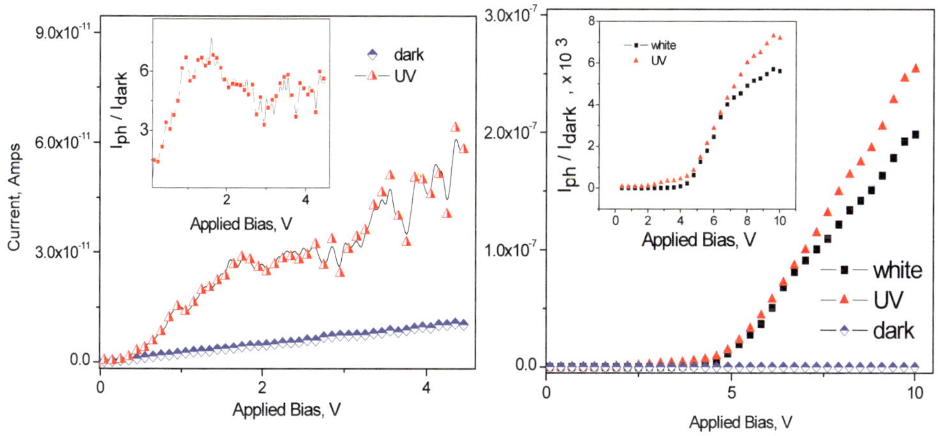

Fig. 5. (left) I-V characteristics of ZnO nanowire for dark and UV-illumination conditions, white-light I-Vs were similar to dark, the inset shows the ratio of photocurrent to dark current as a function of bias, at ~ 8 V the ratio was ~ 6; (right) I-V characteristics of similarly-configured Cu:ZnO nanowire sensors for dark, UV and white light illumination conditions, the inset shows the ratio of photocurrent to dark current as a function of bias for white and UV light. Reproduced with permission, Adv. Mat 20, (2008).

By extrapolating the intercept with the voltage-axis yields a threshold voltage, V_{th}~4.5 eV for both UV and white light photo-IVs as presented in Figure 5. The obtained value of V_{th} (~1.4 E_g of Cu:ZnO) closely corresponds with the threshold energy of 1.5 E_g required for an avalanche multiplication to initiate in this nano-semiconductor (Sze 2002). A multiplication factor, M calculated as a ratio of the avalanche photocurrent at a given voltage to that obtained by extrapolating the low-voltage photocurrent to the same voltage, was ~ 2 $x10^2$ and $7x10^2$ (at a bias of 10V) for UV and white light, respectively, which is taken as evidence that the photo-generated carriers undergo a strong multiplication in this, higher-voltage regime. Since no avalanche multiplication was present in similarly configured, undoped ZnO nanowire-based sensors for voltages of up to 10 V, the introduced Cu centers are likely to play a critical role behind avalanche photocarrier multiplication observed in case of Cu:ZnO nanowires.

In the case of visible excitation, the mechanism responsible for the photoconduction enhancement is likely to be associated with the processes of hole photogeneration, i.e., involving Cu^{2+}/Cu^{+} charge transfer transitions, whereas most of the photogenerated holes are expected to be trapped on the impurity centers. A fraction of these holes will be set free via thermoionic emission at RT, therefore giving rise to a weak, linear-like response observed at a low voltage regime. Yet, as the drift velocity of holes increases with the bias, eventually the holes are to gain enough energy to ionize the host lattice atoms as to initiate avalanche free carrier multiplication effects. This process is held responsible for an exponentially enhanced white light and UV sensitivity of Cu:ZnO nanowires for $V>V_{th}$.

5. Conclusion.

ZnO nanowires have recently emerged as an important class of materials that hold a strong promise for developing low-cost, multi-spectral, and highly sensitive opto-electronic sensors. The intentional introduction of TM impurities such as Cu has been shown to uniquely enhance the photosensitivity in the UV-visible spectral ranges of ZnO nanowires as a result of avalanche photo-multiplication effects. However, additional studies will be required to address the problems with slow response times in order to improve the switching speed of such optical sensors. In general, the highly crystalline character of the core of the Cu:ZnO nanowires grown by high T vapor transport methods is essential to achieving high photocarrier drift velocities in Cu:ZnO nanowire devices, thereby enabling faster photoconduction response without sacrificing gain. While in the near term ZnO photo-detectors are unlikely to replace UV-enhanced Si detectors, they hold many strong prospects in areas such as high-T, radiation-hard, and harsh environment light optical sensing, detection, and monitoring.

6. References.

Gupta, A., M.Omari, et al. (2008). Spectral investigation of carrier recombination processes in ZnO nanorods at high temperatures, J. Appl. Physics. 103: 124312.

Gupta, A., M.Omari, et al. (2009). "Ultraviolet pre-avalanche photoconduction properties of transitional metal-doped ZnO nanowires." Journal of Electronic Materials 38: 596.

Hausmann, A., B. Schallenberger, et al. (1979). "Electronic Structure of Copper in Zinc Oxide" Phys. Rev. B: 34: 129.

Broser, I. and M.Schultz (1969). "EPR studies on ZnO:Cu single crystals" Solid State Communication 7: 651.

Soci, C., A. Z., B. Xiang, S.A. Dayeh, D.P.R. Aplin, J. Park, X.Y. Bao, YH. Lo, D. Wang (2007). "ZnO Nanowire UV Photodetectors with High Internal Gain." Nano Letters 7: 1003.

Lin, D., H.Wu, et al. (2007). "Photoswitches and memories assembled by electrospinning aluminumdoped zinc oxide single nanowires." Adv. Mater. 19: 3968.

Dahan, V., P. Fleurov, et al. (1998). "Properties of the intermediately bound alfpha, betta, and gamma -excitons in ZnO:Cu." J. Phys.: Condens. Matter 10: 2007-2019.

Dietz, R. E., H. Kamimura, et al. (1963). "Electronic Structure of Copper Impurities in ZnO " Phys. Rev. 132: 1559.

Dingle, R. (1969). "Luminescent Transitions Associated With Divalent Copper Impurities and the Green Emission from Semiconducting Zinc Oxide." Physical Review Letters 23(11): 579.

Fan, Z., P. Chang, et al. (2004). "Photoluminescence and polarized photodetection of single ZnO nanowires." Applied Physics Letters 85(25): 6128-6130.

Fan, Z., D. Dutta, et al. (2006). "Electrical and photoconductive properties of vertical ZnO nanowires in high density arrays." Applied Physics Letters 89(21): 213110-3.

Goris, L., R. Noriega, et al. (2008). "Intrinsic and Doped Zinc Oxide Nanowires for Transparent Electrode Fabrication via Low-Temperature Solution Synthesis." Journal of Electronic Materials 34: 586.

Chik, H., J.Liang, et al. (2004). "Periodic Array of uniform ZnO nanorods by Second Order Self-Assembly." Applied Physics Letters 84: 3376

Kind, H., H.Yan, et al. (2002). "Nanowire ultraviolet photodetectors and optical switches." Adv. Mater. 14: 158.

Heitz, R., A. Hoffmann, et al. (1992). "The copper centre: a transient shallow acceptor in ZnS and CdS." J. Phys. : Condes. Matter 4: 157.

Heo, Y., L. Tien, et al. (2004). "Pt/ZnO nanowire Schottky diodes." Appl. Phys. Lett. 85: 3107

Huang, M. H., Y. Wu, et al. (2001). "Catalytic Growth of Zinc Oxide Nanowires by Vapor Transport." Advanced Materials. 13: 113.

I.Broser, L.Podlowsky, et al. (1994). J. Lumin. 60: 588.

Jeong, I. S., J. H. Kim, et al. (2003). "Ultraviolet-enhanced photodiode employing n-ZnO/p-Si structure." Applied Physics Letters 83(14): 2946-2948.

Keem, K., H. Kim, et al. (2004). "Photocurrent in ZnO nanowires grown from Au electrodes." Applied Physics Letters 84(22): 4376-4378.

Kong, Y. C., D. P. Yu, et al. (2001). "Ultraviolet-emitting ZnO nanowires synthesized by a physical vapor deposition approach." Applied Physics Letters 78: 407

L.Wischmeier, T.Voss, et al. (2008). "Correlations between surface-excitonic emission bands in ZnO nanowires." Nanotechnology 19: 135705.

Li, J. Y., X. L. Chen, et al. (2001). "Fabrication of zinc oxide nanorods." Journal of Crystal Growth 233: 5

Lin, C. Y., W. H. Wang, et al. (2009). "Magnetophotoluminescence properties of Co-doped ZnO nanorods." Appl. Phys. Lett 94: 151909.

Omari, M., N.Kouklin, et al. (2008). "Fabrication of Cd3As2 nanowires by direct vapor-solid growth and their infrared absorption properties." Nanotechnology 19 105301.

Garces, N., L.Wang, et al. (2002). "Role of copper in the green luminescence from ZnO crystals." Appl. Phys. Lett. 81: 622.

Gemma, N. (1984). "Electronic states of transitional metal impurities in II-VI and III-V semiconductors." J.Phys. C: Solid State Phys. 17: 2333.

Kouklin,N. (2008). "Cu-doped ZnO nanowires for efficient and multi-spectral detection applications." Adv. Mater. 20: 2190-2194.

Kouklin,N., L.Menon, et al. (2002). "Room Temperature Single Electron Charging in Self-Assembled Quantum Dots and Wires." Applied Physics Letters 80: 1649.

Kouklin,N., L. Menon, et al. (2001). "Giant Photoresistivity and Optically controlled switching in self-assembled nanowires." Applied Physics Letters 79: 4423.

Kouklin,N., S.Sen, et al. (in press, 2008). "Second-Order Self-Assembled Nano-semiconductors and Oxides by Nanometric Alumina Templates: Device Properties and Applications." in "Handbook of Nanoceramics and Their Based Nanodevices", American Scientific Publishers.

Kouklin,N., S.Sen, et al. (2006). "Self-driven formation of Zn3As2 Single Crystal Platelets by CVD." Applied Physics Letters 89: 071901.

Okamoto, H. (2000). "Phase Diagrams for Binary Alloys." ASM International: 314.

Optoelectronics Industry, http://www.oida.org/.

Robbins, D., D. Herbert, et al. (1981). "The origin of the α, β, γ blue no-phonon transitions in ZnO:Cu-A deep-level problem." J. Phys. C: Solid State Phys. 14: 2859.

Rose, A. (1955). "Space-Charge-Limited Currents in Solids." Physical Review 97: 1955.

Sze, S.M. (2002). "Semiconductor Devices: Physics and Technology, 2nd edition " Printed John Wiley and Sons: 79.

Sen, S., D.Chowdhary, et al. (2007). "Negative photoconduction in heterogenous planar network on ZnO-carbon nanotubes." Appl. Phys. Lett. 91: 093125.

Sen, S. and N.Kouklin (in press, 2008). "Nanofabrication Based on Self-Assembled Alumina Templates." in "Nanofabrication: Fundamentals and Applications", World Scientific.

Yong, S. (2006). "Characterization of ZnO metal-semiconductor-metal ultraviolet photodiodes with palladium contact electrodes." Semicond. Sci. Technol. 21: 1507.

Shalish, I., H. Temkin, et al. (2004). "Size-dependent surface luminescence in ZnO nanowires." Physical Review B 69(24): 245401.

Lee, W., M.Jeong, et al. (2004). "Optical characteristics of As-doped ZnO." Appl. Phys. Lett. 85: 6167.

Wang, K., J. J. Chen, et al. (2007). "Fe-doped ZnO Nanowire Arrays Synthesized by Chemical Vapor Deposition." Microsc. Microanal. 13: 736.

Wang, Y. W., L. D. Zhang, et al. (2002). "Catalytic growth of semiconducting zinc oxide nanowires and their photoluminescence properties " Journal of Crystal Growth 234: 171.

Wu, J. J. and S. C. Liu (2002). "Catalyst-Free Growth and Characterization of ZnO Nanorods." Journal Physical Chemistry B 106: 9546.

Xu, C. X., X. W. Sun, et al. (2004). "Photoluminescent properties of copper-doped zinc oxide nanowires." Nanotechnology 15: 856.

Zhao, M., A. H. Li, et al. (2007). "Growth and characterization of Cu-catalyzed ZnO nanowires." Journal of Physics: Conference Series 61: 703.

Zhou, S.-M., X.-H. Zhang, et al. (2004). "The fabrication and optical properties of highly crystalline ultra-long Cu-doped ZnO nanowires." Nanotechnology 15: 1152.

18

Modeling and Performance Analysis of III-V Nanowire Field-Effect Transistors

M. Abul Khayer[1] and Roger K. Lake
University of California, Riverside, CA 92521, USA

1. Introduction

A semiconductor nanowire (NW) is a solid rod with a diameter less than 100-200 nm composed of one or several semiconductor materials. The emergence of bottom-up chemical methods to fabricate one-dimensional semiconductor structures has enabled the synthesis of nanoscale wires that can serve as both devices and interconnects in nanoelectronic circuits.

During the last half century, a dramatic downscaling of electronics has taken place. The miniaturization of the devices found in integrated circuits is predicted by the semiconductor industry roadmap to reach atomic dimensions in few years. The narrowest feature of silicon devices - the gate oxide - should then reach its fundamental physical limit of four to five atoms thickness. At a thickness of less than four layers of silicon atoms, current will penetrate through the gate oxide causing the chip to fail. Intel's innovation of high-k/metal gate (Intel, 2003) has apparently solved the problem, however, to continue the aggressive downscaling of CMOS devices in order to sustain Moore's law beyond 22 nm technology node will require introduction of new materials and device architectures. Intel has been focusing on InSb materials for next generation high-speed, low-power logic applications (Chau et al., 2005).

Efforts have been given to use the self-assembly of one-dimensional semiconductor nanostructures in order to bring new, high-performance logic devices as an add-on to mainstream Si technology. Semiconductor NWs and carbon nanotubes (CNTs) can be the realistic additions.

NWs based on III-V and II-VI compound semiconductor materials have become the most attractive candidates for the next generation field effect transistors (FETs) because of the unique possibilities they offer for the rational control of fundamental properties such as dimension, composition, and doping during growth (Xia et al., 2003). They can be considered promising candidates for future high-speed, low-power electronic devices because of their narrow bandgap (E_g = 0.35 eV for InAs, for example) implying a small electron effective mass, which yields a very high mobility. For undoped InAs, the mobility can be as high as $\mu_n = 33000 cm^2/Vs$, compared with $\mu_n = 1500 cm^2/Vs$ for Si (Lind et al., 2006). Furthermore, the large intravalley separation energy of InAs allows for a high saturated velocity, $v_{sat} = 4 \times 10^7 cm/s$.

Nanowire field effect transistors (NWFETs) have been of particular interest recently, both as vehicles for the investigation of basic carrier-transport behavior and as potential future high-performance electronic devices.

[1] mkhayer@ee.ucr.edu

2. Background and motivation

Nanowires can be synthesized using many techniques, such as molecular/chemical beam epitaxy, vapor phase epitaxy/chemical vapor deposition, and laser ablation. The particular method used influences the potential material compositions, doping and crystal quality, and growth rate.

Most efforts on nanowire devices have focused on making NWFETs. Researchers have typically either studied individual NWs (Xiang et al., 2006) randomly placed on a substrate or have tried to fabricate transistors from arrays of vertical NWs including wrap-gated Si (Schmidt et al., 2006) and InAs NWs (Dayeh, Aplin, Zhou, Yu, Yu & Wang, 2007), even incorporating first attempts at bandgap engineering into the channel of the device. The carrier mobilities of such NWFETs have been shown to match or exceed those of planar silicon, and the devices can achieve gain. Furthermore, p- and n-type nanowires can be assembled into crisscross arrays where the junctions of the crossed wires serve as on-off switches.

A wide range of NW-based devices and systems, including transistors and circuits (Duan et al., 2001; Huang, Duan, Cui, Lauhon, Kim & Lieber, 2001; Zhong et al., 2003), light emitters (Gudiksen et al., 2002; Wang et al., 2001; Huang, Mao, Feick, Yan, Wu, Kind, Weber, Russo & Yang, 2001; Duan et al., 2003), and sensors (Cui et al., 2001) have been explored. NWFETs based on III-V (Dayeh, Soci, Yu, Yu & Wang, 2007; Dayeh, Aplin, Zhou, Yu, Yu & Wang, 2007; Bryllert et al., 2006; Lind et al., 2006; Thelander et al., 2004; Huang et al., 2005; Duan et al., 2001) and II-VI (Goldberger et al., 2005; Ng et al., 2004) compound semiconductor materials have demonstrated promising FET characteristics in various gate geometries, namely, top-gate (Wang et al., 2004; 2003), back-gate (Huang et al., 2005; Duan et al., 2001), wrap-around-gate(Ng et al., 2004; Bryllert et al., 2006), and core-shell (Lauhon et al., 2002) structures.

InSb and InAs NWs, in particular, are attractive candidates for NWFETs due to their high electron mobility at room temperature (Sze, 1981) and low contact resistance (Woodall et al., 1981). The low contact resistance is particularly important in nanoscale systems. Contacts often define performance at the nanoscale. Reports by Intel and Qinetiq on fabricated both *n*- and *p*-type InSb quantum well FETs show that InSb-based quantum well FETs can achieve equivalent high performance with lower dynamic power dissipation (Datta et al., 2005; Ashley et al., 1997; Chau et al., 2005; Radosavljevic et al., 2008). Recent discovery of the observation of Quantum Spin Hall effect in III-V (Liu et al., 2008) and II-VI (König et al., 2007) materials have also motivated the field of spintronics largly due to the fact that there is the possibility of low power logic devices design using the spin degree of freedom of the electron (Wolf et al., 2001; Murakami et al., 2003). Furthermore, III-V materials also have the potential to be used in NW band-to-band tunneling field effect transistors for high-speed, low-power (Khayer & Lake, 2009, in press; Luisier & Klimeck, 2009) and high-speed, high-power (Khayer & Lake, 2009a) applications.

III-V and II-VI NWs tend to be larger (20 - 50 nm diameters) making atomistic modeling difficult or sometimes impossible. However, full band (FB) models are necessary to understand on-off current ratios and radio frequency (RF) power. Although there are series of reports on the experimental realizations of InSb and InAs NWFETs (Datta et al., 2005; Ashley et al., 1997; Chau et al., 2005; Radosavljevic et al., 2008; Lind et al., 2006; Dayeh, Aplin, Zhou, Yu, Yu &Wang, 2007; Bryllert et al., 2006; Thelander et al., 2004; Ng et al., 2004), we find very few attempts to theoretically model them.

To accurately model charge density, spin density. effective mass, and interband tunneling, we have developed a full three dimensional (3D) discretized 8-band **k** · **p** model (Khayer & Lake, 2008a). We applied it to modeling InSb / InP and InAs / InP core-shell NWs. In the following sections, we will present the theory and performance of *n*- and *p*-type InSb and InAs NWFETs.

3. Theory and models

3.1 The InAs / InP and InSb / InP material systems

The choice of InAs and InSb as the base material for the NW devices is motivated by their physical properties: first of all, the bandgap is small, only 0.35 eV for InAs and 0.23 eV for InSb, and problems with electrical contacting of wires should be minimal. Secondly, the low electron effective mass, $m^* = 0.023m_0$ for InAs and $m^* = 0.013m_0$ for InSb, provides strong quantum confinement effects and a large energy level separation in the wires. The mobility is also high due to the low effective mass, which is of interest for high-speed device application. Another feature which is distinctive for these semiconductor materials (InAs) (Noguchi et al., 1991) is that the Fermi level is known to pin in the conduction band at the surface, at least for bulk material. Because of this, there is an accumulation of carriers at the surface. This behavior is in contrast to GaAs, which has a surface depletion that limits the minimum feasible diameter of a wire unless the surface is passivated. In principle, any metal should thus result in a good, Ohmic contact to InAs. However, a disadvantage with the strong pinning in the conduction band might be that it prevents realization of p-type InAs NWs. Most recently, experimental techniques have been developed to unpinning the Fermi level for the design of III-V FET source/drain contacts (Hu et al., 2009). Moreover, the InAs / InP and InSb / InP material systems are also relatively unexplored since the two materials are impossible to combine in bulk growth due to the large lattice mismatch.
The material parameters of these binary semiconductors are listed in Table 1 (Vurgaftman et al., 2001). These binary materials possess a zinc-blende crystal structure.

3.2 Electronic bandstructure calculation

Electronic bandstructure calculation can be performed in various ways (Yu & Cardona, 1999). The *ab initio* methods, such as Hartree-Fock or Density Functional Theory (DFT), calculate the electronic structure from first principles, i.e., without the need for empirical fitting parameters. These methods use a variational approach to calculate the ground state energy of a many-body system where the system is defined at atomic level.
In contrast to *ab initio* method, the empirical methods, such as tight-binding (Chadi & Cohen, 1974) and the **k** · **p** method (Luttinger & Kohn, 1955) involve empirical parameters to fit experimental data. Of them, the **k** · **p** method is widely used for direct bandgap semiconductors due to its simplicity and capability to capture essential physics of the materials in the band extrema. **k** · **p** method is based upon perturbation theory (Kane, 1956; 1957). In this method, the energy is calculated near a band maximum or minimum by considering the wavevector as a perturbation. In our work, we will use an 8-band **k** · **p** method to calculate the electronic bandstructure of InSb and InAs NWs.

3.3 Multiband k · p method

The multiband effective mass equation (Luttinger & Kohn, 1955) is widely used to describe the bandstructure of the low-dimensional structures. It can be expressed as,

$$i\hbar\frac{\partial}{\partial t}\Psi_{\nu}(\mathbf{r},t)=\sum_{\nu'}H_{\nu\nu'}(-i\nabla)\Psi_{\nu'}(\mathbf{r},t)+U(\mathbf{r},t)\Psi_{\nu}(\mathbf{r},t), \tag{1}$$

where ν represents band index and $H_{\nu\nu'}(\mathbf{k})$ is defined as

$$H_{\nu,\nu'}(\mathbf{k})=\begin{cases}\dfrac{\hbar^2k^2}{2m_0}+E_{\nu 0}, & \nu'=\nu\\ \dfrac{\hbar\mathbf{P}_{\nu\nu'}\cdot\mathbf{k}}{m_0}, & \nu'\neq\nu.\end{cases} \tag{2}$$

$P_{\nu\nu'}$ is called the momentum matrix element between bands ν and ν' and is defined as

$$P_{\nu\nu'}=-i\hbar\langle\overline{u}_{\nu,0}\,|\,\nabla\overline{u}_{\nu',0}\rangle. \tag{3}$$

Here $u_{\nu,0}$ represents the zone center Bloch functions for the νth band. The relation between the actual wavefunction and the multiband envelope functions follows readily from,

$$\Psi_0(\mathbf{r},t)=\sum_{\nu}u_{\nu,0}(\mathbf{r},t)\Psi_{\nu}(\mathbf{r},t). \tag{4}$$

Eq. (1) can be solved for the perfect spatially uniform semiconductor using a variety of well known techniques, but in a quantum confined structure, the crystal composition and/or strain varies from region to region and approximations are needed in order to solve Eq. (1). Many such approximate methods are now well known and are extensively used (Sercel & Vahalla, 1990). The choice of how many bands will be needed depends on the details of the problem to be solved.

3.4 Numerical calculations

Fig. 2a shows the schematic diagram of the simulated NWFETs. Square [100] InSb/InP and InAs/InP core-shell NWs with core cross-section of 2 - 60 nm are considered.

The 8-band $\mathbf{k}\cdot\mathbf{p}$ method as described by Gershoni et al. (Gershoni et al., 1993) is used to calculate the electronic bandstructures of the NWs. We include eight basis functions in the set, namely, the spin-up and spin-down s and p atomic orbital-like states. These are arranged in the following order: $|S\uparrow\rangle$, $|X\uparrow\rangle$, $|Y\uparrow\rangle$, $|Z\uparrow\rangle$, $|S\downarrow\rangle$, $|X\downarrow\rangle$, $|Y\downarrow\rangle$ and $|Z\downarrow\rangle$. As a result, the multiband effective mass equation is transformed into eight coupled differential equations for the envelope function F_n

$$\sum_{n'=1}^{8}H_{nn'}(\mathbf{r},\nabla)F_{n'}(\mathbf{r})=EF_n(\mathbf{r}). \tag{5}$$

H is the multiband Hamiltonian matrix and E is the eigenenergy. We use a finite difference method (Mamaluy et al., 2005) with a constant grid spacing of 1 nm to discretize the Hamiltonian. Spurious solutions of the $\mathbf{k}\cdot\mathbf{p}$ equations are eliminated by methods described in Refs. (Foreman, 1997; Kolokolov & Ning, 2003).

To make the discretized $\mathbf{k}\cdot\mathbf{p}$ Hamiltonian look like a nearest neighbor tight-binding Hamiltonian for a layered structure, the NW should be discretized as shown in Fig 1. It should be discretized in planes of sites. The indices of the first plane should run from 1 to N, the indices of the second plane should run from N+1 to 2N, etc. The matrix $[D_1]$ in Fig. 1 is the block of the Hamiltonian matrix of the isolated 1st plane of sites. The matrix $[t_{1,2}]$ is the block of the Hamiltonian matrix that couples plane 1 to plane 2. If the NW consisted only of the 4 planes shown in Fig. 1, then the total Hamiltonian matrix would have the form

$$H = \begin{bmatrix} [D_1] & [t_{1,2}] & 0 & 0 \\ [t_{1,2}]^{\dagger} & [D_2] & [t_{2,3}] & 0 \\ 0 & [t_{2,3}]^{\dagger} & [D_3] & [t_{3,4}] \\ 0 & 0 & [t_{3,4}]^{\dagger} & [D_4] \end{bmatrix} \tag{6}$$

Considering the components of the **k** vector in Eq. (5) to be numbers and diagonalizing the matrix for a bulk crystal for $k = 0$, one can obtain the dispersion relationships, namely, the **k** dependent eigenvalues $E_n(\mathbf{k})$.

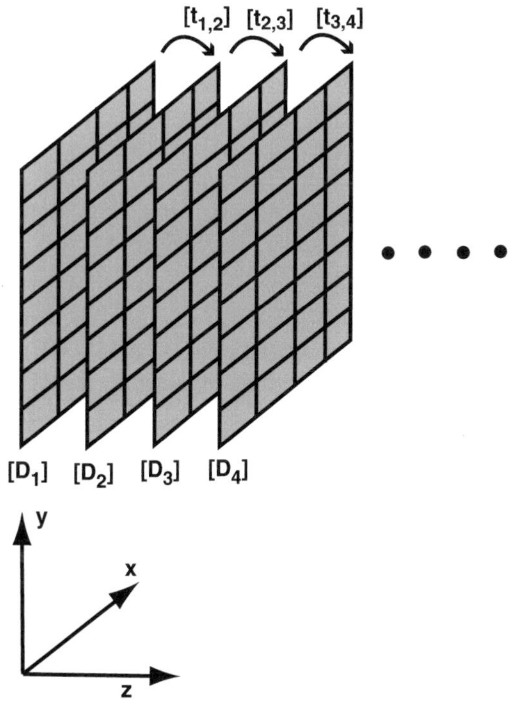

Fig. 1. Discretization scheme for NW.

The *I-V* characteristics of *n*-type InSb and InAs NWFETs are evaluated by using a semiclassical ballistic FET model as described in (Rahman et al., 2003) for ballistic planar MOSFETs and extended by J.Wang (Wang & Lundstrom, 2003) for ballistic high electron mobility transistors. The model is illustrated in Fig. 2(b). It consists of three capacitors, C_G, C_S, and C_D, which describe the electrostatic coupling between the top of the barrier and the gate, the source, and the drain, respectively. The potential at the top of the barrier is calculated as (Rahman et al., 2003)

$$U_{scf} = \left[C_G V_G + C_D V_D + C_S V_S + Q_{top}\right] / C_{\Sigma}, \tag{7}$$

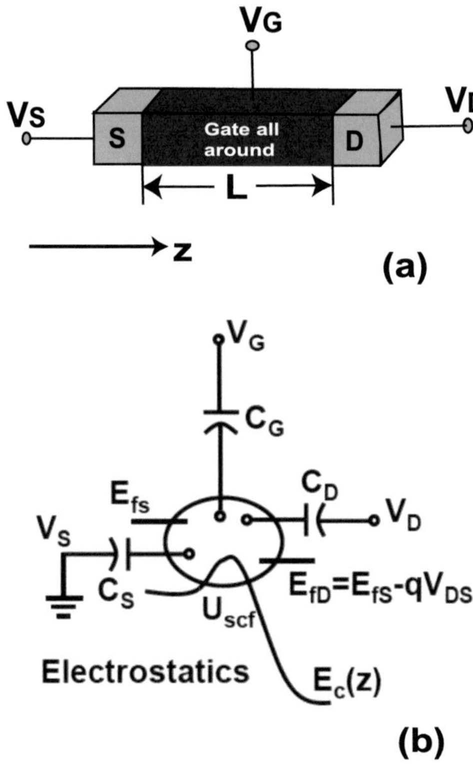

Fig. 2. (a) Schematic diagram of the simulated gate-all-around NW transistor, (b) The electrostatics of the semiclassical ballistic FET model used in the calculation. The diagrams are not to scale.

where V_G, V_D, and V_S are the applied biases at the gate, drain, and source terminals, respectively, $C_\Sigma = C_G + C_D + C_S$, and Q_{top} is the mobile charge at the top of the barrier. Q_{top} is governed by U_{scf}, the source and the drain Fermi levels, E_{FS}, E_{FD}, and the E-k dispersion for the channel material. In terms of energy, Eq. (7) becomes

$$E_{CG} = -q\left[\alpha_G V_{GS} + \alpha_D V_{DS} + Q_{top} / C_\Sigma\right], \tag{8}$$

where E_{CG} is the conduction band-edge under the gate, q is the magnitude of the electron charge, $\alpha_G = \frac{C_G}{C_\Sigma}$, $\alpha_D = \frac{C_D}{C_\Sigma}$, and the source is at ground. α_G and α_D are referred to as the gate and drain control parameters, respectively. They are calculated numerically from $\alpha_G = \frac{C_G}{C_\Sigma} = \left|\frac{\Delta E_{CG}}{\Delta q V_{GS}}\right|_{\Delta V_{DS}=0, \Delta N=0}$ and $\alpha_D = \frac{C_D}{C_\Sigma} = \left|\frac{\Delta E_{CG}}{\Delta q V_{DS}}\right|_{\Delta V_{GS}=0, \Delta N=0}$, with ΔN being the induced charge in the channel, using a three dimensional Laplace solver (Adams, 1989).
The charge under the gate, Q_{top}, for electrons, is calculated from

$$Q_{top} = -(q/2)\int_{\varepsilon_1(0)}^{\infty} dE N_{1D}(E - \varepsilon_1(0))\left[f(E-\eta) + f(E-\eta+U_D)\right], \tag{9}$$

where $U_D = qV_{DS}$, $\eta = E_{FS} - \varepsilon_1(0) + qU_{scf}$, E_{FS} is the source Fermi level, and $\varepsilon_1(0)$ is the 1st subband level at the top of the barrier at zero gate and drain bias. The factor of '1/2" comes from the fact that under the gate in the ballistic limit, the left contact only fills the right moving states under the gate, and the right contact only fills the left moving states under the gate. The density-of-states corresponding to the left or right moving states is one half the full density of states. We use the Green's function method to calculate N_{1D},

$$N_{1D}(E) = \frac{1}{a}\mathrm{tr}\left[\frac{-1}{\pi}\mathrm{Im}\{G(E)\}\right], \tag{10}$$

where a is the discretization length and tr{...} is the trace over all states within a single discretized layer along the transport direction (including spin).
The exact Green's function G of the device is calculated as

$$G(E) = \left[EI - H_D - \Sigma_L - \Sigma_R\right]^{-1}, \tag{11}$$

where H_D is the device Hamiltonian, and $\Sigma_L = t_{1,0}g^L_{0,0}t_{0,1}$, and $\Sigma_R = t_{0,1}g^R_{N,N}t_{1,0}$ are the self-energies to the left and the right contact, respectively. $t_{1,0}$ and $t_{0,1}$ are the coupling matrices between the device and the left and the right contact, respectively. The g's are the surface Green's functions calculated using a decimation technique (Galperin et al., 2002; Sancho et al., 1985).
To evaluate the number of modes contributing to the drive current for the devices, we calculate transmission at each energy point E (Fisher & Lee, 1981),

$$T(E) = \mathrm{tr}\left\{\Gamma_L G \Gamma_R G^\dagger\right\}, \tag{12}$$

where $\Gamma_L = i(\Sigma_L - \Sigma_L^\dagger)$ and $\Gamma_R = i(\Sigma_R - \Sigma_R^\dagger)$, and tr... is the trace over all states within a layer including spin.
After self-consistency between U_{scf} and Q_{top} is achieved, the drain current for the devices is calculated as (Lake et al., 1997)

$$I = \frac{q}{h}\int_0^\infty dE\,T(E)\left[f(E-\eta) - f(E-\eta+U_D)\right]. \tag{13}$$

To compare the performance of different NWFETs, we set the maximum gate bias voltages such that the gate overdrive (V_{OD}) for each device is fixed at 0.2 V, i.e., $(V_{GS} - V_T)_{\max}$=0.2 V, where V_T is the threshold voltage for each device. V_T is determined as follows. For each single-moded n-type device, V_T is taken as the V_{GS} when the conduction band-edge under the gate (E_{CG}) reaches the energy $E_{FS} + kT$. At this V_{GS}, the single-moded n-type devices have the same threshold current, I_{th}. For the p-type devices, V_T is the gate voltage that produces this I_{th}. Thus, by definition, all of our NWFETs have the same current when $V_{GS}=V_T$

3.5 Analytical calculations

To understand the performance metrics of NWFETs, analytical expressions are derived for the current, the charge, the power-delay product, the energy-delay product, the gate delay time, and the cut-off frequency for a single-moded device operating in the quantum capacitance limit (QCL), ignoring thermal broadening, and assuming parabolic dispersion.

Devices are said to operate in QCL when the their quantum capacitance (QC) is lower than their geometric gate capacitance. The expressions for the power-delay product, the energy-delay product, the gate delay time, and the cut-off frequency are fundamental limits for these devices.

In equilibrium, the QC per unit length is defined as $C_Q = \frac{q^2 \partial n}{\partial E_F}$, and under the gate with large $V_{DS} >> k_B T$ applied

$$C_Q = \frac{q^2 \partial n}{\partial (E_{FS} - \varepsilon_1)}, \tag{14}$$

where ε_1 is the energy level of the fundamental mode including the effect of the self-consistent potential, i.e., $\varepsilon_1 = \varepsilon_1(V_{GS} = 0) - qU_{scf}$. The gate capacitance to ground, C_{GS}, is the series combination of the geometric (C_G) and quantum (C_Q) gate capacitance (Burke, 2004). Thus, when $C_Q << C_G$, $C_{GS} \approx C_Q$ and the device is said to be in the QCL. Physically, the QC is the energy broadened density-of-states evaluated at the Fermi level.

The QC affects how the band-edge under the gate responds to the gate bias. Taking the derivative of Eq. (8), we obtain

$$\left| \frac{\partial E_{CG}}{\partial q V_{GS}} \right| = \frac{\alpha_G}{1 + \frac{C_Q}{C_\Sigma}}, \tag{15}$$

where we used the fact that the QC defined in Eq. (14) is also equal to $C_Q = q \frac{\partial Q_{top}}{\partial E_{CG}}$. Eq. (15) shows that in the QCL ($C_Q << C_G$), the gate control of the band-edge under the gate is unaffected by charge in the channel, i.e., there is no screening. E_{CG} moves linearly with gate voltage in proportion to the gate control parameter α_G. However, for $C_Q >> C_G$, the gate voltage has little effect on the band-edge under the gate, i.e., screening is significant.

Under the gate, in the ballistic limit, ignoring thermal broadening, the QC per unit length is given by $C_Q = q^2 \partial n / \partial (E_{FS} - \varepsilon_1) = q^2 N_{1D}^{+}(E_{FS})$ where N_{1D}^{+} is the one dimensional (1D) density-of-states with positive velocity which is just the standard 1D density-of-states divided by 2. N_{1D}^{+} can be written in several ways, two of which will be useful later,

$$N_{1D}^{+}(E_{FS}) = \frac{2}{h v_{FS}} = \frac{1}{h} \sqrt{\frac{2m^*}{(E_{FS} - \varepsilon_1)}}. \tag{16}$$

In Eq. (16), v_{FS} is the velocity of an electron under the gate injected at the source Fermi level, E_{FS}. Therefore, the QC under the gate is,

$$C_Q = \frac{2q^2 L_G}{h v_{FS}} = \frac{q^2 L_G}{h} \sqrt{\frac{2m^*}{(E_{FS} - \varepsilon_1)}}. \tag{17}$$

The maximum drain current occurs when V_{DS} is biased to its maximum value of V_{DD}, and the energy of the fundamental mode under the gate is well below the Fermi level of the source. Under these conditions, for a single-moded, spin-degenerate wire, the second term in Eq. (31) is negligible, and Eq. (31) becomes,

$$I = \frac{2qk_BT}{h}\ln\left(1+e^{(E_{FS}-\varepsilon_1)/k_BT}\right). \tag{18}$$

When the energy of the fundamental mode, ε_1, is pulled several k_BT below the Fermi level of the source, Eq. (18) reduces to

$$I_D \approx \frac{2q}{h}(E_{FS}-\varepsilon_1). \tag{19}$$

Thus, the current is proportional to the energy difference between the source Fermi level and the fundamental mode, and it is independent of any material parameters such as the effective mass or density-of-states. To calculate other quantities such as the power-delay and energy-delay products, quantities such as the current will be needed as a function of voltage rather than energy. The energy scale is converted to a gate voltage using Eq. (15) to write

$$E_{FS}-\varepsilon_1 = q(V_{GS}-V_T)\frac{\alpha_G}{1+\langle C_Q/C_\Sigma\rangle} = q\tilde{\alpha}_G(V_{GS}-V_T), \tag{20}$$

where we define the reduced gate control parameter $\tilde{\alpha}_G \doteq \frac{\alpha_G}{1+\langle C_Q/C_\Sigma\rangle}$ where $\langle C_Q/C_\Sigma\rangle$ is an average value. We will see that for the structures that we consider, the value of $\langle C_Q/C_\Sigma\rangle$ is approximately 0.1. Thus, the current as a function of voltage is

$$I_D \approx \frac{2q^2}{h}\tilde{\alpha}_G(V_{GS}-V_T), \tag{21}$$

with the understanding that $V_{GS} \geq V_T$. The transconductance is defined as $g_m = dI_D/dV_{GS}$ which from Eq. (21) is

$$g_m = \frac{\tilde{\alpha}_G 2q^2}{h}, \tag{22}$$

i.e., g_m is the quantum of conductance times the reduced gate control parameter.
The charge under the gate, Q_{top} is

$$\begin{aligned} Q_{top} = \int_{\varepsilon_1}^{E_{FS}} dE\,N_{1D}^{+}(E) &= \frac{2q\sqrt{2m^*}L_G}{h}\sqrt{E_{FS}-\varepsilon_1} \\ &= \frac{2q\sqrt{2m^*}L_G}{h}\sqrt{q\tilde{\alpha}_G(V_{GS}-V_T)}. \end{aligned} \tag{23}$$

The power-delay product corresponds to the energy required to charge the gate. If one assumes that all of the charge on the gate is imaged by the charge in the channel, one obtains a lower limit for this quantity given by (Knoch et al., 2008)

$$P\cdot\tau_D = \int_0^{V_T+V_{OD}} dV_{GS}\,Q_{top}. \tag{24}$$

Substituting Eq. (23) into (24) and integrating gives

$$P \cdot \tau_D = \frac{4L_G\sqrt{2m^*}}{3h\tilde{\alpha}_G}\left(q\tilde{\alpha}_G V_{OD}\right)^{3/2}. \tag{25}$$

The gate delay time, t_D is defined as (Knoch et al., 2008)

$$\tau_D = \frac{P \cdot \tau_D}{V_{DD} I_D}. \tag{26}$$

Substituting Eqs. (25) and (21) into (26), τ_D is evaluated as,

$$\tau_D = \frac{2L_G\sqrt{2m^*}}{3q\tilde{\alpha}_G V_{DD}}\sqrt{q\tilde{\alpha}_G V_{OD}}. \tag{27}$$

Since we are using a lower limit for $P \cdot \tau_D$, Eq. (27) provides a lower limit for τ_D.
The energy-delay is the product of the power-delay and the delay time. Multiplying Eqs. (25) and (27), $E \cdot \tau_D$ is

$$E \cdot \tau_D = \frac{16qm^* L_G^2}{9hV_{DD}} V_{OD}^2. \tag{28}$$

Again, this expression should be viewed as a lower limit for the energy-delay product.
The intrinsic cut-off frequency is calculated as $f_T = g_m/(2\pi C_{GS})$, where C_{GS} is the series combination of the geometric (C_G) and quantum (C_Q) gate capacitance (Burke, 2004). Assuming that $C_Q << C_G$ so that $C_{GS} \approx C_Q$, one obtains the upper limit of the intrinsic cut-off frequency,

$$f_T \approx \tilde{\alpha}_G \frac{v_{FS}}{2\pi L_G} = \frac{\tilde{\alpha}_G}{2\pi\tau_G} = \frac{\tilde{\alpha}_G}{2\pi L_G}\sqrt{\frac{2(E_{FS}-\varepsilon_1)}{m^*}} = \frac{\tilde{\alpha}_G}{2\pi L_G}\sqrt{\frac{2q\tilde{\alpha}_G V_{OD}}{m^*}}. \tag{29}$$

In Eq. (29), $\tau_G = L_G/v_{FS}$ is the gate transit time. The cut-off frequency has the form of one over the transit time divided by 2π.
The key concept to understanding the high intrinsic performance of single-moded NWFETs is the 'decoupling of the charge and the current.' This is apparent from Eqs. (21) and (23). The current is material independent. It does not depend on the density-of-states or the charge in the channel. The charge in the channel depends on the density-of-states through the effective mass. For a given current, the charge could be anything depending on the value of the effective mass. To optimize performance, one wants maximum current with minimum charge. This is obtained by minimizing the effective mass. Minimizing the effective mass serves two purposes. (i) It minimizes the charge in the channel according to Eq. (23), and (ii) it allows one to achieve a higher Fermi level in the source for a given doping. This maximizes the current by maximizing the source Fermi level, E_{FS} according to Eq. (19).

4. Results and discussion

4.1 The quantum and classical capacitance limit

In each case, the core material contains InSb or InAs surrounded by a cladding layer (shell material) of InP with a thickness of 6 nm (equivalent to an SiO_2 thickness of 2 nm) which is treated as the gate insulator. We first show the effect of confinement on the electronic properties, mode spacing, and quantum capacitance, and then we present calculations of the current, power-delay product, delay times, energy-delay product, and cut-off frequencies. All numerical calculations are performed with T = 300K.

The bandgap vs. NW diameters and the electron effective mass of the lowest conduction band mode vs. NW diameters are found in (Khayer & Lake, 2008a). Fig. 3 shows the NW diameter dependence of (a) the bandgap and (b) the lowest conduction mode electron effective mass at the zone center. The bandgap for the InSb NW is 1.17 eV with the 2 nm wire diameter, and it falls to 0.37 eV with the 12 nm wire diameter. For InAs, the bandgap is 1.23 eV for the 2 nm wire diameter, and it falls to 0.54 eV for the 12 nm wire diameter. The electron effective mass at the zone center for the InSb NW with core cross section of 2 nm is $0.042m_0$, and that with core cross section of 12 nm in $0.023m_0$. For InAs, the mass is $0.052m_0$ with 2 nm core cross section and $0.028m_0$ with core cross section of 12 nm.

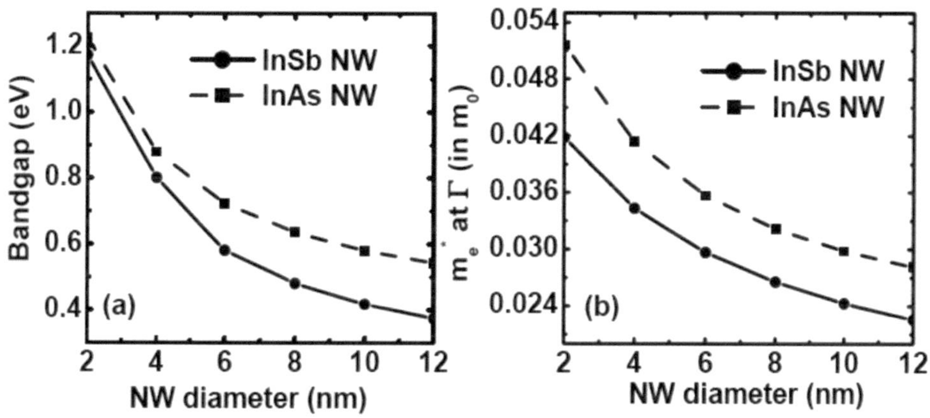

Fig. 3. (a) Bandgap as a function of NW diameter for the simulated InSb and InAs NWs, and (b) theNWdiameter dependence of the lowest conduction band electron effective mass as the zone center. [Reproduced with permission from (Khayer & Lake, 2008a); © 2008 IEEE]

Luryi Luryi (1987). QC in NWFET devices accounts for the fact that the gate field penetrates through the wire since it is not completely screened on the wire surface as is the case in an ideal macroscopic conductor. Devices are said to operate in the quantum capacitance limit (QCL) when their QC is less than their classical capacitance (CC). In highly scaled nanotransistors based on NWs or nanotubes exhibiting 1-D transport, the QCL limit can be reached Rahman et al. (2003); Knoch et al. (2008); Khayer & Lake (2008b). Significant performance improvement in terms of the power-delay product has been predicted in devices operating in the QCL Knoch et al. (2008); Khayer & Lake (2008b;a).

The n-type NWFETs operate in the QCL and the p-type NWFETs operate in the classical capacitance limit (CCL) as shown in Fig. 4(e, f). It is shown that the drive currents at a fixed gate overdrive are well matched for n- and p-type NWFETs (Khayer & Lake, 2009b). This is surprising since the density-of-states of the p-type FET is much larger than the density-of-states of the n-type FET. However, this effect has been observed both theoretically and experimentally by others Rahman et al. (2005); Li et al. (2008). Despite the matched current drive, the p-type devices operating in the CCL have twice the delay times, twice the power-delay products, and 4-5 times the energy-delay products of the n-type devices operating in

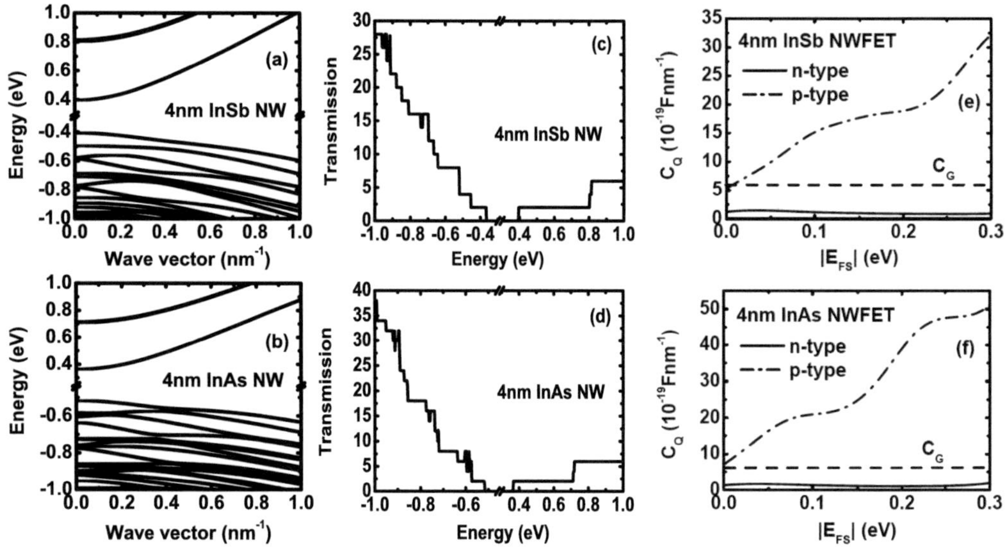

Fig. 4. $E - k$ dispersion relationship and the transmission plots for (a, c) InSb NWs and (b, d) InAs NWs. E=0 corresponds to the lowest conduction band energy of the *bulk* InSb or InAs materials. (e-f) Quantum capacitance under the gate as a function of source energy for the corresponding NWs. Figures are presented for both *n*- and *p*-type devices. Corresponding gate capacitances are also shown. [Reproduced with permission from (Khayer & Lake, 2009b); © 2009 IEEE]

the QCL. These effects have nothing to do with mobility since all transport is assumed ballistic. The effects arise solely from the density-of-states and electrostatics.

Fig. 4(a, b) shows the *E-k* dispersion relations calculated using the 3-D discretized 8-band $\mathbf{k} \cdot \mathbf{p}$ model for the two different materials. In each case, $E = 0$ corresponds to the lowest conduction band energy of the corresponding *bulk* InSb or InAs materials. It is apparent from the $E - k$ plots that the dispersion for the electrons quickly deviates from parabolic becoming significantly more linear above the energy of the mode minimum. Considerable band mixing in the excited hole subbands takes place for both the NWs.

Fig. 4(c, d) shows the transmission plots as a function of energy for the electrons and for the holes calculated from Eq. (12). There are two details worth commenting on for clarification. First, the initial turn-on of the transmission is 2 since the 8-band k·p model explicitly includes spin in the basis. The trace in Eq. (12) traces over all of the basis orbitals which include spin. Second, the non-monotonic behavior of the transmission for the p-type NWs is the result of the finite bandwidth and non-monotonic nature of the energy versus wavevector relations of the hole modes as seen in Fig. 4(a, b). Some modes have multiple regions of positive velocity, and as the energy moves down below a local extremum, a mode can turn off. The same effect has been observed in Si NWs (Zheng et al., 2005). Fig. 4(c, d) gives insight into the number of modes contributing to the drive current. For the n-type InSb and InAs NWs, the second set of modes occur at 0.35 eV and 0.4 eV above the fundamental mode, respectively. In all cases for the *n*-type devices, with a maximum gate overdrive of 0.2 V, the NWFETs are single-moded (2 spins) as shown in Fig. 4(c, d). For the *p*-type InSb and InAs NWs, the second set of modes occur at 57 meV and 89 meV below the fundamental mode, respectively. Thus, the p-type devices are not single-moded, and there is a contribution from the higher modes to the current, charge and, in particular, to the quantum capacitance that we discuss next.

Fig. 4(e, f) shows the quantum capacitance under the gate as a function of the source Fermi level for both *n*- and *p*-type devices. The quantum capacitance is calculated from Eq. (14) with the charge, $-qn = Q_{top}$ calculated from Eq. (9). The geometrical gate capacitance, C_G, for the corresponding NWFET is also shown in each figure. C_G is calculated from $C_G = (2\pi\epsilon_r\epsilon_0)/(ln((2T_{ox} + T_{NW})/T_{NW}))$ assuming a coaxial gate geometry (Ramo et al., 1994), where T_{NW} is the diameter of the NW core and T_{ox} is the thickness of the shell. The effect of discrete, well-spaced modes on the quantum capacitance is clearly depicted in Fig. 4(e, f). For all *n*-type devices, $C_Q << C_G$ for the entire energy range considered and the devices are operating in the QCL. For all *p*-type devices, however, C_Q is comparable to or larger than C_G and the devices are operating in the CCL.

Fig. 5(a, b) shows the log(I_{DS}) vs. V_{GS} transfer characteristics of both the *n*- and *p*-type InSb and InAs NWFETs calculated from Eq. (31). For the *p*-type devices, the polarity of $V_{GS} - V_T$ has been reversed. For all devices, V_{DS} is chosen to be large enough so that there is no back injection from the drain. The currents are balanced for the *n*- and *p*-type devices. The *n*- and *p*-type device currents are well-matched. This is surprising since the *n*- and *p*-type devices have very different densities-of-states and they fall into different capacitance regimes. Naively, one would expect that, in the ballistic limit, the device with the larger density-of-states and more closely spaced modes would carry more current.

Fig. 5(c, d) shows the carrier density, Q_{top} vs. V_{GS} for the InSb and InAs NWFETs. The carrier density is higher for the *p*-type devices. The *p*-type devices not only have more occupied modes, but the dispersion of each mode is considerably flatter than those in the conduction band (see Fig. 4(a, b)). This results in a larger density-of-states associated with each valence band mode. The net result is more charge in the channel for the *p*-type devices for a given gate overdrive and current.

Fig. 5(e, f) shows how the band-edges under the gate change with applied gate bias. These curves are explained by Eq. (15). Initially, the channel in both the *n*- and *p*-type devices is empty, there is a significant source-channel barrier to electron or hole flow, and, thus, the quantum capacitance, $|\partial Q_{top}/\partial(E_{FS} - \varepsilon_1)|$, is exponentially reduced. Therefore, initially, the band-edges for both *n*- and *p*-type devices move identically with gate voltage with a slope whose magnitude is given by the gate control parameter α_G. When the first set of excited modes under the gate are shifted by the gate bias such that they start to become populated by the source Fermi level at $V_{GS} = V_T$, the quantum capacitance turns on and is as shown in Fig. (4(e, f)). The single-moded, *n*-type devices remain in the QCL for the entire range of gate bias. At threshold, for the *n*-type devices, the magnitude of the slope $\left|\frac{\Delta E_{CG}}{\Delta q V_{GS}}\right|$ reduces slightly from $\alpha_G = 0.8$ to $\tilde{\alpha}_G \approx 0.7$. When the first set of excited modes of the *p*-type devices are shifted by the gate bias such that they start to become populated by the source Fermi level, the quantum capacitance becomes larger than the classical capacitance and the *p*-type devices move into the CCL with $C_Q > C_\Sigma$ as shown in Fig. 4(e, f). When this happens, the denominator of Eq. (8) increases, and the magnitude of the slope $|\partial E_{CG}/\partial q V_{GS}|$ decreases as shown in Fig. 5(e, f). In the ideal classical limit with an infinite density-of-states, C_Q becomes infinite, the slope $|\partial E_{CG}/\partial q V_{GS}|$ goes to zero, and the band-edge under the gate becomes

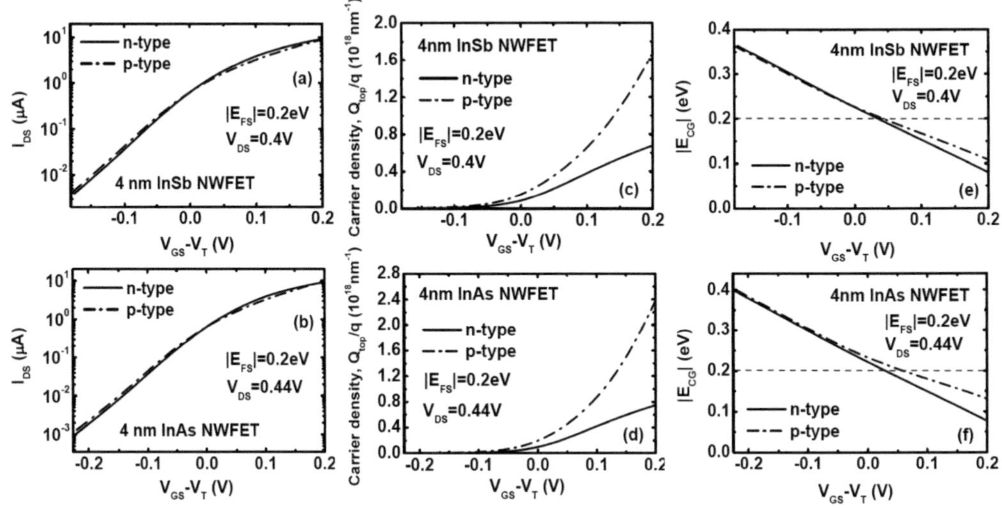

Fig. 5. log(I_{DS}) vs. V_{GS} transfer characteristics and the carrier density, Q_{top} vs. V_{GS} plots for (a, c) InSb NWs and (b, d) InAs NWs. The drain bias voltages are fixed at half of the energy bandgap for each NW as shown. For the *p*-type devices, the polarity of the voltage is reversed. (e, f) band-edges under the gate as a function of the gate bias for the devices. The source Fermi energy E_{FS} is at 0.2 eV for all devices and is shown by the horizontal dashed line. [Reproduced with permission from (Khayer & Lake, 2009b); © 2009 IEEE]

fixed, independent of the gate voltage. This decrease in $|\partial E_{CG}/\partial qV_{GS}|$ resulting from charging of the channel is the expected behavior of devices operating in the CCL.

For an *n*-type device, the positive gate bias lowers the band-edge in the channel which results in charging of the channel. The charge results in a self-consistent potential, U_{scf}, which works against the gate bias to raise the band-edge. The larger the density-of-states, the larger the negative feedback will be. This negative feedback is absent, or much reduced, in the QCL. Therefore, the gate bias moves the band-edge less in the *p*-type devices than in the *n*-type devices, and the band-edge in the channel presents a larger barrier to the source in the *p*-type devices than in the *n*-type devices for the same gate overdrive. Therefore, even though the density-of-states and carrier density is higher in the *p*-type channel than in the *n*-type channel, the barrier to source injection is also higher. For a NW in the ballistic limit, ignoring thermal broadening, the current resulting from a single mode m is $\frac{2q}{h}$ $(E_{FS} - \varepsilon_m)$ where ε_m is the mode energy in the channel. For a *p*-type channel, a larger U_{scf} results in smaller energy differences $|(E_{FS} - \varepsilon_m)|$. Thus, although there are more modes carrying current, each mode is carrying less current than the single mode in the *n*-type device. As a result, the currents of the *n*-type and *p*-type devices tend to be similar for the same gate overdrive.

The negative feedback due to charging affects how far the bands move under the gate, which, in turn, determines how large the source Fermi level must be to avoid source injection saturation. At the maximum gate overdrive of 0.2V, the conduction band edge under the gate is pushed 0.12 eV below the Fermi level of the source for both the InAs and InSb NWFETs. For the p-type devices with greater charging, the band edges move less. At the maximum gate overdrive, the InSb and InAs bandedges under the gate are pushed 91 meV and 69 meV above the Fermi level of the source, respectively. Therefore to ensure that the current is not limited by source injection saturation, the source Fermi level should be at least ~ 0.15 eV above the conduction band edge of the source for the 4nm n-type NWFETs and at least ~ 0.1 eV below the valence band edge of the source for the 4nm p-type NWFETs. Finally, we compare to the numbers predicted or projected for other materials and dimensionalities. For *n*-type InSb and InAs NWFETs with 4 nm NW diameter, the values of $P \cdot \tau_D$ are half the value of 5×10^{-20} J, and the values of $P \cdot \tau_D$ with 10 nm NW diameter are close to the value of 5×10^{-20} J predicted in Ref. (Knoch et al., 2008) for a 3 nm diameter Si NWFET with a 10 nm gate length. The corresponding gate delay times for the *n*-type devices with 4 nm NW diameter are close to the value of 8 fs and the gate delay times with 10 nm NW diameter are twice the value of 8 fs predicted in Ref. (Knoch et al., 2008) for a 3 nm diameter Si NWFET with 10 nm long gate. The energy-delay product for *n*-type devices with 4 nm NW diameter is found to be 10 - 100 times lower than the projected experimental curve for a III-V planar *n*-channel MOSFET with a 10 nm channel width (Chau et al., 2005). For *n*-type devices with 10 nm NW diameter, however, the energy-delay product falls on the projected experimental curve for a III-V planar *n*-channel MOSFET with a 10 nm channel width (Chau et al., 2005). For *p*-type devices with 4 nm NW diameter, $E \cdot \tau_D$ is found to be 10 - 100 times lower than the projected experimental curve and with 10 nm NW diameter, $E \cdot \tau_D$ falls on the projected experimental curve for *p*-channel CNTFET and Si NWFET with 10 nm channel width (Chau et al., 2005).

4.2 The diameter dependent performance

To investigate diameter dependent performance, we choose the *n*-type NWFETs composed of InSb or InAs as core and InP as shell materials. The diameters are varied from 10 nm to 60 nm. To calculate the mode energies and E – k dispersions for various diameter NWs, we use an analytical 2-band dispersion relation,

$$E(1+\alpha E) = \frac{\hbar^2 k^2}{2m^*}, \tag{30}$$

where $\alpha = \frac{1}{E_G}(1 - \frac{m^*}{m_0})^2$. $E_G = 0.23$ eV is the bulk bandgap, $m^* = 0.013m_0$ is the bulk electron effective mass, and m_0 is the bare electron mass.

The semiclassical ballistic model as described in sec. 3.4 is used to obtain the charge density, the self consistent potential and the current. After self consistency between the potential and the charge is achieved, the drain current for the *n*-type InSb NWFETs is calculated as,

$$I = \sum_m M \frac{qk_BT}{2\pi\hbar}\left[\Im_0(\eta_m) - \Im_0(\eta_m - U_D)\right], \tag{31}$$

where M stands for spin degeneracy, m is the mode index, k_B is the Boltzmann constant, T is the temperature. $U_D = V_{DS}/(k_BT/q)$ and $\eta_m = (E_{FS} - \varepsilon_m(0) + qU_{scf})/k_BT$, where E_{FS} is the source Fermi level and $\varepsilon_m(0)$ is the m-th subband level at the top of the barrier. The function $\Im_i(x)$ is the Fermi-Dirac integral.

The drain bias voltage is fixed at 0.5 V for all devices. To compare the effect of different cross-sections and doping, we fix the gate overdrive to 0.2 V for all devices. The threshold voltages are determined from the linear I_D-V_{GS} curves. For all diameters, two different Fermi levels in the source are modeled, $E_F - E_c = 0.1$ eV and 0.2 eV.

It is shown that relatively larger diameter (≤60 nm) InSb NWFETs (Khayer & Lake, 2008b) operate in the QCL. This is a result of the small effective mass, and hence the smaller density of states of these material. Both the energy-delay and power-delay products are reduced as the diameter is reduced, and optimum designs are obtained for diameters in the range of 10 - 40 nm (Khayer & Lake, 2008b).

Fig. 6(a-c) presents the number of spin degenerate modes as a function of mode energy for 10nm, 30nm, and 60nm diameter InSb NWs, and Fig. 6(d-f) shows the corresponding quantum capacitances vs. the source Fermi energy. We consider InSb NWFETs with Fermi levels of up to 0.2eV in the source. In all cases, the NWFETs contain multiple modes as shown in Fig. 6(a-c). The relationship between the number of populated modes and the quantum capacitance is obvious from Figs. 6. Figs. 6(a) and 6(b) show the effect of discrete, well-spaced modes on the quantum capacitance, C_Q. The dashed lines in Figs. 6(e-f) show the geometrical gate capacitance, C_G. For source Fermi energy such that $E_F - E_C$=0.1eV, all of the InSb NWs with diameters ≤60nm have $C_Q < C_G$, and thus operate in the QCL.

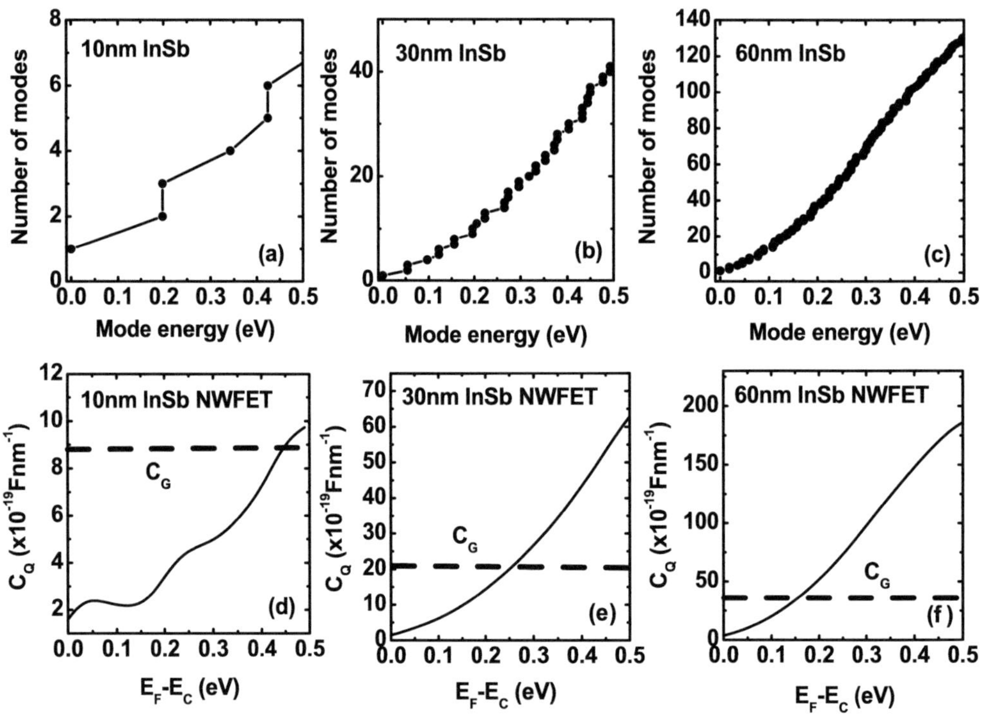

Fig. 6. (a-c) Number of conduction modes (without spin) as a function of the mode energy for the simulated InSb NWFETs with 10 nm, 30 nm, and 60 nm NW diameters. All the devices contain multiple modes within the energy range considered. (e-f) Quantum capacitance as a function of the source Fermi level for the corresponding NWFETs. The geometrical capacitance, C_G is also shown. C_G is calculated from $C_G = (2\pi\epsilon_r\epsilon_0)/(ln((2d_{ox} + d_{NW})/d_{NW}))$ assuming a coaxial gate geometry. All the devices are operating in the quantum capacitance limit within source Fermi level ($E_F - E_c = E_{FS}$) range of 0.1-0.2 eV. [Reproduced with permission from (Khayer & Lake, 2008b); © 2008 IEEE]

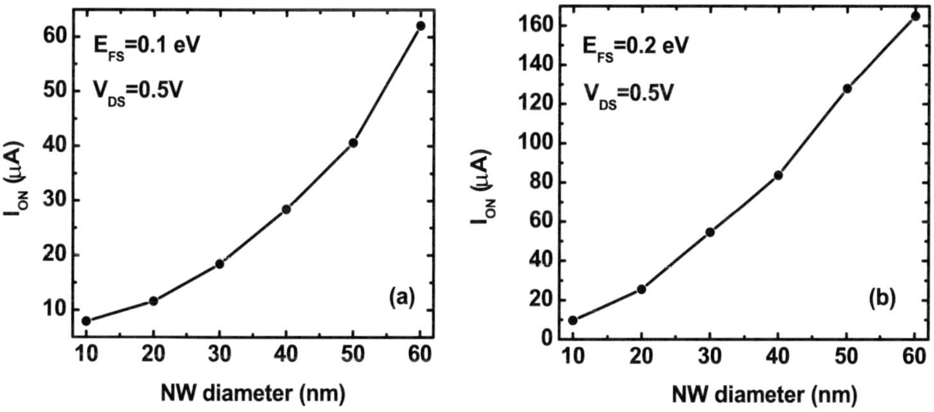

Fig. 7. ON-current as a function of NW diameter with source Fermi level ($E_{FS} = E_F - E_c$) of 0.1 eV (a), and 0.2 eV (b). The drain bias voltage is fixed at 0.5 V as shown. The gate voltage is fixed at a gate overdrive of 0.2V, i.e., $V_{GS} - V_T = 0.2$ where V_T is the threshold voltage for each device. [Reproduced with permission from (Khayer & Lake, 2008b); © 2008 IEEE]

within a particular Fermi energy, larger number of conduction modes become populated as the NW diameter increases (Fig. 6(a-c)), and the drain current increases. Fig. 7 shows the ON-current as a function of the NW diameter for two source Fermi energies, 0.1 eV (a), and 0.2 eV (b). The drain bias voltage is fixed at 0.5 V for each device, and the gate bias voltage is fixed at a gate overdrive of 0.2 V, $V_{GS} - V_T$=0.2 V, where V_T is the threshold voltage and has been calculated from the linear I_{DS}-V_{GS} characteristics for each device. The ON-current for all devices varies from 7 - 165 μA within a source Fermi energy variation of 0.1 - 0.2 eV.

With an increase of the NW diameter, as shown in Fig. 8(a, b), the power-delay product increases and the gate delay decreases . The power-delay product, $P \cdot \tau$, is calculated from $\int QdV_{GS}$, where Q is the magnitude of the charge in the channel. The gate delay is obtained from $\tau = \int QdV_{GS}/(V_{DD} I_{ON})$. The gate delay time τ for all devices varies from 4 – 16 fs within a source Fermi level range of 0.1-0.2 eV and decreases as the NW diameter increases (Figs. 8(a, b)). This is due to the higher ON-current for the larger diameter NWs (Figs. 7). $P \cdot \tau$ varies from 2×10^{-20} J to 68×10^{-20} J for all devices with a source Fermi level range of 0.1 - 0.2 eV. For $E_F - E_c = 0.1$ eV, the values of $P \cdot \tau$ for diameters 10 - 50 nm all match closely the value of 5×10^{-20} J predicted in Ref. (Knoch et al., 2008) for a 3 nm diameter Si NW FET with a 10 nm gate.

Fig. 8(c, d) presents the energy-delay product as a function of NW diameter. As diameter and Fermi energy increase, the energy-delay product also increases. The energy-delay is the product of the power-delay and the delay time shown in Fig. 8(a, b). It is interesting that the 1.4×10^{-33} Js energy-delay product of the 10 nm NW FET with $E_F - E_c = 0.2$ eV falls on the projected experimental curve for a III-V planar HEMT with a 10 nm channel width shown in Fig. 7 of (Chau et al., 2005).

Both the energy-delay and power-delay products are optimal for diameters in the range of 10 - 40 nm with source Fermi levels of 0.1 eV. Increasing the source Fermi level increases C_Q which causes a rapid degradation of the energy-delay product. Approximating τ as $\frac{C_Q V_{DD}}{I_D}$, the energy-delay product is proportional to $\frac{C_Q^2}{I_D}$. In this case, the increase in C_Q overrides the increase in I_D.

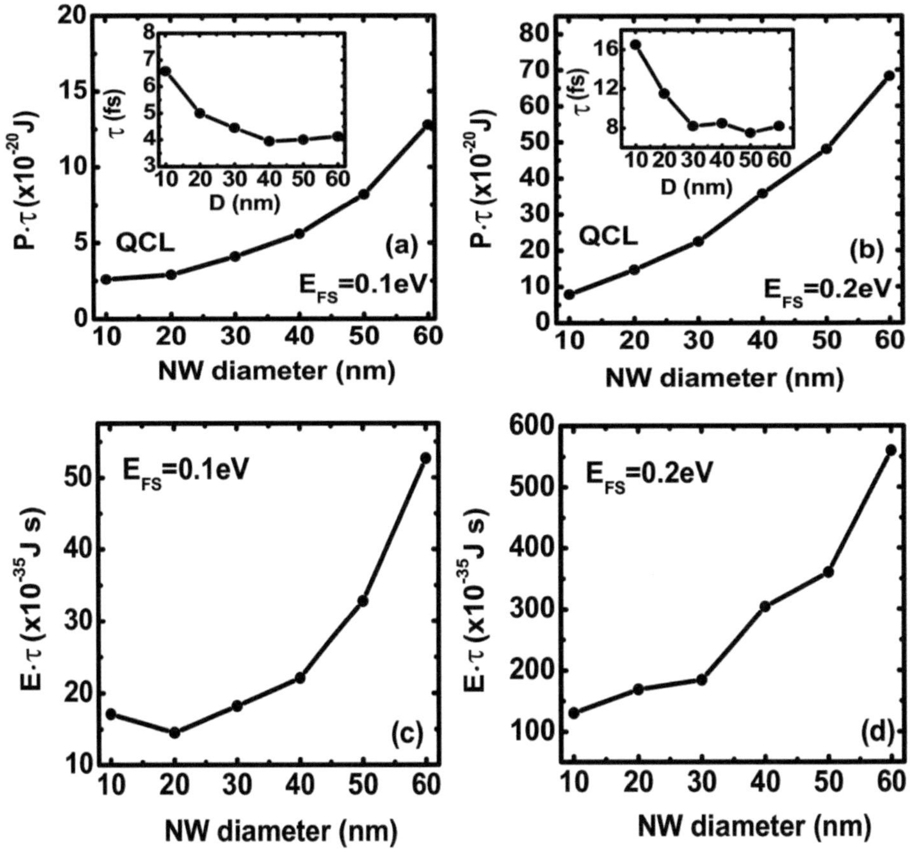

Fig. 8. Simulated $P \cdot \tau$ (power-delay product) (a, b) and $E \cdot \tau$ (energy-delay product) (c, d) as a function of NW diameter with source Fermi level ($E_{FS} = E_F - E_C$) of 0.1 eV , and 0.2 eV. $P \cdot \tau$ is calculated as $\int Q dV_{GS}$, where Q is the total channel charge extracted from the simulations. Accordingly the gate delay time $\tau = \int Q dV_{GS} / (V_{DD} I_{ON})$. $E \cdot \tau$ is the product of the $P \cdot \tau$ and the gate delay τ. Device operating regime, quantum capacitance limit (QCL), is indicated in the plots. These curves show a significant performance improvement in terms of the power-delay product and energy-delay product for the devices scaled towards the quantum capacitance limit with E_{FS} =0.1 eV. Inset in Figs. (a, b) shows the gate delay time as a function of NW diameter extracted from the simulations. A gate length of 10 nm was considered to calculate these values. [Reproduced with permission from (Khayer & Lake, 2008b); © 2008 IEEE]

5. Conclusion

In conclusion, we have presented a full 3-D discretized 8-band **k** · **p** model to calculate the electronic bandstructures of III-V and II-VI compound semiconductor materials in quantum confined structures. In particular, we have investigated high speed, low power InSb and InAs NWFETs in two operating regimes, the quantum capacitance limit and the classical capacitance limit. These materials are believed to be strong candidates for extending high performance logic beyond the 22 nm technology node.

6. References

Adams, J. (1989). Mudpack: multigrid fortran software for the ef_cient solution of linear elliptic partial differential equations, *Applied Math. and Comput.* 34: 113.

Ashley, T., Barnes, A. R., Buckle, L., Datta, S., Dean, A. B., Emeny, M. T., Fearn, M., Hayes, D. G., Hilton, K. P., Jefferies, R., Martin, T., Nash, K. J., Phillips, T. J., Ting,W. H. A. & Chau, R. (1997). InSb quantum well transistors for high-speed, low power applications, *IEEE Int. Electron Devices Meeting Tech. Dig.* p. 751.

Bryllert, T., Wernersson, L. E., Fr ¨oberg, L. E. & Samuelson, L. (2006). Vertical high-mobility wrap-gated inas nanowire transistor, *IEEE Electron Device Lett.* 27(5): 323.

Burke, P. J. (2004). Carbon nanotube devices for ghz to thz applicatons, *Proc. SPIE* 5593: 52.

Chadi, D. J. & Cohen, M. L. (1974). Tight-binding calculations of the valence bands of diamond and zincblende crystals, *Physica Status Solidi (b)* 68: 405.

Chau, R., Datta, S., Doczy, M., Doyle, B., Jin, B., Kavalieros, J., Majumder, A., Metz, M. & Radosavljevic, M. (2005). Benchmarking nanotechnology for high-performance and low-power logic transistor applications, *IEEE Trans. Nanotechnol.* 4(2): 153.

Cui, Y., Wei, Q., Park, H. & Lieber, C. M. (2001). Nanowire nanosensors for highly sensitive and selective detection of biological and chemical species, *Science* 293(5533): 1289.

Datta, S., Ashley, T., Brask, J., Buckle, L., Doczy, M., Emeny, M., Hayes, D., Hilton, K., Jefferies, R., Martin, T., Phillips, T. J., Wallis, D., Wilding, P. & Chau, R. (2005). 85nm gate length enhancement and depletion mode InSb quantum well transistors for ultra high speed and very low power digital logic applications, *IEEE Int. Electron Devices Meeting Tech. Dig.* .

Dayeh, S. A., Aplin, D. P. R., Zhou, X., Yu, P. K. L., Yu, E. T. & Wang, D. (2007). High electron mobility inas nanowire _eld-effect transistors, *Small* 3(2): 326.

Dayeh, S. A., Soci, C., Yu, P. K. L., Yu, E. T. & Wang, D. (2007). Transport properties of inas nanowire field effect transistore: The effects of surface states, *J. Vac. Sci. Technol. B* 25(4): 1432.

Duan, X., Huang, Y., Agarwal, R. & Lieber, C. M. (2003). Single-nanowire electrically driven lasers, *Nature* 421(5523).

Duan, X., Huang, Y., Cui, Y.,Wang, J. & Lieber, C. M. (2001). Indium phosphide nanowires as building blocks for nanoscale electronic and optoelectronic devices, *Nature* 409: 66.

Fisher, D. S. & Lee, P. A. (1981). Relation between conductivity and transmission matrix, *Phys. Rev. B* 23(12): 6851.4.

Foreman, B. A. (1997). Elimination of spurious solutions from eight-band k · p theory, *Phys. Rev. B* 56(20): 748.

Galperin, M., Toledo, S. & Nitzan, A. (2002). Numerical computation of tunneling fluxes, *J. Chem. Phys.* 117: 10817.10826.

Gershoni, D., Henry, C. H. & Baraff, G. A. (1993). Calculating the optical properties of multidimensional heterostructures: application to the modeling of quaternary quantum well lasers, *IEEE J. Quantum Electron.* 29(9): 2433.

Goldberger, J., Sirbuly, D., Law, M. & Yang, P. (2005). Zno nanowire transistors, *J. Phys. Chem. B* 109(1): 9.

Gudiksen, M., Lauhon, L. J.,Wang, J., Smith, D. C. & Lieber, C. M. (2002). Growth of nanowire superlattice structures for nanoscale photonics and electronics, *Nature* 415(6872): 617.

Hu, J., Guan, X., Choi, D., Harris, J. S., Saraswat, K. &Wong, H. S. P. (2009). Fermi level depinning for the design of iii-v fet source/drain contacts, *Int. Symp. on VLSI Technology, Systems, and Applications* p. 123.

Huang, M. H., Mao, S., Feick, H., Yan, H., Wu, Y., Kind, H., Weber, E., Russo, R. & Yang, P. (2001). Room-temperature utraviolet nanowire nanolasers, *Science* 292(5523): 1897.

Huang, Y., Duan, X., Cui, Y., Lauhon, L. J., Kim, K. H. & Lieber, C. M. (2001). Logic gates and computation from assembled nanowire building blocks, *Science* 294(5545): 1313.

Huang, Y., Duan, X., Cui, Y. & Lieber, C. M. (2005). Gallium nitride nanowire nanodevices, *Nano Lett.* 2: 101.

Intel (2003). www.intel.com/research/silicon .

Kane, E. O. (1956). *J. Phys. Chem. Solids* 1: 82.

Kane, E. O. (1957). *J. Phys. Chem. Solids* 1: 249.

Khayer, M. A. & Lake, R. K. (2008a). Performance of n-type InSb and InAs nanowire field effect transistors, *IEEE Trans. Electron Devices* 55(11): 2939.

Khayer, M. A. & Lake, R. K. (2008b). The quantum capacitance limit of high-speed, lowpower InSb nanowire field effect transistors, *IEEE Int. Electron Devices Meeting Tech. Dig.* p. 193.

Khayer, M. A. & Lake, R. K. (2009, in press). Drive currents and leakage currents in InSb and InAs nanowire and carbon nanotube band-to-band tunneling FETs, *IEEE Electron Dev. Lett.* .

Khayer, M. A. & Lake, R. K. (2009a). Performance analysis of InP nanowire band-to-band tunneling field-effect transistors, *Appl. Phys. Lett.* 95(7): 073504.

Khayer, M. A. & Lake, R. K. (2009b). The quantum and classical capacitance limits of InSb and InAs nanowire FETs, *IEEE Trans. Electron Devices* 56(10): 2215.

Knoch, J., Riess,W. & Appenzeller, J. (2008). Outperforming the conventional scaling rules in the quantum capacitance limit, *IEEE Electron Device Lett.* 29(4): 372.

Kolokolov, K. I. & Ning, C. Z. (2003). k · p hamiltonian without spurious-state solutions, *Phys. Rev. B* 68(16): 161308.

König, M., Wiedmann, S., Br ¨ une, C., Roth, A. & Buhmann, H. (2007). Quantum spin hall insulator state in HgTe quantum wells, *Science* 318: 766.

Lake, R., Klimeck, G., Bowen, R. C. & Jovanovic, D. (1997). Single and multiband modeling of quantum electron transport through layered semiconductor devices, *J. Appl. Phys.* 81(12): 7845.

Lauhon, L. J., Gudiksen, M. S., Wang, D. & Lieber, C. M. (2002). Eptaxial core-shell and coremultishell nanowire heterostructures, *Nature* 420(7): 57.

Li, N., Harmon, E. S., Hyland, J., Salzman, D. B., Ma, T. P., Xuan, Y. & Ye, P. D. (2008). Properties of InAs metal-oxide-semiconductor structure with atomic-layer-deposited Al_2O_3 dielectric, *Appl. Phys. Lett.* 92: 143507.

Lind, E., Persson, A. I. & nd L. E.Wersersson, L. S. (2006). Improved subthreshold slope in an InAs nanowire heterostructure field-effect transistor, *Nano Lett.* 6(9): 1842.

Liu, C., Hughes, T. L., X. L. Qi, K. W. & Zhang, S. C. (2008). Quantum spin hall effect in inverted type *ii* semiconductors, *arXiv:0801.2831v1[cond-mat.mes-hall]* .

Luisier, M. & Klimeck, G. (2009). Atomistic full-band design study of InSb and InAs bandtoband tunneling field-effect transistors, *IEEE Electron Dev. Lett.* 30(6): 602.

Luryi, S. (1987). Quantum capacitance devices, *Appl. Phys. Lett.* 52(16): 501.

Luttinger, J. M. & Kohn, W. (1955). Motion of electrons and holes in perturbed periodic fields, *Phys. Rev.* 97(4): 869.

Mamaluy, D., Vesileska, D., Sabathil, M., Zibold, T. & Vogi, P. (2005). Contact block reduction method for ballistic transport and carrier densities of open nanostructures, *Phys. Rev. B* 71(24): 245321.

Murakami, S., Nagaisa, N. & Zhang, S. C. (2003). Dissipationless quantum spin current at room temperature, *Science* 301: 1348.

Ng, H. T., Han, J., Yamada, T., Nguyen, P., Chen, Y. P. & Meyyappan, M. (2004). Single crystal nanowire vertical surround-gate field-effect transistor, *Nano lett* 4(7): 1247.

Noguchi, M., Hirakawa, K. & Ikoma, T. (1991). Intrinsic electron accumulation layers on reconstructed clean inas (100) surfaces, *Phys. Rev. Lett.* 66(17): 2243.

Radosavljevic, M., Ashley, T., Andreev, A., Coomber, S. D., Dewey, G., Emeny, M. T., Fearn, M., Hayes, D. G., Hilton, K. P., Hudait, M. K., Jefferies, R., Martin, T., Pillarisetty, R., Rachmady,W., Rakshit, T., Smith, S. J., Uren, M. J.,Wallis, D. J.,Wilding, P. J. & Chau, R. (2008). High-performance 40nm gate length insb p-channel compressively strained quantum well field effect transistors for low-power (v_{CC}=0.5v) logic applications, *IEEE Int. Electron Devices Meeting Tech. Dig.* p. 727.

Rahman, A., Guo, J., Datta, S. & Lundstrom, M. S. (2003). Theory of ballistic nanotransistors, *IEEE Trans. Electron Devices* 50(9): 1853.

Rahman, A., Klimeck, G. & Lundstrom, M. (2005). Novel channel materials for ballistic nanoscale mosfets - bandstructure effects, *IEEE Int. Electron Devices Meeting Tech. Dig.*

Ramo, S., Whinnery, J. R. & Duzer, T. V. (1994). *Fields and Waves in Communication Electronics*, 3 edn, New York, NY:John Wiley and Sons Inc., New York.

Sancho, M. P. L., Sancho, J. M. L. & Rubio, J. (1985). Highly convergent schemes for the calculation of bulk and surface green functions, *J. Phys. F* 15: 851.858.

Schmidt, V., Riel, H., Senz, S., Karg, S. & Gösele, W. R. U. (2006). Realization of a silicon nanowire vertical surround-gate field-effect transistor, *Small* 2(1): 85.

Sercel, P. C. & Vahalla, K. J. (1990). Analytical formalism for determining quantum wire and quantum dot band structure in the multi-band envelope function approximation, *Phys. Rev. B* 42: 3690.

Sze, S. M. (1981). *Physics of Semiconductors*, 2 edn, John Wiley and Sons, New York.

Thelander, C., Björk, M. T., Larsson, M. W., Hansen, A. E., Wallenberg, L. R. & Samuelson, L. (2004). Electron transport in inas nanowires and heterostructure nanowire devices, *Solid State Communications* 131: 573.

Vurgaftman, I., Meyer, J. R. & Ram-Mohan, L. R. (2001). Band parameters for iii-v compound semiconductors and their alloys, *J. Appl. Phys.* 89(11): 5815.

Wang, D., Chang, Y., Wang, Q., Cao, J., Farmer, D., Gordon, R., Dai, H. & Saraswat, K. (2004). Surface chemistry and electrical properties of germanium nanowires, *J. Am. Chem. Soc.* 126(37): 11589.

Wang, J., Gudiksen, M. S., Duan, X., Cui, Y. & Lieber, C. M. (2001). Highly polarized photoluminescence and photodetection from single indium phosphide nanowires, *Science* 293(5534): 1455.

Wang, J. & Lundstrom, M. (2003). Ballistic transport in high electron mobility transistors, *IEEE Trans. Electron Devices* 50(7): 1604.

Wang, Q., Javey, A., Tu, R., Dai, H., Kim, H. H., Mclntyre, P., Krishnamohan, T. & Saraswat, K. (2003). Germanium nanowire field-effect transistors with SiO_2 and high-k HfO_2 gate dielectrics, *Appl. Phys. Lett.* 83(12): 2432.

Wolf, S. A., Awschalom, D. D., Buhrman, R. A., Daughton, J. M., Mòlnar , S. V., Roukers, M. L., Chtchelkanova, A. Y. & Treger, D. M. (2001). Spintronics: A spin-based electronics vision for the future, *Science* 294: 1488.

Woodall, J. M., Freeouf, J. L., Pettit, G. D., Jackson, T. & Kirchner, P. (1981). Ohmic contacts to *n*-GaAs using graded band gap layers of $Ga_{1-x}In_xAs$ grown by molecular beam epitaxy, *J. Vac. Sci. Technol* 19: 626.

Xia, Y., Yang, P., Sun, Y., Wu, Y., Mayers, B., Gates, B., Yin, Y., Kim, F. & Yan, H. (2003). One dimensional nanostructures: synthesis, characterization, and applications, *Adv. Mater.* 15(5): 353.

Xiang, J., Lu, W., Hu, Y., Wu, Y., Yan, H. & Lieber, C. M. (2006). Ge/Si nanowire heterostructures as high-performance field effect transistors, *Nature* 441: 489.

Yu, P. Y. & Cardona, M. (1999). *Fundamentals of semiconductors*, Springer-Verlag, Berlin.

Zheng, Y., Rivas, C., Lake, R. K., Alam, K., Boykin, T. B. & Klimeck, G. (2005). Electronic properties of Si nanowires, *IEEE Trans. Electron Dev.* 52(6): 1097.

Zhong, Z.,Wang, D., Cui, Y., Bockrath, M.W. & Lieber, C. M. (2003). Nanowire crossbar arrays as address decoders for integrated nanosystems, *Science* 302(5649): 1377.